Einstein Studies

Editors: Don Howard John Stachel

Published under the sponsorship
of the Center for Einstein Studies,
Boston University

Volume 1: Einstein and the History of General Relativity
Don Howard and John Stachel, editors

Volume 2 : Conceptual Problems of Quantum Gravity
Abhay Ashtekar and John Stachel, editors

Volume 3 : Studies in the History of General Relativity
Jean Eisenstaedt and A. J. Kox, editors

IN PRESS :

Volume 4 : Recent Advances in General Relativity
Allen I. Janis and John R. Porter, editors

Jean Eisenstaedt A. J. Kox
Editors

Studies in the History of General Relativity

Based on the
Proceedings of the
2nd International Conference
on the History of General Relativity,
Luminy, France, 1988

Birkhäuser
Boston · Basel · Berlin

Jean Eisenstaedt
Centre National de la
Recherche Scientifique
Université Pierre et Marie Curie
75231 Paris Cédex 05
France

A. J. Kox
Instituut voor Theoretische Fysica
Universiteit van Amsterdam
1018 XE Amsterdam
The Netherlands
and
The Collected Papers of Albert Einstein
Boston University
Boston, MA
U.S.A.

Library of Congress Cataloging-in-Publication Data

International Conference on the History of General Relativity (2nd : 1988 : Luminy, Marseille, France)
Studies in the history of general relativity : proceedings of the Second International Conference on the History of General Relativity, Luminy, Marseille, France, 6-9 September 1988 / Jean Eisenstaedt, A.J. Kox, editors.
p. cm. -- (Einstein studies : v. 3)
Summaries in French.
Includes bibliographical references and index.
ISBN 0-8176-3479-7 (alk. paper)
1. General Relativity (Physics) --History--Congresses.
I. Eisenstaedt, Jean, 1940- . II. Kox, Anne J. III. Title.
IV. Series.
QC173.6 I572 1988 92-20275
530.1'1'09--dc20 CIP

Printed on acid-free paper.

The Einstein Studies series is published under the sponsorship of the Center for Einstein Studies, Boston University.

ISBN 0-8176-3479-7
ISBN 3-7643-3479-7

Typeset in TeX by ARK Publications, Inc., Newton Centre, MA.
Printed and bound by Edwards Brothers, Inc., Ann Arbor, Michigan.
Printed in the United States of America.

9 8 7 6 5 4 3 2 1

Preface to Volume Three

In May 1988, the Osgood Hill Conference Center in North Andover, Massachusetts, housed a conference that was unique in its kind: the first international conference exclusively devoted to the history of general relativity. As John Stachel, who organized the meeting, wrote in the introduction to its proceedings (*Einstein and the History of General Relativity*, Don Howard and John Stachel, eds., Boston: Birkhäuser, 1989; *Einstein Studies*, Vol. 1), the history of general relativity—as well as general relativity itself—has in the past been a much neglected field of research. In recent years, however, a new interest in general relativity and its history has arisen, among historians as well as philosophers of science. Inspired by the success of the Osgood Hill Conference, its participants decided to try to make it the first in a regular series of meetings. Thus, the Second International Conference on the History of General Relativity was held September 6–9, 1988, at the International Center of Mathematical Research (CIRM) at Luminy, France. It was organized by Jean Eisenstaedt (Institut Henry Poincaré, Paris), with the assistance of an organizing committee consisting of S. Goldberg (Smithsonian Institution, Washington) A. J. Kox (University of Amsterdam and The Collected Papers of Albert Einstein), M. de Maria (University of Rome), J. Renn (The Collected Papers of Albert Einstein), and J. Stachel (The Collected Papers of Albert Einstein). The organizing committee was advised by a scientific committee, which consisted of the following persons: P. G. Bergmann (New York University), J. Goldberg (Syracuse University), P. Havas (Temple University), W. Israel (University of Alberta), M. Jammer (The Hebrew University of Jerusalem), A. Lichnerowicz (Collège de France), M. A. H. MacCallum (Queen Mary College), J. Merleau-Ponty (Université Paris XI) J. D. North (University of Groningen and International Academy of the History of Science), A. Papapetrou (Institut Henri Poincaré), J. C. Pecker (Collège de France), E. Schucking (New York University), J. Stachel

(The Collected Papers of Albert Einstein), and V. P. Vizgin (Academy of Sciences of the USSR).

At the end of the Luminy conference, it was decided to continue the series with a third meeting, to be organized by Professor John Norton and to be held June 27–30, 1991, in Pittsburgh, Pennsylvania. The proceedings of this conference will appear in a future volume of *Einstein Studies*.

The names of the participants at the Luminy meeting are listed in the back matter of this volume, along with the addresses of the contributors and French summaries of their papers.

Acknowledgments

The editors gratefully acknowledge the organizational and financial help of the following institutions and organizations: The Division of Mathematics and Physics and the Division of Philosophy of the CNRS, the Université Pierre et Marie Curie, the French Ministries of Education and of Foreign Affairs, the Philosophical Division of UNESCO, the Center for Einstein Studies, the International Committee on General Relativity and Gravitation, the Société Mathématique de France, and the Laboratoire de Physique Théorique de l'Institut Henri Poincaré. We would also like to thank the staff of the CIRM and Emmanuelle Cesari, who extended invaluable organizational assistance in Luminy.

We are much indebted to Don Howard and John Stachel for their assistance in the editing of the papers and to Jim Ritter for his help in translating some of the papers.

A Note on Sources

In view of the frequent citations of unpublished correspondence or other items in the Einstein Archive, the editors have adopted a standard format for such citations. For example, a designation such as "EA 26-107" refers to item number 26-107 in the Control Index to the Einstein Archive. Copies of the Control Index can be consulted at the Jewish National and University Library (The Hebrew University), Jerusalem, where the Archive is housed; and at Mudd Manuscript Library, Princeton University, and Mugar Memorial Library, Boston University, where copies of the Archive are available for consultation by scholars.

Contents

Introduction

Jean Eisenstaedt and A. J. Kox

This volume of the *Einstein Studies* series presents the proceedings of the Second International Conference on the History of General Relativity, which was held in Luminy, France, in the fall of 1988. In spite of our efforts, we are not able to include in this volume all of the papers presented at the meeting: personal circumstances prevented Peter Havas from finishing his paper dealing with the history of the controversy between Einstein and Ludwik Silberstein before this volume went to press. It is planned to include his paper in a future volume of *Einstein Studies*. However, not all of the papers in this volume were delivered at Luminy. Joshua Goldberg's paper on the history of the US Air Force support for research in general relativity was actually presented at the Osgood Hill conference on the history of general relativity (see Preface), and the paper by Thibault Damour and Gerhard Schäfer on Levi-Civita and the problem of motion in general relativity was delivered at the Fifth Marcel Grossmann Conference at the University of Western Australia. Gennady Gorelik's history of the first stages of quantum gravity is included, even though the author was unable to attend the meeting.

Several of the papers presented here are devoted to the reception and development of the general theory of relativity. David Lopes Gagean and Manuel da Costa Leite concentrate on the situation in Portugal, Hubert F. M. Goenner offers a survey of the reception of the theory of relativity in Germany through the books published between 1908 and 1945, and in addition, he provides an extensive bibliography of this period. One of us (AJK) presents an account of the reception and further development of general relativity in the Netherlands. Finally, José M. Sánchez-Ron gives a detailed description of the reception of Einstein's theory of gravity among British physicists and mathematicians between 1915 and 1930.

On a more personal level, Joshua N. Goldberg tells the intriguing story of the relativity research group at Wright-Patterson Air Force Base,

Ohio, of which he was a leading member; André Lichnerowicz analyzes mathematical research in the field during the years 1937–1945 and discusses his own work in this context; and André Mercier offers us a description and a personal testimony about the origins of the international meetings on general relativity (GRG), in particular of the first one, which was held in Bern in 1955.

The discussion of conceptual issues in general relativity constituted a second theme of the conference. Julian B. Barbour discusses in a historical perspective the question of the "Machian" character of Einstein's theory of gravitation, while Don Howard explores the role in Einstein's thinking of what he calls the *Eindeutigkeit* principle. József Illy gives an account of the discussion on general covariance as it took place in the correspondence between Einstein and Gustav Mie, whereas Ludwik Kostro treats another fundamental topic in Einstein's thinking through his concise history of Einstein's relativistic ether conception. Finally, John Norton analyzes the question of Einstein's attitude to the physical content of general covariance and reviews Kretschmann's objection to Einstein's use of covariance principles.

A third set of papers deals with the more technical aspects of general relativity. Arthur I. Miller treats the prehistory of general relativity through the analysis of Einstein's 1907 *Jahrbuch* paper, and Gunnar Berg sketches the origin of the Einstein space. Michel Janssen analyzes Lorentz's 1916 geometrical approach to general relativity. There are two papers on quantum gravity: Peter G. Bergmann presents a short review of the quantization of the gravitational field from 1930 to the present, and Gennady E. Gorelik concentrates on the first steps toward quantum gravity, in particular on Matvey P. Bronsteyn's work in this field. S. Kichenassamy's study focuses on the history of the description of the coupling of an electron with a gravitational field. Thibault Damour and Gerhard Schäfer give a technical analysis of Levi-Civita's approach to the problem of motion in general relativity, while the work of the Polish physicist Myron Mathisson on the problem of motion is analyzed by Bronisław Średniawa. In addition, John Stachel retraces the early history of the Cauchy problem in general relativity, and Christiane Vilain gives an overview of the question of spherical coordinates in general relativity.

Finally, we should mention two papers on cosmology: Odon Godart gives a review of the work of Lemaître, and Helmuth Urbantke surveys Schrödinger's contributions to cosmology.

PART I

The Reception and Development of General Relativity

General Relativity and Portugal: A Few Pointers Towards Peripheral Reception Studies

David Lopes Gagean and Manuel da Costa Leite

1. Introduction

Some recent studies on the history of relativity have surpassed mere conceptual and theoretical reconstructions (e.g., its universal expression) and have extended to comparative studies of its early reception in various countries. These studies, though of a second order, can obviously help in the understanding and characterization of its early discussion, particularly with regard to prejudices of an epistemological and psychosociological nature.

Our paper intends to make some contributions to the comparative historical study of relativity and to the understanding of the diffusion processes, discussing aspects of the history of general relativity with reference to Portugal from its earliest days until 1950.

Why have we assumed that it may be interesting to know about the reception of general relativity in a culturally and scientifically peripheral area such as Portugal? We think there are two main reasons: (1) those that involve an immediate historical interest, i.e., occurrences or events in a particular country that are strictly related to the history of the theory, which we call *reasons of fact*; and (2) those that involve the mere possibility of reflecting on the local reception and the local diffusion of the theory; these reasons support and give legitimacy to comparative studies (no matter how passive the reception of the theory) and are called *general reasons*.

On this subject, Thomas Glick (1987c) considered five general assumptions:

> One is to posit that the differential receptivity of cultural or social groups to specific ideas reveals something significant about social and cultural processes.
>
> A second is to suppose that such differentiation illuminates science as a social and cultural phenomenon.
>
> A third is to challenge perceived views of what is normative in science by expanding the number of cases examined and broadening the basis of study to include scientific communities thought peripheral to mainstream science.
>
> A fourth is to change the focus of study from an internal development of ideas to the social and cultural contexts of their development and diffusion.
>
> A fifth is to cast a social definition of science as ideology; to do so is to evade the question of what is normative experience, to rid the research program of ethnocentric bias, and to even out the distinction between invention and innovation, discovery and reception.

In addition to Glick's notes, we would like to emphasize the second-order character of the comparative studies and their philosophical and cognitive significance. From a factual point of view, there are several items relating general relativity to Portugal before 1950. These are:

1. Eddington's expedition to the Isle of Principe to observe the total eclipse of the sun in 1919,
2. The papers by Mira Fernandes published in *Rendiconti della Accademia dei Lincei*, and
3. Einstein's nomination by the Lisbon Academy of Sciences.

Among these factual items, only the papers by Mira Fernandes have the distinction of being regarded as an active participation in the internal, technical development of the general theory of relativity. However, the three items supported by reasons of fact, added to others, result in some interesting provisional conclusions.

2. Philosophical and Popular Perspectives

The first book to mention and focus on the (special) theory of relativity was not written by a physicist or a mathematician, but by a philosopher, Leonardo Coimbra, at the beginning of 1912 (Coimbra 1912). This is not at all surprising, first, because of the conceptual "radicalness" of the theory and its natural attraction for an attentive philosopher, and second, because in a case characterized by a generally passive reception, it seems highly probable that the "strictly scientific" discussions within the scientific community were mixed ambiguously with philosophical or even popular levels of approach. Leonardo Coimbra knew the philosophical theories of that time very well, and he knew of the revolutionary changes physics was undergoing. Coimbra often quoted the works of Henri Poincaré, especially *La science et l'hypothèse* (Poincaré 1902), and those of Lucien Poincaré, such as *La physique moderne. Son évolution* (Poincaré 1906).

Although Coimbra was influenced by Auguste Comte, as well as by other philosophers of his time, he was not a positivist. In fact, in *O criacionismo* (*Creationism*, Coimbra 1912), he criticized several philosophical schools of that time, including materialism, Ostwald's energetics, and Kant's a priorism. Let us look at some of his criticisms.

> One of the benefits that philosophical thought owes to positivism is the relevance of metaphysics today. Philosophical thought has been shamed by scientific thought because it has always been progressive and because of its prolific achievements. (Coimbra 1983, Vol. 1, p. 5)

For Leonardo Coimbra, "science is considered the right perspective, whereas metaphysics is reduced to a mere complement of science" (Morujão 1983). In his idiosyncratic style Coimbra wrote:

> Studying the mental being that takes itself for granted more readily and accurately. Analyzing it in such a manner that its rationalism can be shown to be its criterion of truth, and then showing that rationalism is never hollow, but on the contrary, always results from an activity that functions on oppositions. Thus, the spirit stands—the sensitive and free spirit; sensitive because its life does not come from isolation, *free*... because it dominates and coordinates the sensitive flux rather than mixes with it. That mental being is science....
>
> Therefore, we have to demonstrate that science is real and rational. It has a real content, even though it has an ideal order. Science is absolute and concrete in so far as its legitimate domains are concerned; ... [science] is made of notions and not of things. (Coimbra 1983, Vol. 1, p. 8)

We think the dynamics of *notion* and that of *notionness* (as opposed to *thingness*) are the most significant ideas in Coimbra's philosophy of knowledge as far as relativity is concerned. Science develops by dynamical processes of successive and integrating synthesis, by "a rationalization that Leonardo calls notion" (Freitas 1960). Opposite to notion, there is the *thing*, which results from a stagnant tendency of reason. The thing and thingness lead to a static representation of the world. "Thus we cannot admit the existence of things. The thing is the biggest philosophical sin" (Coimbra 1983, Vol. 1, p. 18).

We can say that Coimbra's philosophy of knowledge and science inevitably led him to a world of events rather than a world of things, to a dynamic, differentially geometrized world rather than to a world of successive pictures. "Once the vice of thingness is dismantled, creationism is the only philosophy that remains" (Coimbra 1983, Vol. 1, p. 327). Therefore, creationism is the process of the creation of new notions within the progressive development of reason.

The philosophical apparatus of Coimbra and his attention to physics led him to Lorentz's electron theory and the problems associated with it; Poincaré's criticism of Newton's principles; and the principle of relativity as enunciated by Poincaré. In 1912 Leonardo Coimbra wrote:

> Langevin, in a notable and audacious thesis, demonstrates that the Principle of Relativity stands if we change the meaning of the notions of space and time. (Coimbra 1983, Vol. 1, p. 103)

Following this passage, Coimbra formulated the principles and presented some calculations of the (special) theory of relativity. One of his bibliographical sources was an article from the *Bulletin de la Société Française de Philosophie* (January 1912), which described the debate at the Société on October 19, 1911, when the introductory talk was Langevin's "Temps, l'espace et la causalité en physique moderne." It should be stressed that not only was this one of the first papers on relativity in France (which demonstrates Coimbra's awareness of the literature), it was also published in a periodical of a philosophical nature. Thus, it is not surprising that philosophers rather than scientists (Portuguese or French) read it.

It is significant that at this stage Coimbra confused the names of Einstein and Langevin, which seems to support our own assumption that in peripheral regions the strictly scientific, philosophical, and popular levels can easily be confused. However, Coimbra was an exception within the Portuguese cultural and academic environments in the 1920s.

Let us briefly look at the situation at that time. Physics was one of the least developed or taught scientific disciplines at the University of Coimbra or at the two other Portuguese institutions of higher scientific studies at that time (Escola Politécnica de Lisboa and Academia Politécnica do Porto). At that time, we had no physicists (either theoretical or experimental) with an international reputation. Until 1861, physics was confined to only one chair in the Faculty of Philosophy at the University of Coimbra. In the Faculty of Mathematics, there was a chair in Mathematical Physics, which was essentially mathematics applied to constructions, where the theory of elasticity was taught. At the aforementioned institutions of higher scientific studies in Lisbon and Oporto, a course on experimental physics was offered.

In 1921, the first Joint Congress of the Portuguese and Spanish Association for the Development of Science was held in Oporto, Portugal. The opening lecture of the mathematics section was delivered by the Spanish professor at the Universidad Central de Madrid, José Maria Plans y Freire; its title was "Processo histórico del calculo diferencial absoluto y su importancia actual" ("The Historical Development of Absolute Calculus and Its Present Importance," Plans y Freire 1921a). General relativity, which until then had been diffused informally and with little response in academic environments, subsequently rose in importance. This was the first formal introduction to the general theory of relativity in Portugal.

Another work by Plans y Freire that was widely read in Portugal was *Nociones fundamentales de mecanica relativista* ("Fundamental Notions of Relativistic Mechanics," Plans y Freire 1921b), for which Plans y Freire won the prize offered in 1919 by the Real Academia de Ciencias Exactas, Fisicas e Naturais (Royal Academy of Sciences) of Madrid. Though the work is mainly devoted to the special theory, it has two chapters concerning general relativity, particularly, the field equations of gravitation and the motion of the perihelion of Mercury and the deflection of light. It is important to know that the manuscript was written and presented to the Academy during 1919 before the famous joint meeting of the Royal Society of London and the Royal Astronomical Society, but the book was not published until 1921, in Madrid[1] (Cunha 1921; Glick 1987b). This circumstance—an interesting one, as Glick wrote (Glick 1987b, p. 235)—explains why it did not prepare the Portuguese scientists for Eddington's expedition.

The lecture and the works by Plans y Freire impressed the Portuguese academic world so much that the Lisbon Academy of Sciences nominated Freire as a corresponding member on March 18, 1922. This was the first time the Lisbon Academy of Sciences had officially heard about relativity (either special or general).

In the academic year 1922–1923, the Professor of Mathematical Physics at the Faculty of Sciences in Lisbon, António dos Santos Lucas, gave a course on both Einstein's special and general relativity, which became the first academic course on general relativity. Also during 1922–1923, A. Cyrilo Soares, in a dissertation entitled *O conceito de matéria e o desenvolvimento das teorias físicas* (*The Concept of Matter and the Development of Physical Theories*, Soares 1922), included a chapter on (special) relativity.

Leonardo Coimbra expanded his dissertation of 1912 (where he included for the first time Einstein's equations of special relativity) into a book. In this book, *A razão experimental* (*Experimental Reason*, Coimbra 1923), Coimbra described with mathematical detail the general theory of relativity, including (though in less detail) the cosmological solutions to the gravitational equations of Einstein and De Sitter, as well as the first unified theories of Weyl (1918) and Eddington (1921). Then, in 1925, Vitor Hugo de Lemos, Professor of Mathematical Physics at the University of Lisbon, wrote *Cálculo tensorial* (Lemos 1925), which was a book on tensor calculus as an introduction to general relativity.

Therefore, we can say that the turning point for the diffusion of general relativity, in both scientific and philosophical terms, was between 1922 and 1923. Nevertheless, it was a nonactive diffusion (nonactive in Glick's sense) (Glick 1987c). In an active manner, general relativity in Portugal centered around two important mathematicians: Mira Fernandes and Ruy Luís Gomes. We shall refer only to Fernandes, as Gomes's work extended into the 1950s.

3. The Mathematician Mira Fernandes: The Active Response

Mira Fernandes, who was perhaps the greatest Portuguese geometrician of his time, was Professor of Infinitesimal Calculus and Rational Mechanics at the Instituto Superior Técnico, which was the most important school of engineering in Portugal. He wrote about 60 mathematical papers on analysis, rational mechanics, differential geometry, quantum mechanics, the general theory of relativity, and unified theories. Beginning in 1934, Fernandes included the general theory of relativity in his lectures on rational mechanics. We would like to emphasize that he was a mathematician and not a physicist[2] and that he corresponded with the Italian mathematician Tullio Levi-Civita and the French mathematician Elie Cartan.[3] From 1928 to 1938, his most important papers were published in *Rendiconti della Accademia dei Lincei*, undoubtedly at the suggestion of Levi-Civita. In 1938, Levi-Civita was forced to retire from the University of Rome for political reasons, and Fernandes ceased to be a contributor to *Rendiconti della Accademia dei Lincei*, though he continued writing (in Italian) for two Portuguese journals of an international character: *Portugaliae Mathematica* and *Revista da Faculdade de Ciências de Lisboa* (*Journal of the Faculty of Sciences of Lisbon*).

The following are the relevant papers Fernandes wrote on general relativity and unified theories and their mathematical basis.

1. "Fundamentos da geometria diferencial dos espanços lineares" ("Fundamentals of Differential Geometry in Linear Spaces," 1927). This paper methodologically followed J.A. Schouten's book *Der Ricci-Kalkül* (Schouten 1924). As far as we know, besides Schouten's book, this is one of the best works in the literature of geometry.
2. "Proprietà di alcune connessioni lineari" (1931). In this paper, Fernandes analyzed the properties of some linear connections and made generalizations of some formulas from *Der Ricci-Kalkül*.

3, 4. "Sulla teoria unitaria dello spazio fisico" (1932, 1933). These two papers by Fernandes are on unified theories. He gives special attention to the unified theory proposed by the Swiss physicist Straneo in 1931 and 1932.

5. "La teoria unitaria dello spazio fisico e le equazioni relativiste della meccanica atomica" (1934). This article tries to unify general relativity, electromagnetism, and wave mechanics. This attempt is of unique historical interest.
6. "Mecànica e geometria" ("Mechanics and Geometry," 1936). In various papers, E.A. Milne from Oxford and S. Milne from the University of Sheffield maintained that it ought to be possible to choose arbitrarily any kind of geometry for cosmological theories. Fernandes used this idea as a starting point for a new unifying attempt.

7, 8. "Connessioni finite" (1945) and "Trasporti finiti" (1950). In both of these articles, Fernandes discussed the theory of Einstein–Bergmann (Einstein and Bergmann 1944; Einstein 1944), pointing out that there are dispensable axioms in it. At the same time, Fernandes applied some formalism from *Der Ricci-Kalkül* to the Einstein–Bergmann theory in an attempt to make an approximation to his own theory of 1933.[4]

4. Three Other Peripherally Significant Facts

Peripheral regions tend to have no clear contours; strictly scientific, philosophical or popular movements of ideas tend to be intermingled. Events that occur in a peripheral region and belong to different levels of importance appear to be part of a homogeneous (though a very rarefied) field because of their imprecise footprints on the historical process of the development of scientific ideas. Some of these events seem to be more like *faits-divers* than actual significant facts.

Let us look at three of these peripherally significant facts: Eddington's expedition to the Isle of Principe, anti-relativistic reactions, and Einstein's nomination by the Academy of Sciences of Lisbon.

1. *Eddington's expedition to the Isle of Principe*: According to the report "A Determination of the Deflection of Light by the Sun's Gravitational Field, from Observations Made at the Total Eclipse of May 29, 1919," published in the *Philosophical Transactions of the Royal Society of London* (Dyson, Eddington, and Davidson 1920, p. 292), we can be sure that there were some contacts with the Portuguese scientific community, namely, with the Director of the National Observatory of Lisbon.

> Vice-Admiral CAMPOS RODRIGUES and Dr. F. Oom of the National Observatory, Lisbon, had kindly given us introductions.... (Dyson, Eddington, and Davidson 1920, p. 312)

However, these contacts produced no tangible results. In contrast to the Brazilian attitude, no Portuguese scientist (astronomer, physicist, or mathematician) was sent to Principe.

2. *Anti-relativism*: Before 1950, there were three anti-relativistic reactions in Portugal. Fundamentally, they were of the kind that M. Biezunski called a "conservative position"; that is, they did not represent "a straightforward attack, but an effort to avoid a change of theory" (Biezunski 1987). However, these three appear to have been of two very different types if we use a more precise analytical view. Two of them were produced in the traditional academic environment. The other was a curious mixture of popular and scientific criticisms. The academic refutations came from two professors of physics: namely, José Maria de Almeida Lima and Francisco Miranda da Costa Lobo.

José Maria de Almeida Lima (1859–1939) was Professor of Physics at the Faculdade de Ciências da Universidade de Lisboa. He was one of the professors mainly responsible for the improvement of the teaching of physics during the first two decades of the 20th century in Portugal. He published several scientific articles on thermodynamics and a very reasonable textbook on (general) physics. Almeida Lima advocated a Newtonian and mechanistic view of the world and of physics. Therefore, Einstein's theories were not exactly in harmony with his views. So, between 1920 and 1923, he published some articles against relativity: "Reflexões sobre o princípio de inércia" ("Reflections on the Principle of Inertia"), "O critério de verdade: racionalismo e dogmatismo" ("The Criterion of Truth: Rationalism and Dogmatism"), and "A física perante as teorias de Einstein" ("Physics Before the Theories of Einstein") (Lima 1920, 1922, 1923).

Francisco Miranda da Costa Lobo (1864–1945) was Professor of Astronomy in the Faculdade de Ciências da Universidade de Coimbra, as well as the director of its astronomical observatory. Strong Newtonian as he was, he proposed a "radiant" theory aimed at explaining all kinds of physical phenomena essentially on the basis of the corpuscular theory of gravitation by Le Sage. A radical critic of relativity and quantum theories, Costa Lobo announced and propagated his theory in various articles, as well as at the meeting of the British Association for the Advancement of Science in September 1931 in London. Costa Lobo and his theory were unwillingly responsible for a very curious and significant case of support for relativity. In fact, in the same year as the above-mentioned meeting of the British Association, Costa Lobo published an article in the *Revista da Faculdade de Ciências da Universidade de Coimbra* (Vol. II, 1931–1932, p. 71–73), where he developed exactly the same ideas as in his paper at the meeting. Curiously, the same journal published a collective article against Lobo and his radiant theory signed by lecturers and professors of Lobo's own faculty. It is, perhaps, a unique case of an *intra muros* controversy on relativity. The counterattack against Costa Lobo was a dismantling of his main attack against the contradictions and criticisms of recent discoveries in physics.

Finally, the third criticism came from an interesting Portuguese personality, Admiral Carlos Viegas Gago Coutinho (1869–1959), a very well known personality, mainly because he and Sacadura Cabral were the first to fly between Europe (Lisbon) and South America (Rio de Janeiro), in 1922. He was an expert in air and astronomical navigation, a geographer, a hydrographer, and a historian of his discoveries. The professional biography of Gago Coutinho makes him an unusual critic of (special) relativity. His main criticism is based on the common sense notions of absolute space and time. He wrote articles and books against relativity from 1926 to 1942. However, it should be noted that although Gago Coutinho's criticism was more common sense oriented than strictly scientific, his opinions were so well considered that the Portuguese mathematician Ruy Luís Gomes argued against them in an attempt to put relativity in its proper place. We could say that, just as anti-relativism is obviously of interest to historians of relativity, this case certainly would be interesting for a historian of cases of prejudices and hesitations in the spreading of ideas (both in central and peripheral processes). Costa Lobo and Gago Coutinho, albeit for different reasons, are two structurally different, but nevertheless good examples of such cases.

3. *Einstein's nomination by the Academy of Sciences of Lisbon*: During the presidency of Dr. Egas Moniz, the Academy of Sciences of Lisbon nominated Einstein and Tullio Levi-Civita as foreign corresponding members at its meeting on March 17, 1932. By that time, Einstein was a well known celebrity, so the Lisbon Academy of Sciences nomi-

nation was not surprising. This is a clear example of what we meant by an almost *fait-divers* rather than a significant scientific fact.

Let us look at the very concise, formal, and polite letter of acknowledgment by Einstein (similar to others received by other academies).[5]

Berlin 2. den 9. April 1932
Haberlandstr. 5

An die Akademie der Wissenschaften
Lissabon
Portugal

Ich nehme die Ernennung zum korrespondieren Mitglied der Akademie der Wissenschaften mit freundlichem Dank an.

Mit ausgezeichneter Hochachtung

A. Einstein

There are some more brief comments we should like to make about this: (l) the nomination occurred during the presidency of Egas Moniz, the Portuguese Nobel Laureate in Medicine (to some extent, the Portuguese counterpart of Ramón y Cajal); (2) the nomination was proposed by the above-mentioned Portuguese mathematician Fernandes; and (3) Einstein was nominated at the same time as Levi-Civita, also a mathematician.

5. Some Tentative Conclusions

What can we conclude from this collection of facts? There are several significant issues that may endorse the results of some comparative studies already done, especially in Italy, France, and Spain.

1. In Portugal, as in Spain and, to some extent, France, "only mathematics had sufficient weight as a discipline to interpret" (Glick 1987b) general relativity and to produce active scientific research.
2. For Portugal, as for Spain, Italy was its main mathematical source. In the case of Portugal, however, there were no institutional links; the mathematical link always existed on a very fragile personal basis, for instance, through Tullio Levi-Civita.
3. Two influences, from Spain and France, occurred through individual links: through the Spaniard José Maria Plans y Freire and the Frenchman Paul Langevin.
4. Plans y Freire was responsible for creating the general level of interest; Langevin helped the philosophical and popular diffusion of relativity.

5. Only Levi-Civita, although on an individual basis, succeeded in producing an active diffusion of the general theory of relativity.

In conclusion, we are tempted to say that in peripheral regions the diffusion process seems to stand on a very fragile basis. Individual links, contingent occurrences, indirect connections (which mean dependence), and a rarefied field of events are the components of such fragility.

Acknowledgments: The authors would like to thank the Gabinete de Filosofia do Conhecimento for supporting the research project "História da Cultura Científica: Introdução das Ideias Quânticas e Relativistas em Portugal na Primeira Metade do Século XX," of which this paper is a small part. They would also like to thank the Fundação Calouste Gulbenkian and the Academia das Ciências de Lisboa. Finally, the authors would like to acknowledge and express their gratitude for the valuable questions, comments, and suggestions made by Don Howard, José Manuel Sánchez-Ron, J.D. North, and Michel Biezunski.

Notes

[1] The book *Nociones de calculo diferencial absoluto* (Madrid, 1924), by Plans y Freire, was known in the Portuguese scientific community, but we cannot be sure to what extent.

[2] In Spain, it is also significant that the mathematicians rather than the physicists were the first to actively receive relativity.

[3] Cf. Cartan, Elie, *Oeuvres complètes.* Part III, Vol. 2, p. 1392 (Paris: Gauthier-Villars).

[4] See references in *Zentralblatt fur Mathematik*, 40, as well as in *Mathematical Reviews*, 1946 and 1951.

[5] Archives of the Academia das Ciências de Lisboa.

References

Biezunski, Michel (1987). "Einstein's Reception in Paris in 1922." In Glick, 1987a, pp. 169–188.

Coimbra, Leonardo (1912). *O criacionismo.* Oporto: Biblioteca da Renascenca Portuguesa.

——— (1923). *A razão experimental.* Oporto: Biblioteca da Renescenca Portuguesa.

——— (1983). *Obras, Vol. 1.* Oporto: Lello & Irmão.

Cunha, Pedro José da (1921). *Parecer acerca da candidatura do Sr. D. José Maria Plans y Freire a sócio correspondente estrangeiro.* Lisbon: Academia das Ciências de Lisboa.

Dyson, Frank W., Eddington, Arthur S., and Davidson, Charles (1920). "A Determination of the Deflection of Light by the Sun's Gravitational Field." *Royal Society of London. Philosophical Transactions A* 220: 291–333.

Einstein, Albert (1944). "Bivector Fields, II." *Annals of Mathematics* 45: 15–23.

Einstein, Albert, and Bergmann, Valentin (1944). "Bivector Fields, I." *Annals of Mathematics* 45: 1–14.

Eisenstaedt, Jean (1986). "La relativité générale à l'étiage: 1925–1955." *Archive for History of Exact Sciences* 35: 115–185.

Freitas, Manuel de (1960). "Leonardo Coimbra, incidências positivistas na sua filosofia." *Revista Portuguesa de Filosofia* 16: 160–175.

Glick, Thomas, F. ed. (1987a). *The Comparative Reception of Relativity.* Dordrecht and Boston: Reidel.

——— (1987b). "Relativity in Spain." In Glick 1987a, pp. 231–263.

——— (1987c). "Cultural Issues in the Reception of Relativity." In Glick 1987a, pp. 381–400.

Lemos, Vitor Hugo de (1925). *Cálculo Tensorial.* Lisbon.

Lima, José Maria de Almeida (1920). "Reflexões sobre o principio da inércia." *Academia das Ciências de Lisboa. Jornal das Sciencias Mathematicas, Fisicas e Naturaes* 7: 132–145.

——— (1922). "Os critérios da verdade: racionalismo e dogmatismo." *Academia das Ciências de Lisboa. Jornal das Sciencias Mathematicas, Fisicas e Naturaes* 9: 1–12.

——— (1923). "A fisica perante as teorias de Einstein." *Academia das Ciências de Lisboa. Jornal das Sciencias Matematicas, Fisicas e Naturaes* 15: 79–96.

Morujão, Alexandre Fradique (1983). "O sentido da filosofia em Leonardo Coimbra." *Revista Portuguesa de Filosofia* 34: 345–364.

Plans y Freire, José Maria (1921a). "Processo histórico del calculo diferencial absoluto y su importancia actual." In *Asociación Espanõla para el Progreso de las Ciências,* 8. Congresso, Oporto. Madrid: E. Arias.

——— (1921b). *Nociones fundamentales de mecanica relativista.* Madrid: Real Academia de Ciências Exactas, Fisicas e Naturais.

Poincaré, Henri (1902). *La science et l'hypothèse.* Paris: Flammarion.

Poincaré, Lucien (1906). *La physique moderne. Son évolution.* Paris: Flammarion.

Schouten, Jan A. (1924). *Der Ricci-Kalkül.* Berlin: Springer.

Soares, A. Cyrilo (1922). *O conceito de matéria e o desenvolvimento das teorias físicas.* Lisbon.

The Reception of the Theory of Relativity in Germany as Reflected by Books Published Between 1908 and 1945

Hubert F. M. Goenner

1. Method and Empirical Data

Any program attempting to assess the reception of the theory of relativity in the 1920s and 1930s must collect the printed sources with a reasonable degree of completeness. From Lecat's bibliography (Lecat 1924) containing more than 2000 entries for the period 1919–1924, it is obvious that a single person cannot cope with this wealth of scientific papers, books, pamphlets, and newspaper articles. The Berlin physicist E. Gehrke wrote in 1924 that he had collected more than 5000 newspaper clippings and essays in journals concerning relativity (Gehrke 1924). If private letters by relativists and anti-relativists and relevant public documents of the time, many of which cannot be easily reached are added, a whole research institute is needed for taking on the task.

In view of the situation, I decided to restrict myself to *books on relativity in the German language* published between 1908 and 1945. In Lecat's statistics, Germany takes first place in both the number of authors (30%) and publications (38%). Thus, it seemed to me that the German book publication statistics should more or less reproduce Lecat's (Fig. 1) and for the other years give a first insight into the more than lively discussion induced by Einstein's relativity in all circles and classes of Germany and German speaking countries.

Besides the uncertainty in the definition of what constitutes a book, a number of *methodological* problems arise. During the period studied, articles in scientific journals often were not just reprinted as they are today, but they were also offered under separate cover through the bookstores in the form of booklets. In the statistics, I included these as books *except*

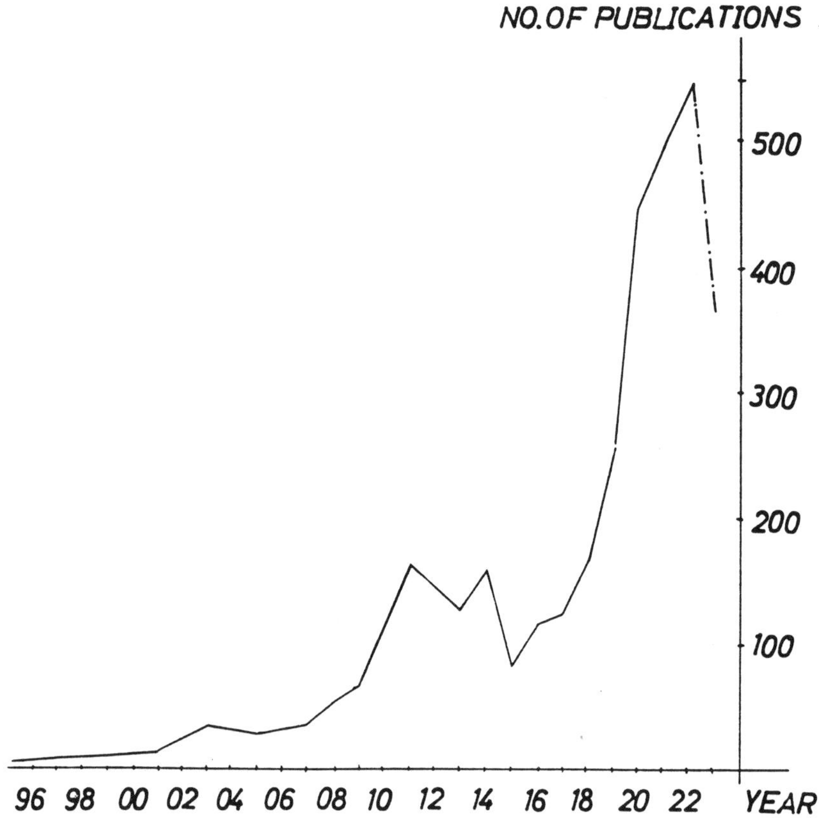

FIGURE 1. Lecat's statistics on relativity publications of all sorts.

if the publisher sold them *in Kommission* (not at his own risk but at the risk of the learned society editing the journal, e.g., *the Sitzungsberichte der Preussischen Akademie der Wissenschaften*). Furthermore, issues of a journal covering a single author and subject and sold separately are included. Quite a few anti-relativists published *im Selbstverlag* (author and publisher are the same). I took them into account because this is the only way unorthodox opinions can be voiced. However, pamphlets by obvious cranks directed against Newton's and other prerelativistic theories have been excluded. In addition, dissertations on the subject are generally not included.

So far, I have dated and identified a total of 312 books without actually having had in hand more than *half of them.*[1] They are listed in the appended bibliography. About 15 titles still wait to be dated, or their relevancy to relativity must still be determined. Only the first edition of a book is noted, with one or two exceptions when the first edition falls outside the period studied. My impression is that the problem of the *Sonderdrucke* mentioned will not greatly affect the shape of the resulting plot given in Fig. 2.

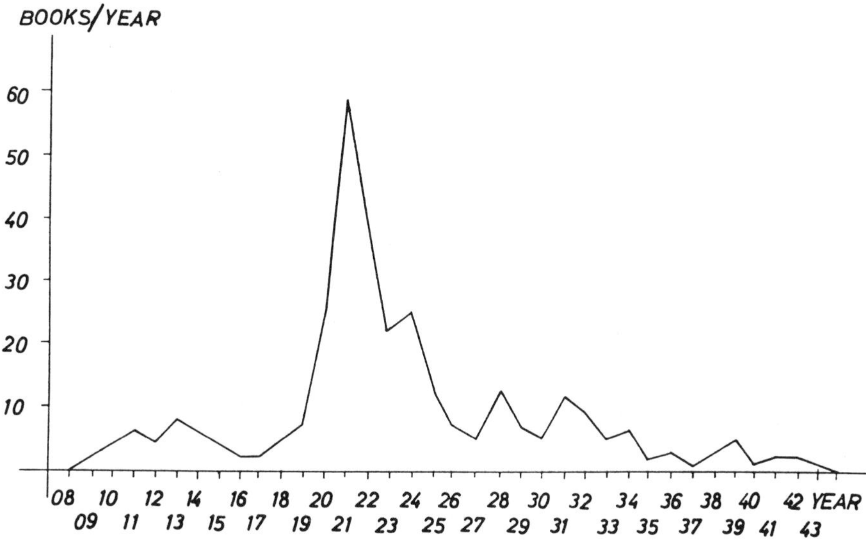

FIGURE 2. Yearly distribution of German books on relativity, total.

I included four articles in encyclopedias (e.g., Pauli's or Kottler's), but omitted the various monographs on theoretical physics with chapters on relativity. The 12 mathematics books that appear were included mostly because of their sections on Riemannian geometry. An example of a book the title of which tries to cash in on Einstein's fame but does not contain anything directly related to relativity is Blaschke's three volume set. When pro-relativity and anti-relativity titles are plotted separately, both curves follow the same general trend (Figs. 3 and 4). Among the 149 anti-relativity books and brochures, less than 10% are from serious critics. The 135 relativity books, besides the encyclopedic articles mentioned, comprise 46 monographs or lectures, 35 popularizations and applications to high school teaching, and 50 books on natural philosophy or philosophy proper. With relativity being just one among other items in some of the last category, their inclusion in the bibliography is thus debatable. In Figure 2 there are also 22 books of a general educational purpose (for an educated public), partly published by book clubs.

In order to check the influence of the development of quantum mechanics on the book publication activity in relativity, I also made statistics of books on the Bohr–Sommerfeld atom model and quantum theory

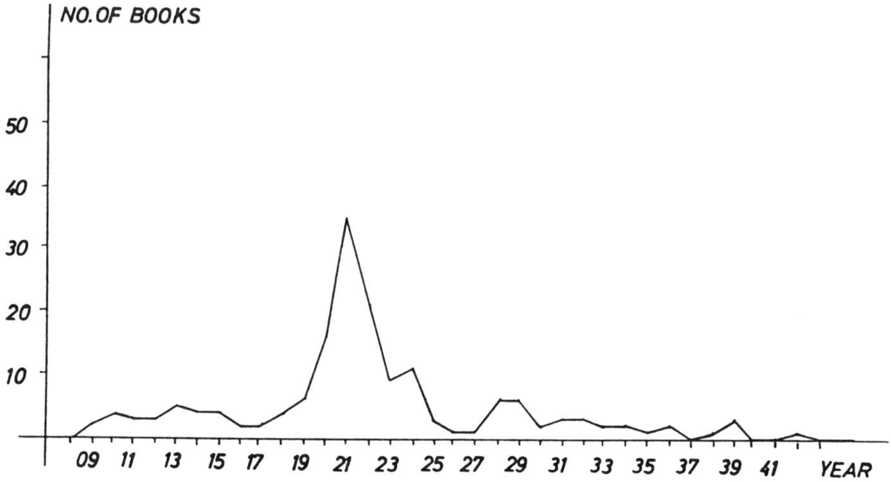

FIGURE 3. Yearly distribution of German books on relativity by relativists.

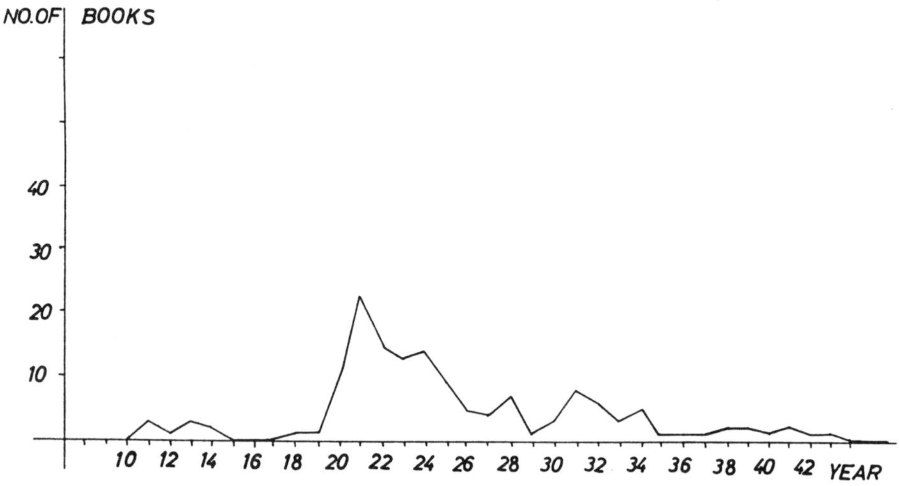

FIGURE 4. Yearly distribution of German books on relativity by anti-relativists.

(1919–1945). Again, there are problems, because in the 1930s, the field developed rapidly with applications to atomic, molecular, and nuclear physics; in the 1940s, even elementary particle physics appears. In my statistics (Fig. 5), I included books on atomic physics *before* 1926, but discarded those published afterward. The effect of nuclear physics books and some atomic physics books after 1926 on the statistics is also shown in Figure 5.

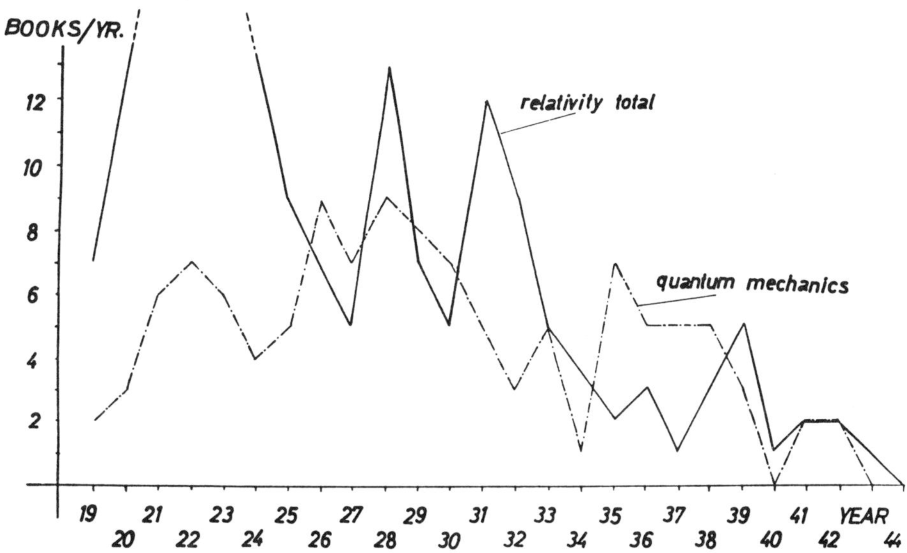

FIGURE 5. Comparison of book publication activity in relativity and quantum mechanics.

2. Interpretation

It is clear from the various plots that the upsurge in book publication on relativity with its peak in 1921 arose *after* the announcement of the light deflection measurements at the sun in November 1919. Thus, we are led to accept, as a reason for the 1921 peak, Einstein's fame, dating from the beginning of 1920, and the public response to it. A closer look at the years 1920–1924 confirms this interpretation: popular presentations of relativity theory, natural philosophy texts concerned with relativity, and anti-relativity books add up to three quarters of the total number of titles published during 1920–1922. This percentage grows in the following years. However, the number of monographs or lecture notes on relativity rises from four in 1920 to nine in 1921 and then falls to six in 1922. After 1925 and until 1933, no further monograph on relativity theory appeared. The only monograph published during the Nazi rule in Germany was Heckmann's *Theories of Cosmology* (1942). As shown in Table 1, until 1921 the major textbooks on relativity had already reached their third (or higher) edition. The Nobel prize won by Einstein for 1921 and announced in 1922 came too late to be of relevance for the peak in Figures 2–4. This assessment agrees with Missner's conclusion (Missner 1985) that the crucial period for the development of Einstein's fame in the United States was bracketed by November 1919 and August 1920.

TABLE 1. Editions of well-known monographs on relativity
1. Über die spezielle und allgemeine Relativitätstheorie.
2. Collection of papers edited by O. Blumenthal.

		Editions					
Author		1	2	3	4	5	12
A. Einstein		1917	1917	1918			1921
M. v. Laue	ST	1911	1913	1919	1923	1952	
	GRT	1921	1923			1965	
M. Born		1920	1921	1922	1964		
H. Weyl		1921	1923				
Lorenz–Einstein–Minkowski		1913		1919	1921	1923	

In addition to this explanation, we must not omit the influence of the general political and economical environment. The end of World War I may also explain some of the growing book publishing activity. In particular, the first sharp decrease in book publication is well correlated with the *monetary* crisis in Germany during 1922–1923. Because of the cost of the lost war, including the reparation payments Germany had to make to the victorious allies, the fiscal budgets of the German states became very tight. This resulted in drastic losses in buying power of the public libraries, especially university libraries. Publishers had to adjust quickly.[2] The downward slope thus coincides with the monetary inflation in Germany culminating in 1923. It ended in November 1923 with the introduction of a stable currency. In order to find out if the downward slope stands for a real decrease of interest in relativity or is due to economic factors, a comparison should be made with book publication activity in Germany in other scientific fields.[3]

Studies of the relativity publication activity in other countries or on an international scale would also be of interest. As for the latter, in using Combridge's bibliography (Combridge 1965) of research papers in relativity and gravitation, I did not reach a conclusive result (Fig. 6). The not-so-marked peaks of 1923 and 1933 rise from a rather high average not showing a steady decline. The plot selects as relevant only 918 of the 1707 entries in Combridge's bibliography, which contains many papers on quantum theory and pure mathematics. It is very likely that the plot reflects more of Combridge's interests and available reading time than of the actual publishing activity. The peak at the beginning of the 1930s might be connected with the discussions of cosmological questions and alternative theories of gravitation. It is invisible though in the German book publication activity if we regard only relativity books.

My interpretation of Combridge's bibliography is in complete accord with the investigation concerning publications on general relativity abstracted by *Physikalische Berichte* and *Physics Abstracts* and made previously by J. Eisenstaedt (Eisenstaedt 1986). He found a peak shortly after 1922, a rapid decline from 1925–1926 to 1930, and a not very pronounced second high in the beginning of the 1930s. Eisenstaedt's survey covers all countries. Together, our data imply that the publication rate of books on relativity in Germany genuinely reflects a very hot phase of public interest in relativity between 1920 and 1922 and a wane of interest beginning in 1924–1925, the particular economic factors in Germany notwithstanding.

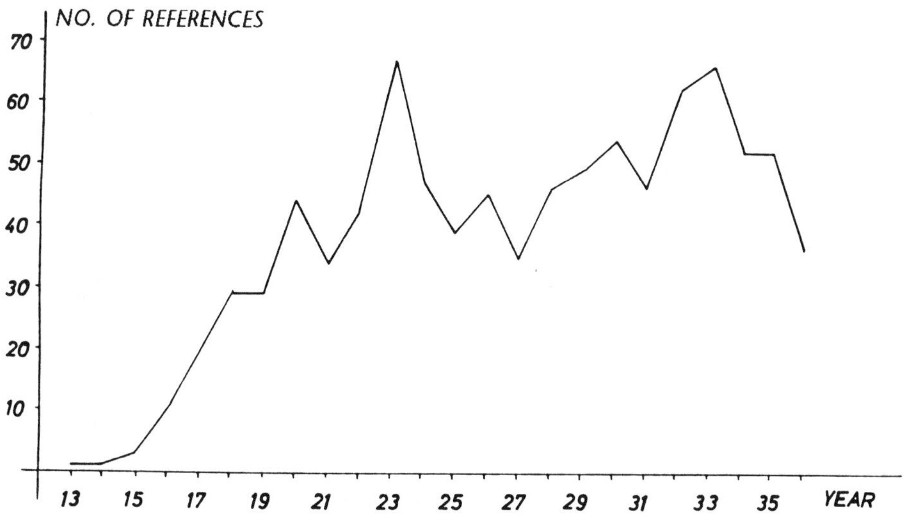

FIGURE 6. Combridge's statistics concerning research papers on relativity and differential geometry.

It is not so obvious from the statistics that with the advent of quantum mechanics in 1925–1926 interest in relativity lessened; any such effect was masked by the ongoing anti-relativist publishing activity that reached a second climax at the beginning of the 1930s; 1931 was the year in which the preposterous title *Hundert Autoren gegen Einstein (A Hundred Authors Against Einstein)* was published. The comparison of the graphs for relativity and quantum theory books shows the overwhelming impact of relativity. Quantum mechanics in Germany never received as much public response as relativity, nor did it encounter as much resistance and resiliency.

Although these results may not be surprising, it is gratifying to me that they are backed by empirical data, something I have rarely seen after Lecat's statistics. It was astonishing for me that relatively few books in German concerning quantum theory and its application in this period exist. However, this fits well with the decline of the number of science publications in Germany during the 1930s.[4] I do not wish to imply that a subject such as the reception of relativity theory in Germany can be dealt with properly by the presentation of only quantitative material.[5] Nevertheless, before an attempt is made to explain the singular time scale of the impact of relativity theory on the general public (e.g., in terms of the influence exerted by the media), a solid empirical foundation must be laid.

Acknowledgments. I thank the participants of the conference, notably John Stachel, Boston; G. Berg, Uppsala; and S. Sigurdsson, for interesting comments.

NOTES

[1] The sources for the titles that I do not own or have copies of are: *Deutsches Bücherverzeichnis*, Verlag des Börsenvereins der deutschen Buchhändler zu Leipzig; Naturae Novitates, R. Friedländer u. Sohn, Berlin; *Bibliografie der schweizerischen naturwissenschaftlichen Literatur*, Bern; book reviews, notably in *Physikalische Zeitschrift*; advertisements of the publishers in the books seen.

[2] I am grateful to Professor M. E. Bergmann, New York, for the information concerning the situation of university libraries.

[3] Professor J. D. North kindly informed me that from a bibliography of works in relativity that he compiled in connection with his book on cosmology (North 1965) that comprises an international list of books and articles a marked peak in 1922 resulted.

[4] Cf. the figures given on p. 185 of Beyerchen (Beyerchen 1977).

[5] For a critique of the quantitative evaluation of concepts in the history of science, see the article of M. J. Nye on scientific decline (Nye 1984).

REFERENCES

Beyerchen, Alan D. (1977). *Scientists under Hitler.* New Haven, Connecticut: Yale University Press.

Combridge, J.T. (1965). *Bibliography of Relativity and Gravitation Theory 1921–1937.* London: King's College.

Eisenstaedt, Jean (1986). La relativité générale à l'étiage: 1925–1955. *Archive for History of Exact Sciences* 35: 115–185.

Gehrcke, Ernst (1924). *Die Massensuggestion der Relativitätstheorie.* Berlin: Meusser.

Lecat, Maurice (1924). *Bibliographie de la relativité.* Bruxelles: Lamertin.

Missner, Marshall (1985). "Why Did Einstein Become Famous in America?" *Social Studies of Science* 15: 267–291.

North, John David (1965). *The Measure of the Universe.* London and New York: Oxford University Press (Clarendon).

Nye, Mary Jo (1984). "Scientific Decline. Is Quantitative Evaluation Enough?" *Isis* 75: 697–708.

APPENDIX

Bibliography of Books on Relativity in the German Language Between 1908 and 1945.

Adler, Friedrich (1920). *Ortszeit, Systemzeit, Zonenzeit und das ausgezeichnete Bezugssystem der Elektrodynamik.* Wien: Volksbuchhandlung.

Alliata, Giulio (1921). *Verstand contra Relativität.* Locarno: Vito Carminati.

——— (1922). *Das Weltbild der Äthermechanik.* Leipzig: Hillman.

——— (1922). *Die Planetenanomalien im Weltbild der Äthermechanik.* Leipzig: Hillman.

——— (1923). *Missverständnisse zu den Grundlagen der Einsteinschen Relativitätstheorie, zu de Sitter's Einwand zum Impulsprinzip, zum Dopplereffekt.* Leipzig: Hillmann.

Angersbach, Adam (1920). *Das Relativitätsprinzip.* Leipzig: Teubner.

Aster, von Ernst (1922). *Raum und Zeit in der Geschichte der Philosphie und Physik.* Berlin: Paetel.

Auerbach, Felix (1921). *Raum und Zeit, Materie und Energie.* Leipzig: Dürrsche Buchhandlung.

Bäcker, Hans (1923). *Die Relativitätslehre und der Mensch.* Potsdam: Der weiße Ritter Verlag.

Balster, Wilhelm (1928). *Der Fehler in der Einsteinschen Relativitätstheorie, gemeinverständlich dargestellt.* Leipzig: Hillmann.

Barnewitz, Friedrich (1920). *Einsteins Relativitätstheorie. Versuch einer volkstümlichen Zusammenfassung.* Rostock: Leopolds Universitäts-Buchhandlung.

Bauer, Hans (1922). *Mathematische Einführung in die Gravitationstheorie Einsteins nebst einer exakten Darstellung ihrer wichtigsten Ergebnisse*. Leipzig, Wien: Deuticke.

Bavink, Bernhard (1921). *Ergebnisse und Probleme der Naturwissenschaft, eine Einführung in die heutige Naturphilosophie*. Leipzig: Hirzel.

Becher, Erich (1915). *Weltgebäude, Weltgesetze, Weltentwicklung. Ein Bild der unbelebten Natur*. Berlin: Reimer.

Beck, Guido (1929). "Allgemeine Relativitätstheorie." In *Handbuch der Physik, Band 4*. Berlin: Springer.

Becker, Walther (1921). *Die Relativitätstheorie gemeinverständlich dargestellt*. Leipzig: Hachmeister u. Thal.

Beer, Fritz (1919). *Die Einsteinsche Relativitätstheorie und ihr historisches Fundament*. Wien und Leipzig: Perles.

Benedicks, Carl A. F. (1923). *Raum und Zeit. Eines Experimentalphysikers Auffassungen von diesen Begriffen und von deren Veränderung*. Zürich: Orell Füssli.

Berg, Otto (1910). *Das Relativitätsprinzip der Elektrodynamik*. Sonderabdruck aus Abhandl. Friesschen Schule Göttingen (2), 3: 333–382. Göttingen: Vandenhoeck and Ruprecht.

Bernays, Paul (1913). *Über die Bedenklichkeit der neueren Relativitätstheorie*. Göttingen: Vandenhoeck and Ruprecht.

Binder, Erich (1923). *Elementare Einführung in die spezielle Relativitätstheorie*. Lübeck: Coleman.

Blaschke, Wilhelm (1921). *Vorlesungen über Differential-Geometrie und geometrische Grundlagen von Einstein's Relativitätstheorie*. Band 1, *Elementare Differentialgeometrie* (1923); Band 2, *Affine Differentialgeometrien* (1929); Band 3, *Differentialgeometrie der Kreise und Kugeln*. Berlin: Springer.

Bloch, Werner (1918). *Einführung in die Relativitätstheorie*. Leipzig and Berlin: Teubner.

Bollert, Karl (1921). *Relativitätstheorie und ihre Stellung im System der Gesamterfahrung*. Dresden and Leipzig: Steinkopf.

Borel, Emile (1931). *Zeit und Raum*. Stuttgart: Francksche Verlagshandlung.

Born, Max (1920). *Die Relativitätstheorie Einsteins und ihre physikalischen Grundlagen*. Berlin: Springer.

Botezadu, N. (1939). *Die Wahrheit in der Relativitätstheorie, Erklärung des Michelsson-Versuchs*. Cluj: Produktivgenossenschaft Buchdruckerei Pallas.

Brack-Warstade, J. (1924). *Materie und Weltraum in ihrem gegenseitigen physischen und metaphysischen Verhältnis auf Grund der Teilbarkeit des Raumes*. Stade i. Hann.: Zwei-Welten-Verlag.

Bresler, Johannes (1922). *Jenseits von Klug und Blöde, 1. Bezugslehre (Relativitätstheorie), 2. Psychiatrie und Psychoanalyse*. Halle A.S.: Marhold.

Brill, Alexander W. von (1909). *Vorlesungen zur Einführung in die Mechanik raumerfüllter Massen*. Berlin and Leipzig: Teubner.

——— (1912). *Das Relativitätsprinzip*. Sonderabdruck aus Band 21 der Deutschen

Mathematischen Vereinigung. Leipzig and Berlin: Teubner.

Brösske, Ludwig (1931). *Der Sturz der Irrlehre Einsteins und der bisherigen Auslegungen der Aberration, des Airy-und des Fizeauschen Versuchs sowie Lösung.* Düsseldorf: Industrieverlag and Druckerei.

Brühlmann, Otto (1924). *Wille und Licht, 1. Teil: Licht und Kraft in der Physik.* Bern: Haupt.

——— (1931). *Möglichkeiten und Deutung der absoluten Konstanz der Lichtgeschwindigkeit.* Leipzig: Hillmann.

——— (1932). *Licht gestaltet Physik. Erkenntniskritische Untersuchung der physikalischen Gestaltung grundsätzlichen Sicherung der Lorentz-Transformation und Aufhebung der speziellen Relativitätstheorie.* Wien: Braumüller Universitäts-Verlagsbuchhandlung.

Brüll, Erhard (1929). *Erkenntniskritische Grundprobleme der Relativitätstheorie, Quanten- und Wellenmechanik.* Breslau: Borgmeyer.

Bucherer, Heinrich (1923). *Die Planetenbewegung auf Grund der Quantentheorie und eine Kritik der Einsteinschen Gravitationsgleichungen.* Bonn: Röhrscheid.

Büchler, Robert (1923). *Die Gesetze der Natur.* Aachen: Aachener Verlags- und Druckereigesellschaft.

——— (1922). *Über die Einsteinsche Relativitätstheorie.* Aachen: Selbstverlag.

——— (1921). *Lehrsätze über das Weltall mit Beweis in Form eines offenen Briefes an Prof. Einstein.* Aachen: Aachener Verlags- und Druckereigesellschaft.

Busam, Theodor (1921). *Der Irrtum Einsteins. Der Begriff Raum und Zeit. Relativität. Der Irrtum. Der Ausweg.* Baden-Baden: Selbstverlag.

Carlebach, Joseph (1912). *Die Geschichte des Trägheitssatzes im Lichte des Relativitätsprinzips.* Berlin: Wiedemann.

Carnap, Rudolph (1922). *Der Raum. Ein Beitrag zur Wissenschaftslehre.* Kantstudien: Ergänzungsheft 57. Berlin: Reuther and Richard.

Cassirer, Ernst (1921). *Erkenntnistheoretische Betrachtungen zur Einsteinschen Relativitätstheorie.* Berlin: Bruno Cassirer.

Christiansen, Hans (1920). *Absolut und Relativ! Eine Ablehnung des "Relativitätsprinzips" Einsteins auf Grund einer reinen Begriffsmathematik.* Wiesbaden: Hofbuchhandlung Staadt.

Christoph, Hans (1925). *Die Fahrt in die Zukunft. Ein Relativitätsroman.* Stuttgart: Deutsche Verlagsanstalt.

Clauss, Karl (1937). *Der Äther eine logische Notwendigkeit.* Krefeld: Greven.

Cohn, Emil (1913). *Physikalisches über Raum und Zeit.* Leipzig: Teubner.

Dieck, Wilhelm (1924). *Die Relativitätslehre und ihre Stellung zur zeitgenössischen Philosophie.* Sterkrade: Osterkamp.

Dingler, Hugo (1921). *Physik und Hypothese. Versuch einer induktiven Wissenschaftslehre nebst einer kritischen Analyse der Relativitätstheorie.* Berlin: de Gruyter.

——— (1921). *Kritische Bemerkungen zu den Grundlagen der Relativitätstheorie.*

Wiederabdruck aus Physikalische Zeitschrift 23: 668–675. Leipzig: Hirzel.

——— (1922). *Relativitätstheorie und Oekonomieprinzip.* Leipzig: Hirzel.

——— (1923). *Das Problem des absoluten Raums in historisch-kritischer Behandlung.* Sonderabdruck aus Jahrbuch der Radioaktivität und Elektronik 19: 165–214. Leipzig: Hirzel.

——— (1938). *Die Methode der Physik.* München: Reinhardt.

Driesch, Hans A. E. (1930). *Relativitätstheorie und Weltanschauung.* (2. umgearbeitete Auflage von *Relativitätstheorie und Philosophie*). Leipzig: Quelle und Meyer.

Duschek, Adalbert, und Mayer, Walther (1930). *Lehrbuch der Differentialgeometrie. Band 2: Riemannsche Geometrie.* Leipzig und Berlin: Teubner.

Düsing, Karl (1922). *Einstein's Relativitätstheorie.* Leipzig: Jänecke Verlagsbuchhandlung.

Eddington, Arthur S. (1923). *Raum, Zeit und Schwere. Ein Umriss der allgemeinen Relativitätstheorie.* Braunschweig: Vieweg.

——— (1925). *Relativitätstheorie in mathematischer Behandlung.* (Grundlehren der mathematischen Wissenschaft, Band 18). Berlin: Springer.

——— (1935). *Die Naturwissenschaft auf neuen Bahnen.* Braunschweig: Vieweg.

——— (1939). *Das Weltbild der Physik und ein Versuch seiner philosophischen Deutung.* Braunschweig: Vieweg.

Ehrenfest, Paul (1912). *Zur Krise der Lichtäther-Hypothese.* Antrittsrede Leiden. Leiden: IJdo; Berlin: Springer, 1913.

Eidlitz, Otto (1925). *Vom Syllogismus in der Relativitätstheorie. Die Gravitation und die Lösung eines Welträtsels.* Wien: Selbstverlag.

Einstein, Albert (1915). "Die Relativitätstheorie." In *Die Kultur der Gegenwart. 3. Teil, 3. Abtl. 1. Band.* Leipzig and Berlin: Teubner.

——— (1916). *Die Grundlagen der allgemeinen Relativitätstheorie.* Leipzig: Barth.

——— (1917). *Über die spezielle und die allgemeine Relativitätstheorie.* Braunschweig: Vieweg.

——— (1920). "Aether und Relativitätstheorie." *Rede gehalten am 5. Mai 1920 an der Reichsuniversität zu Leiden.* Berlin: Springer.

——— (1921). *Geometrie und Erfahrung.* Erweiterte Fassung des Festvortrages an der Akademie der Wissenschaften. Berlin: Springer.

——— (1922). *Vier Vorlesungen über Relativitätstheorie.* (Gehalten im Mai 1921 an der Princeton University.) Braunschweig: Vieweg.

——— (1934). *Mein Weltbild.* Amsterdam: Querido.

Einstein, Albert, und Grossmann, Marcel (1913). *Entwurf einer verallgemeinerten Relativitätstheorie und einer Theorie der Gravitation.* Sonderabdruck aus Zeitschrift für Mathematik und Physik 62: 225–261. Leipzig and Berlin: Teubner.

Einstein, Albert, und Infeld, Leopold (1938). *Physik als Abenteuer der Erkenntnis.* Leiden: Sijthoff.

Elsbach, Alfred C. (1924). *Kant und Einstein.* Berlin: de Gruyter.

Engelhardt, Victor (1922). *Einführung in die Relativitätstheorie.* Charlottenburg: Volkschochschulverlag.

Exner, Franz (1919). *Vorlesungen über die physikalischen Grundlagen der Naturwissenschaft.* Leipzig und Wien: Deuticke.

Farsky, Hermann (1925). *Vom Relativen zum Absoluten, Kritik der Einsteinschen Theorie.* Berlin: Taurus.

Felke, Georg N. (1928). *Einstein für Jedermann. Die Relativitätstheorie in ihren Grundgedanken für Laien.* Berlin: Verlag der neuen Gesellschaft.

Fischer, Franz X. (1921). *Das Einsteinsche Relativitätsprinzip und die philosophischen Anschauungen der Gegenwart.* Wissen und Glauben 19: 129–159, Mergentheim a.d.T.: Ohlinger.

Flamm, Ludwig (1915). *Die neuen Anschauungen über Raum und Zeit. Das Relativitäts-Prinzip.* Wien: Braumüller.

Frank, Philipp (1920). *Relativitätstheorie.* Leipzig: Teubner.

Freundlich, Erwin (1916). *Die Grundlagen der Einsteinschen Gravitationstheorie.* Berlin: Springer.

——— (1924). *Drei Vorlesungen über Relativitätstheorie.* London: Methuen.

Fricke, Hermann (1919). *Eine neue und einfache Deutung der Schwerkraft und eine anschauliche Erklärung der Physik des Raumes.* Wolfenbüttel: Heckner.

——— (1920). *Der Fehler in Einstein's Relativitätstheorie.* Wolfenbüttel: Heckner.

——— (1920). *Die neue Erklärung der Schwerkraft.* Wolfenbüttel: Heckner.

——— (1934). *Die im Innern erdähnliche Sonne.* Weimar: Borkmann.

——— (1939). *Weltätherforschung.* Weimar: Borkmann.

Friedländer, Salomo (1932). *Kant gegen Einstein. Fragelehrbuch.* Berlin: Verlag Der neue Geist.

Friedrichs, Gustav (1920). *Die falsche Relativität der Sinne. Keine Mathematik nur gewöhnliches Rechnen.* Osnabrück: Baumert.

Gartelmann, Henri (1920). *Zur Relativitätslehre. Eine kritische Betrachtung.* Berlin: Verlag der neuen Weltanschauung.

——— (1934). *Der Fall Einstein.* Dresden: Pierson.

Gawronsky, Dimitry (1924). *Die Relativitätstheorie Einsteins im Lichte der Philosophie—Ein neuer Beweis der Lorentz-Transformation.* Bern: Haupt.

——— (1924). *Das Trägheitsgesetz und der Aufbau der Relativitätstheorie.* Stuttgart: Engelhorns Nachfahren.

Gehrcke, Ernst (1924). *Kritik der Relativitätstheorie. Gesammelte Schriften über absolute und relative Bewegung.* Berlin: Meusser.

——— (1924). *Die Massensuggestion der Relativitätstheorie. Kulturhistorisch-psychologische Dokumente.* Berlin: Meusser.

Geiger, Moritz (1921). *Die philosophische Bedeutung der Relativitätstheorie.* Halle a. S.: Niemeyer.

Geissler, Friedrich J.K. (1921). *Gemeinverständliche Widerlegung des formalen Relativismus (von Einstein und Verwandten) und zusammenhängende Dar-*

stellung einer grundwissenschaftlichen Relativität. Leipzig: Hillmann.

Geppert, Hermann (1923). *Ist die Welt absolut oder relativ? Vollständige Widerlegung der Relativitätstheorie.* Karlsruhe: Reiff.

Gimmerthal, Armin (1926). *Die Irrtümer und Trugschlüsse in Einsteins Relativitätstheorie.* Langendreer: Selbstverlag.

Gilbert, Leo (1914). *Das Relativitätsprinzip, die jüngste Modenarrheit der Wissenschaft. Und die Lösung des Fizeau Problems.* Brackwede: Breitenbach.

Gleich, Gerold von (1930). *Einsteins Relativitätstheorie und physikalische Wirklichkeit.* Leipzig: Barth.

Goldschmidt, Ludwig (1923). *Gegen Einstein's Metaphysik. (Eine kritische Befreiung.)* Lübeck: Coleman.

Goldstein, Emil (1932). *Der Relativitätsgedanke und Relativität und Entwicklung.* Berlin: Konzentration, A. G. Abt. Buchverlag.

Graetz, Leo (1923). *Der Aether und die Relativitätstheorie. Sechs Vorträge.* Stuttgart: Engelhorns Nachfahren.

Gruner, Paul (1910). *Elementare Darlegung der Relativitätstheorie.* Sonderabdruck aus Mitteilung der Naturforschenden Gesellschaft in Bern. Bern: Wyss.

——— (1922). *Das moderne physikalische Weltbild und der christliche Glaube. (Stimmen aus der Deutschen Christlichen Studentenbewegung, Heft 13.)* Berlin: Furche.

——— (1922). *Elemente der Relativitätstheorie.* Bern: Haupt.

Haas, Arthur (1934). *Kosmologische Probleme der Physik.* Leipzig: Akademische Verlagsgesellschaft.

Hack, Karl (1912). *Eine neue Aetherhypothese auf Grund des mechanischen Satzes von der Durchkreuzbarkeit der Gasmolekülwege.* Würzburg: Physiochemische Vertriebsanstalt.

Haedicke, Johannes (1932). *Die physikalische Unhaltbarkeit der Relativitätstheorie Einsteins.* Leipzig: Hillmann.

Haering, Theodor L. (1923). *Philosophie der Naturwissenschaften.* München: Rösl and Cie., Berlin: Paetel.

Hasse, Max (1920). *Einsteins Relativitätstheorie. Versuch einer volkstümlichen Darstellung.* Magdeburg: Hasse.

Heckmann, Otto (1942). *Theorien der Kosmologie. Fortschritte der Astronomie, Band 2.* Berlin: Springer.

Heffter, Lothar (1921). *Über eine vierdimensionale Welt.* Freiburg i.Br.: Speyer and Kerner.

Hegedusić, Mladen (1928). *Charakteristik der Gravitationsfelder.* Berlin: Friedländer; Zagreb: Naklada "Astra" Klub.

Heinsohn, Johannes (1933). *Einstein Dämmerung: Kritische Betrachtungen zur Relativitätstheorie.* Leipzig: Hillmann.

Henning, Hans (1922). *Einsteins Relativitätstheorie im Lichte der experimentellen Psychologie und des philosophischen Realismus.* Leipzig: Barth.

Hentschel, Willibald (1920). *Das Relativitätsprinzip im Rahmen einer gesamt-*

ansicht von Welt und Mensch. Leipzig: Malthers.

Hesse, Paul (1922). *Licht und Weltäther als Ursachen der Planetenumläufe und Rotationen. Die astrologische Forschung und die Einsteinsche Theorie im Lichte einer neuen Himmelsmechanik und Naturphilosophie.* Berlin-Pankow: Bücher-Verlag.

Hopf, Eberhard (1921). *Einstein's Relativitätstheorie.* Leipzig: Verlag für Kunst und Wissenschaft, A. O. Paul.

Hopf, Ludwig (1931). *Die Relativitätstheorie* (Verständliche Wissenschaft, Band 14). Berlin: Springer.

Horvath, Clemens von (1921). *Raum und Zeit im Lichte der speziellen Relativitätstheorie.* Berlin: Springer.

Huber, Franz (1924). *Newton oder Einstein? Die Grundprobleme der Relativitätstheorie in historisch-genetischer Entwicklung, klärender Beleuchtung und neuer Lösung nach Newtonschen Prinzipien.* Leitmeritz: Selbstverlag.

Isenkrahe, Caspar (1921). *Zur Elementaranalyse der Relativitätstheorie* (Sammlung Vieweg: 51.) Braunschweig: Vieweg.

Israel, Hans (1929). *Beweis, weshalb die Einsteinsche Relativitätstheorie ad acta zu legen ist.* Leipzig: Hillmann.

Israel, Hans, Ruckhaber, Erich, und Weinmann, Rudolf (1931). *100 Autoren gegen Einstein.* Leipzig: Voigtländer.

Jaeger, Wilhelm (1925). *Zeit, Raum, Stoff, Äther, Kraft, Masse, Relativitätstheorie.* Plauen: Bredel and Jost.

Jaffe, Georg (1931). *Zwei Dialoge über Raum und Zeit.* Leipzig: Akademische Verlagsgesellschaft.

Jeans, James H. (1937). *Der Weltenraum und seine Rätsel.* (Übersetzung von The Mysterious Universe.) Stuttgart: Deutsche Verlagsanstalt.

Jontschow, Th. (1928). *Die Offenbarung. Theorie und Wirklichkeit. 3. Widerlegung der von A. Einstein verfassten Relativitätstheorie.* Wien: Minerva.

Jordan, Pascual (1936). *Die Physik des 20. Jahrhunderts.* Braunschweig: Vieweg.

Jovičić, Milorad Z. (1924). *Über den Wert der Relativitätstheorie Einsteins.* Leipzig and Wien: Hölder-Pichler-Tempsky AG.

Karollus, Franz (1921). *Wo irrt und was übersieht Einstein? Ein neuer Versuch zur Lösung einer strittigen physikalischen Frage.* Brünn: Winkler.

Kirchberger, Paul (1920). *Was kann man ohne Mathematik von der Relativitätstheorie verstehen?.* Karlsruhe: Müllersche Hofbuchhandlung.

Keller, Hugo (1924). *Haltlosigkeit der Relativitätstheorie.* Leipzig: Hillmann.

——— (1925). *Gegenbeweise gegen die Relativitätstheorie.* Leipzig: Hillmann.

Klein, Felix (1927). *Vorlesungen über die Entwicklung der Mathematik im 19. Jahrhundert. Band 2.* Berlin: Springer.

Kleiner, Hermann (1942). *Kritische Betrachtung der Begriffe: Schwere, Gewicht, Masse, Kraft, Energie, Äther, Zeit und Raum. Die Widerlegung der Newtonschen Gravitationstheorie.* Leipzig: Diskus-Verlag.

Kleinert, Heinrich (1923). *Die Prüfungsmöglichkeiten der Einsteinschen Relativitätstheorie.* Bern and Leipzig: Bircher.

Kleinschrod, Franz (1920). *Das Lebensproblem und das Positivitätsprinzip in Zeit und Raum und das Einsteinsche Relativitätsprinzip in Raum und Zeit* (Frankfurter zeitgemässe Broschüren: 40), Frankfurt: Hamm.

Kneser, Adolf (1918). *Mathematik und Natur. Von der Schwere.* Zwei Akademische Reden. Breslau: Trewendt und Granier.

König, Edmund (1929). *Ist Kant durch Einstein widerlegt?.* Sondershausen: Eupel.

Kohler, Karl-Maria (1921). *Das Exzentrizitätsprinzip als Korrelat zur Relativitätstheorie. Versuch zur Postulierung eines Kriteriums der absoluten Bewegung und einer Erklärung des Michelsonschen Effektausfalles.* Leipzig, Wien: Deuticke.

Koller-Aeby, Hermann (1927). *Die Universalität der Gravitation in den grössten und kleinsten Systemen.* Basel: Schwabe.

—— (1931). *Der Grundirrtum Newton's als Ursache des Einsteinschen Grundirrtums.* Leipzig: Hillmann.

—— *Die Einsteinsche Relativitätstheorie und das Problem der Kausalität.* Leipzig: Hillmann.

Kopf, August (1920). *Die Einsteinsche Relativitätstheorie.* Leipzig: Greszner.

—— (1921). *Grundzüge der Einsteinschen Relativitätstheorie.* Leipzig: Hirzel.

Kottler, Friedrich (1922). *Newton'sches Gesetz und Metrik.* Wien: Hölder-Pichler-Tempsky AG.

—— (1922). "Gravitation und Relativitätstheorie." In *Encyklopädie der Mathematischen Wissenschaften, Band 6, Teil 2, 2. Hälfte.* Leipzig: Teubner.

Kraus, Oskar (1925). *Offene Briefe an A. Einstein und M. von Laue.* Wien and Leipzig: Braumüller.

Kremer, Josef (1921). *Einstein und die Weltanschauungskrisis.* Graz and Wien: Styria.

Kržiwanek, Karl (1915). *Gravitation und Lichtgeschwindigkeit.* Teschen: Prochaska.

Lämmel, Rudolf (1921). *Die Grundlagen der Relativitätstheorie populärwissenschaftlich dargestellt.* Berlin: Springer.

—— (1921). *Wege zur Relativitätstheorie.* Stuttgart: Francksche Verlagsbuchhandlung.

—— (1930). *Relativitätstheorie Lieferung 1. Technische Selbst-Unterrichtsbriefe.* Potsdam: Bonness and Hachfeld.

Lagally, Max (1928). *Vorlesungen über Vektorrechnung.* Leipzig: Akademische Verlagsgesellschaft.

Lamberty, Paul (1924). *Wie der Äther Stoff und Kraft erzeugt. Neues vom Weltäther.* Haag: Selbstverlag.

—— (1925). *Die Ursache von allem erkannt. Das Ende der Relativitätstheorie.* Haag: Selbstverlag.

Laue, Max von (1911). *Das Relativitätsprinzip.* (Sammlung Die Wissenschaft, Band 38). Braunschweig: Vieweg.

—— (1921). *Die Relativitätstheorie. Band 2: Die allgemeine Relativitätstheorie*

und Einstein's Lehre von der Schwerkraft (Sammlung Die Wissenschaft, Band 68). Braunschweig: Vieweg.
——— (1921). "Das physikalische Weltbild." Ein Vortrag. Karlsruhe: Müller.
Lenard, Philipp (1910). *Über Äther und Materie*. Heidelberg: Winter.
——— (1918). *Über Relativitätsprinzip, Äther, Gravitation*. Leipzig: Hirzel.
——— (1921). *Über Aether und Uräther*. Leipzig: Hirzel.
Levi-Cività, Tullio (1924). *Fragen der klassischen relativistischen Mechanik*. Berlin: Springer.
——— (1928). *Der absolute Differentialkalkül (und seine Anwendungen in Geometrie und Physik)*. Berlin: Springer.
Liesegang, Paul F. (1921). *Ein Schaubild zur Darstellung der Zeit-Raum-Verhältnisse in der speziellen Relativitätstheorie*. Sonderabdruck aus Zeitschrift für mathematischen und naturwissenschaftlichen Unterricht 52: 193–201. Leipzig: Teubner.
Lipsius, Friedrich R. (1918). *Naturphilosophie und Weltanschauung*. Leipzig: Kröner.
——— (1927). *Wahrheit und Irrtum in der Relativitätstheorie*. Tübingen: Mohr.
Lodge, Oliver J. (1911). *Der Weltäther*. Sammlung Die Wissenschaft, Band 41. Braunschweig: Vieweg.
Lorentz, Hendrik A. (1914). "Das Relativitätsprinzip." *Drei Vorlesungen gehalten in Teylers Stiftung zu Haarlem*. Bearbeitet von W.H. Keesom. Leipzig: Teubner.
——— (1929). *Vorlesungen über Theoretische Physik an der Universität Leiden*. Band 3, *Äthertheorien und -modelle*, Band 4, *Die Relativitätstheorie für gleichförmige Translation*. Leipzig: Akademische Verlagsgesellschaft.
Lorentz, Hendrik A., Einstein, Albert, und Minkowski, Hermann (1913). *Das Relativitätsprinzip*. Leipzig: Teubner.
Maag, Ernst, und Reihling, Karl (1921). *Vom Relativen zum Absoluten. Das Aetherrätsel und seine Lösung*. Stuttgart: Nägele-Schweizerbart.
Madary, Carl (1921). *Einstein, E.H. Schmitt und das Ende der "Philosophie." Versuch einer Synthese*. Berlin: Alberti.
Marcus, Ernst (1926). *Kritik des Aufbaus der speziellen Relativitätstheorie und Kritik der herrschenden Hypothese der Lichtausbreitung*. Berlin: Verlag Der Sturm.
May, Eduard (1941). *Am Abgrund des Relativismus*. Berlin: Lüttke.
Maximilian, Herbert (1931). *Contra Einstein*. Elbing: Hohmann.
Mercier, André (1939). *Theoretisch-physikalische Fragen der Kosmologie*. Bern and Leipzig: Haupt.
Merz, Karl (1922). *Die Relativitätstheorie in philosophischer Bedeutung mit Hinweis auf die Kategorienlehre*. Sonderabdruck aus Jahresbericht der Naturforschenden Gesellschaft Graubündens 61. Chur: Schuler.
Mewes, Rudolf (1921). *Raumzeitlehre oder Relativitätstheorie in Geistes- und Naturwissenschaft und Werkkunst*. Berlin: Selbstverlag.
Michelson, Albert A. (1911). *Lichtwellen und ihre Anwendungen*. Leipzig: Barth.

Mie, Gustav (1910). *Lehrbuch der Elektrizität und des Magnetismus.* Stuttgart: Enke.
——— (1921). *Die Einsteinsche Gravitationstheorie.* Leipzig: Hirzel.
Minkowski, Hermann (1909). "Raum und Zeit." *Vortrag auf der 80. Naturforscherversammlung Köln.* Leipzig: Teubner.
——— (1910). *Zwei Abhandlungen über die Grundlagen der elektromagnetischen Vorgänge in bewegten Körpern.* Leipzig: Teubner.
Mitis, Lothar (1930). *Einsteins Grundirrtum.* Leipzig: Hillmann.
Mohorovičič, Stjepan (1923). *Die Einsteinsche Relativitätstheorie und ihr mathematischer, physikalischer und philosophischer Charakter.* Berlin: de Gruyter.
Mokrzycky, Gustav (1922). *Relativisierung des Kausalitätsbegriffes.* Leipzig: Hillmann.
Moszkowski, Alexander (1921). "Einstein." *Einblicke in seine Gedankenwelt.* Hamburg: Hoffmann and Campe; Berlin: Fontane.
Müller, Aloys (1922). *Die philosophischen Probleme der Einsteinschen Relativitätstheorie.* Vorlesungen an der Universität Bonn. Braunschweig: Vieweg.
——— (1922). *Der Gegenstand der Mathematik mit besonderer Beziehung auf die Relativitätstheorie.* Braunschweig: Vieweg.
Müller, Paul J. (1914). *Das Rätsel der Schwerkraft gelöst durch die Raumenergetik.* Wien, Teschen, Leipzig: Prochaska.
Neuberg, Arthur (1939). *Das naturwissenschaftliche Weltbild der Gegenwart. 1. Das neue Weltbild der Physik.* Göttingen: Vandenhoeck und Ruprecht.
Neumann, Ernst R. (1922). *Vorlesungen zur Einführung in die Relativitätstheorie.* Jena: Fischer.
Newest, Th. (Goldzier, Hans) (1921). *Gegen Einstein. Die Erfahrung im Weltall. Einige Weltprobleme. Allgemeinverständliche Abhandlungen.* Wien: Frisch-Verlag.
Niedermöller, Heinrich A. (1922). *Die relativ veränderliche Lichtgeschwindigkeit.* Dahle i.W.: Ossenberg-Engels.
Nordmann, Charles (1922). *Einstein und das Weltall.* Stuttgart: Hoffmann.
Oppenheim, Samuel (1921). "Die Gravitation." In *Die Kultur der Gegenwart, 3. Teil, 3. Abteilung, Band 3.* Leipzig: Teubner.
Orthner, Rudolf (1931). *Der Michelson-Versuch, seine Deutung und seine Konsequenzen.* Linz a.D.: Winkler.
——— (1922). *Der Indizienbeweis in der Physik. Die Einsteinsche Relativitätstheorie als Beispiel eines solchen.* Cöthen: van Halem.
Palágyi, Melchior (1914). *Die Relativitätstheorie in der modernen Physik. Vortrag auf dem 85. Naturforschertag in Wien.* Berlin: Reimer.
——— (1925). *Zur Weltmechanik. Beiträge zur Metaphysik der Physik.* Leipzig: Barth.
Passarge, Hans (1928). *Die Gravitation. Wesen und Ursprung.* Leipzig: Hillmann.
Patschke, Arthur (1920). *Umsturz der Einsteinschen Relativitätstheorie. Einführung in die einheitliche Erklärung und Mechanik der Naturkräfte. Kreuzi-*

gung und Auferstehung des Lichtäthers. Berlin-Wilmersdorf: Selbstverlag.
Pauli, Wolfgang (1921). *Relativitätstheorie.* Leipzig: Teubner.
Pécsi, Gusztáv (1923). *Kritik der Relativitätstheorie Einsteins.* Innsbruck: Tyrolia.
——— (1925). *Liquidierung der Relativitätstheorie. Berechnung der Sonnenwindgeschwindigkeit.* Regensburg: Verlagsanstalt vormals Manz.
Petraschek, Karl O. (1922). *Der Grundwiderspruch in der speziellen Relativitätstheorie und seine Folgen.* Leipzig: Hillmann.
Petzoldt, Joseph (1921). *Die Stellung der Relativitätstheorie in der geistigen Entwicklung der Menschheit.* Dresden: Sibyllen Verlag.
Pfaff, Alfred (1921). *Für und gegen das Einsteinsche Prinzip. Für Laien geschrieben von einem Laien.* München: Huber.
Pfeil, Joachim (1932). *Einstein ist erledigt. Die Haltlosigkeit der Relativitätstheorie.* Rathenow: Selbstverlag.
Pflüger, Alexander W. (1920). *Das Einsteinsche Relativitätstheorie.* Bonn: Cohen.
Phalén, Adolf (1922). *Über die Relativität der Raum- und Zeitbestimmungen.* Uppsala: Akad. Bokhandeln; Leipzig: Harrassowitz.
Planck, Max (1910). "Allgemeine Dynamik. Prinzip der Relativität." In *Acht Vorlesungen über theoretische Physik.* Leipzig: Hirzel.
——— (1925). *Vom Relativen zum Absoluten.* Gastvorlesung. Leipzig: Hirzel.
Platte, Konrad F. (1940). *Das Weg-Zeitproblem bei Transversalwellen. Versuche mit stehenden Wellen zum Nachweis der Ätherbewegung.* Weidhofen a.d. Ybbs: Selbstverlag.
Quint, Heinz (1922). *Die Relativitätstheorie. Ein Blick in die Welt Einstein's.* Wein: Anzengruber.
Rawitz, Bernhard (1923). *Raum, Zeit, Gott. Eine kritische-erkenntnistheoretische Untersuchung auf der Grundlage der physikalischen Relativitätstheorie.* Leipzig: Hillmann.
Reichenbach, Hans (1920). *Relativitätstheorie und Erkenntnis a priori.* Berlin: Springer.
——— (1924). *Axiomatik der relativistischen Raum-Zeit-Lehre.* (Reihe Die Wissenschaft, Band 72.) Braunschweig: Vieweg.
——— (1928). *Philosophie der Raum-Zeit-Lehre.* Berlin and Leipzig: de Gruyter.
Reichinstein, David (1934). *Albert Einstein. Ein Bild seines Lebens und seiner Weltanschauung.* Prag: Stella Publishing House.
Richter, Gustav (1912). *Bewegung, die 4. Dimension (Philosophische Grundlagen der Naturwissenschaft.)* Leipzig: Hillmann.
——— (1921). *Kritik der Relativitätstheorie Einsteins.* Leipzig: Hillmann.
——— (1921). *Die Entwicklung der Begriffe: "Kraft, Stoff, Raum, Zeit" durch die Philosophie mit Lösung des Einsteinschen Problems.* Leipzig: Hillmann.
——— (1923). *Vom Heiligen zum Gravitationsgesetz.* Leipzig: Hillmann.
Riebesell, Paul (1921). *Einführung in die Relativitätstheorie.* Hamburg.
——— (1926). *Die Relativitätstheorie im Unterricht. Beihefte der Unterrichtsblätter für Mathematik und Naturwissenschaften* 5. Berlin: Salle.
Ripke-Kühn, Lenore (1920). *Kant contra Einstein. Beiträge zur Philosophie des*

deutschen Idealismus. Veröffentlichungen der Deutschen Philosophischen Gesellschaft, Heft 7. Erfurt: Keysersche Buchhandlung.

Ritz, Walter (1911). "Die Gravitation." In *Gesammelte Werke*. Ursprünglich in Scientia 5: 241–255. Paris: Gauthier-Villars.

Roderich-Stoltheim, Ferdinand (1921). *Einsteins Truglehre*. Leipzig: Hammer-Verlag.

Ruckhaber, Erich (1928). *Die Relativitätstheorie widerlegt durch das Widerspruchsprinzip und die natürliche Erklärung des Michelson-Versuchs. Das dreidimensionale Raum-Zeit System*. Leipzig: Hillmann.

—— (1929). *Relativita, der Roman eines Propheten (Satire)*. Berlin-Spandau: Kuntz.

Rudolf, Heinrich (1911). *Die Stellung der Physik und Naturphilosophie zur Weltätherfrage*. Berlin: Allgemeine medizinische Verlagsanstalt.

—— (1913). *Die hydrodynamische Äthertheorie und die Berechnung von Naturkonstanten aus der Lichtgeschwindigkeit*. Koblenz: Groos.

Russel, Bertrand (1928). *Das ABC der Relativitätstheorie*. München: Drei Masken Verlag.

Sandgathe, Franz (1934). "Die absolute Zeit in der Relativitätstheorie. Ein raumzeitlicher Umbau der Relativitätstheorie." *Bibliothek für Philosophie, Band 30*. Berlin: Heymann.

—— (1934). *Das Ende der Einsteinschen Zeittheorie*. Bonn: Röhrscheid.

Sauer, Friedrich (1943). *Naturgesetzlichkeit und Relativismus*. München: Reinhardt.

Schiffner, Viktor (1931). *Relativitätsprinzip und Gravitations-Problem*. Leipzig: Voigtländer.

—— (1932). *Das Wesen des Alls und seiner Gesetze. Grundzüge einer neuen, einheitlichen Weltanschauung auf physikalischer Grundlage*. Leipzig: Voigtländer.

—— (1934). *Die Probleme des Raumes und der Zeit und die Vorstellung der realen Unendlichkeit*. Leipzig: Voigtländer.

Schimank, Hans (1920). *Gespräch über die Einsteinsche Theorie. Versuch einer Einführung in den Gedankenkreis*. Berlin: Seemann.

Schlesinger, Ludwig (1920). *Raum, Zeit und Relativitätstheorie*. Leipzig: Teubner.

Schlick, Moritz (1917). *Raum und Zeit in der gegenwärtigen Physik*. Berlin: Springer.

Schmidt, Harry (1920). *Das Weltbild der Relativitätstheorie. Allgemeinverständliche Einführung in die Einsteinsche Lehre von Raum und Zeit*. Hamburg: Halting.

—— (1921). *Allgemeinverständliche Einführung in die Grundgedanken der Einsteinschen Relativitätstheorie*. Altona: Hammerich and Lesser.

—— (1921). *Weltäther, Elektrizität, Materie*. Physikalische Fragen der Gegenwart. Hamburg: Hartung.

Schneider, Ilse (1921). *Das Raum-Zeit-Problem bei Kant und Einstein*. Berlin: Springer.

Schouten, Jan A. (1919). *Die direkte Analyse zur neueren Relativitätstheorie.* Sonderabdruck aus Koninklijke Akademie van Wetenschappen. Verhandelingen. 1 Sec., Teil 12, n° 6. Amsterdam: Müller.

——— (1924). *Über die Entwicklung der Begriffe des Raumes und der Zeit und ihre Beziehungen zum Relativitätsprinzip.* Leipzig: Teubner.

——— (1924). *Der Ricci-Kalkül.* Berlin: Springer.

Schrempf, Christoph (1934). *Der Weltäther als Grundlage eines einheitlichen Weltbildes.* Leipzig: Hillmann.

Schuster, Friedrich (1913). *Die moderne theoretische Physik und der Äther. Eine Verteidigung des materiellen Äthers.* Karlsruhe: Braunsche Hofbuchdruckerei.

Schwassmann, Arnold (1921). *Relativitätstheorie und Astronomie. Auf wissenschaftlicher Grundlage allgemein verständlich dargestellt.* Hamburg: Grand.

Schwinge, Otto (1921). *Eine Lücke in der Terminologie der Einsteinschen Relativitätstheorie. Ihre Beseitigung schlichtet automatisch den Streit über das Thema: absolut oder relativ?* Berlin-Steglitz: Puhle.

Schworetzky, Gustav (1922). *Weltäther und Weltall.* Stuttgart: Steinkopf.

Sellien, Ewald (1919). *Die Erkenntnistheoretische Bedeutung der Relativitätstheorie.* Kantstudien Nr. 48. Berlin: Reuther & Richard.

Sepp H. (Josef Hiemesch) (1921) *Stoff und Kraft in Verbindung mit Raum und Zeit nach den neuesten Forschungen über den Atomzerfall, sowie nach der Quanten- und der Einsteinschen Relativitätstheorie.* Lüdenscheid: Heimatverlag Hiemesch.

Siebert, Otto (1921). *Albert Einsteins Relativitätstheorie und ihre kosmologischen und philosophischen Konsequenzen.* Langensalza: Beyer & Söhne.

Stark, Johannes (1922). *Die gegenwärtige Krisis in der deutschen Physik.* Leipzig: Barth.

Stickers, Joe (1922). *Die wahre Relativitätstheorie der Physik und die Mißgriffe Einsteins. Allgemeinverständlich, systematisch und grundlegend dargestellt.* Bielefeld: Breitenbach.

Strasser, Hans (1922). *Die Grundlagen der Einsteinschen Relativitätstheorie. Eine kritische Untersuchung.* Bern: Haupt.

——— (1923). *Einstein's spezielle Relativitätstheorie; eine Komödie der Irrungen.* Bern and Leipzig. Bircher AG.

——— (1924). *Transformationsformeln von Lorentz und die Transformationsformeln der Einsteinschen speziellen Relativitätstheorie.* Leipzig: Bircher.

Struik, Dirk J. (1922). *Grundzüge der mehrdimensionalen Differentialgeometrie in direkter Darstellung.* Berlin: Springer.

Tamari, Leo (1926). *Das Wesen und Wirken der Gravitation philosophisch-naturwissenschaftlich begründet und dargestellt.* Berlin-Charlottenburg: Foerster und Mewis.

Thedinga, Eddo (1922). *Einstein und wir Laien. Wege zu einer neuen Lichttheorie.* Leipzig: Hillmann.

——— (1927). *Einsteins Wunderglaube.* Leipzig: Hillmann.

Thirring, Hans (1921). *Die Idee der Relativitätstheorie.* Berlin: Springer.

Thomas, Bruno (1933). *Axiome und Dogma in der Relativitätstheorie.* Wien: Braumüller.

Thürring, Bruno (1941). "Albert Einsteins Umsturzversuch der Physik." *Sonderabdruck aus Forschungen zur Judenfrage, Band 4.* Berlin: Lüttke.

Tobien, Waldemar (1938). *Der Einstein-Putsch als Werkzeug zur Verewigung der Jahweherrschaft. Seelenkundlich und naturwissenschaftlich gesehen.* Mülheim-Ruhr: Fabri.

Tummers, Josef (1929). *Die spezielle Relativitätstheorie Einsteins und die Logik.* Leipzig: Hillmann.

Uller, Karl (1935). *Das Grundgesetz der Wellenfortpflanzung aus bewegter Quelle in bewegtem Mittel: der Michelson-Versuch und die Raum-Zeit-Lehre von Einstein.* Berlin: Oldenbourg.

Valentiner, Richard W. S. (1921). *Wie stellt sich die Philosophie zum Prinzip der Relativität? Rede zur Rektoratsübergabe, Clausthal am 7.11.1921.* Clausthal: Piepersche Buchdruckerei.

Varičak, Vladimir (1924). *Darstellung der Relativitätstheorie im dreidimensionalen Lobatschefskijschen Raume.* Zagrab: Zaklade Tishare Narodnih Novina.

Veblen, Oswald (1933). *Projektive Relativitätstheorie.* Berlin: Springer.

Vogt, Johann G. (1914). *Die Relativitätstheorie der Physik.* Humboldt-Bibliothek, Heft 12. Brackwede und Bielefeld: Breitenbach.

Vogtherr, Karl (1923). *Wohin führt die Relativitätstheorie. Kritische Betrachtungen vom physikalischen und erkenntnistheoretischen Standpunkt aus.* Leipzig: Hillmann.

——— (1926). *Ist die Schwerkraft relativ? Kritische Betrachtungen über den Relativismus in der neuesten Physik.* Karlsruhe: Macklotsche Druckerei.

——— (1933). *Das Problem der Gleichzeitigkeit.* München: Reinhardt.

Vortisch, Hermann (1921). *Die Relativitätstheorie und ihre Beziehung zur christlichen Weltanschauung.* Hamburg: Agentur des Rauhen Hauses.

Walte, Wilhelm (1921). *Einstein, Michelson, Newton, die Relativitätstheorie. Wahrheit und Irrtum.* Hamburg: Gente.

Wantoch, Rudolf (1921). *Die Einsteinsche Relativitätstheorie. Kurze, für jedermann verständliche Besprechung.* Wien: Hölder.

Weinberg, Arthur B. von (1922). *Die Grenzen der Relativität.* Frankfurt: Senckenbergische Naturforschende Gesellschaft.

Weinman, Rudolf (1922). *Gegen Einstein's Relativierung von Zeit und Raum.* München und Berlin: Odenburg.

——— (1923). *Anti-Einstein.* Leipzig: Hillmann.

——— (1925). *Widersprüche und Selbstwidersprüche der Relativitätstheorie.* Leipzig: Hillmann.

——— (1926). *Versuch einer endgültigen Widerlegung der speziellen Relativitätstheorie.* Leipzig: Hillmann.

—— (1930). *Die Unhaltbarkeit der speziellen Relativitätstheorie.* Innsbruck, Wien and München: Tyrolia.

Weinstein, Max B. (1913). *Die Physik der bewegten Materie und die Relativitätstheorie.* Leipzig: Barth.

Weiss-Jonak, Witpold (1919). *Energie und Entwicklungslehre im Lichte der Relativitätstheorie.* Vortrag. München: Deutscher Monistenbund.

Weizenböck, Roland (1929). *Der vierdimensionale Raum.* Braunschweig: Vieweg.

Wenzl, Aloys (1924). *Das Verhältnis der Einsteinschen Relativitätstheorie zur Philosophie der Gegenwart mit besonderer Berücksichtigung der Philosophie des Als-Ob.* München and Leipzig: Rösl and Cie.; Berlin: Patel.

—— (1929). *Das naturwissenschaftliche Weltbild der Gegenwart.* Leipzig: Quelle und Meyer.

Weyl, Hermann (1918). *Raum-Zeit-Materie. Vorlesungen über die allgemeine Relativitätstheorie.* Berlin: Springer.

—— (1927). "Philosophie der Mathematik und Naturwissenschaft." In *Handbuch der Philosophie.* München and Berlin: Oldenburg.

Weyland, Paul (1920). *Betrachtungen über Einsteins Relativitätstheorie und die Art ihrer Einführung.* (Schriften der Arbeitsgemeinschaft Deutscher Naturforscher zur Erhaltung reiner Wissenschaft, Band 2). Leipzig: Köhler.

Wiechert, Emil J. (1915). "Die Mechanik im allgemeinen Rahmen der Physik." In *Kultur der Gegenwart, 3. Teil, 3. Abt., 1. Band.* Leipzig und Berlin: Teubner.

—— (1921). *Der Äther in Weltbild der Physik.* Sonderabdruck aus Königliche Gesellschaft der Wissenschaften zu Göttingen. Mathematisch-physikalische Klasse. Nachrichten: 29–70. Berlin: Weidmann.

Wien, Wilhelm K. (1919). *Neuere Entwicklungen der Physik und ihrer Anwendungen.* Leipzig: Barth.

—— (1921). *Die Relativitätstheorie vom Standpunkt der Physik und Erkenntnislehre.* Leipzig: Barth.

Wildhagen, Eduard (Herausgeber) (1932). *Materie und Energie. Vom naturwissenschaftlichen Weltbild der Gegenwart.* Berlin: Deutsche Buch-Gemeinschaft.

Winternitz, Josef (1923). *Relativitätstheorie und Erkenntnislehre.* Leipzig and Berlin: Teubner.

Witte, Hans (1914). *Raum und Zeit im Lichte der neueren Physik.* Sammlung Vieweg, Heft 17. Braunschweig: Vieweg.

Wittig, Hans (1921). *Die Geltung der Relativitätstheorie. Eine Untersuchung ihrer naturwissenschaftlichen Bedeutung.* Berlin: Sack.

Wolff, Oskar (1920). *Natur und Mathematik. Hinter den Kulissen von Raum und Zeit.* Seestadt (Tschechoslowakei): Druck E. Weigand.

Wulf, Theodor (1921). *Einsteins Relativitätstheorie gemeinverständlich dargestellt.* Innsbruck: Tyrolia.

—— (1929). *Lehrbuch der Physik.* Freiburg: Herder.

Zbořil, J. (1924). $\sqrt{1 - v^2/c^2}$ *oder das Relativitätsprinzip im Lichte der Wahrheit.* Pressburg: Selbstverlag.

Zehnder, Ludwig (1928). *Die Entwicklung des Weltalls aus mechanischen Grundlagen.* Tübingen: Laupp.

——— (1933). *Der Aether im Lichte der klassischen Zeit und Neuzeit.* Tübingen: Laupp.

Ziegler, Franz (1924). *Die Zahlendisziplin natürlicher Grössen. Kritische Gedanken zu dem Relativitätsprinzip von A. Einstein.* Pokrovsk: Staatsverlag der Autonomen Sowjetrepublik der Wolga-Deutschen.

Ziegler, Johann H. (1923). *"Das Ding an Sich" und das Ende der sogenannten Relativitätstheorie.* Zürich: Weltformel-Verlag.

Zimmermann, Hermann (1924). *Zur Relativitätslehre. Gedanken eines Technikers.* Berlin: Ernst and Sohn.

Zlamal, Heinrich (1924). *Das Verhältnis der Einsteinschen Relativitätstheorie zur exakten Naturforschung.* Wien und Leipzig: Braumüller.

——— (1932). *Die Lehre von der Zustandsbeeinflussung des Weltäthers durch die Anwesenheit und Bewegung ponderabler Körper als klassisches Correlatum der Einsteinschen Relativitätstheorie.* Wien: Selbstverlag.

General Relativity in the Netherlands, 1915–1920

A. J. Kox

1. Introduction

In the years immediately following Einstein and Grossmann's first steps toward a satisfactory generalization of special relativity to a relativistic theory of gravitation (Einstein and Grossmann 1913), a group of Dutch scientists and mathematicians, all of them working in Leiden, became very interested in general relativity and actively contributed to the further development of the new theory. In the years 1915–1920, nine Leiden authors published a total of more than 30 research papers on general relativity. After 1920, however, Dutch activity diminished rapidly, as was the case all over the world.[1] Only two members of the Leiden group remained actively interested in general relativity: the physicist Adriaan D. Fokker, who wrote two textbooks and several research papers, and the astronomer Willem De Sitter, who published works on cosmological problems until his death in 1934.

In this paper, I will give an overview of the work on general relativity produced in the Netherlands from 1915 to 1920, and I will try to assess the impact of this work on the development of general relativity. I will also address the question of why the activities of the Leiden group did not give rise to a lasting tradition of research in general relativity in the Netherlands in spite of the fact that in the years 1915–1920 the members of this group clearly formed a coherent and authoritative "college." Finally, in an appendix, I will comment on the general context in which the Dutch reception of general relativity took place, and I will briefly discuss the reactions to relativity in the Netherlands, in particular the negative reactions.

2. Reception

From the very beginning, with the Einstein–Grossmann paper of 1913, an enthusiastic attitude toward general relativity prevailed in Leiden. Both Hendrik Antoon Lorentz and Paul Ehrenfest, the two leading Leiden physicists of the time,[2] became interested in the theory and studied Einstein's work carefully. The correspondence between Lorentz, Einstein, and Ehrenfest shows this interest very clearly, and, in particular, Lorentz's scientific notebooks record his concentrated efforts to gain an understanding of the theory. In addition, Lorentz and his student Johannes Droste published scientific papers that dealt with the Einstein–Grossmann theory. The frequent contact, both in the correspondence and in person, between Lorentz and Ehrenfest on the one hand, and Einstein on the other, undoubtedly stimulated the interest.[3]

Another factor was that Lorentz's student Adriaan D. Fokker spent the winter semester of 1913–1914 with Einstein at the ETH in Zürich. Einstein and he wrote a joint paper analyzing the rival (scalar) theory of the Finnish physicist Gunnar Nordström from the point of view of absolute differential calculus (Einstein and Fokker 1914). A little later, in January 1915, Fokker published a summary of the Einstein–Grossmann theory for English readers in *Philosophical Magazine* (Fokker 1915), which was written during a stay with Bragg in Leeds.

After the publication of the final form of the field equations of general relativity in the late fall of 1915,[4] activities in Leiden became more intense. Lorentz lectured on the theory in the winter and spring of 1916, and through these lectures and the accompanying discussions, a group of scientists that included Ehrenfest, Fokker, Droste, and De Sitter obtained a working knowledge of general relativity and a deep insight in the new theory.[5] Lorentz's lectures were instrumental in training the younger generation in the methods of general relativity, and the fact that he chose this topic for his famous Monday morning lecture series must have given instant respectability to this difficult and at the time still fairly esoteric work.

3. The Dutch Work

3.1 General

If we look at the papers written in the Netherlands in the period from 1915 to 1920, several characteristics are noted. First, we see that many papers show an emphasis on purely geometrical, rather than differential geometrical, methods. This holds in particular for Lorentz's series of four papers that appeared during 1916 and 1917 in which he reviewed the theory and put it on the basis of a Hamiltonian formalism that incorporated gravitation as well as electromagnetism (Lorentz 1916a, 1916b, 1916c, 1917a).[6] These papers also exhibit the second characteristic trait: an emphasis on the development and use of variational principles. This approach was initiated by Lorentz (1915) in the framework of the version of Einstein's theory that was published in 1914 (Einstein 1914) and in which general covariance had not yet been achieved.[7] Third, we see the emergence of a strong interest on the part of the Leiden mathematicians (in particular of Jan Schouten and Dirk Struik) in the mathematical foundations of the theory, specifically in the development of a general and very rigorous formalism, with a completely new and rather inaccessible notation. This line of research quickly separated itself from the more physically oriented work done by the physicists and developed into a branch of mathematics in its own right. A discussion of this work falls outside the scope of this paper. Let me now turn to an overview of the work of the various authors.

3.2 Johannes Droste

In a series of papers published in 1915 and 1916 (Droste 1915a, 1915b, 1916a, 1916b) and in his dissertation, which he wrote under the guidance of Lorentz and which he defended in December 1916 (Droste 1916c), Droste analyzed the problem of the gravitational field, first of a single center, and later of n slowly moving bodies. The first two papers still use the Einstein–Grossmann theory; in the later ones, the final covariant field equations serve as a starting point. Independently of Schwarzschild—and in a simpler way—Droste derived what is known today as the Schwarzschild solution. Droste's work did not remain unnoticed: Einstein expressed his admiration for it in a letter to Ehrenfest,[8] and Droste's choice of coordinates for the field of a single center soon became the standard choice for this problem. The work on the field of n bodies was taken up by others as well, in particular by De Sitter, who applied it to astronomical problems.[9] In spite of its having been influential, Droste's work gradually receded into the background in the following decades. Only in recent years has its importance been recognized in the historical literature.[10]

Droste's work on the n-body problem was expanded in a set of two papers written in collaboration with Lorentz, in which a variational principle was developed for this problem (Lorentz and Droste 1917a, 1917b). Remarkably, these papers remained largely unnoticed, in spite of the novelty and importance of the results presented in them. The fact that of these papers only a Dutch version appeared and that there was no additional English translation (as was usual for papers published by the Dutch Academy of Sciences) must have played a role here.[11] The result was that when Einstein, Infeld, and Hoffmann did their work on the problem of motion in the 1930s, they unknowingly duplicated some of the work by Lorentz and Droste.[12]

3.3 Adriaan D. Fokker

Fokker's first paper on general relativity after his work in connection with the Einstein–Grossmann theory (see Section 2) gave a reformulation and an extension of earlier work by Lorentz on the Hamiltonian formulation of general relativity (Fokker 1917). Notable in this paper was a derivation of the contracted Bianchi identities (as a consequence of the variational principle) and a discussion of its implications for the field equations.[13] In his next paper (Fokker 1918a), the emphasis was on geometrical considerations. Fokker's main contribution to general relativity was the work he did on the elucidation of the phenomenon of geodesic precession (Fokker 1920). Geodesic precession, sometimes called Fokker precession, is the phenomenon of the precession of a freely falling gyroscope (or any other object the direction of which remains fixed in the inertial frame that moves along with it) in a gravitational field. This phenomenon might almost be called a typical Dutch subject; in addition to Fokker, De Sitter and Schouten worked on it as well (see Sections 3.7 and 3.8). As Fokker first calculated, the effect is minute: for the earth in its orbit around the sun, the yearly precession would amount to $0.019''$ (Fokker 1920).[14] Still, Fokker valued the effect so much that in the two books he wrote on general relativity he listed geodesic precession as a fourth test of the theory, along with the three "classic" tests.[15]

Fokker's first textbook was published in 1929 (Fokker 1929), but it was not very influential because of its geometrical approach, the late date of its appearance, and the fact that it was written in Dutch. His second book, eccentric in some respects, appeared in 1960 (Fokker 1960) and was not very influential either. In this book, as well as in other publications from his later years, Fokker advocated the use of the term "chronogeometry," instead of general relativity, reflecting his conviction that one should not treat space and time in the same way.[16] In spite of geodesic precession, however, in the field of relativity Fokker is better remembered for his work on special relativity (Fokker action, dynamical definition of the center of mass) than on general relativity.

3.4 Hendrik A. Kramers

Kramers' contribution to general relativity consists of two papers (Kramers 1920a, 1920b) that deal with the stationary gravitational field, in particular, with rotating bodies in such fields (including an alternative treatment of geodesic precession). Strictly speaking, Kramers was not part of the Leiden group when he wrote these papers: at the time, he was working with Bohr in Copenhagen. Nevertheless, I have included Kramers because during these years he was still working on his dissertation under the supervision of Ehrenfest, and he was in frequent contact with Leiden. Kramers' papers made little impact and are mainly remarkable because when he wrote them, problems in quantum theory were his main interest. It is striking that he had the time and energy to spare for research so far removed from his other work.[17]

3.5 Hendrik Antoon Lorentz

In Lorentz's work, which has already been mentioned in Section 3.1 and discussed in more detail elsewhere,[18] three lines of development can be distinguished: the work on a geometrical, coordinate-free formulation of general relativity, which essentially failed; the important work on variational principles; and the work, together with Droste, on the problem of motion (see Section 3.2). Lorentz's last paper on general relativity (Lorentz 1923) dealt with the problem of determining the metric tensor for the case when a set of world lines of light rays and material particles is given and discussed the theories of Weyl and Eddington.

3.6 Gunnar Nordström

Although not Dutch, Nordström was a member of the Leiden group during the years 1916–1919.[19] After abandoning his own (scalar) theory of gravitation in 1916, he worked on various problems in general relativity, in particular on the problem of the localization of the gravitational field energy (Nordström 1917, 1918a, 1918b, 1918c). In passing, he derived an expression for the gravitational field of an electrically charged center (Nordström 1918b), generalizing earlier work by Reissner on the field of a massless electric charge. The resulting metric is known today as the Reissner–Nordström metric.

3.7 Jan A. Schouten

The mathematician Schouten has to be credited with the first published discussion of geodesic precession (Schouten 1918a). His results, however, were inaccurate, and subsequently corrected and generalized by

Fokker (see Section 3.3). Schouten's interest shifted toward the formal mathematical side of general relativity, and partly in collaboration with his student Dirk Struik, he started to develop a formalism that is as mathematically rigorous as it is abhorrent to the average physicist.[20] Schouten made no further contributions to the physics of general relativity.

3.8 Willem De Sitter

The astronomer De Sitter is a special case, if only because of his choice of topics. Between 1916 and 1920, he wrote 10 papers on general relativity and its applications to astronomy and cosmology.[21] For several reasons, De Sitter's work was very important and influential. First, De Sitter was instrumental in spreading the "gospel" of general relativity across the North Sea to England: his three papers in the *Monthly Notices of the Royal Astronomical Society* (De Sitter 1916c, 1916e, 1917c) reviewed Einstein's work (which was inaccessible to the English because of World War I) and contained important new results.[22] Second, De Sitter discovered that the Einstein equations admitted a nonflat metric as their solution, even in the absence of matter (De Sitter 1917a). This important result, for which De Sitter is still mentioned in textbooks, was obtained in the course of a discussion between De Sitter and Einstein on fundamental issues in general relativity, such as the relativity of rotation and the cosmological consequences of the theory.[23] In later years, De Sitter continued to publish articles on cosmology, and he became one of the world's leading experts in this field.

3.9 Others

In 1917, in a short paper (Tresling 1917), Ehrenfest's student Jan Tresling gave an extension to Lorentz's earlier variational derivation of the generally relativistic Maxwell equations. Lorentz's last dissertation student, Willem van den Berg, obtained his doctorate in 1921 on a thesis (Van den Berg 1920) dealing with three topics in general relativity: the propagation of light in a gravitational field, including the study of effects such as double refraction (all on the basis of the Maxwell–Einstein equations as derived by Lorentz); the problem of the energy-momentum of a small particle (such as an electron) moving in a gravitational field; and the calculation of the field associated with a stationary, spherically symmetric energy-momentum tensor. As far as I know, none of the thesis was ever published in a journal.

4. Conclusion

One of the characteristic traits of the reception of general relativity in the Netherlands is the speed with which it was accepted and integrated into the ongoing research of the Leiden group. Important in this respect is, of course, the personal factor: the already existing personal links between Einstein, on the one hand, and Lorentz and Ehrenfest (and their students) on the other. General relativity entered the Netherlands through the physics community and not through the efforts of the mathematicians, as was, for instance, the case in Portugal and to a certain extent in France.[24] It should be noted, however, that in the Netherlands, too, there was a strong interest in general relativity on the part of the mathematicians, which soon developed into a separate and much more abstract line of research.

It is well known that in those days, the Netherlands was by no means a peripheral community in the world of physics. Does this mean that the Dutch work was pivotal in the development of general relativity? Summarizing what has been said above, it seems fair to say that the Dutch work on general relativity was of sizable contemporary importance, not only because of the concrete contributions to the further development of the theory (such as variational principles, geodesic precession, the problem of motion, and De Sitter's solution), but also because of the respectability that arose from the fact that a prestigious group of physicists such as the Leiden group occupied itself with general relativity. At the same time, surprisingly little of lasting value has come out of this concerted effort by a group of highly gifted physicists: geometrical methods are out of fashion, De Sitter's solution is not often discussed, and variational principles are much more sophisticated. There are two exceptions; both Droste's own work and his joint work with Lorentz are recognized as important in the historical literature, and geodesic precession has been revived from oblivion because of plans to use it as a test of general relativity.

It is interesting to ask why a tradition of general-relativistic research did not emerge in the Netherlands after this period. Of course, the malaise into which general relativity slipped after the three classic tests had been carried out with a positive result is a factor here. But, there is still a striking difference with, for instance, England, where there was a continuing interest in general relativity, even during the period of its lowest ebb.[25]

A general factor in this development is the fact that Paul Ehrenfest was the only one of the Leiden physicists who really created a school, in the sense of a group of students with a common background and a shared view of physics, including an opinion on what the important problems were. It is Ehrenfest who was responsible for the emergence in the Netherlands of a school of statistical physics, and in that field his influence is noticeable even after two generations. But Ehrenfest, although very interested in general relativity and actively participating in discussions on the subject, never encouraged his students to take up research in the field, nor did he publish anything after a short paper in 1913 (Ehrenfest 1913). None of the other major figures in Leiden, such as Lorentz or Fokker, had a close-knit group of students like the one Ehrenfest had. It was Fokker in particular who could have kept a tradition of general relativity alive; as was already pointed out in Section 3.3, he remained interested in general relativity throughout his life. The somewhat unorthodox character of his approach to general relativity may be partly responsible for the fact that he attracted no students.

Clearly then, the interest in the Netherlands in general relativity resembled the behavior of a comet in the sky: highly visible, exciting to watch, and a cause for much lively discussion for a short period, but rapidly disappearing out of sight afterwards.

Appendix: The Dutch Anti-Relativists

When one looks at the reception of relativity outside of the scientific community in the Netherlands, one is struck by the contrast between the attitudes in the Netherlands and those in other countries, in particular Germany and England. Although the general public was interested in the theory and fascinated by its remarkable predictions, especially after the results of the light-bending measurements had been announced in the fall of 1919, there was no wild excitement, as in other countries, neither were there large-scale attempts to refute the theory and attack its proponents. An important factor was, without a doubt, the authority and prestige enjoyed by the leading Dutch scientists among the general public. Lorentz in particular was revered almost as a demigod, and his word carried much weight. Several Dutch scientists, including Lorentz himself, wrote popular expositions of the theory of relativity in newspapers and various other periodicals.[26]

The few anti-relativists in Holland came from the ranks of the philosophers or were not professional scientists.[27] It is striking that, even though they voiced objections against relativity, they remained polite and were careful to avoid any impression of mounting a personal attack against Einstein and his fellow relativists. In some instances, Einstein is even praised for the brilliant way in which he developed his—unfortunately misguided—ideas. In addition, no trace of anti-Semitism can be found. A general (and not typically Dutch) characteristic of the arguments of the anti-relativists is the emphasis on the counterintuitiveness of the theory of relativity.[28] As an illustration, let me briefly describe two "major" attempts to refute the theory of relativity.

The most active of the Dutch anti-relativists was undoubtedly the engineer M.W. Polak. In a booklet (Polak 1918), in a polemic in the periodical *De Ingenieur* (Polak 1919a, 1919b), and in later papers, (e.g., Polak 1922), he tried to refute the theory of relativity, concentrating on the special theory. His arguments strike us as rather naive, centering around the view that the theory lacks a solid logical foundation and that concepts such as the relativity of simultaneity are counterintuitive and absurd. An example of Polak's way of reasoning was his objection against the statement that relativity reduces to classical mechanics in the limit of an infinite speed of light. He argued that it was absurd to take such a limit since the speed of light has a fixed value! One of Polak's opponents in the discussion was Fokker, but his attempts (Fokker 1919a, 1919b) to refute Polak's arguments were unsuccessful.

In 1921, the well-known philosopher Gerard Heymans listed a series of more philosophically oriented objections against the relativistic view of space and time (Heymans 1921). His main objections, framed rather as questions posed by a layman, were directed toward what he called the positivistic foundation of the theory and, in particular, against the relativization of time and simultaneity.[29] In a reaction, Fokker responded to some of Heymans's arguments (Fokker 1922).

In conclusion, it appears that there was no strong anti-relativistic movement in the Netherlands and that the few Dutch anti-relativists who made their objections public did not cause much commotion and were not very influential.

Acknowledgments: I have greatly profited from comments made after the presentation of this paper at the Luminy meeting, in particular from those of Peter Havas, József Illy, and John Stachel. I am indebted to Jean Eisenstaedt for his moral support and for many valuable suggestions and to Henriette Schatz for linguistic improvements.

NOTES

[1] See Eisenstaedt 1986, 1988b for more details.

[2] Although Ehrenfest had succeeded Lorentz as professor of theoretical physics in the summer of 1912 and Lorentz had moved from Leiden to Haarlem,

Lorentz was still an influential factor in Leiden physics through his fame and prestige, as well as through his position as an Extraordinary Professor ("Buitengewoon Hoogleraar").

[3] See Kox 1988 and Illy 1989 for more details on Lorentz, and Klein 1970 for more on the contacts between Ehrenfest, Lorentz, and Einstein.

[4] See Einstein 1915.

[5] In a footnote to his first paper in the *Monthly Notices*, De Sitter made the following comment in connection with Lorentz's lectures: "Many of the results contained in the present paper are wholly or partially derived from these lectures" (De Sitter 1916c, p. 707, footnote). He also mentioned "much free interchange of ideas" between Lorentz, Ehrenfest, Droste, and himself.

[6] See Janssen's paper in this volume for a detailed discussion of this work.

[7] Lorentz was the first to cast the 1914 version of general relativity in Hamiltonian form (Lorentz 1915). His work on the 1915 version generalized earlier work by Hilbert (1915).

[8] "Droste's Dissertation ist ausserordentlich schön," Einstein to Ehrenfest, May 25 1917.

[9] See, for instance, De Sitter 1916e.

[10] See in particular Eisenstaedt 1982, 1987, 1988a.

[11] An English translation did not appear until 1937, in the fifth volume of Lorentz's *Collected Papers* (The Hague: Nijhoff).

[12] See Havas 1988 and the paper by Damour and Schäfer in this volume for more details.

[13] This must be one of the first derivations of the contracted Bianchi identities, if not the first; Pais (1982, p. 275) credits Weyl for the first derivation and dates it at August 1917, whereas Fokker's paper was submitted in January 1917.

[14] The first calculation is in fact due to Schouten (1918a), but his value was off by a factor of $2/3$. See also Schouten 1921.

[15] In fact, geodesic precession may still become a test of general relativity. It has been proposed to measure the precession of a gyroscope on board a satellite orbiting the earth. Strictly speaking, this would not be a determination of geodesic precession only, since other effects also contribute to the total precession. See Will 1981, pp. 208–212, for more details.

[16] As Fokker puts it: "Space is nothing but the mere possibility of events taking place here-and-there, while time is merely the possibility of events occurring before-and-after each other." (Fokker 1965, p. 2).

[17] See a similar comment by Kramers' biographer Max Dresden (1987, p. 98). Dresden only mentions the papers in passing.

[18] See note 3; see also Janssen's paper in this volume.

[19] He was in Leiden on a three-year stipend. In 1917, his ties with Leiden became more personal when he married Nelly van Leeuwen, one of Ehrenfest's students. See Isaksson 1985 for more details on Nordström's work and life.

[20] See, for instance, Schouten 1918b and Schouten 1924.

[21] At an early stage, De Sitter was interested in the possible consequences

on astronomy of the principle of relativity; see De Sitter 1911. His first two papers on general relativity (De Sitter 1916a, 1916b) were concerned with this topic as well.

[22] See the paper by Sánchez-Ron in this volume for more details on the reception of general relativity in England.

[23] The discussion took place in a series of papers as well as in the correspondence between De Sitter and Einstein; see, e.g., De Sitter 1916d, 1917a, 1917b, 1918, 1920 and Einstein 1918. It is outside the scope of this paper to go into more detail; see Kerzsberg 1988, 1989 for an extensive discussion.

[24] See, e.g., Glick 1987 and the paper by Lopes Gagean and da Costa Leite in this volume.

[25] See Eisenstaedt 1986, 1988b.

[26] See, e.g., Fokker 1918b; Kuenen 1917; Lorentz 1917b, 1919; Schouten 1920; and Van der Waals Jr. 1923.

[27] I limit myself to the more serious anti-relativists and leave out of consideration a few attacks by people such as the writer Lodewijk van Deyssel (1922), whose arguments betrayed a profound lack of knowledge of the theory.

[28] A typical example was a statement by Greeve (1922): "What is called *simultaneity* by Einstein has nothing to do with the nature of this concept. [...] Everyone *knows* what simultaneity is." ("Wat door Einstein *gelijktijdigheid* wordt genoemd is totaal bezijden den aard van dit begrip. [...] Ieder *weet* wat gelijktijdigheid is." Emphasis by Greeve.)

[29] His view was shared by the physics teacher Wigersma (1922). A typical objection raised by Wigersma was the claim that Einstein first established the constancy of the speed of light as a fact and then used it as a starting point for the development of a theory, instead of taking Lorentz's approach and using the existing theory to find a theoretical explanation for the outcome of experiments, such as the Michelson–Morley experiment.

References

De Sitter, Willem (1911). "On the Bearing of the Principle of Relativity on Gravitational Astronomy." *Monthly Notices of the Royal Astronomical Society 71* (1910–1911): 388–415.

——— (1916a). "De planetenbeweging en de beweging van de maan volgens de theorie van Einstein." *Koninklijke Akademie van Wetenschappen te Amsterdam. Verslagen van de Gewone Vergaderingen der Wis- en Natuurkundige Afdeeling* 25 (1916–1917): 232–245; English translation: "Planetary Motion and the Motion of the Moon According to Einstein's Theory." *Koninklijke Akademie van Wetenschappen te Amsterdam. Proceedings of the Section of Sciences* 19 (1916–1917): 367–381.

——— (1916b). "Space, Time, and Gravitation." *Observatory* 39: 412–419.

——— (1916c). "On Einstein's Theory of Gravitation, and its Astronomical Consequences. First Paper." *Monthly Notices of the Royal Astronomical Soci-*

ety 76 (1915–1916): 699–728; 77 (1916–1917): 481 (Errata).

——— (1916d). "De relativiteit der rotatie in de theorie van Einstein." *Koninklijke Akademie van Wetenschappen te Amsterdam. Verslagen van de Gewone Vergaderingen der Wis- en Natuurkundige Afdeeling* 25 (1916–1917): 499–504; English translation: "On the Relativity of Rotation in Einstein's Theory." *Koninklijke Akademie van Wetenschappen te Amsterdam. Proceedings of the Section of Sciences* 19 (1916–1917): 527–532.

——— (1916e). "On Einstein's Theory of Gravitation, and Its Astronomical Consequences. Second Paper." *Monthly Notices of the Royal Astronomical Society* 77 (1916–1917): 155–184; 481 (Errata).

——— (1917a). "Over de relativiteit der traagheid: beschouwingen naar aanleiding van Einstein's laatste hypothese." *Koninklijke Akademie van Wetenschappen te Amsterdam. Verslagen van de Gewone Vergaderingen der Wis- en Natuurkundige Afdeeling 25* (1916-1917): 1268–1276; English translation: "On the Relativity of Inertia. Remarks Concerning Einstein's Latest Hypothesis." *Koninklijke Akademie van Wetenschappen te Amsterdam. Proceedings of the Section of Sciences* 19 (1916–1917): 1217–1225.

——— (1917b). "Over de kromming der ruimte." *Koninklijke Akademie van Wetenschappen te Amsterdam. Verslagen van de Gewone Vergaderingen der Wis- en Natuurkundige Afdeeling* 26 (1917–1918): 222–236; English translation: "On the Curvature of Space." *Koninklijke Akademie van Wetenschappen te Amsterdam. Proceedings of the Section of Sciences* 20 (1917–1918): 229–243.

——— (1917c). "On Einstein's Theory of Gravitation and its Astronomical Consequences. Third Paper." *Monthly Notices of the Royal Astronomical Society* 78 (1917–1918): 3–28.

——— (1918). "Nadere opmerkingen omtrent de oplossingen der veldvergelijkingen van Einstein's gravitatie-theorie." *Koninklijke Akademie van Wetenschappen te Amsterdam. Verslagen van de Gewone Vergaderingen der Wis- en Natuurkundige Afdeeling* 26 (1917–1918): 1472–1475; English translation: "Further Remarks on the Solutions of the Field-Equations of Einstein's Theory of Gravitation." *Koninklijke Akademie van Wetenschappen te Amsterdam. Proceedings of the Section of Sciences* 20 (1917–1918): 1309–1312.

——— (1920). "Over de mogelijkheid van statisch evenwicht van het heelal." *Koninklijke Akademie van Wetenschappen te Amsterdam. Verslagen van de Gewone Vergaderingen der Wis- en Natuurkundige Afdeeling* 29 (1920–1921): 651–653; English translation: "On the Possibility of Statistical Equilibrium of the Universe." *Koninklijke Akademie van Wetenschappen te Amsterdam. Proceedings of the Section of Sciences* 23 (1920–1922): 866–868.

Dresden, Max (1987). *H. A. Kramers: Between Tradition and Revolution.* Berlin and New York: Springer-Verlag.

Droste, Johannes (1915a). "Over het veld van een enkel centrum in Einstein's theorie der zwaartekracht." *Koninklijke Akademie van Wetenschappen te Amsterdam. Verslagen van de Gewone Vergaderingen der Wis- en Natuurkundige Afdeeling 23* (1914–1915): 968–981; English translation: "On the Field of a Single Centre in Einstein's Theory of Gravitation." *Koninklijke Akademie van Wetenschappen te Amsterdam. Proceedings of the Section of Sciences* 17 (1914–1915): 998–1011.

—- (1915b). "Het veld van twee bolvormige rustende centra in Einstein's theorie der zwaartekracht." *Koninklijke Akademie van Wetenschappen te Amsterdam. Verslagen van de Gewone Vergaderingen der Wis- en Natuurkundige Afdeeling* 24 (1915–1916): 749–757; English translation: "On the Field of Two Spherical Fixed Centres in Einstein's Theory of Gravitation." *Koninklijke Akademie van Wetenschappen te Amsterdam. Proceedings of the Section of Sciences* 18 (1915–1916): 760–769.

——— (1916a). "Het veld van een enkel centrum in Einstein's theorie der zwaartekracht, en de beweging van een stoffelijk punt in dat veld." *Koninklijke Akademie van Wetenschappen te Amsterdam. Verslagen van de Gewone Vergaderingen der Wis- en Natuurkundige Afdeeling* 25 (1916–1917): 163–180; English translation: "The Field of a Single Centre in Einstein's Theory of Gravitation and the Motion of a Particle in That Field." *Koninklijke Akademie van Wetenschappen te Amsterdam. Proceedings of the Section of Sciences* 19 (1916–1917): 197–215.

——— (1916b). "Het veld van n bewegende centra in Einstein's theorie der zwaartekracht." *Koninklijke Akademie van Wetenschappen te Amsterdam. Verslagen van de Gewone Vergaderingen der Wis- en Natuurkundige Afdeeling* 25 (1916–1917): 460–467; English translation: "The Field of n Moving Centres in Einstein's Theory of Gravitation." *Koninklijke Akademie van Wetenschappen te Amsterdam. Proceedings of the Section of Sciences* 19 (1916–1917): 447–455.

——— (1916c). *Het zwaartekrachtsveld van een of meer lichamen volgens de theorie van Einstein.* Leiden: Brill.

Ehrenfest, Paul (1913). "Over Einstein's theorie van het stationaire gravitatieveld." *Koninklijke Akademie van Wetenschappen te Amsterdam. Verslagen van de Gewone Vergaderingen der Wis- en Natuurkundige Afdeeling* 21 (1912–1913): 1234–1239; English translation: "On Einstein's Theory of the Stationary Gravitation Field." *Koninklijke Akademie van Wetenschappen te Amsterdam. Proceedings of the Section of Sciences* 15 (1912–1913): 1187–1191.

Einstein, Albert (1914). "Die formale Grundlage der allgemeinen Relativitätstheorie." *Königlich Preussische Akademie der Wissenschaften* (Berlin). *Sitzungsberichte*: 1030–1085.

——— (1915). "Die Feldgleichungen der Gravitation." *Königlich Preussische Akademie der Wissenschaften* (Berlin). *Sitzungsberichte*: 844–847.

——— (1918). "Kritisches zu einer von Herrn De Sitter gegebenen Lösung der Gravitationsgleichungen." *Königlich Preussische Akademie der Wissen-*

schaften (Berlin). *Sitzungsberichte*: 270–272.

Einstein, Albert, and Fokker, Adriaan Daniël (1914). "Die Nordströmsche Gravitationstheorie vom Standpunkt des absoluten Differentialkalküls." *Annalen der Physik* 44: 321–328.

Einstein, Albert, and Grossmann, Marcel (1913). *Entwurf einer verallgemeinerten Relativitätstheorie und einer Theorie der Gravitation.* Leipzig: Teubner. Also, with some additional remarks, in *Zeitschrift fur Mathematik und Physik* 62 (1914): 225–261.

Eisenstaedt, Jean (1982). "Histoire et singularités de la solution de Schwarzschild (1915–1923)." *Archive for History of Exact Sciences* 27: 157–198.

—— (1986). "La relativité générale à l'étiage: 1925–1955." *Archive for History of Exact Sciences* 35: 115–185.

—— (1987). "Trajectoires et impasses de la solution de Schwarzschild." *Archive for History of Exact Sciences* 37:275–357.

—— (1988a). "The Early Interpretation of the Schwarzschild Solution." In *Einstein and the History of General Relativity.* Don Howard and John Stachel, eds. Boston: Birkhäuser, pp. 213–233.

—— (1988b). "The Low Water Mark of General Relativity, 1925–1955." In *Einstein and the History of General Relativity.* Don Howard and John Stachel, eds. Boston: Birkhäuser, pp. 277–292.

Fokker, Adriaan Daniël (1915). "A Summary of Einstein and Grossmann's Theory of Gravitation." *Philosophical Magazine* 29: 77–96.

—— (1917). "De virtueele verplaatsingen van het elektromagnetische en van het zwaartekrachtsveld bij de toepassing van het variatieprincipe van Hamilton." *Koninklijke Akademie van Wetenschappen te Amsterdam. Verslagen van de Gewone Vergaderingen der Wis- en Natuurkundige Afdeeling* 25 (1916–1917): 1067–1084; English translation: "The Virtual Displacements of the Electro-Magnetic and Gravitational Field in Applications of Hamilton's Variation Principle." *Koninklijke Akademie van Wetenschappen te Amsterdam. Proceedings of the Section of Sciences* 19 (1916–1917): 968–984.

—— (1918a). "Over hetgeen in niet-Euclidische ruimten beantwoordt aan eene verplaatsing evenwijdig aan zichzelf, en over de Riemanniaansche kromtemaat." *Koninklijke Akademie van Wetenschappen te Amsterdam. Verslagen van de Gewone Vergaderingen der Wis- en Natuurkundige Afdeeling* 27 (1918–1919): 363–376; English translation: "On the Equivalent of Parallel Translation in Non-Euclidean Space and on Riemann's Measure of Curvature." *Koninklijke Akademie van Wetenschappen te Amsterdam. Proceedings of the Section of Sciences* 21 (1918–1919): 505–517.

—— (1918b). "Over de natuurlijke verkorting van lichamen, die zich bewegen, en de paradox der relativiteitstheorie, die ermede samenhangt." *Vragen van den Dag* 33: 395–407.

—— (1919a). "Het anti-relativisme van ir. M.W. Polak weerlegd." *De Ingenieur* 34: 297, 299–300.

——— (1919b). "Het relativiteitsbeginsel in de mechanica. Zakelijk." *De Ingenieur* 34: 651–652.

——— (1920)."De geodetische precessie; een uitvloeisel van Einstein's gravitatietheorie." *Koninklijke Akademie van Wetenschappen te Amsterdam. Verslagen van de Gewone Vergaderingen der Wis- en Natuurkundige Afdeeling* 29 (1920–1921): 611–621; English translation: "The Geodesic Precession: A Consequence of Einstein's Theory of Gravitation." *Koninklijke Akademie van Wetenschappen te Amsterdam. Proceedings of the Section of Sciences* 23 (1920–1922): 729–738.

——— (1922). "Relativistische studie. Proeve van antwoord aan Prof. Dr. G. Heymans." *De Gids* 86 (IV): 244–271.

——— (1960). *Tijd en ruimte, traagheid en zwaarte.* Zeist: De Haan. English translation: Fokker 1965.

——— (1965). *Time and Space, Weight and Inertia: A Chronogeometrical Introduction to Einstein's Theory.* Oxford: Pergamon.

Glick, Thomas F. (1987). "Cultural Issues in the Reception of Relativity." In *The Comparative Reception of Relativity.* Thomas F. Glick, ed. Dordrecht: Reidel, pp. 381–400.

Greeve, Max (ps. of P. Hekmeyer) (1922). De *onhoudbaarheid der relativiteitstheorie.* Den Haag: Haga.

Havas, Peter (1988). "The Early History of the 'Problem of Motion' in General Relativity." In *Einstein and the History of General Relativity.* Don Howard and John Stachel, eds. Boston: Birkhäuser, pp. 234–276.

Heymans, Gerard (1921). "Leekenvragen ten opzichte van de relativiteitstheorie." *De Gids* 85 (II): 85–108.

Hilbert, David (1915). "Die Grundlagen der Physik. (Erste Mitteilung)." *Königliche Gesellschaft der Wissenschaften zu Göttingen. Mathematisch-physikalische Klasse. Nachrichten:* 395–407.

Illy, József (1989). "Einstein Teaches Lorentz, Lorentz Teaches Einstein: Their Collaboration in General Relativity, 1913–1920." *Archive for History of Exact Sciences* 39: 247–289.

Isaksson, Eva (1985). "Der finnische Physiker Gunnar Nordström und sein Beitrag zur Entstehung der allgemeinen Relativitätstheorie Albert Einsteins." *NTM-Schriftenreihe für die Geschichte der Naturwissenschaft, Technik und Medizin* 22: 29–52.

Kerszberg, Pierre (1988). "The Einstein–De Sitter Controversy of 1916–1917 and the Rise of Relativistic Cosmology." In *Einstein and the History of General Relativity.* Don Howard and John Stachel, eds. Boston: Birkhäuser, pp. 325–366.

——— (1989). *The Invented Universe: The Einstein–De Sitter Controversy (1916–1917) and the Rise of Relativistic Cosmology.* London and New York: Oxford University Press (Clarendon).

Klien, Martin J. (1970). *Paul Ehrenfest: The Making of a Theoretical Physicist.* Amsterdam: North-Holland.

Kox, A. J. (1988). "Hendrik Antoon Lorentz, the Ether, and the General Theory of Relativity." *Archive for History of Exact Sciences* 38: 67–78. Also in *Einstein and the History of General Relativity.* Don Howard and John Stachel, eds. Boston: Birkhäuser, pp. 201–212.

Kramers, Hendrik Anthony (1920a), (1920b). "Over de toepassing van Einstein's gravitatietheorie op een stationair zwaarteveld I, II." *Koninklijke Akademie van Wetenschappen te Amsterdam. Verslagen van de Gewone Vergaderingen der Wis- en Natuurkundige Afdeeling* 29 (1920–1921): 409–418; 419–433; 834 ("Naschrift"); English translation: "On the Application of Einstein's Theory of Gravitation to a Stationary Field of Gravitation." *Koninklijke Akademie van Wetenschappen te Amsterdam. Proceedings of the Section of Sciences* 23 (1920–1922):1052–1073.

Kuenen, J. P. (1917). "Relativiteitstheorie." *De Gids* 81 (I): 462–481, 81 (II): 96–123.

Lorentz, Hendrik Antoon (1915). "Het beginsel van Hamilton in Einstein's theorie der zwaartekracht." *Koninklijke Akademie van Wetenschappen te Amsterdam. Verslagen van de Gewone Vergaderingen der Wis- en Natuurkundige Afdeeling* 23 (1914–1915): 1073–1089; English translation: "On Hamilton's Principle in Einstein's Theory of Gravitation." *Koninklijke Akademie van Wetenschappen te Amsterdam. Proceedings of the Section of Sciences* 19 (1916–1917): 751–765.

——— (1916a), (1916b), (1916c), (1917a). "Over Einstein's theorie der zwaartekracht I, II, III, IV." *Koninklijke Akademie van Wetenschappen te Amsterdam. Verslagen van de Gewone Vergaderingen der Wis- en Natuurkundige Afdeeling* 24 (1915–1916):1389–1402; 1759–1774; 25 (1916–1917): 468–486; 1380–1396; English translation: "On Einstein's Theory of Gravitation I, II, III, IV." *Koninklijke Akademie van Wetenschappen te Amsterdam. Proceedings of the Section of Sciences* 19 (1916–1917): 1341–1354; 1354–1369; 20 (1917–1918): 2–19; 20–34.

——— (1917b). "De gravitatietheorie van Einstein en de grondbegrippen der natuurkunde." *De Ingenieur* 32:649–655.

——— (1919). "De zwaartekracht en het licht. Een bevestiging van Einstein's gravitatietheorie." *Nieuwe Rotterdamsche Courant* November 19, 1919.

——— (1923). "De Bepaling van het g-veld in de algemeene relativiteitstheorie met behulp van de wereldlijnen van lichtsignalen en stoffelijke punten, met eenige opmerkingen over de lengte van staven en den duur van tijdsintervallen en over de theorieën van Weyl en Eddington." *Koninklijke Akademie van Wetenschappen te Amsterdam. Verslagen van de Gewone Vergaderingen der Wis- en Natuurkundige Afdeeling* 32: 383–402. English translation: "The Determination of the Potentials in the General Theory of Relativity, with Some Remarks about the Measurement of Lengths and Intervals of Time and about the Theories of Weyl and Eddington." *Koninklijke Akademie van Wetenschappen te Amsterdam. Proceedings of the Section of Sciences* 29 (1926): 383–399.

Lorentz, Hendrik Antoon, and Droste, Johannes (1917a), (1917b). "De beweging van een stelsel lichamen onder den invloed van hunne onderlinge aantrekking, behandeld volgens de theorie van Einstein I, II." *Koninklijke Akademie van Wetenschappen te Amsterdam. Verslagen van de Gewone Vergaderingen der Wis- en Natuurkundige Afdeeling* 26 (1917–1918): 392–403; 649–660.

Nordström, Gunnar (1917). "De gravitatietheorie van Einstein en de mechanica der continua van Herglotz." *Koninklijke Akademie van Wetenschappen te Amsterdam. Verslagen van de Gewone Vergaderingen der Wis- en Natuurkundige Afdeeling* 25 (1916–1917): 836–843; English translation: "Einstein's Theory of Gravitation and Herglotz's Mechanics of Continua." *Koninklijke Akademie van Wetenschappen te Amsterdam. Proceedings of the Section of Sciences* 19 (1916–1917): 884–891.

—— (1918a). "Iets over de massa van een stoffelijk stelsel volgens de gravitatietheorie van Einstein." *Koninklijke Akademie van Wetenschappen te Amsterdam. Verslagen van de Gewone Vergaderingen der Wis- en Natuurkundige Afdeeling* 26 (1917–1918): 1093–1108; English translation: "On the Mass of a Material System According to the Gravitation Theory of Einstein." *Koninklijke Akademie van Wetenschappen te Amsterdam. Proceedings of the Section of Sciences* 20 (1917–1918): 1076–1091.

—— (1918b). "Een en ander over de energie van het zwaartekrachtsveld volgens de theorie van Einstein." *Koninklijke Akademie van Wetenschappen te Amsterdam. Verslagen van de Gewone Vergaderingen der Wis- en Natuurkundige Afdeeling* 26 (1917–1918):1201–1208; English translation: "On the Energy of the Gravitation Field in Einstein's Theory." *Koninklijke Akademie van Wetenschappen te Amsterdam. Proceedings of the Section of Sciences* 20 (1917–1918): 1238–1245.

—— (1918c). "Berekening voor eenige bijzondere gevallen volgens de gravitatietheorie van Einstein." *Koninklijke Akademie van Wetenschappen te Amsterdam. Verslagen van de Gewone Vergaderingen der Wis- en Natuurkundige Afdeeling* 26 (1917–1918): 1577–1589; English translation: "Calculation of Some Special Cases, in Einstein's Theory of Gravitation." *Koninklijke Akademie van Wetenschappen te Amsterdam. Proceedings of the Section of Sciences* 21 (1918–1919): 68–79.

Pais, Abraham (1982). *'Subtle is the Lord. . .': The Science and the Life of Albert Einstein.* London and New York: Oxford University Press (Clarendon).

Polak, M.W. (1918). *Bezwaren tegen de opvattingen der relativisten.* Deventer: Kluwer.

—— (1919a). "Is de relativiteitstheorie te aanvaarden?" *De Ingenieur* 34: 21–26.

—— (1919b). "Relativiteitsbeginsel in de mechanica. Weerlegging vernietigd." *De Ingenieur* 34: 439–441.

—— (1922). "De relativiteitstheorie en het denken." *Tijdschrift voor Wijsbegeerte* 16: 209–219.

Schouten, Jan Arnoldus (1918a). "Over het ontstaan eener praecessiebeweging tengevolge van het niet euklidisch zijn der ruimte in de nabijheid van de zon." *Koninklijke Akademie van Wetenschappen te Amsterdam. Verslagen van de Gewone Vergaderingen der Wis- en Natuurkundige Afdeeling* 27 (1918–1919): 214–220 (pp. 219–220: addendum by W. De Sitter); English translation: "On the Arising of a Precession-Motion Owing to the Non-Euclidian Linear Element of the Space in the Vicinity of the Sun." *Koninklijke Akademie van Wetenschappen te Amsterdam. Proceedings of the Section of Sciences* 21 (1918–1919): 533–539 (pp. 538–539: addendum by W. De Sitter).

—— (1918b). "Die direkte Analysis zur neueren Relativitätstheorie." *Koninklijke Akademie van Wetenschappen te Amsterdam. Verhandelingen (Eerste Sectie)* 12, No. 6.

—— (1920). *Over de ontwikkeling der begrippen ruimte en tijd in verband met het relativiteitsbeginsel.* Rotterdam: Nijgh en Van Ditmar. (Based on an article in the newspaper *Handelsblad,* January 8, 1920.)

—— (1921). "Over de geodetische praecessie." *Koninklijke Akademie van Wetenschappen te Amsterdam. Verslagen van de Gewone Vergaderingen der Wis- en Natuurkundige Afdeeling* 29 (1920–1921): 1150–1165; English translation: "On Geodesic Precession." *Koninklijke Akademie van Wetenschappen te Amsterdam. Proceedings of the Section of Sciences* 23 (1920–1922): 1108–1112.

—— (1921). *Der Ricci Kalkül.* Berlin: Springer, 1924.

Tresling, Jan (1917)."De vergelijkingen der electronentheorie in een gravitatieveld van Einstein afgeleid uit een variatieprincipe. De principale functie der electronenbeweging." *Koninklijke Akademie van Wetenschappen te Amsterdam. Verslagen van de Gewone Vergaderingen der Wis- en Natuurkundige Afdeeling* 25 (1916–1917): 844–848; English translation: "The Equations of the Theory of Electrons in a Gravitation Field of Einstein Deduced from a Variation Principle. The Principal Function of the Motion of the Electrons." *Koninklijke Akademie van Wetenschappen te Amsterdam. Proceedings of the Section of Sciences* 19 (1916–1917): 892–896.

Van den Berg, Willem (1920). *Vraagstukken uit Einstein's gravitatietheorie.* Haarlem: Loosjes.

Van der Waals Jr., Johannes D. (1923). *De relativiteitstheorie.* Haarlem: Bohn.

Van Deyssel, Lodewijk (1922). "Einstein's relativiteitstheorie. Een enkele opmerking." *De Nieuwe Gids* 37 (I): 153–158.

Wigersma, B. (1922). *Natuurkunde en relativiteitstheorie. Hun uitkomst en hun doel.* Haarlem: Boissevain.

Will, Clifford M. (1981). *Theory and Experiment in Gravitational Physics.* London and New York: Cambridge University Press.

The Reception of General Relativity Among British Physicists and Mathematicians (1915–1930)

José M. Sánchez-Ron

1. Introduction

The theory of general relativity was discovered, as is well known, by Albert Einstein when he was working in Berlin and being supported by the Prussian Academy of Sciences. Although this momentous discovery took place in the middle of World War I, when relations between many countries were interrupted, it can hardly be said that the early history of general relativity is exclusively, or even mainly, a chapter of the history of German physics. Many non-German scientists and philosophers contributed to the understanding and development of Einstein's theory. In this paper, I discuss the contributions made to the understanding and development of the general theory of relativity by British scientists.[1] I limit my discussion to the period between 1915 and 1930, that is, to the period beginning with the formulation of general relativity and ending with Eddington's rediscovery and acceptance of Lemaître's model of an expanding universe. At least one of the episodes of this period, the joint 1919 Royal Society and Royal Astronomical Society Eclipse expeditions to Sobral in Brazil and to Principe off the west coast of Africa,[2] was of considerable importance to the subsequent history of Einstein's new theory of gravitation, but it was by no means the only one; in other words, no history of the general theory of relativity would be complete without taking into account the contributions made by British scientists.

Before I proceed, let me point out that I do not include the topic, important as it is, of the reception given to Einstein's theory by British society at large. However, I would like to mention that I have never been completely satisfied with the tentative explanations of the "Einstein phenomenon" some authors have put forward. For Lewis Elton (Elton 1986), for instance, Einstein's fame arose from controversies surrounding the verification of his general theory of relativity and from the myth of its incomprehensibility, both of which originated in the Anglo-Saxon world.[3] For Marshall Missner (Missner 1985), to give another example, the American press was the instrument that made Einstein a celebrity.[4]

It might be that these explanations, among others, reveal a part of the real mechanisms behind the public reception of relativity in different national communities, but I still have to confess that I do not really understand why Einstein became so popular. Perhaps there are still some important elements missing that would allow us to solve that puzzle. What I want to point out in passing is that perhaps we have not paid sufficient attention to what could be termed the "institutional history of science." To show what I mean, let me quote from a leader that appeared in the British newspaper *The Morning Post* on September 27, 1922 (it was entitled "The Scientific Front"):

> One of the results of the war, in which the scientific brains of this country were mobilized to such good purpose, is an appreciable increase of public interest in the achievements of science, whether theoretical or practical. In prewar days, neither the man in the street nor the man at the club window could have been persuaded to read articles about the Einstein versus Newton controversy.

If we accept the point of view expressed in this leader, then the initial reaction of the British public to Einstein and his theories, a reaction that no doubt was enhanced by the results of the already-mentioned 1919 Eclipse Expedition, ought to be looked at, at least in part, from a new perspective. Such a reaction would then be, not just the more-or-less straightforward response of a substantial part of the public when informed about the confirmation of a striking prediction in which the science of Britain's great hero, Isaac Newton, was confronted by that of a German Jew, Albert Einstein; rather, it would be to some extent, the fortunate coincidence of results obtained in a British-led scientific expedition with the end of a war that had definitely alerted British government and society to the values of science for the well-being and prosperity of the nation.

In this sense it would not be very difficult, although perhaps somewhat exaggerated, to attach some kind of symbolic meaning to the 1921 publication of Richard Haldane's book, *The Reign of Relativity* (Haldane 1921); which had an extraordinary reception: it was sold out in the first week of publication, and within three months, there was a third edition.[5] Haldane was for many years one of the most active promoters of science in Britain.[6] Most of his efforts on behalf of science and education were made in public: within the framework of extraparliamentary organizations, in the press, at meetings, on commissions of inquiry, and on the founding committees of new institutions (the Imperial College of Science and Technology, the National Physical Laboratory, and the Department of Scientific and Industrial Research are important examples in this sense). But, they also took place behind the scenes, in negotiations and discussions with politicians and civil servants in which important decisions relating to science policy were made.

One of Haldane's last prominent services to science was precisely this book on relativity, of which he said: "I projected [it]... on the day of my release from office as Lord Chancellor in 1915."[7] A release that was in fact a resignation, brought on by his admiration for the German system of education and science; after August 1914, such admiration marked him as a friend of Germany, and consequently he was exposed to the hostility and defamation of his fellow countrymen. Fortunately, by 1921 when his book appeared things had changed. Now it is time to abandon these passing comments and start with the real theme of my paper.

2. Physicists and Astronomers

2.1 The Introduction of General Relativity in Great Britain Before 1919

By the time (late 1915) Einstein finally succeeded in formulating a satisfactory relativistic theory of gravitation, special relativity was already well known in Great Britain.[8] There were, of course, scientists (physicists, astronomers, and mathematicians) who did not accept Einstein's 1905 new world view. In 1914, two influential monographs wholly dedicated to special relativity were published, namely, Ludwik Silberstein's *The Theory of Relativity,* and Ebenezer Cunningham's *The Principle of Relativity* (Silberstein 1914, Cunningham 1914). This fact, i.e., that Einstein's "relativity world" was by no means unknown to an important number of British scientists, should not be forgotten when one deals with the history of the reception of general relativity in Britain; it is indeed because of this fact that we can avoid the temptation of reducing the 1915–1919 period to a few events: the three papers published by Willem De Sitter in the *Monthly Notices of the Royal Astronomical Society* (De Sitter 1916a, 1916b, 1917),[9] the role played by Arthur Eddington and Frank Dyson in the preparation and development of the 1919 Eclipse Expedition and Eddington's famous work, *Report on the Relativity Theory of Gravitation* (Eddington 1918a). There is no doubt that these events are very important, but they are only elements of a richer history.

A more complete history has to take into account, for instance, that Eddington's interest in Einstein's approaches to gravitation, as John Stachel has pointed out (Stachel 1986), had been aroused well before De Sitter appeared on the horizon in 1916. In fact, in 1912, he participated in a Brazilian solar eclipse expedition that included in its research program a test of Einstein's 1911 prediction of the apparent position of a star in the gravitational field of the sun. The hypothesis that Eddington knew—presumably from C.D. Perrine, an American astronomer who had become the director of the Argentine National Observatory in 1909—of Einstein's 1911 work is supported by the fact that he referred to it in print in an article published in February 1915 (Eddington 1915).

As a matter of fact, Eddington's reference to Einstein's 1911 theory came too late, even for British scientists, since in the January 1915 issue of *Philosophical Magazine* Adriaan D. Fokker, one of the better known students of Hendrik A. Lorentz, had already published a paper explaining Einstein and Marcel Grossmann's 1913 theory of gravitation (Fokker 1915). Fokker, who had collaborated with Einstein for a few months during 1913–1914 in a problem dealing with Gunnar Nordström's 1912 theory of gravitation (Einstein and Fokker 1914), submitted his paper in July 1914 when he was visiting William H. Bragg in Leeds.[10] The Dutch physicist, who at that time was very much interested in Einstein's ideas on gravity,[11] spent some time in England in 1914, making the acquaintance of Paul Dirac, Alfred A. Robb, and Ernest Rutherford, as well as of Bragg. In fact, it seems quite likely that he could have discussed Einstein's ideas with some of them.

Almost as soon as he began to receive information on Einstein's final theory,[12] through the good services of De Sitter, Eddington helped to organize a discussion on gravitation that was held at the annual meeting of the British Association for the Advancement of Science.[13] In 1916 (September 5–9), the British Association, which was overshadowed by the war,[14] met at Newcastle-on-Tyne. Unfortunately, the report of the meeting mentions only that the discussion took place and that it was opened by Ebenezer Cunningham[15]; we also know that Eddington discussed Einstein's new theory of gravitation.[16] Thus, as early as the summer of 1916, many British scientists had the opportunity of hearing about general relativity.[17]

Soon afterward, and well before the famous 1919 eclipse, the Lindemanns, father and son, published a paper (submitted on December 4, 1916, and published in the December issue of *Monthly Notices of the*

Royal Astronomical Society) dealing with daylight photography of stars as means of testing the equivalence postulate of the new theory of relativity (Lindemann and Lindemann 1916). They indicated methods for testing by photography Einstein's hypothesis "of the refraction of light by a gravitational field," as they called it; they also discussed the most favorable conditions for photographing stars in the daytime, describing experiments that showed that stars may be photographed in daytime using red filters and red sensitive plates. Moreover, they described an instrument by means of which the transparency of the air at different times and places may be compared quantitatively. As their conclusion, the Lindemanns suggested that "experiments in daylight photography of stars be undertaken by some observatory possessing a suitable instrument, and enjoying a fine climate, with a view to testing Einstein's theory." It was, as far as I know, the only contribution made by the Lindemanns to the development of general relativity, but it shows that the early reception of Einstein's theory in Great Britain included more protagonists than the ones usually mentioned.

During 1917, things began to speed up. Eddington prepared a council note[18] that the Council of the Royal Astronomical Society included in its report to the 97th Annual General Meeting, held on February 9, 1917. Eddington's note, entitled "Einstein's theory of gravitation" (Eddington 1917a), left no doubt as to how he viewed general relativity. He stated

> A Council Note in 1910 begins with the sentence, "Celestial mechanics, which has hitherto been based on the Newtonian laws of motion, is profoundly affected by the discoveries which have been made in recent years regarding measurements of space, time, and force" [*Monthly Notices,* 70: 363]. These words were of the character of a prediction, for the new ideas had at that time scarcely reached the principal force with which astronomy is concerned—gravitation. But, remarkable progress has been made, especially in the last two years, and now there emerges a complete and self-consistent new system of mechanics, which can be applied without ambiguity to the problems of gravitational astronomy.

In the same volume of *Monthly Notices*, Frank Dyson, the Astronomer Royal, wrote about the opportunity that the coming eclipse of May 29, 1919, offered for verifying Einstein's theory of gravitation (Dyson 1917). Besides speaking of the possibilities this eclipse would offer, Dyson mentioned that he had already tried to use four photographs taken with the astrographic equatorial at the 1905 eclipse by C.R. Davidson, an astronomer of the Royal Observatory in Greenwich who had been one of Eddington's assistants in the 1912 Brazil eclipse expedition and who would also be on the 1919 expedition (he went to Sobral with

Crommelin). The results did not support Einstein's theory, but, as Dyson pointed out, the observations had not been designed with the purpose of testing a gravitational theory. Dyson's conclusion was that the observations show, however, that astrographic object-glasses are very suitable for this important observation.

Another event that helped introduce general relativity to Great Britain during 1917 and 1918, was a discussion that took place in the pages of *Philosophical Magazine*, with Oliver Lodge and Arthur Eddington as the main protagonists and a small contribution from G.W. Walker (Walker 1918). The discussion arose out of a paper published by Lodge in which he analyzed some astronomical consequences of what he called "the electrical theory of matter" (Lodge 1917a). As one would have expected of him, Lodge's starting points were expressions of the "electromagnetic world view." Thus, he supposed that (1) the motion of matter through the ether has a definite meaning; (2) an extra inertia due to this motion is to be expected at high speeds, in accordance with the FitzGerald–Lorentz contraction; and (3) this extra or high-speed inertia is not part of the mass but is dependent on the ether and hence not subject to gravity. What he thought is that from these "reasonable hypotheses," rooted in the tradition of the works of J.J. Thomson and Oliver Heaviside, some astronomical consequences would follow, in particular, that "the outstanding discrepancy in the theory of the perihelion of Mercury would be accounted for by attributing a certain value to a component of the true solar motion through the ether in the direction of the planet's aphelion path." What is important for us to realize is that although Lodge did not share Einstein's special relativity approach, his work shows that he could not ignore the kind of problems raised by relativity. As a matter of fact, it was not only the special theory of relativity that appeared as a sort of rival formulation behind the scenes of Lodge's work, the general theory also contributed to this situation. Indeed, in his paper, Lodge showed that he was already informed of and worried by Einstein's new theory. He also wanted to show that his "electrical theory of matter" could cope with astronomical problems. Thus, he wrote:

> Professor Einstein's genius enabled him in 1915 to deduce astronomical and optical consequences (some not yet verified) from the Principle of Relativity.[19] ... I wish to show that one of them [the anomalous motion of the perihelion of Mercury] at least can be deduced without reference to that principle.

In fact, Lodge also considered the question of whether "the gravitative deflection of a ray of light, predicted by Einstein, also follows from ether theory." However, to get the same result as Einstein, he had to resort to what he freely admitted were very speculative assumptions. But, the important lesson to be learned is that by the summer of 1917 there were scientists in England who felt obliged to enter into a dialectic on general relativity at the same time that they developed their own theories.

Soon after Lodge's paper, Eddington, who had been corresponding with Lodge even before the publication of this article, wrote two notes for *Philosophical Magazine* (Eddington 1917b, c). Apparently, Eddington took Lodge's ideas very seriously. Thus, he stated that:

> His explanation [of the motion of the perihelion of Mercury] is comparatively simple, and on that account will be widely preferred to the recent theory of Einstein, which introduces very revolutionary conceptions, provided that it meets certain other astronomical requirements which seem necessary.

In particular, what Eddington wanted was to discover whether or not in removing the discordances for Mercury, Lodge's theory introduced discordances for Venus and Earth. "If the explanation breaks down under this further test," Eddington went on, "the discussion will make prominent a feature of the success of Einstein's theory which has perhaps not been sufficiently emphasized." Fortunately for general relativity, Eddington's results did not favor the electrical theory of matter, and the Plumian Professor of Astronomy at the University of Cambridge finally concluded that:

> It is disappointing to find that this interesting suggestion [i.e., Lodge's], which gives a simple explanation of the most celebrated discordance of gravitational theory, is apparently unable to satisfy the most stringent test proposed. (Eddington 1917b)

The discussion did not end at that point. Lodge accepted Eddington's comments but could not refrain from declaring that although he did not see how to disprove Eddington's conclusion, he felt "that the last word has not been said on the subject" (Lodge 1917b). Indeed, the next year he continued the discussion (Lodge 1918), but once more his hopes were dashed by Eddington who openly stated that "Einstein's generalized relativity theory gives the precise value required without any arbitrary constant" (Eddington 1918b).

That year also saw the publication of Eddington's famous article, *Report on the Relativity Theory of Gravitation* (Eddington 1918a). Prepared at the request of the Physical Society of London, the report was the first complete account of general relativity to appear in the English language. Its success was great, especially after the results of the 1919

Eclipse became known, and another edition, slightly modified, appeared in 1920.

The scene was, therefore, ready for the great year, the year of the eclipse—a year that marked the irrefutable success in Great Britain of Einstein's theory. To some extent, it was, as we have seen, both the end of and the sequel to a three-year period that saw a significant number of discussions, analyses, and publications.[20]

To end this section, 1918, the year previous to the eclipse, was the first year that the contributions of Ludwik Silberstein to the analysis of general relativity were published (Silberstein 1918a, 1918b). I will return to Silberstein later; let me just say that he was most probably at that time, with the possible exceptions of Eddington and De Sitter, the physicist publishing in Britain with the broadest and deepest knowledge of general relativity, although it is true that at that time he did not fully accept Einstein's theory.

2.2 1919: The Year of the Eclipse

Much has been written about the famous joint meeting of the Royal Astronomical Society and the Royal Society, held on Thursday, November 6, 1919, in the rooms of the Royal Society at Burlington House. A large audience, consisting mainly of the fellows of both societies, assembled with the president of the Royal Society, J.J. Thomson, to hear the Astronomer Royal (Dyson) and Eddington's accounts of the eclipse expedition.[21] The meeting was in fact a "meeting for discussion,"[22] and although the conclusion put forward by Dyson and Eddington, namely, that the results obtained were in full agreement with the prediction for the bending of light in Einstein's general relativity, was generally accepted, one speaker, Silberstein, dissented. F.A. Lindemann reported on a pilot experiment that aimed at detecting the deflection of infrared radiation by the sun outside eclipses.

The announcement of the results obtained in the eclipse expedition caused a tremendous sensation in England. Eddington, the real champion of relativity in England, wrote to Einstein on December 1[23] that since the meeting of November 6, "all England has been talking about your theory." Headlines, such as the one appearing in the London *Times* of November 7, "Revolution in Science/New Theory of the Universe/ Newtonian Ideas Overthrown," were not infrequent. But, it was not only the daily press or the scientific journals that paid attention to the news; magazines, such as *The Athenaeum* and *The Nineteenth Century and After*, also included in their pages commentaries dedicated to the new gravitational theory.[24]

The meeting and discussion of November 6 was not the only scientific discussion organized in England soon after the official announcement of the eclipse results. The Council of the Royal Astronomical Society decided that its own meeting of December 12 should also be devoted to a discussion of general relativity;[25] the Royal Society also organized a discussion that took place on February 5, 1920; the next month (March 26), it was the turn of the Physical Society of London.[26]

In a sense, the least interesting of these meetings was the one organized by the Physical Society of London. Indeed, it was but a sort of dialogue between Eddington, who gave a general presentation of the theory, and those in the audience, who wanted to ask him some questions. Eventually, 10 persons did so.[27]

As for the other two meetings, several speakers, Eddington, J. Jeans, Dyson, Silberstein, and F.A. Lindemann, spoke at both, with Lodge, H. Jeffreys, and Larmor (the latter with a contributed paper) participating only in the Royal Astronomical Society meeting, and R.H. Fowler, Cunningham, and H.F. Newall in the one organized by the Royal Society.

There were some common traits between the Royal Astronomical Society and Royal Society meetings; Eddington and Jeans gave introductory talks,[28] and the Astronomer Royal made brief comments dealing with observational matters. Of all the speakers participating in both events, the most critical was Silberstein, discounting Larmor and Lodge. Thus, at the Royal Astronomical meeting, Silberstein questioned the British astronomers' conclusions, observing "displacements from the rectangular components quoted by the Astronomer Royal at the joint meeting of November 6." Although he was "aware of the difficulty of these measurements and of the circumstance that the direction is a too severe test for them," he could not avoid concluding that if "we were not prejudiced by Einstein's theory, we should not have said that the figures strongly indicated a radial law of displacement." As a matter of fact, Silberstein argued that even if the "displacements were ideally radial, we may have to account for them in some other, non-Einsteinian way."

One of those non-Einsteinian ways was considered by Silberstein in a paper that appeared in the February 1920 issue of *Philosophical Magazine* (Silberstein 1920). In his article, Silberstein stated that it was premature to interpret the bending of light effect found by the British astronomers as a verification of Einstein's theory, "not merely in view of the small outstanding discrepancies, but chiefly in view of the failure of detecting the spectrum shift predicted by the theory, with which the whole theory stands or falls." To Silberstein, the eclipse result proved that there was an alteration, a change of light-velocity all around the sun.

It was at this stage that he turned to the Stokes–Planck ether:

> The condensation claimed by Planck's modification of Stokes's theory, for the sun as well as for the Earth and for all other material bodies, is no longer devoid of influence on observable phenomena. It suddenly acquires physical life, so to speak. In other words, the discovery made in Brazil naturally suggests the idea that the observed deflection *is due to the condensation of the ether* around the sun, [n.b.] and that although one has been an implacable enemy of any ether at all for the last 15 years, one does not hesitate to point out this possibility—a last glimpse of hope, perhaps, for the vanished medium.

Silberstein's revival of the Stokes–Planck's ether gave Lodge (Lodge 1920) the opportunity to call attention to an old paper (1907) of his, but nothing else came out of it.

Coming back to the Royal Astronomical Society meeting, we note that Silberstein's most important point there was his insistence on the lack of any observation confirming Einstein's theoretical prediction of a gravitational displacement toward the red of the solar spectral lines. "Einstein's whole theory," he remarked "stands or falls with this effect."

Although agreeing on the importance of the red-shift prediction, Lindemann did not think that the absence of such a shift had been proved; to support his contention, he quoted the conflicting values found by a number of researchers (C.E. St. John, L. Grebe and A. Bachem, and K. Schwarzschild).[29] However, Harold Jeffreys' reading of the observed spectral shifts was somewhat different from Lindemann's; comparing Einstein's theoretical value with those measured by John Evershed and Charles St. John, he concluded that "while they do not show any uniform agreement with the theory [i.e., general relativity], they are at least sufficiently in agreement with it to suggest that it cannot be condemned on the spectroscopic evidence alone." At the Royal Society meeting, Alfred Fowler took another view, pointing out that different "interpretations may possibly be put on the observations, but my own opinion is that they are decidedly unfavorable to the existence of the Einstein field."[30]

As a matter of fact, the question of the spectral shift, the third Einstein test, became a topic frequently discussed in England, as elsewhere, during the 1920s. The problem was that the testing of this prediction was a very difficult matter, not so much because the small displacement of 0.008Å was difficult to measure, but because it was masked by larger displacements of an unknown amount. Scientists, such as Duffield (Duffield 1919–1920), and Dodwell and Davidson (Dodwell and Davidson 1922–1923), studied the effect, but it was mostly because of the efforts of John Evershed, a FRS British astronomer, and of Charles E. St. John, of the

Mt. Wilson staff,[31] that it became accepted, although not without a certain amount of doubt, that the observations agreed with the theoretical prediction derived from general relativity. Briefly, the story is as follows: from his observations at Kodaikanal in India, Evershed had convinced himself by 1921 that the solar spectrum yielded evidence for the gravitational displacement of lines to the red in accordance with the requirements of Einstein's theory (Evershed 1921). But, St. John with his magnificent equipment at Mt. Wilson was still dubious in 1922[32] and outlined the difficulties and conflicting evidence in his address at the Centenary Meeting of the Royal Astronomical Society on May 30, 1922, with Eddington in the chair as president.[33] However, one year later, St. John (St. John 1922–1923) thought he had confirmed the displacement to the red within 14% of the Einstein value. The chief cause of his change of attitude was his recognition of the very small pressure in the reversing layer of the solar atmosphere, so that it was no longer necessary to confine the study to lines that were insensitive to pressure shifts. In view of these results, the report of the Council of the Royal Astronomical Society at its 104th General Meeting[34] included the statement, "The measures of the solar spectral lines no longer offer any difficulty to the existence of the Einstein shift, even to those who hesitate about adopting this explanation of the observed facts." It was, nevertheless, an overly optimistic assessment; thus, during his address as the new president of the Royal Astronomical Society in 1926, Jeans (1925–1926) pointed out that "Even today the problem can hardly be said to be solved with absolute finality so far as solar light is concerned, although probably Evershed, St. John, and others have done all that it is possible to do." The new president thought that the case was different, however, with the light coming from one of the faintest of stars, Sirius, because of the fact that Sirius has a smaller and far fainter star as a companion, describing an orbit around it. The corresponding measurements were undertaken at Mt. Wilson, and Jeans remarked, "The observed shift was found to agree almost exactly with that predicted by theory. This experiment not only established the validity of Einstein's theory of the gravitational displacement of spectral lines, but also showed its usefulness as an instrument to be utilized by the practical astronomer in his everyday work." Again, not everybody would have subscribed to Jeans's words, although, certainly, the available observations were regarded as weighting the evidence for Einstein's theory against that of the sceptics.[35]

Before concluding this section, I would like to mention a rather insignificant episode that shows from the announcement of the eclipse results, the impact that general relativity produced among British physicists; to further the point, we can distinguish between physicists and astronomers. The episode involves Ernest Rutherford, the same Rutherford who, during a conversation held in 1910 with Wilhelm Wien, replied to Wien's comment that "no Anglo-Saxon can understand relativity," with the remark: "No, they have too much sense."[36] By 1919, Rutherford's feelings as regarding relativity, at least as far as general relativity was concerned, seem to have changed; thus, the December 25 issue of *Nature* contained a note entitled "Radioactivity and Gravitation," co-authored by Rutherford and Arthur H. Compton (Rutherford and Compton 1919), in which they reported the results of some experiments performed in the Cavendish Laboratory to test whether the rate of decay of radioactive substances is affected by subjecting them to the high centrifugal acceleration at the rim of a spinning disc.[37] The motivation of this investigation was to test whether the rate of transformation of a radioactive substance was influenced by the intensity of gravitation,[38] a problem that became particularly interesting in view of the fact that, as they put it, "according to Einstein's theory, a gravitational acceleration is in no sense different from a centrifugal acceleration."[39] Rutherford and Compton's contribution was, however, purely experimental; they did not make any effort to see whether or not their experimental results agreed with the predictions of general relativity, most probably because they were unable to deal with the complexity of the new theory.

2.3 General Relativity and British Physicists (1920–1930)

General relativity in Great Britain during the period 1920–1930 was dominated by the towering figure of Arthur Eddington. Although it is not my intention to analyze his works and ideas here,[40] I can hardly proceed without mentioning some of his contributions.

Eddington's activities in support of general relativity can be divided into several groups. As far as books are concerned, after publishing *Report on the Relativity Theory of Gravitation*, he saw that there was a need for a less mathematical account, which would explain as many as possible of the underlying ideas and results of relativity theory using a minimum of mathematical symbols. Thus, in 1920, he published his third book: *Space, Time and Gravitation* (Eddington 1920).[41] Soon, a French edition was published (Eddington 1921a); it included a mathematical supplement that was in fact a draft of Eddington's masterpiece, *The Mathematical Theory of Relativity* (Eddington 1923).[42] For many years, this treatise was the standard textbook from which many physicists and mathematicians learned the special and general theories of relativity.[43] Eddington's book was not only important because of its "didactic dimension" (i.e., because it could be regarded as a textbook in which the basic elements of Einstein's gravitational theory were put forward), it also contained quite a number of what were then advanced topics. Of course, most of these advanced topics had been previously published in specialized journals,[44] but their inclusion in *The Mathematical Theory of Relativity* increased their popularity among a larger audience, thus aiding in their development.

Turning now toward Eddington's own original contributions to the development of general relativity, we can say that during the period considered here his most important works were those dealing with gravitational waves and with the generalization of Weyl's unified theory of electromagnetic and gravitational fields.[45] Indeed, both topics received attention from some British scientists: for example, the papers that O.R. Baldwin and G.B. Jeffery of University College, London, published dealing with the relativity theory of plane and divergent waves[46] and the articles related to unified field theories.[47] Let me make clear, however, that the number of research papers published in this process of assimilation and development of general relativity in Great Britain is rather small, especially when one considers what the situation was like in other fields of physics. Thus, if we review the papers published in British scientific journals during the period 1920–1930,[48] we find that the number of articles dealing with general relativity published by British physicists, astronomers, and mathematicians[49] was almost insignificant compared with the number of papers with quantum physics as their subject.[50] A possible way of illustrating this is by looking at the reports of the British Association for the Advancement of Science meetings. If we take into account the meetings held between 1921 (Edinburgh) and 1930 (Bristol), we find that, according to the corresponding reports, only two papers dealing with general relativity were read, namely, Silberstein's "Determination of the Curvature Radius of Space-Time" (Toronto 1924),[51] and S. Brodetsky's "The Equations of the Gravitational Field in Two and in Three Dimensions of Space-Time" (Leeds 1927),[52] the latter being read within the Department of Mathematics of Section A (Mathematical and Physical Sciences). To these two papers, we ought to add the lecture delivered by the mathematician Edmund Whittaker (Whittaker 1927), as president of Section A of the 1927 Leeds meeting. Whittaker's lecture, and the one delivered in 1921 by Owen Richardson (also in Section A) were the only addresses, including the presidential ones, delivered during those years that mentioned Einstein's theory (in Whittaker's case, that theory was the theme of the lecture; in Richardson's case, Einstein's theory was only briefly mentioned).[53] When one compares this with the number of addresses, papers, or discussions delivered, read, or held at the British Association forum that dealt with quantum physics issues (for example, with the analysis of crystal structure by x-rays, or with the analysis of line spectra), one sees how poorly represented general relativity was, or what amounts to the same thing, where the real strengths and potentialities of British physics lay. Whether similar conclusions apply to other countries is beyond the scope of this paper, although I suspect, considering the global history of physics during this period, that indeed they do apply.

If we now look at the topics treated in the general relativity articles published by British scientists, setting aside those papers considered in the last section dealing with the classical experimental tests of the theory, we find that most dealt with electromagnetic phenomena (the electromagnetic field of the electron, in particular) in a gravitational field,[54] a theme with which many classical scientists could feel relatively at home. Following in the list of often-treated topics was what we could term mathematical aspects of general relativity, but as the authors of these works were usually mathematicians, I will delay my discussion of them until the next section. There were also papers dealing with other topics (with the problem of motion, for example), but the numbers involved were rather small.

As far as journals are concerned, and although *Philosophical Magazine*, the *Proceedings of the Royal Society of London*, and even the *Proceedings of the Cambridge Philosophical Society* included in their pages an important number of general relativity papers, in my opinion it was the *Monthly Notices of the Royal Astronomical Society* that played the most important role in the introduction and development of Einstein's gravitational theory in Great Britain.[55] Besides the articles published in these notices, there were also the council notes dealing with general relativity matters that were included in the Royal Astronomical Society journal. We have discussed some of them, but now I will mention another one that coincides with the end of the period I consider in this paper. In many ways, Eddington's discovery or, better, rediscovery and acceptance, of Lemaître's 1927 model of an expanding universe constituted the end of a period in the history of general relativity in Great Britain; the universe was then to be seen, especially in Eddington's case, in a new light, a light that opened up fresh problems that a new generation of young physicists (such as McVittie, McCrea, and Chandrasekhar),[56] who had grown up with relativity as a well-established theory, could tackle. Eddington's new convictions were evident in his 1930 paper, "On the Instability of Einstein's Spherical World" (Eddington 1930a),[57] but a more public pronouncement was the council note he prepared early the following year, under the attractive title "The expansion of the Universe" (Eddington 1931).[58] Finally, it is worth pointing out that in the index of the 80th volume (1919–1920) of the *Monthly Notices* there was a section dedicated to relativity; this fact clearly shows the status attained by Einstein's theory in the minds of the authorities of the Royal Astronomical Society.[59]

Turning to monographs dedicated to general relativity, we see that Eddington's books were not alone. Indeed, the first half of the 1920s saw the publication of a large number of books, usually textbooks or general introductions, that had Einstein's relativity as their leitmotiv. Leaving aside the English translations of foreign books,[60] as well as pamphlets and books of a too general nature or of a philosophical character,[61] we can mention a relatively small number of books,[62] Cunningham's *Relativity, the Electron Theory, and Gravitation* (Cunningham 1921),[63] Rice's *Relativity: A Systematic Treatment of Einstein's Theory* (Rice 1923),[64] Dingle's *Relativity and Gravitation* (Dingle 1923), Nunn's *Relativity and Gravitation: An Elementary Treatise upon Einstein's Theory* (Nunn 1923),[65] Norman Campbell's "Relativity" (a supplementary chapter to his 1907–1913 *Modern Electrical Theory*[66] (Campbell 1923), and Silberstein's *The Theory of Relativity* (Silberstein 1924)[67] and *The Size of the Universe* (Silberstein 1930).

As we have seen, Ludwik Silberstein was one of the most prolific physicists[68] to publish papers or books dealing with general relativity in British journals and publishing houses (I would say that he could be considered "the second," after Eddington). He was not, however, a British subject. A native of Warsaw, and educated in Cracow, Heidelberg, and Berlin, he was a lecturer, successively in Lemberg, Bologna, and Rome during the period from 1895 to 1920. However, during 1912–1920, he also held, in addition to his lectureship in Rome, positions at the research department of Adam Hilger, Ltd., London (a firm, founded in 1874, known for the high quality of its optical works), and the University of London.[69] (We have already seen that he was one of the participants in the discussions that took place in London shortly after the 1919 eclipse.) Finally, in 1920, he moved to Rochester, New York, to work for the Eastman Kodak Co., where he worked until his retirement in 1929, remaining in the United States until his death in 1948. Silberstein, therefore, cannot be considered as a representative of the British scientific *milieu*; however, because his many papers and books were widely circulated in Britain, he is necessarily included in this study. Again, Silberstein was initially somewhat critical of Einstein's theory; he never lost his independent way of looking at general relativity problems,[70] but we should not confuse such an independence with a rejection of the theory, at least when referring to him around 1922. The preface he included in his book *The Theory of General Relativity and Gravitation* (Silberstein 1922) is expressive in this sense:

> Some of my readers will miss, perhaps, in this volume the enthusiastic tone which usually permeates the books and pamphlets that have been written on the subject (with a notable exception of Einstein's own writings). Yet the author is the last man to be blind to the admirable boldness and the severe architectonic beauty of Einstein's theory. But it has seemed that beauties of such a kind are rather enhanced than obscured by the adoption of a sober tone and an apparently cold form of presentation.

3. Mathematicians

3.1 General Relativity and British Mathematicians

The mathematicians' response to the general theory of relativity is a particularly interesting topic, inasmuch as Einstein's theory requires some mathematical tools that are not usually needed in other fields of physics. Indeed, it is a well-known fact that the contributions of many people who had been educated as mathematicians proved to be very important for the development of the general theory. As a matter of fact, many physicists, and philosophers by the way, seemed to perceive Einstein's theory as belonging in some sense to the domain of mathematics. Thus, Oliver Lodge's remark, made during the December 1919 "Discussion on the Theory of Relativity" held at the Royal Astronomical Society, could be considered as representative of the feelings of at least some of his colleagues:[71]

> The surprising thing is that this theory has arrived at verifiable results; it is marvelous, and it represents very brilliant mathematical work. It has been done by using an out-of-the-way calculus developed by pure mathematicians—a ponderous kind of tool which only a few people can use. I do not pretend to be able to use it; I only with difficulty follow it; but the principle must coincide with some kind of reality, for by writing down equations on that principle you can get results. The theory is not dynamical. There is no apparent aim at real truth. It is regarded as a convenient mode of expression.

Of course, we know that Lodge was not a sympathetic commentator; yet, he was not totally unfair to Einstein's gravitational theory. It is true that at that time few people mastered the absolute calculus needed to understand the theory, and its experimental dimension, although impressive, was rather slim. Therefore, there were reasons that could justify the physicists' discomfort with the theory, especially in the case of physicists not too worried with fundamental principles.

With mathematicians, the reaction was very different. General relativity, a scientific theory that was even discussed in newspapers, gave them new opportunities; some mathematicians were able to use their abilities to contribute to the development of a theory about which scientists and the general public were speaking. General relativity offered them a "feeling of life" so to speak, which they could hardly find in other domains of mathematics or mathematical physics. Many physicists had to radically change, when confronted with Einstein's theory, their education and (conservative) ideas of what physics was really about; however, with mathematicians, it was almost the opposite. People such as Andrew Russell Forsyth, Cayley's successor in the chair of Pure Mathematics at the University of Cambridge (later on, in 1910, he moved to Imperial College, London);[72] Edward T. Copson, lecturer in mathematics at the University of Edinburgh and later Regius Professor of Mathematics at the University of St. Andrews; and Harry Edward James Curzon, an obscure mathematician, as far as I know, who was lecturer in mathematics at Goldsmiths' College, London, could participate (or thought they did) in the spectacular new quest to know the structure of the universe.[73]

A particularly illustrative case in this sense is that of John Edward Campbell (1862–1924). A mathematician attached for most of his life to Hertford College, Oxford, Campbell worked on miscellaneous topics. By 1897, he was writing about matters connected with Lie's transformation groups, which were his main concern for several years, culminating in his book *Introductory Treatise on Lie's Theory of Finite Continuous Transformation Groups* (Campbell 1903), a book that somewhat dampened John Theodore Merz's criticism of British mathematics, as expressed in his classic *A History of European Scientific Thought in the Nineteenth Century.*[74] The geometrical applications of Lie's contact transformations gradually led Campbell to the differential geometry of surfaces, and papers on Bäcklund's transformation, the application of quaternions to the deformation of a surface, and cyclic congruencies were written by him during the 10 years after the publication of his book. Then, his research came to an end. He only recovered his interest in mathematics after the war; it was tensor calculus and the applications of differential geometry to Einstein's theory that appealed to him, although he never had shown any interest either in mathematical physics or in experimental science.[75] In 1922 and 1923, he published two articles dealing with purely mathematical aspects of general relativity (Campbell 1922, 1923). Moreover, elected president of the London Mathematical Society in 1920, he chose as the subject of his presidential address (read on November 11, 1920) "Einstein's Theory of Gravitation as an Hypothesis of Differential Geometry" (Campbell 1920). He also wrote a book that was posthumously published by Clarendon (Oxford University) Press under the title *A Course of Differential Geometry* (Campbell 1926); unfortunately, as his son (J.M.H. Campbell) wrote in the preface, he died before completing a chapter, or appendix, that was to deal with the connection between the rest of the book and Einstein's theory.

General relativity served another obvious purpose for British mathematicians: through it some of them became acquainted with new areas of mathematics, such as parallelism and teleparallelism (or distant parallelism),[76] a topic to which Edmund Whittaker (Whittaker 1930) dedicated his 1929 presidential address to the London Mathematical Society. In Great Britain, during the period 1915–1930, there were no mathematicians of the category of Weyl, Levi-Civita, Cartan, Schouten, or Veblen, who could promote the development of differential geometry in a significant way;[77] at this time, British mathematics was stronger in other fields, such as analysis, mathematical logic, and applied mathematics, its principal names including Hardy, Littlewood, Whitehead, Russell, Whittaker, Forsyth, Watson, Lamb, Love, and Pearson. There is no doubt that the accomplishments of British mathematicians in differential geometry were not terribly impressive (especially when compared with the achievements of mathematicians in other countries),[78] but the rather small number of papers dealing with different aspects (differential invariants, tensor calculus, curves in Riemannian spaces, and so on) of this area of mathematics, papers—it is my contention—that were published at least in part because of general relativity, helped in the diversification of British mathematics. In this way, general relativity contributed to the development of mathematics in Great Britain, a phenomenon not exclusive to the British case, but felt more intensively in Great Britain than in other countries.

As far as individual British mathematicians are concerned, it was probably Edmund Taylor Whittaker, Professor of Mathematics at the University of Edinburgh from 1912, who achieved the greatest distinction for his contributions to the development of general relativity. (It could be argued that John L. Synge's contributions were more numerous and important than Whittaker's, but one has to remember that Synge was an Irishman, he worked in Dublin, and finally, from 1920 until 1925, he was an assistant professor of mathematics at the University of Toronto, in Canada; actually, many of Synge's papers during the 1920s appeared in American journals. Synge's influence in the development of general relativity in Great Britain would be more notable later, both through his papers and through people such as Felix A.E. Pirani, who spent some time with him in Dublin.)[79]

Whittaker's case is remarkable not only because of his general relativity papers, all of them concerned in some way with electromagnetism,[80] but also because at Edinburgh he did what his colleagues in the physics or, better, natural philosophy department (Charles G. Barkla and Charles G. Darwin were the professors there) did not—he lectured on Einstein's theory of relativity to senior year undergraduates, and postgraduates. (In the late 1920s, for example, he lectured on the unification of gravitation and electromagnetism. In the academic year 1927–1928, a graduate student who followed this course was McVittie: that education served him well when he went to Cambridge to work on his master's thesis under Eddington. His Ph.D. thesis, which he submitted in 1930, dealt with various modifications and exact solutions of Einstein's 1929 theory of teleparallelism.)[81] However, there were not many British mathematicians or physicists who contributed so actively to the development and assimilation of general relativity as did Whittaker.

On the whole, the great majority of the scientists who in one way or another contributed to the reception and development of general relativity in Great Britain during the period 1915–1930, did so from time to time; Einstein's theory constituted a small part of their professional interests, something, of course, that should be understood in the light of the fact that general relativity was a completely new theory, created by the genius of essentially one person, Albert Einstein. Consequently, the process of its institutionalization as a distinctive branch of physics, with its own experts (scientists working almost exclusively on its development), took some time, especially in a scientific community with the characteristics of the British one. During 1919 and 1920, the discussions could easily be intense because of the novelty and experimental dimension that the eclipse results brought to the theory, but afterward, the situation changed radically; new experimental effects were difficult, if not impossible, to find or deal with, and the development of general relativity relied mainly on its more theoretical and mathematical aspects. Few British physicists and mathematicians had the abilities, technical skills, opportunities for discussion with other colleagues, and time to work systematically and exclusively on the development of general relativity, especially in view of the many opportunities offered at the time by the successful atomic theory. By 1930, however, the situation began to change, with a new generation of scientists who had grown up with Einstein's theory as an already established theory.

NOTES

[1] I shall discuss the case of British philosophers in another publication.

[2] See, for instance, Earman and Glymour 1980.

[3] Concerning the myth of the incomprehensibility of general relativity in the Anglo-Saxon world, it is worth recalling the presumably apocryphal exchange that had Arthur Eddington as its main protagonist: "Professor Eddington," someone asked him, "is it true that only three people understand general relativity?" "Oh, who is the third?" was Eddington's reply. Independently of its apocryphal or even jocular character, this anecdote sums up the prevailing opinion among a large number of individuals.

[4] For more about the cultural issues in the reception of relativity, see Glick 1987.

[5] See, in this sense, Carr 1921.

[6] See, in this regard, Alter 1987.

[7] Haldane 1921, p. xi.

[8] See Sánchez-Ron 1987.

[9] A few aspects related to De Sitter's three papers in the *Monthly Notices* were discussed by A.D. Crommelin (Crommelin 1931) in the address he delivered on the occasion of the award of the Gold Medal of the Royal Astronomical Society to De Sitter.

[10] Actually, it was Bragg who communicated Fokker's paper to *Philosophical Magazine*.

[11] See the interview conducted by John L. Heilbron with Fokker, as part of the "Archive for History of Quantum Physics. Sources for History of Quantum Physics" project. For descriptions and some locations of this archive, see Kuhn, Heilbron, Forman, and Allen 1967.

[12] This was before De Sitter sent Eddington his first paper for the *Monthly Notices* [De Sitter's paper was completed during August 1916, and in July (see Stachel 1986), Eddington was writing to the Dutch astronomer that "I feel sure you will allow me to make use of the papers you send"].

[13] Whether or not Eddington was the moving spirit in the organization of that discussion is something I do not know, although it seems likely he was (see in this regard the letter Eddington sent to De Sitter on July 14, quoted in Stachel 1986).

[14] Actually, no meetings were held in 1917 and 1918.

[15] See p. 364 of *Report of the Eighty-Sixth Meeting of the British Association for the Advancement of Science: Newcastle-On-Tyne: 1916* (London 1917).

[16] See Eve 1939, p. 254, and the account of the BAAS meeting published in *Nature* 98: 120 (1916). Eddington's talk at Newcastle-On-Tyne presumable coincides with his first published paper devoted to the general theory of relativity (Eddington 1916).

[17] Among those present at Newcastle-On-Tyne were F.W. Dyson, E. Rutherford, J.C. McLennan, H.R. Hassè, W.H. Hicks, H.H. Turner, and A.N. Whitehead, who presided over Section A (mathematics and physics).

[18] The council notes, or, more properly, the "Reports on the Progress of Astronomy," were commissioned by the Council of the Royal Astronomical Society. It was only from 1938 that the authors' names were spelled out, although the owners of the initials used until then were easily identifiable.

[19] Here, Lodge referred to an "interesting account" published by Eddington in *Nature* (Eddington 1916).

[20] This view of the 1915–1918 history of general relativity in Great Britain, although not essentially compatible with the one put forward by Earman and Glymour (1980), offers new perspectives and insights concerning the early reception of Einstein's 1915 theory in Britain. Compare, in this sense, what I have said in the text with the following statements made by Earman and Glymour (1980 p. 50): "Yet there is no evidence that before 1919 many British physicists had warmed to general relativity. What little literature appeared on the theory was generally critical, or concerned with presenting alternatives. Before 1919 there did not appear in British scientific journals a single article, other than De Sitter's, that applied or extended the new theory. British physicists had to take cognizance of the theory, primarily because Eddington and De Sitter had made them aware that Einstein had succeeded in explaining the long-standing anomaly in the motion of Mercury, but prior to the eclipse expedition they were not disposed to accept the new relativistic account of gravitation or even to trouble much to understand it."

To further my point, I will recall the case of William Wilson, of King's College, London, who, as can be seen in bibliographies such as the ones prepared by Combridge 1965 and Lecat 1924 or in Whittaker's encyclopedic history (Whittaker 1953), contributed to the development of general relativity during the 1920s and 1930s. [(He also helped in the development of the old quantum theory). By 1918, Wilson was already discussing specific points of Einstein's theory at the Physical Society of London (see Wilson 1919); in this paper, received at the Physical Society on October 21, 1918, Wilson showed that the equations of motion in a gravitational field can be put in a Hamiltonian form.

[21] The full account of the eclipse expedition results was published by Dyson, Eddington, and Davidson 1920 in the *Philosophical Transactions of the Royal Society of London*. It might be worth pointing out at this stage that a preliminary discussion of part of the eclipse results (the photographs taken at Principe) was presented by Eddington and Cottingham 1920 at the 87th meeting of the British Association for the Advancement of Science, held at Bournemouth on September 9–13, 1919.

[22] See, for instance, the note published in *The Athenaeum*, November 21, 1919 (p. 1229), as well as Tayler, ed., 1987, p. 26.

[23] Quoted in Stachel 1986, p. 234.

[24] See *The Athenaeum*, "Einstein's Theory of Gravitation," November 14, 1919, p. 1189; "A Matter of Evidence," November 21, 1919, pp. 1128–1129; and Lodge 1919. In his paper, published in the magazine *The Nineteenth Century and After*, Lodge accepted without any reservation the measurements taken by the

British expedition, pointing out also that "Before Einstein's prediction nothing of the kind had been seen, nothing of the kind had been looked for, nor, so far as is known, has such an amount of deflection been suspected." However, he could not refrain from warning of the dangers of accepting Einstein's theory. "The present writer holds it dangerous," he wrote, "to base such far-reaching consequences [i.e., Einstein's ideas of space and time as expressed in his gravitational theory], even if anything like them can legitimately be drawn—which is doubtful—on a predicted effect which may after all be accounted for and expressed in simpler fashion. Our admiration for the brilliant way in which the fact was arrived at must not make us too enthusiastically ready to assimilate the whole complicated theory out of which it arose. For, independent of the above speculation, it is indubitable that the mathematical theory of relativity—very different from what philosophers have hitherto meant by the term—is almost inevitably complicated." Lodge also took this opportunity to mention his "Electrical Theory of Matter," to which I have already referred.

[25] Usually, the ordinary meetings of the Royal Astronomical Society were devoted to the reading of papers. However, in view of the widespread interest in the theory of relativity caused by the publication of the results of the eclipse observations, the council thought they would meet the wishes of the fellows by giving the whole time of the meeting of December 12 to the consideration of Einstein's theory. Regarding this, see Dreyer and Turner, eds., 1923, p. 238.

[26] See "Discussion on the Theory of Relativity," *Monthly Notices of the Royal Astronomical Society* 80: 96–119 (1919); "Discussion on the Theory of Relativity," *Proceedings of the Physical Society of London* 32: 245–251 (1920).

[27] Joseph Larmor sent a communication that was read by the secretary of the society; in it, Larmor criticized Einstein's theory, of which he said "How far it is from being a determinate theory of the universe will appear to anyone who dips into the writings of its developers." At the same time, Larmor took the opportunity to point out the compatibility of general relativity with his beloved principle of least action. Indeed, Larmor was quite active in the discussions that followed the 1919 announcement; being, however, a truly 19th century scientist, he could not fully accept Einstein's theory. A representative sentence in this sense is the following one, which appears in the paper Larmor wrote in collaboration with W.J. Johnston for the 1919 meeting at the British Association for the Advancement of Science (Johnston and Larmor 1920): "If physical science is to evolve on the basis of relations of permanent matter and its motions, time must continue to be thrown on to the material observing system in the form of slight modification of its structure." For a list of Larmor's papers connected with issues aroused by general relativity, see Lecat 1924 and Combridge 1965. I will not discuss Larmor's ideas in this article.

[28] At the royal Astronomical Society meeting, Jeans confined practically all of his comments to the special theory of relativity; perhaps his knowledge of general relativity was still rather superficial.

[29] For the exact references, see Lecat 1924.

[30] The observations referred to by Fowler were those of Evershed 1914 and 1918; Schwarzschild 1914; St. John 1917; and Grebe and Bachem 1919.

[31] Probably, St. John's most important work was his revision of the Rowland's table of solar wavelengths undertaken during the 1920s in collaboration with other Mt. Wilson colleagues.

[32] St. John's first observations and discussions were made in 1917.

[33] See Douglas 1956, pp. 44–45.

[34] "The Astronomical Tests of Einstein's Law," *Monthly Notices of the Royal Astronomical Society* 84: 292–295 (1924). The author of this report answered to the initials A.C.D.C., most probably A.C.D. Crommelin, who was the Society's secretary during the period 1917–1923.

[35] The problem continued to be investigated (see, for instance, Evershed 1928, 1931). Eisenstaedt 1986 has also considered some aspects of this test of general relativity.

[36] This story was told by Eve 1939.

[37] Rutherford and Compton's note dealt, as they acknowledged, with the same sort of problems that the chemist Frederick George Donnan was studying at the time (see Donnan 1919). Indeed, their results were "not in disaccord with the relation deduced by Prof. Donnan."

[38] Arthur Schuster had suggested this problem to Rutherford before the outbreak of the war.

[39] Of course, this was in fact a test of the principle of equivalence.

[40] See in this regard Douglas 1956, Stachel 1986, and Sánchez-Ron 1987b.

[41] "One of the literary gems of English scientific literature," Rice 1923 called this book.

[42] A second edition was published in 1924. Also, a German translation appeared in 1925; it contained an appendix written by Einstein.

[43] As a matter of fact, during these years, Eddington gave a course of lectures on general relativity each Easter term. According to Douglas (1956), "a few senior undergraduates and research students and a few of the younger fellows took advantage of these lectures. About eight attended the course in 1922.

[44] This is the case, for instance, with Chapter VII (World Geometry), which includes Eddington's generalization of Weyl's unified theory of the electromagnetic and gravitational fields, which he had put forward in 1921 (Eddington 1921b). I have considered a few of the more advanced topics included in *The Mathematical Theory of Relativity* in Sánchez-Ron 1987b.

[45] See Eddington 1921b, 1922.

[46] See Baldwin and Jeffrey 1926, and Baldwin 1929.

[47] See, for instance, items numbers 954, 955, 1673, and 1692 of Cambridge 1965, corresponding to papers of McVittie, Whittaker, and Wirtinger.

[48] As far as journals are concerned, my discussion throughout this paper is based on the analysis of the following journals: *Proceedings of the Royal Society of London* (series A), *Philosophical Magazine, Monthly Notices of the Royal Astronomical Society, Philosophical Transactions of the Royal Society of Lon-*

don, Proceedings of the Royal Society of Edinburgh, The Journal of the London Mathematical Society, Proceedings of the Cambridge Philosophical Society, The Quarterly Journal of Pure and Applied Mathematics (after Vol. 50, its name was *The Quarterly Journal of Mathematics)*, and *The Memoires and Proceedings of the Manchester Literary and Philosophical Society.* I have also been helped by the splendid bibliographies prepared by Lecat 1924 and Combridge 1965.

[49] It really makes no difference to consider also non-British scientists.

[50] Of course, there is nothing strange about this; the problems of quantum physics at the time absorbed the energies of an important part of theoretical and experimental physicists. In England, the school, or influence, of Rutherford was dominant.

[51] See p. 373 of the *Report* (London 1925).

[52] See p. 315 of the *Report* (London 1927).

[53] Indeed, even when he was mentioning general relativity, Richardson could not avoid referring to quantum physics in very favorable terms. Thus, he said (Richardson 1922), "Relativity is the revolutionary movement in physics which has caught the public eye, perhaps because it deals with familiar conceptions in a manner which for the most part is found pleasantly incomprehensible. But is is only one of a number of revolutionary changes of comparable magnitude. Among these we have to place the advent of the quantum, the significance of which I hope we shall thoroughly discuss early next week."

[54] See, in this regard, for example, items numbers 277, 655, 667, 731, 734, 1000, 1671, and 1677 of Combridge 1965, corresponding to papers of Conway, Cunningham, Hargreaves, Jeffery, Meksyn, E.T. Whittaker, and J.M. Whittaker. Of course, the list is not complete.

[55] Of course, *Nature* was also important (the February 17, 1921, issue, for example was totally dedicated to relativity; it included articles by Einstein, Cunningham, Dyson, Crommelin, St. John, Mathews, Jeans, Lorentz, Lodge, Weyle, Eddington, Campbell, Wrinch and Jeffreys, and Carr). Also important was *The Observatory*, the monthly review that aimed at presenting in a popular form the progress of astronomy. [For a list of the articles dealing with general relativity that appeared in *The Observatory*, see the entry "Relativity" in *The Observatory. General Index, Vols. 1–72 (1877–1955)*, pp. 108–109.] However, I have in general excluded *Nature* and *The Observatory* from my considerations because they were more scientific magazines than scientific journals, and I am only considering the latter ones here.

[56] McVittie was a student of Eddington's (see in this regard McVittie 1987).

[57] See also Eddington 1930b.

[58] An English translation of Lemaître's 1927 paper (Lemaître 1927) was published in *Monthly Notices* in 1931 (Lemaître 1931).

[59] It is interesting to recall that Eddington was the president of this society during the period from 1921 to 1923.

[60] A partial list can be found in the "Books and Pamphlets" of the "Bibliography of Relativity" published in the February 17, 1921, *Nature* issue dedicated

to relativity (pp. 811–813). A more complete bibliography can be found in Sánchez-Pérez 1934.

[61] The number of these philosophical books was important. Let us recall, for instance, the works of Whitehead, Broad, Wildon Carr, Haldane, and Russell.

[62] Indeed, the number of books dealing with relativity written by British scientists is much lower than, for example, what one has in the German case. The comparison is also unfavorable in other cases, such as the French one. A large list of relativity books that covers the period considered here is included in Sánchez-Pérez 1934.

[63] The first edition, which did not include general relativity, had been published in 1915.

[64] Rice was the senior lecturer in Physics at the University of Liverpool.

[65] Nunn was Professor of Education at the University of London. I have included him here, in spite of his not being really a scientist, because his book shows a considerable technical knowledge of relativity.

[66] The first edition of *Modern Electrical Theory* (Cambridge University Press) appeared in 1907; the second one, really a new book, in 1913. Only one chapter (Ch. XVI) of the four basic chapters of Campbell's *Relativity* (Campbell 1923) dealt with general relativity.

[67] The first edition had appeared in 1914.

[68] As a matter of fact, he also published mathematical books. See, for instance, Silberstein 1919a,b.

[69] In 1912 and 1913, for example, he delivered a course of lectures dealing with Einstein's special theory of relativity at University College, London. This course constituted the base of his book *The Theory of Relativity* (Silberstein 1914).

[70] Another expression of his independent,and often unsuccessful,approaches can be found in his controversial conception of cosmology, which he summed up in *The Size of the Universe* (Silberstein 1930), written before Lemaître's model became known. When Eddington 1930b reviewed Silberstein's book, he called attention to Lemaître's "very substantial advance" which "renders obsolete the contrast between Einstein's and De Sitter's cosmologies" on which Silberstein based his work (see in this regard Kragh 1987).

[71] See note 26; Lodge's remark appears on p. 107.

[72] In 1912, Forsyth 1912 had published a book on the differential geometry of curves and surfaces.

[73] See Forsyth 1920, 1921. In these papers, the author of the *Theory of Differential Equations* applied his considerable knowledge of differential equations to solve some of the equations appearing in general relativity. (Copson 1928) and Curzon 1924a,b. Curzon's two papers in the *Proceedings of the London Mathematical Society*, a society of which he was a member, contained the so-called "Curzon Exact Solution" of Einstein's field equations, a solution still often discussed in some of the modern general relativity texts. As far as I know, these two important papers of Curzon (the first is really only an abstract) constitute his

only contributions to the development of general relativity.

[74] See pp. 690–691 of Merz 1904. There, Merz wrote the following: "The 'Theory of Groups' has now grown into a very extensive doctrine which according to the late Prof. Marius Sophus Lie... is destined to occupy a leading and central position in the mathematical science of the future.... But though it is an undoubted fact that the largest systematic works on the subject emanate from the great Norwegian mathematician, and that his ideas have won gradual recognition ... the epoch-making tract which pushed the novel conception into the foreground was Prof. F. Klein's 'Erlangen Programme' (1872).... From that date onward, the different kind of groups have been defined and systematically studied, notably by Klein and Lie and their pupils. In this country [Great Britain], although many of the relevant ideas were contained in the writings notably of Cayley and Sylvester, the systematic treatment of the subject was little attended to before the publication (1987) of Prof. Burnside's 'Theory of Groups of Finite Order,' and latterly of his article on the whole theory of groups in the 29th volume of the 'Ency. Brit.' It has been remarked by those who have studied most profoundly the development of the two great branches of mathematical tactics—viz. 'The Theory of Invariants' and the 'Theory of Groups'—that the progress of science would have been more rapid if the English School had taken more notice of the general comprehensive treatment of Lie."

[75] For more information about Campbell, see the obituary note (signed by H. H.) that appeared in the *Proceedings of the London Mathematical Society* 23: lxx-lxxi (1924).

[76] Here is Whittaker's definition of 'teleparallelism': "When the vector obtained at the terminus is the same whatever be the path by which the journey from the initial point has been made, we say that there is *teleparallelism*" (Whittaker 1930)

[77] Of course, Eddington did contribute, but he was not a mathematician.

[78] Of course, we are not taking Eddington into account.

[79] The list of Synge's publications is included in O'Raifeartaigh, ed. 1972. There, one can see the papers dealing with general relativity, or with mathematical problems related to Einstein's theory, that Synge published in British journals during the 1920s.

[80] Whittaker's contributions to relativity are discussed in Synge 1958, as well as in Temple 1956.

[81] See McVittie 1987. See also McVittie's papers referred to in note 47.

REFERENCES

Alter, Peter (1987). *The Reluctant Patron. Science and the State in Britain 1850–1920.* Oxford: Berg.

Baldwin, O. R. (1929). "The Relativity Theory of Divergent Waves." *Royal Society of London. Proceedings A* 123: 119–133.

Baldwin, O. R., and Jeffery, George B. (1926). "The Relativity Theory of Plane Waves." *Royal Society of London. Proceedings A* 111: 95–104.

Campbell, John E. (1903). *Introductory Treatise on Lie's Theory of Finite Continuous Transformation Groups.* Oxford: Clarendon Press; reprint Bronx, New York: Chelsea, 1966.

—— (1920). "Einstein's Theory of Gravitation as an Hypothesis in Differential Geometry." *London Mathematical Society. Proceedings* 20: 1–14.

—— (1922). "On a Class of Surfaces in Euclidean Space which Generate an Expression for the Space Time Interval in Einstein's Geometry of a Particular Form." *London Mathematical Society. Proceedings* 21: 317–324.

—— (1923). "On the Generation of the Ground Form of the Statical Gravitation Field in Einstein's Theory of Gravitation from the Ground Form of any Surface in Ordinary Space." *London Mathematical Society. Proceedings* 22: 92–103.

—— (1926). *A Course of Differential Geometry.* Oxford: Clarendon Press.

Campbell, Norman Robert (1923). *Modern Electrical Theory.* Cambridge: Cambridge University Press.

Carr, H. Wildon (1921). "Review of: Viscount Haldane, The Reign of Relativity." *Mind* 30: 462–467.

Combridge, J. T. (1965). *Bibliography of Relativity and Gravitation Theory 1921 to 1937.* London: King's College.

Copson, Edward T. (1928). "On Electrostatics in a Gravitational Field." *Royal Society of London. Proceedings A* 118: 184–194.

Crommelin, Andrew C. D. (1931). "The President's Address." *Royal Astronomical Society. Monthly Notices* 91: 422–434.

Cunningham, Ebenezer (1914). *The Principle of Relativity.* Cambridge: Cambridge University Press.

—— (1921). *Relativity, the Electron Theory, and Gravitation.* New York: Longmans, Green.

Curzon, Harry E. J. (1924a). "Bipolar Solutions of Einstein's Gravitation Equations." *London Mathematical Society. Proceedings* 23: xxix.

—— (1924b). "Cylindrical Solutions of Einstein's Gravitation Equations." *London Mathematical Society. Proceedings* 23: 477–480.

De Sitter, Willem (1916a). "On Einstein's Theory of Gravitation and its Astronomical Consequences." *Royal Astronomical Society. Monthly Notices* 76: 699–728.

—— (1916b). "On Einstein's Theory of Gravitation and Its Astronomical Consequences. Second Paper." *Royal Astronomical Society. Monthly Notices* 77: 155–184, 481.

——— (1917). "On Einstein's Theory of Gravitation and Its Astronomical Consequences. Third Paper." *Royal Astronomical Society. Monthly Notices* 78: 3–28.

Dingle, Herbert (1923). *Relativity and Gravitation.* London: Methuen.

Dodwell, G.F., and Davidson, Charles R. (1922–1923). "Determination of the Deflection of Light by the Sun's Gravitational Field from Observations Made at Cordillo Downs, South Australia, during the Total Eclipse of 1922 September 21." *Royal Astronomical Society. Monthly Notices* 84: 150–162.

Donnan, Frederick G. (1919). "Heat of Reaction and Gravitational Field." *Nature* 104: 392–393.

Douglas, A. Vibert (1956). *The Life of Arthur Stanley Eddington.* London: Thomas Nelson.

Dreyer, John, L.E., and Turner, Herbert M., eds. (1923). *History of the Royal Astronomical Society, 1820–1920.* London: Royal Astronomical Society; reprint Oxford: Blackwell, 1987.

Duffield, W.G. (1919–20). "The Displacement of Spectrum Lines and the Equivalence Hypothesis." *Royal Astronomical Society. Monthly Notices* 80: 262–272.

Dyson, Frank W. (1917). "On the Opportunity Afforded by the Eclipse of 1919 May 29 of Verifying Einstein's Theory of Gravitation." *Royal Astronomical Society. Monthly Notices* 77: 445–447.

Dyson, Frank W.; Eddington, Arthur S.; and Davidson, Charles (1920). "A Determination of the Deflection of Light by the Sun's Gravitational Field." *Royal Society of London. Philosophical Transactions A* 220: 291–333.

Earman, John, and Glymour, Clark (1980). "Relativity and Eclipses: The British Eclipse Expeditions of 1919 and their Predecessors." *Historical Studies in the Physical Sciences* 11: 49–85.

Eddington, Arthur S. (1915). "Gravitation." *Observatory* 38: 93.

——— (1916). "Gravitation and the Principle of Relativity." *Nature* 98: 328–331.

——— (1917a). "Einstein's Theory of Gravitation." *Royal Astronomical Society. Monthly Notices* 77: 377–382.

——— (1917b). "Astronomical Consequences of the Electrical Theory of Matter. Note on Sir Oliver Lodge's Suggestions." *Philosophical Magazine* 34: 163–167.

——— (1917c). "Astronomical Consequences of the Electrical Theory of Matter. Note on Sir Oliver Lodge's Suggestions, II." *Philosophical Magazine* 34: 321–327.

——— (1918a). *Report on the Relativity Theory of Gravitation.* London: Physical Society of London.

——— (1918b). "Electrical Theories of Matter and Their Astronomical Consequences with Special Reference to the Principle of Relativity." *Philosophical Magazine* 35: 481–487.

——— (1920). *Space, Time and Gravitation.* Cambridge: Cambridge University Press.

——— (1921a). *Espace, temps et gravitation.* Paris: Hermann.

——— (1921b). "A Generalisation of Weyl's Theory of the Electromagnetic and Gravitational Fields." *Royal Society of London. Proceedings A* 99: 104–122.

——— (1922). "The Propagation of Gravitational Waves." *Royal Society of London. Proceedings A* 102: 268–282.

——— (1923). *The Mathematical Theory of Relativity.* Cambridge: Cambridge University Press.

——— (1930a). "On the Instability of Einstein's Spherical World." *Royal Astronomical Society. Monthly Notices* 90: 668–678.

——— (1930b). "Space and Its Properties." *Nature* 125: 849–850.

——— (1931). "The Expansion of the Universe (Council Note)." *Royal Astronomical Society. Monthly Notices* 91: 412–416.

Eddington, A. S., and Cottingham, E.T. (1920). "Photographs Taken at Principe during the Total Eclipse of the Sun, May 29th." In *Report of the Eighty-Seventh Meeting of the British Association for the Advancement of Science.* London: John Murray, pp. 156–157.

Einstein, Albert, and Fokker, Adriaan D. (1914). "Nordströmsche Gravitationstheorie vom Standpunkt des absoluten Differentialkalküls." *Annalen der Physik* 44: 321–328.

Eisenstaedt, Jean (1986). "La relativité générale à l'étiage: 1925–1955." *Archive for History of Exact Sciences* 35: 115–185.

Elton, Lewis (1986). "Einstein, General Relativity, and the German Press, 1919–1920." *Isis* 77: 95–103.

Eve, Arthur S. (1939). *Rutherford. Being the Life and Letters of the Rt. Hon. Lord Rutherford, O. M.* New York: Macmillan.

Evershed, John (1921). "The Relativity Shift in the Solar Spectrum." *Observatory* 44: 243–245.

——— (1928). "The Solar Rotation and the Einstein Displacement Derived from Measures of the H and K Lines in Prominences." *Royal Astronomical Society. Monthly Notices* 88: 126–134.

——— (1931). "The Shift towards Red on the Calcium, Aluminium, and Iron Lines in the Solar Spectrum." *Royal Astronomical Society. Monthly Notices* 91: 260–270.

Fokker, Adriaan D. (1915). "A Summary of Einstein and Grossmann's Theory of Gravitation." *Philosophical Magazine* 29: 77–96.

Forsyth, Andrew R. (1912). *Lectures on the Differential Geometry of Curves and Surfaces.* Cambridge: Cambridge University Press.

——— (1920). "Note on the Central Differential Equation in the Relativity Theory of Gravitation." *Royal Society of London. Proceedings A* 97: 145–151.

——— (1921). "Note on the Path of a Ray of Light in the Einstein Relativity Theory of Gravitational Effect." *Royal Astronomical Society. Monthly Notices* 82: 2–11.

Glick, Thomas F. (1987). "Cultural Issues in the Reception of Relativity." In *The*

Comparative Reception of Relativity. Thomas F. Glick, ed. Dordrecht: Reidel, pp. 381–400.

Haldane, Viscount (1921). *The Reign of Relativity.* London: John Murray.

Jeans, James H. (1925–1926). "President's Address." *Royal Astronomical Society. Monthly Notices* 86: 262–269.

Johnston, W. J., and Larmor, Joseph (1920). "The Limitations of Relativity." In *Report of the Eighty-Seventh Meeting of the British Association for the Advancement of Science* London: John Murray, pp. 15–159.

Kragh, Helge (1987). "The Beginning of the World: Georges Lemaître and the Expanding Universe." *Centaurus* 32: 114–139.

Kuhn, Thomas S., et al. (1967). *Sources for History of Quantum Physics. An Inventory and Report.* Philadelphia: The American Philosophical Society.

Lecat, Maurice (1924). *Bibliographie de la relativité.* Bruxelles: Maurice Lamertin.

Lemaître, Georges (1927). "Un univers homogène de masse constante et de rayon croissant, rendant compte de la vitesse radiale des nébuleuses extragalactiques." *Société Scientifique de Bruxelles. Annales A* 47: 49–59.

——— (1931). "Homogeneous Universe of Constant Mass and Increasing Radius Accounting for the Radial Velocity of Extra-Galactic Nebulae." *Royal Astronomical Society. Monthly Notices* 91: 483–490. [English translation of Lemaître 1927.]

Lindemann, Adolphus F., and Lindemann, Frederick A. (1916). "Daylight Photography of Stars as a Means of Testing the Equivalence Postulate in the Theory of Relativity." *Royal Astronomical Society. Monthly Notices* 77: 140–151.

Lodge, Oliver (1917a). "Astronomical Consequences of the Electrical Theory of Matter." *Philosophical Magazine* 34: 81–94.

——— (1917b). "Astronomical Consequences of the Electrical Theory of Matter. Supplementary Note." *Philosophical Magazine* 34: 517–521.

——— (1918). "Continued Discussion of the Astronomical and Gravitational Bearings of the Electrical Theory of Matter." *Philosophical Magazine* 35: 141–156.

——— (1919). "The New Theory of Gravity." *The Nineteenth Century and After* 86: 1189–1201.

——— (1920). "Note on a Possible Structure for the Ether." *Philosophical Magazine* 39: 170–174.

Merz, John Theodore (1904). *A History of European Thought in the Nineteenth Century, Vol. 2.* London: William Blackwood. Reprinted under the title *A History of European Scientific Thought in the Nineteenth Century,* Vol. 2. New York: Dover, 1965.

McVittie, George C. (1987). "An Anglo-Scottish University Education." In *The Making of Physicists.* Rajkumari Williamson, ed. Bristol: Adam Hilger, pp. 66–70.

Missner, Marshall (1985). "Why Einstein Became Famous in America." *Social*

Studies of Science 15: 267–291.

Nunn, T. Percy (1923). *Relativity and Gravitation. An Elementary Treatise On Einstein's Theory.* London: University of London Press.

O'Raifeartaigh, Lochlainn, ed. (1972). *General Relativity. Papers in Honour of J.L. Synge.* Oxford: Oxford University Press (Clarendon).

Rice, James (1923). *Relativity. A Systematic Treatment of Einstein's Theory.* London: Longmans, Green.

Richardson, Owen W. (1922). "Problems of Physics." In *Report of the Eighty-Ninth Meeting of the British Association for the Advancement of Science.* London: John Murray, pp. 25–35.

Rutherford, Ernest, and Compton, Arthur H. (1919). "Radio-activity and Gravitation." *Nature* 104: 412.

Sánchez-Pérez, José A. (1934). "Bibliografia sobre relatividad." In *Discurso leido en el acto de su recepción.* Madrid: Academia de Ciencias Exactas, Fisicas y Naturales, pp. 31–71.

——— (1987a). "The Reception of Special Relativity in Great Britain." In *The Comparative Reception of Relativity.* Thomas F. Glick, ed. Dordrecht: Reidel, pp. 27–58.

——— (1987b). "The Role Played by Symmetries in the Introduction of Relativity in Great Britain." In *Symmetries in Physics (1600–1980).* Manuel G. Doncel et al., eds. Barcelona and Singapore: Universitat Autònoma de Barcelona/ World Scientific Publishers, pp. 165–184.

Silberstein, Ludwik (1914). *The Theory of Relativity,* 2nd. ed. London: Macmillan.

——— (1918a). "Bizarre Conclusion Derived from Einstein's Gravitation Theory." *Royal Astronomical Society. Monthly Notices* 78: 465–467.

——— (1918b). "General Relativity without the Equivalence Hypothesis." *Philosophical Magazine* 36: 94–128.

——— (1919a). *Projective Vector Algebra.* London: G. Bell and Sons.

——— (1919b). *Elements of Vector Algebra.* London: Longmans, Green.

——— (1920). "The Recent Eclipse Results and Stokes–Planck's Aether." *Philosophical Magazine* 39: 161–170.

——— (1922). *The Theory of General Relativity and Gravitation.* New York: Van Nostrand-Reinhold.

——— (1924). *The Theory of Relativity,* 2nd ed. London: Macmillan.

——— (1930). *The Size of the Universe.* London: Oxford University Press.

Stachel, John (1986). "Eddington and Einstein." In *The Prism of Science.* E. Ullmann-Margalit, ed. Dordrecht: Reidel, pp. 225–250.

St. John, Charles E. (1922–1923)."On Gravitational Displacement of Solar Lines." *Royal Astronomical Society. Monthly Notices* 84: 93–96.

Synge, John L. (1958). "Whittaker's Contributions to the Theory of Relativity." *Edinburgh Mathematical Society. Proceedings* 11: 39–55.

Tayler, Roger J., ed. (1987). *History of the Royal Astronomical Society, 1920–1980.* Oxford: Blackwell.

Temple, George F. (1956). "Edmund Taylor Whittaker, 1873–1956." *Biographical Memoirs of Fellows of the Royal Society* 2: 299–325.

Walker, George W. (1918). "Relativity and Electrodynamics." *Philosophical Magazine* 35: 327–338.

Whittaker, Edmund T. (1927). "The Outstanding Problems of Relativity." *Report of the Ninety-Fifth Meeting of the British Association for the Advancement of Science*. London: Office of the British Association, pp. 16–26.

—— (1930). "Parallelism and Teleparallelism in the Newer Theories of Space." *Journal of the London Mathematical Society* 5: 68–80.

—— (1953). *A History of the Theories of Aether and Electricity. The Modern Theories, 1900–1926*. London and Edinburgh: Thomas Nelson and Sons.

Wilson, William (1919). "Relativity and Gravitation." *Physical Society of London. Proceedings* 31: 69–78.

US Air Force Support of General Relativity: 1956–1972

Joshua N. Goldberg

1. Introduction

In 1955, the Aeronautical Research Laboratories (ARL) at Wright-Patterson Air Force Base in Ohio consisted of three groups: physics, chemistry, and applied mathematics. The applied mathematics people were mostly civilians. About half of these were brought into the United States under operation "Paperclip." The physics and chemistry labs were staffed mostly by people who were in the Air Force through Reserve Officer Training Corps (ROTC) programs at various universities. Most of these people were interrupted in their graduate studies and were to serve a two-year term at ARL. In the physics department, there were two main projects. One had developed methods of growing good crystals, particularly CdS, and was interested in studying their electrical and luminescent properties. The other had a 4-MeV van de Graaff accelerator for studying nuclear energy levels. Civilians headed both efforts.

Dr. Max Scherberg, a mathematician who was the senior scientist at ARL, decided to initiate support in general relativity at Indiana University with Vaclav H'lavaty and at Syracuse University with Peter Bergmann. The lieutenant colonel who was the commandant of ARL at the time believed that a strong in-house research effort that supervised an external support program was an important means for keeping Air Force abreast of new scientific developments and in a position to exploit new technology. Therefore, he encouraged hiring new people in the solid state and nuclear physics groups. In addition, he considered general relativity as a field of growing interest that would be important to the Air Force. Nuclear physics had already found its niche, and the work on nuclear energy levels was clearly related to it.

As a result, Dr. Sherberg was asked to explore the existence of interest in Air Force support and the possibility of hiring someone for ARL who would take over that responsibility. He interviewed Peter Bergmann and Vaclav H'lavaty in the United States and Pascual Jordan, Nathan Rosen, and Julius Géhéniau abroad. Bergmann suggested that I might be interested in the position as I was at that time planning to leave Armour Research Foundation. After some consideration, I accepted the position, not only because it allowed me to return full time to research in general relativity, but also because I found the idea of supporting the research of others attractive.

Therefore, in September 1956, I joined the General Physics Laboratory of the Aeronautical Research Laboratories to begin a research and support program that would grow and last for 16 years. In the fall in 1958, a student of H'lavaty came; a year later, Roy Kerr joined the group; and in the fall of 1961, Stuart Fickler, a student of Richard Arnowitt, joined under a program of postdoctoral research associateships administered by the National Academy of Science. Although Roy Kerr and I left ARL after Sputnik in 1963 for the University of Texas at Austin and Syracuse University respectively, gravitational research flourished at ARL under Stu Fickler until 1972. It was terminated when the US Congress passed the Mansfield amendment, which said that the Defense Department should not be engaged in basic research, but should restrict its research efforts to projects that had application to military needs. To compensate partially for the resulting loss of basic research support, the National Science Foundation (NSF) was strengthened. During the latter period, Lou Tamburino and Jeff Winicour were permanent members of ARL, while David Robinson, Ed Glass, and Richard Isaacson, among others, spent two years with the group as research associates.

In the following sections, I briefly describe the activities and some of the results of both the in-house and supported research. Since this period introduced and developed our understanding of gravitational radiation, that topic will form the focus for the discussion.

2. Conferences

Shortly after I arrived at ARL, Bryce DeWitt called to discuss his and Cecile's interest in organizing a conference on general relativity at Chapel Hill, North Carolina. After he submitted a formal proposal, a small amount of money ($5000?) was made available. The Air Force also offered travel from Europe by the Military Air Transport Service (MATS) for a few people. A very successful conference took place January 18–23, 1957. It was the first such conference in which postwar students of general relativity were able to participate, and it was a marvelous experience for us. We were able to meet intimately with people who were just names on papers and books: P.G. Bergmann, J.A. Wheeler, V. Bargmann, N. Rosen, A. Lichnerowicz, M.A. Tonnelat, S. Mavrides, Y. Choquet, H. Bondi, T. Gold, A. Schild, F. Pirani, among many others. My most vivid recollection is long discussions with Bondi and Gold, who took the position that gravitational radiation did not exist. I no longer recall the arguments they used, but I presume that those discussions prompted Bondi to develop his analysis at null infinity. Obviously, he had to show mathematically that reasonable boundary conditions inhibited gravitational radiation. A wide variety of topics were discussed with emphasis on gravitational radiation and quantum gravity. In particular, DeWitt gave a very long lecture that introduced many of the ideas associated with his development of the covariant approach to quantum gravity.

It was a most stimulating experience, and all agreed that with a growing international effort regular conferences should take place. A. Lichnerowicz agreed to host the next conference in France, and a date was set for 1959. Thus, was born the General Relativity and Gravitation (GR) series of conferences.

The Air Force did not give direct support to any of the later GR conferences. However, it did offer MATS transportation to a large number of American participants in the 1959 meeting in Royaumont, the 1962 meeting in Warsaw, and the 1965 meeting in London. The 1968 meeting took place in Tblisi, USSR, and the US Air Force declined to offer air travel to that destination. By 1971, the Mansfield amendment had been proposed, and the Air Force did not offer MATS transport to friendly Denmark. I believe that many of us have fond memories of meeting at McGuire Air Force Base in New Jersey for our flights to London and Paris.

Together with the University of Cincinnati (Louis Witten was then the head of the physics department), ARL sponsored a conference that took place June 2–6, 1969, in Cincinnati. This conference was memorable because in his presentation, Joe Weber made his first public claim to have detected gravitational radiation—a claim that has not yet been substantiated, but has stimulated significantly increased activity in the design and building of gravity wave antennas.

3. In-House Research

When I arrived at ARL, my interests were in studying the derivation of the equations of motion from the Einstein equations. I followed Einstein's attitude that one did not want to introduce explicitly the matter field. Rather, one was to find singular solutions. The singularities were to be associated with matter whose properties (motion) were to be determined by the gravitational field itself. Therefore, I drew the conclusion that while the fields may be singular, the Bianchi identities should not be singular. At the Chapel Hill conference, V. Bargmann pointed out that this was equivalent to the use of delta-function sources and that Leopold Infeld had already used this idea in a published paper. My study then split into two parts. One involved the relation of the conservation laws to motion without approximations. The other developed the "fast" approximation method, which led to the fruitful collaboration with Peter Havas. The goal was to exhibit radiation damping in the equations of motion.

This work did not progress beyond the lowest approximation because of mathematical difficulties. At this level of approximation, the results were, to say the least, ambiguous. Only in relatively recent times have good results showing radiation damping been obtained.

In the summer of 1957, Ted Newman came to ARL for three months as a consultant. Although we discussed a variety of topics, the focus of our summer's work was on the nature of observables in a curved space-time. We formulated a definition of distance in terms of the equation of geodesic deviation along null rays. This was the precursor to the introduction of the optical parameters ρ, σ, and ω—the rates of contraction, shear, and twist—by R.K. Sachs. The latter was more useful as it involved an area measure, whereas our definition only used the linear expansion.

Joseph Schell, a recent student of H'lavaty, came to ARL in the fall of 1958. This was shortly after the appearance of Pirani's paper applying the Petrov classification to the definition of gravitational radiation. We began to discuss the holonomy group as another method to probe the properties of a curved space-time. Schell was led to consider a classification scheme for the holonomy group. The characteristic bivectors in the various Petrov classes are invariant under subgroups of the Lorentz group. The holonomy group is defined by parallel propagation of a vector around a closed path. Therefore, it induces a curvature-dependent Lorentz transformation of a local tetrad. Thus, the holonomy group yields a refinement of the Petrov classification. However, the Petrov classes have been very useful in the construction of solutions to the Einstein equations, and the refinement has not yet been applied. Nonetheless, Roy Kerr and I did base our null solutions on the holonomy classes.

After spending a postdoctoral year with Peter Bergmann at Syracuse University, Roy Kerr joined our group in the fall of 1959. He had been interested in the equations of motion and had prepared three papers that unfortunately did not go beyond the point Havas and I had achieved. At ARL, he and I discussed the tetrad formulation of the Einstein equations as a means of applying the holonomy classification that Schell was developing. This led to the null solutions mentioned previously. In addition, he developed a novel approach to the construction of invariants. Having constructed a four-dimensional invariant, one constructs a second invariant in its level surfaces. A third is constructed in the level surfaces of the preceding two, and the fourth invariant, in the level surfaces of the preceding three. Thus, one has a hierarchical structure for the invariants. While this was very interesting at the time, it did not lead to any progress in the problem of quantum gravity, which was always a topic for discussion.

In the fall of 1960, I went to King's College, London, as an NSF Fellow. At the same time, Ray Sachs and Alfred Schild were there. Sachs gave a series of lectures on gravitational radiation in asymptotically flat space-times. These lectures formed the basis of his long paper that generalized the impressive work by Bondi, van der Burg, and Metzner. They led to our collaboration on the paper that showed the existence of a shear-free congruence of null geodesics is both necessary and sufficient for the existence of a vacuum space-time that is degenerate in the Petrov classification. During this period, I developed a set of integrals over the conformal tensor that could be identified with the multipole structure. However, it did not occur to me to try to define the energy in terms of such an integral because I thought of energy as defined by the pseudotensor and its corresponding superpotential. Indeed, I was able to show that the superpotential did define the energy and the pseudotensor the energy transport for Bondi's asymptotic solution. Roger Penrose came to King's College in the fall of 1961 (I was there for the fall term) and showed me his construction of the energy interval in terms of the component Ψ_2 of the conformal tensor. With hindsight, it was a natural component to consider in a surface integral at null infinity.

When I returned to ARL, Roy Kerr and I collaborated on paper analyzing the peeling of the electromagnetic field. We assumed an arbitrary charge-current distribution of compact spatial support. The electromagnetic field defined with the retarded Green's function exhibits the behavior at null infinity found in the asymptotic analysis. Kerr was invited to spend 1962–1963 at the University of Texas in Austin. While there, he found the axially symmetric metric known as the Kerr solution and began his collaboration with Schild on the Kerr–Schild metrics. In the fall of 1963, I left ARL for Syracuse University, and Kerr accepted a position at Texas.

Stuart Fickler, who came to ARL while I was abroad, became head of the group after we left. He was interested in quantum gravity and focused on the Yang–Mills SU(2) theory as an example. In the next few years, the activities and the membership of the group expanded. Jeff Winicour and Lou Tambourino joined Stu Fickle in 1964 as permanent members, as did Moshe Carmeli in 1967. David Robinson was a research associate from 1966 to 1968, Edward Glass from 1967 to 1969, and Richard Isaacson from 1967 to 1968, and there were others, whom I did not know.

Winicour worked on a broad range of problems with different collaborators. With Tambourino, he studied the propagation of the gravitational field out from an inner cylinder that presumably contained the matter. In this context, they also developed the idea of flux linkages to define energy-momentum and angular momentum. This idea has been elaborated by Winicour himself. With Robinson, he tested the positive energy theorem (a conjecture at the time) with various examples and test assumptions. In collaboration with Isaacson, he looked for effects on the age of the universe due to gravitational radiation, and he looked at other aspects of the gravitational radiation field. With Ed Glass, he worked out the theory of elastic deformation in general relativistic systems. For him, the Air Force ambiance was exceedingly fruitful.

During his stay at ARL, Moshe Carmeli continued his work on the equations of motion, although the greater part of his research was concerned with understanding the role of group theory in general relativity. This latter work resulted in the publication of the book *Group Theory and General Relativity* after he left ARL.

Undoubtedly, much more research was conducted by the relativity group at ARL. What I have briefly described is that which I know about from personal experience or have been able to find in my files. The amount and high level of research generated in-house alone justified the existence of the relativity group at ARL. At the same time, members of the group were asked to comment and advise on other areas of interest to the Air Force. More important here, they carried out a major support program in general relativity at other institutions. This will be discussed in the next section.

4. The Research Support Program

Dr. Max Scherberg had begun to determine the interest and the need for support of research in general relativity before I arrived at ARL. Shortly after I arrived, however, I visited Peter Bergmann at Syracuse and John Wheeler at Princeton to introduce myself in my new position and to find out from them where there was need for support. I was received warmly, and each gave generously of his time for both scientific and administrative discussions. The preliminary visits by Dr. Scherberg and my visits set the direction of the subsequent support program.

As noted in Section 2, the first support given by ARL was for the Chapel Hill Conference. At the conference, I had a chance to talk with Hermann Bondi and Felix Pirani about support for King's College. Unsolicited proposals were received from P. Jordan, J. Géhéniau, and N. Rosen, among others. The first contracts were let either toward the end of 1957 or early 1958. I did not keep a diary, and I have not been able to get administrative information from the Air Force archives. Therefore, dates and financial records are unavailable to me. Contracts were written with Syracuse University (P.G. Bergmann) in the United States and through the Air Force European Office in Brussels, with P. Jordan in Hamburg and the Technion (N. Rosen) in Israel.

I had requested that a contract be written with Géhéniau in Brussels, but even with my strong insistence, the Air Force refused to do so for political reasons. One source of pleasure for me is that, in spite of that refusal by Air Force procurement, I was able to maintain good relations with J. Géhéniau, P. Debever, and M. Cahen in Brussels. There was no corresponding hesitation to support Jordan, although he had been close to the Nazi government. Personally, I had some misgivings about that situation, but because that support to some extent helped his students Wolfgang Kundt and Jürgen Ehlers, I believe it was the right thing to do.

In the next few years, the support program grew and contracts were let with Peter Havas, Ted Newman, Vaclav H'lavaty, Richard Arnowitt, Jim Anderson, Ralph Schiller, and Alfred Schild. The last contract that I recommended was with William Fairbank at Stanford to support his design of a gyroscope to measure the Lense–Thirring effect, making use of the flux quantization of magnetic fields threading currents in superconductors. While much work was done in preparation for the experiment, it has not yet been scheduled, although it is part of NASA's program.

5. Selective Description of Research Supported

The first report published with ARL support was by Andrzej Trautman, *Lectures on General Relativity*, issued by King's College in 1958. In these lectures, Trautman introduced what have come to be called the Trautman boundary conditions for a space-time with gravitational radiation. He also discussed the propagation of discontinuities in the gravitational field and the problem of gravitational radiation damping in the equations of motion. These lecture notes were distributed both by King's College and by ARL.

The first major report of contract activity was from the Jordan group in 1961. This report was divided into two parts. The first was written by Jordan and was concerned with observational evidence for the Dirac hypothesis of the dependence of fundamental constants on the age of the universe. The second part was written by Ehlers and Kundt. It discussed the geometry of null rays, a similar analysis of hydrodynamic flow, and began the analysis of exact solutions, which later appeared in a chapter of *Gravitation* (John Wiley, New York, 1962), edited by Lou Witten. Three years later, a second report was issued that continued the discussion of the Dirac hypothesis, introduced the algebra of skew lattices, and had a third section authored by students on problems closer to the interests of relativists—an introduction to the differential calculus of spinor forms (the importance of which had been brought to our attention the previous year by Roger Penrose), a three-dimensional formulation of the Bianchi identities in vacuum space-times, and a presentation of thermodynamic equilibrium in stationary space-times.

Peter Bergmann's first comprehensive report in 1963 to ARL was published by the Air Force as the first in a series of ARL reports that bound together a number of reprints, preprints, discussions, and notes prepared during the contract period. These reports were distributed by ARL and were greatly valued by the recipients. The Bergmann report contains a review of canonical quantization and the work by Bergmann and Komar on the construction of observables for general relativity; an analysis of space-like infinity that exhibits the supertranslations geometrically; the shear-free solutions of Robinson and Trautman; the Newman–Penrose paper introducing their spinor formalism; and a long-unpublished paper by Roger Penrose on null hypersurface initial data for classical fields of arbitrary spin. His later reports contain further work on quantum gravity, with emphasis on the Hamilton–Jacobi theory; an analysis of spherically symmetric systems; a successive approximation approach to the scattering of gravitational waves; and a first application of fiber bundles in general relativity by Andrzej Trautman.

During the 1950s and the 1960s, there was a great deal of activity at King's College due to the stimulation and creativity of Bondi and Pirani. Visitors came with their own money, and support was available from British sources and ARL. A major collection of their work was distributed by ARL in 1966. It contains most of the important developments of our current understanding of gravitational radiation theory in asymptotically flat space-times: the work by Bondi and his co-workers; the work by Ray Sachs; a review paper by Pirani; the Robinson–Trautman and Newman–Penrose (NP) papers, again; a paper by Friedlander analyzing the initial value problem on a characteristic surface; the first paper by Penrose with the conformal description of null infinity; a geometrical interpretation of the Weyl tensor by Schild and Pirani; the paper on shear-free null geodesic congruences by Sachs and Goldberg; some important papers by Bondi on gravitational collapse; and several other works of less direct interest here, but of importance in their place.

The group at the University of Pittsburgh under E.T. Newman and A. Janis was remarkably productive. Their work under the sponsorship of ARL had been distributed in three ARL reports, 1964, 1969, and 1970. The first report contains initial work by Newman and his students using the NP formalism to study the gravitational field in the vicinity of null infinity. These papers laid the foundation for most of the work that followed in the analysis of the properties of asymptotically flat space-time at null infinity. Included in this collection is an interesting paper by Janis that studies the equivalence problem in space-time with symmetries. The study of the scattering of gravitational waves is the main topic of the next report. This study begins with the linearized equations and continues into the second order. One finds that with an incoming wave of a finite duration, space-time can be Minkowskian at an early time or a late time, but not both. There follows a general discussion of the existence or nonexistence of tails in solutions of hyperbolic equations in two dimensions. Finally, there is a paper by Janis and Halliday making full use of the NP formalism and a multipole expansion of the asymptotic gravitational field. This collection also includes the paper by Newman and Penrose, introducing their new constants of the motion at null infinity.

The final ARL report mainly contains a novel approach to the equations of motion. This approach is based on studying the change in a sphere surrounding a singularity in linearized theory and then moving these results to null infinity. The results give equations of motion with radiation damping. However, the limitation is that the motion is not related to the sources of the gravitational field but is a global result. Therefore, its application to the interaction of particles or even the motion of point sources in an external field has not yet been determined. This report also contains two papers discussing infinite dimensional representations of the Lorentz group on the sphere.

The 1964 report issued for the Schild group in Austin, Texas, contains work emphasizing the conformal group in general relativity. First, there is a paper with Pirani on the physical interpretation of the conformal tensor; then, there is a paper with I. Robinson extending the theorem that the existence of a shear-free null congruence implies Petrov degeneracy. The collection also includes a very important paper by Roy Kerr, which derives the Kerr solution for a rotating mass. During this period, Schild also worked on relativistic interactions of particles as a means of understanding elementary particles and on alternate theories of gravitation, particularly of the Whitehead type. After this, came the development of the Kerr–Schild metrics.

The report from Peter Havas in 1964 is concerned with the problem of motion. Havas himself worked on the relation between conservation laws and equations of motions in affine spaces and on both relativistic and Newtonian particle interactions. Restricting his attention to position and velocity-dependent forces, he showed that these must be composed only of invariants. J. Plebanski calculated the geodesics for the Schwarzschild solution, and W. Tulczyjew studied the motion of particles with internal structure.

The work reported by R. Schiller in 1966 studied fundamental questions in quantum electrodynamics and linearized general relativity. There was an attempt to understand problems that arose in the effort to quantize gravity, but this work did not influence the principal developments in understanding quantum gravity. A long report on the work of V. H'lavaty issued in 1966 is mathematical in its focus and did not have any influence on work of interest to most relativists.

Some of the work by Arnowitt, Deser, and Misner was supported by ARL through a contract with Richard Arnowitt when he was at Syracuse University and then continued when he moved to Northeastern University in Boston. The early work was not distributed in an ARL report, but it has been reviewed by Arnowitt and Deser in *Gravitation*. They developed the canonical formalism using the spatial metric and extrinsic curvature as fundamental and introducing the remaining metric components as a lapse function and shift vector that act as Lagrange multipliers. To analyze the asymptotic behavior of the gravitational field, they then decomposed the metric into longitudinal, transverse, and transverse–transverse (also traceless) parts. Through this decomposition, they arrived at a quantity that is now known as the ADM (for Arnowitt, Deser, and Misner) energy. The ARL report of 1965 gives further developments using these ideas. Arnowitt and Deser used the equation of geodesic deviation to show that from the quantity they used to identify the energy, they obtained an invariant function that can be identified with the Newtonian potential in an appropriate limit at spatial infinity. Then, they showed that the requirement of a $1/r$ approach to the Minkowski metric at spatial infinity was sufficient to produce a wave front beyond which all derivatives of the metric had a falloff faster than $1/r^{3/2}$. Essentially, this resulted from the requirement of a finite total energy. Infinite energy would not lead to the $1/r$ behavior of the metric itself. There follows a proof of the Lorentz invariance of their definition of energy momentum. The report is completed with papers by Pirani and Schild extending their study of the conformal tensor and Pirani and Deser criticizing a proposed new theory by Hoyle and Narlikar. Pirani was on leave from King's College visiting jointly at Brandeis University and Northeastern.

While the first report issued discusses only classical problems, the second report, issued in 1971, stresses quantum gravity. The techniques employed were taken from quantum electrodynamics with consideration of attempts to quantize the Yang–Mills non-Abelian gauge theory. Two problems were reported. One concerned the question of whether the canonical approach, which intrinsically is not Lorentz covariant, can give a Lorentz covariant S matrix. The analysis primarily used the linearized equations and therefore difficult questions about factor ordering did not arise. The second problem specifically faced the factor ordering question. It showed that if one wishes to obtain quantum Bianchi identities by application of a quantum Noether's theorem, the quantum Lagrangian, and hence the quantum Einstein equations, must be modified to order $\hbar^2$. It is not clear to what extent this result depended on the initial ordering assumed. There follows a paper that analyzes the Poincaré covariance of angular momentum as was done earlier for energy momentum. The result obtained was coordinate dependent and showed possible difficulties associated with supertranslations. The final paper discusses the classical scattering of gravitational waves by a massive particle. This was done to first order and for low energy. It agrees with a similar quantum calculation. In the second order, the structure of the source became important and the calculation was unreliable.

6. Conclusion

From the above description of the activities of the relativity group of the Aerospace Research Laboratories, it is evident that the Air Force support of research in general relativity had a positive effect on the development of the field. This does not mean that the field would not have developed without this support. However, by support of international meetings, as well as of individual research groups in the United States and abroad, the Air Force encouraged and stimulated the exchange of people and ideas. For those of us who participated in this activity from the Air Force side, it was a rewarding and interesting experience. Those who benefited from the support appreciated the vision of leaders who recognized that technological developments cannot be predicted and that basic research must be supported. While many of us were saddened by the passage of the Mansfield amendment, which terminated the relationship between the Air Force and the relativity community, we were pleased that the net result was a strengthened National Science Foundation.

Nonetheless, there are many people who are deeply concerned about the involvement of the military in pure science. However, one must understand the context in which that involvement began. At the end of World War II, there was no civilian agency whose mission was to support scientific research. Shortly after the end of World War II, it was the Navy that began to support basic research at US universities, including general relativity research. The Navy was very interested in this research. The military had seen in the course of two world wars that basic science was important to the development of new weapons. They stepped into what was a vacuum to insure that American science remained strong so that the military could remain strong. The specific motives for supporting general relativity, particularly by the Air Force, cannot be analyzed in the absence of the written documents authorizing the programs. On the other hand, one made many jokes about the interest in antigravity devices, which no doubt would be of great value to the Air Force. However, it should be recognized that, in the United States, the Department of Defense played an essential role in building a strong scientific community without widespread encroachment on academic values.

Acknowledgments: This report is more in the form of a memoir than a true history. I tried to get the administrative reports concerning this project from the US Air Force. Special thanks are due to Mr. Walter Blados, Deputy for Scientific and Technical Information, Department of the Air Force, in pulling together whatever information he could locate.

This information was mainly an outline of research performed under many of the contracts that were let without giving a dollar figure for those contracts. However, the detailed documents that would give the administrative justification of the program together with its budget and personnel could not be found. Preparation of this report was preceded by discussions with Stuart Fickler, Jeff Winicour, and Max Scherberg. A draft was read and commented on by Winicour and Scherberg. Preparation was supported in part by the National Science Foundation through grant number PHY-870065 1.

NOTES

Operation "Paperclips" was a program by which German scientists, engineers, and technicians came to the United states after WWII to work on various scientific and technical projects.

REFERENCES

While the description of the in-house research is based on my memory and the discussions with Fickler and Winicour, the description of the contract-supported research was prepared on the basis of the ARL Technical Reports, which were prepared by the principal investigators and printed and distributed by ARL. These contain a descriptive introduction followed by reprints and preprints of the research performed during the contract period covered. According to the inside cover, copies may be obtained by qualified contractors from the Defense Documentation Center, Cameron Station, Alexandria, VA 22314. It also says that the documents have been released for sale through the Clearing House, US Department of Commerce, Springfield, VA 22151. Below is a chronological list of the ARL reports that are concerned with general relativity.

ARL 61-228, "Problems of Gravitation," Pascual Jordan, Wolfgang Kundt, and Jürgen Ehlers.

ARL 63-56, "Quantization of Covariant Field Theories," Peter G. Bergmann.

ARL 64-2, "Research on General Relativity and other Gravitational Theories," Alfred Schild.

ARL 64-3, "Studies in the Relativistic Theory of Interacting Particles," Peter Havas.

ARL 64-22, "Research in Gravitational Radiation and Space Time," Ezra T. Newman.

ARL 65-24, "Contributions to Actual Problems of General Relativity," Pascual Jordan.

ARL 65-254, "Research on Elementary Particles and Relativity Theory," Richard Arnowitt.

ARL 66-0075, "Research Program in Relativity Physics," Hermann Bondi.

ARL 66-0194, "Research on Riemannian Spaces," Vaclav H'lavaty.
ARL 66-0206, "Problems Relating Quantum Theory to the General Theory of Relativity," Ralph Schiller.
ARL 67-0053, "Research on Solutions of the Gravitational Field Equations," Ezra T. Newman and Allen I. Janis.
ARL 69-0113, "Research in Gravitational Theory," Ezra T. Newman and Allen I. Janis.
ARL 70-0003, "A Study of Null Surfaces in Gravitational Theory," Joshua N. Goldberg.
ARL 70-0066, "Quantization of the Gravitational Field," Peter G. Bergmann.
ARL 70-0120, "Experiments on General Relativity at Liquid Helium Temperatures and in Zero Magnetic Fields," William M. Fairbank and William O. Hamilton.
ARL 70-0136, "Research in General Relativity," Ezra T. Newman and Allen I. Janis.
ARL 71-0323, "Representations of the Lorentz Group: Recent Developments," Moshe Carmeli and Shimon Malin.
ARL 72-0086, "Group Theory and General Relativity," Moshe Carmeli.
ARL 73-0151, "Studies in Gravitational Theory and General Relativity," Joshua N. Goldberg.

Mathematics and General Relativity: A Recollection

André Lichnerowicz

I worked in the field of relativity from 1937 until 1967. In 1937, the relativity community had a strange mix. There was a small group of specialized physicists, Einstein's friends or students, such as W. Pauli, L. Infeld, B. Hoffman, and V. Fock (USSR); and a small group of specialized astronomers, such as G. Lemaître; and a small group of mathematicians, such as T. Levi-Civita, T. de Donder, and G. Darmois. Their interests ranged from differential geometry to rational mechanics and the theory of partial differential equations. The physics community of the time, passionately involved with quantum mechanics, considered the relativists to be marginal. P. Dirac alone was an exception. This state of affairs lasted until 1955 and the Berne Colloquium presided over by Pauli.

The development of differential geometry, which evolved into a real research program by Élie Cartan in 1922, owed much to Cartan's reflections on general relativity and efforts toward a unified field theory so dear to Einstein (Cartan 1923, 1924, 1938, 1952–1955). After having devoted several important articles to these problems (and here one should note the recent rebirth of interest in Einstein–Cartan theory), Cartan turned toward global problems concerning Lie groups and then in 1937 to the teaching of spinor theory, a notion introduced by him in 1913 in a study of irreducible representations of orthogonal groups. This lecture course was published in 1938, remarkably written up by our colleague A. Mercier, to be followed by a study of the status of spinors in general relativity. This study, although generally misunderstood at the time, clarified and corrected Fock's approach and criticized the point of view of Infeld and B. Van der Waerden.

We should note that at this time, although there was an interest in global groups, no intrinsic global theory of manifolds was available. It was only around 1936, with the studies of H. Whitney, that such a theory appeared. To Whitney are attributable the global definition of manifolds in terms of what we today call an atlas, made up of local charts, and the existence of embeddings of such manifolds in a space $\mathbb{R}^N$ with N sufficiently large. Several years later, thanks to C. Ehresmann, a great mathematician and student of Cartan's, and to Whitney and E. Stiefel, a student of H. Hopf's, definitions of the concept of the *fiber bundle* became available. Thus was achieved an axiomatization of one of Cartan's profound intuitions, of which he had made such great use. The appearance of the fiber bundles resulted in only a limited and polite interest among mathematicians—and not even that among physicists.

Such were the events that awaited me. My recollections can now only be partial and incomplete, but I shall try not to deform the perspective. In 1936 and 1937, having received a solid training in differential geometry from Élie Cartan, I became interested, upon the advice of George Darmois and with the blessing of Cartan, in the theory of relativity. In 1926 in Belgium, Darmois had given a course of four lectures on "the equations of Einsteinian gravitation" in the presence of de Donder. The monograph version (Darmois 1927), appearing in the series *Mémorial des Sciences Mathématiques* became my bedside reading. In this book, although treated in Gaussian coordinates, which to my mind present some serious problems, is the first rigorous analysis of the hyperbolic nature of the Einstein equations, i.e., the foundation of the relativistic theory of gravitation as a theory of wave propagation. With profound understanding, the splitting of the Einstein equations relative to the Cauchy problem into two sets is clearly discussed: one treats the initial conditions, and the other deals with time evolution. A geometric interpretation of the de Donder coordinate conditions is now known as the *harmonic* conditions.

At that time, I began by thinking about the differentiable structure of the space-time manifold, so that I could attack what interested me—the global problems of relativity, the keys to a real understanding of the theory. In 1939, feeling the war approaching, I published a rather imperfect monograph called *Problèmes globaux en mécanique relativiste* (*Global Problems in Relativistic Mechanics*, Lichnerowicz 1939) in the series *Actualités Scientifiques et Industrielles.* After having completely analyzed the formal Cauchy problem for arbitrary local coordinates, I gave the first (partial) proof of a result that had been sought by Einstein and Pauli: *an exterior stationary space-time, everywhere regular and asymptotically Euclidean, must be locally flat,* i.e., without gravitation. I gave a complete demonstration of this theorem in 1945—that earned me the congratulations of Pauli. In the meantime, I had analyzed the system of constraints on the initial conditions and introduced the "conformal" method that has remained standard.

In two beautiful articles appearing in Germany in 1938, K. Stellmacher analyzed in detail the propagation of singularities in the gravitational field along isotropic geodesics (Stellmacher 1938). Although my junction conditions for the field were present in the 1939 monograph, they were not yet well presented. They were to find their definitive form in my book of 1955 (Lichnerowicz 1955). The differentiable structure of space-time must be C^2 (piecewise C^4), the metric being C^1 (piecewise C^3), and any supplementary precision in the degree of differentiability is devoid of physical meaning. In particular, for total coherence, an analysis must fall outside of this framework, which explains my permanent perplexity concerning the Kruskal analytic prolongation.

In 1952, my student Y. Choquet-Bruhat gave, in the differential framework thus defined, the first proof of existence and uniqueness of the local Cauchy problem for the Einstein equations (Choquet-Bruhat 1952). She made use of harmonic coordinates and the generalized Kirchoff formulae, at the time a *tour de force* of functional analysis. She has since given a simpler and more sophisticated proof based on the technique of Sobolev spaces. These demonstrations have been the source of many works by contemporary mathematical physicists. At the same time, J. Synge and S. O'Brien used a more complicated, though perhaps more intuitive, approach to my junction conditions. All of these results were confronted and diffused at the International Congress at Berne (1955), which marked the true renaissance of interest in general relativity on the part of physicists, astronomers, and mathematicians. There, I personally met the young members of the English school.

One remark that I think is important is that on several occasions, and in particular at the Berne Congress, I made the following announcement: local coordinates of the space-time manifold are useful only for cartography, the placement of events. The fundamental space is, in fact, the fiber bundle of time-like directions tangent to the space-time, to each of which is attached the orthogonal space-like 3-plane. Given a system of observers, i.e., a local field of time-like directions, the decomposition into squares of the corresponding ds^2 provides, by integration along an arc of the world line, a standard time and standard length relative to the system of observers, which are directly comparable to their homologous classical counterparts. This interpretive key is sufficient for all needs.

Only Carlo Cattaneo has really taken seriously this suggestion, and during the period 1958–1961, he developed a differential geometry and formulas that permit a direct and systematic comparison with classical mechanics with respect to terrestrial observers. The astronomers, I believe, would benefit from the use of his formulas given in the notes of a course taught at the Collège de France (Cattaneo 1961) and in a book in Italian.

I then became interested in *shock waves* in general relativity. Starting from a new technique for tensor distributions on space-time that I developed in 1964, I analyzed the relations between electromagnetic shock waves and gravitational shock waves, which presupposed a rewriting of the Einstein equations in terms of tensor distributions. It followed that any electromagnetic shock wave (discontinuity in the electromagnetic field) would create a gravitational shock wave (discontinuity in the connection), and one would have an energy balance. Using these techniques, I also studied (1967–1970) magnetohydrodynamic shock waves necessarily present, for example, around a pulsar.

The quantization of a physical field on a given globally hyperbolic, curved space-time poses difficult, but very interesting, problems. For a tensor or spinor field, it is possible to construct, starting from elementary solutions of the field equations, an "antisymmetric" propagator solution of the homogeneous equations (1960–1963). The definition of creation–annihilation operators reduces to finding a "symmetric" propagator solution, with positive norm, of a quadratic integral equation. The existence of such propagators has been shown by Carlos Moreno for spins of 0 and $1/2$ but, except for the stationary case, there is no uniqueness. The problem of giving precise mathematical meaning to the creation of particles by the expansion of the universe, as L. Parker has tried to do, remains completely open after 15 years.

Thus, general relativity remains a privileged field of mathematical physics. In the course of the last 15 years, remarkable works concerning the global study of space-times from different points of view (S. Hawking, Choquet-Bruhat, J. York) have shown the difficulty of the undertaking. They have also shown that the theory of general relativity is a remarkably rigid theory and, therefore, an excellent theory with practically no arbitrariness. It must, no doubt, be surpassed and enveloped in a more comprehensive theory, but no one yet knows the road that must be traveled.

REFERENCES

Cartan, Élie (1923). "Sur les variétés à connexion affine et la théorie de la relativité généralisée." *École Normale. Annales* 40: 325–412. Reprinted in Cartan 1952–1955, Vol. 3, pp. 659–746.

——— (1924). "Sur les variétés à connexion affine et la théorie de la relativité généralisée." *École Normale. Annales* 41: 1–25. Reprinted in Cartan 1952–1955, Vol. 3, pp. 799–823.

——— (1938). *Leçons sur la théorie des spineurs. II. Les spineurs de l'espace à $n \geq 3$ dimensions. Les spineurs en géométrie riemannienne.* Actualités Scientifiques et Industrielles, No. 701. Paris: Hermann.

——— (1952–1955). *Oeuvres complètes.* 3 vols. Paris: Gauthier-Villars.

Cattaneo, Carlo (1961). "La formulation relative des lois physiques en relativité générale." Cours au Collège de France. Unpublished.

Choquet-Bruhat [Fourès], Yvonne (1952). "Théorème d'existence pour certains systèmes d'équations aux dérivées partielles non linéaires." *Acta Mathematica* 88: 141–225.

Darmois, Georges (1927). "Les *équations de la gravitation einsteinienne.*" Mémorial des Sciences Mathématiques No. 25. Paris: Gauthier-Villars.

Ehresman, Charles (1941). "Espaces fibrés associés." *Académie des Sciences* (Paris). *Comptes Rendus* 213: 762–764.

——— (1942). "Espaces fibrés de structures comparables." *Académie des Sciences* (Paris). *Comptes Rendus* 214: 144–147.

Lichnerowicz, André (1939). *Sur certains problèmes globaux relatifs au système des équations d'Einstein.* Paris: Hermann. Also issued as *Problèmes globaux en mécanique relativiste.* Actualités Scientifiques et Industrielles, No. 833. Paris: Hermann, 1939.

——— (1941a). "Sur l'intégration des équations d'Einstein." *Académie des Sciences* (Paris). *Comptes Rendus* 213: 516–518.

——— (1941b). "Sur l'intégration des équations d'Einstein." *Académie des Sciences* (Paris). *Comptes Rendus* 213: 549–551.

——— (1944). "L'Intégration des équations de la gravitation relativiste et le problème des n corps." *Journal des Mathématiques Pures et Appliquées* 23: 37–63.

——— (1946). "Sur le caractère euclidien d'espaces-temps extérieurs statiques partout réguliers." *Académie des Sciences* (Paris). *Comptes Rendus* 222: 432–436.

——— (1955). *Théories relativistes de la gravitation et de l'électromagnétisme.* Paris: Masson.

——— (1961). "Propagateurs et commutateurs en relativité générale." *Institut des Hautes Etudes Scientifiques. Publications Mathématiques.* 10: 1–55.

——— (1964). "Champs spinoriels et propagateurs en relativité générale." *Société Mathématique de France. Bulletin* 92: 11–100.

——— (1973). "Ondes de choc gravitationnelles et électromagnétiques." In *Relatività. Convegno del Febbraio del 1972.* Istituto Nazionale di Alta Mathematica. Symposia Mathematica: 12. San Diego: Academic Press, pp. 92–110.

Pauli, Wolfgang (1956). "Ansprach durch den Präsident der Konferenz." In *Fünfzig Jahre Relativitätstheorie.* Bern, 11–16, July, 1955. *Verhandlungen.* André Mercier and Michel Kervaire, eds. (*Helvetica Physica Acta Suppl. 4*). Basel: Birkhäuser, p. 27.

Stellmacher, Karl (1938). "Zum Anfangswertproblem der Gravitationsgleichungen." *Mathematische Annalen* 115: 136–152.

General Relativity at the Turning Point of Its Renewal

André Mercier

1. General Relativity Zero or the Turning Point

The organizers of the Marseilles Second International Conference on the History of General Relativity (September 6–9, 1988), especially Professor Eisenstaedt, had, I believe, among other wishes, wanted to see some of the representatives of the older generation who are still among relativists taking an active part and being given the opportunity to recall the main steps in the development of the physics of the general relativity (GRG) at a time when it was not quite so evident that GRG was one main chapter in physical research. By saying so, I am perhaps putting in their real intention some interpretation of my own that was not theirs. Still, since it appears so to me, let me explain what I mean. To do so, I must not only go back to the preparation of the Berne Jubilee Conference of 1955, but I must also be very personal, more than is perhaps pertinent when writing on a scientific matter.

When, during the years 1953–1954, I began putting together in my mind some main ideas that would constitute the justification for holding such a conference, friends working in fields of physics other than (general) relativity theory expressed serious doubts about the desirability of such an undertaking, even though they recognized that 1955 was a jubilee year, a golden one indeed, and should be celebrated in the very town where Einstein had lived and worked when he published his first paper on (special) relativity. Their argument was that if, on the one hand, special relativity was no longer a "problem," and even some spectacular observations (perihelion motion, deviation of light rays) had shown the immense advantage of imposing general relativity as the correct explanation of the behavior of massive bodies, including particles and photons, further developments of that very theory of general relativity had, on the other hand, become so much of a game consisting of playing with problems of a mathematical nature, that an acceptable connection with "real" physical problems had been lost and practically most of the scientists working in that field were not or were no longer attached to physical laboratories, but to mathematical institutions. Of course, I was an exception, but this was surely not a sufficient argument. Moreover, some were afraid that the Bohr–Einstein controversy might unduly be thrown into the discussion, which, owing to my well known propensity to philosophize instead of posing and solving matter-of-fact problems, might lead to unfruitful discussions.

It is true that I began to doubt that the current epistemological interpretations of relativity were the best and that I personally became engaged in such philosophical speculations (I shall come back to this later), but my ideas in connection with the Berne Jubilee Conference were, on the contrary, very matter of fact. Actually, I personally felt that physics was at a turning point and that some extremely bold attempts within the field of general relativity might, and probably would in fact, have a great impact on physics proper, and when I say physics proper, I mean (and meant) as much "particle physics" as what had become known as "astrophysics." I had a double feeling that, on the one hand, the very attribute that particles were elementary or nonelementary was a bet (in other words, perhaps the very idea of elementary was false) and, on the other hand, that if particles could be both sources of fields, themselves quantifiable, and so be the possible "go-between" endowed with the task of mediating interactions, then gravitation, in spite of classical mechanics having made that old concept the paradigm of all forces, should not be confused with a new concept called interaction. Gravitation has plainly made room for the very structure of the universe, but not for an interaction between particles; it is therefore a universal (even the only universal) aspect of the world, so to speak, identical with the material world. Finally, one could in no way avoid the task of elaborating a detailed theory of matter as consisting of either fields or particles on a background that would have to identify the whole world with its being endowed with a universal property that only classically was called gravitation.

Worried by the reluctance on the side of certain colleagues, I decided to go to Zurich and pay a vist to Wolfgang Pauli, telling him what I had in mind and asking him the double question: would he recommend having a jubilee conference in Berne, and if so, would he accept not simply to chair it but also to help with counsel and instructions toward its realization?

In scientific discussions, Pauli was well known for being capable of making a younger physicist, such as myself, look like a fool. It was for him a test of the latter's capacities. However, I was prepared. He knew me when I was an assistant to Professor H. Favre in the chair of mechanics at the Federal Institute of Technology. There, I also attended a course of lectures on higher analytical dynamics that I had personally suggested Pauli should teach (the formal reason being that Favre, who had just been appointed professor—he himself specialized in theoretical hydraulics—fell ill and could not assume such a course). I had on one occasion brought Pauli a copy of *Leçons sur la théorie des spineurs*, which I had written three years before for Élie Cartan from his lectures at the Institut Henri Poincaré. Another young theoretical physicist who was present asked, "Who is Cartan?" Pauli answered (in German, of course), "Ein grosser Bonze," and from then on, he had liked me, even though always teasing me whenever possible (especially when we met in Copenhagen). Astonishingly to many, though not to me, he was immediately enthusiastic about the idea of the Jubilee Conference; he not only agreed to chair it, but began conceiving a plan for its realization. So, I went home with concrete instructions.

My guess is that, having at a certain time in his career been very interested in general relativity (remember that he wrote the chapter on it in the celebrated *Encyklopädie der mathematischen Wissenschaften*) he felt, of course, much more strongly that the time had come for a new start in general relativistic physics that would not just solve mere mathematical problems, however profound they might be, but would make the tremendous attempt to become the true cosmology that would include the origin of all things and explain the world of today, eventually from hypotheses about its origin. This, perhaps, is more important than usually thought, for on the whole, Pauli has always been considered a typical particle physicist. (No one has ever received the Nobel Prize for an achievement in general relativistic research.) There is, in that respect, a similarity between him and Dirac.

In those days, one did not yet talk about "big science." Still, people began to plan huge laboratories. It was, e.g., the time of the origin of CERN, in which I also became involved. Big science has since, it is true, often been understood only in terms of staff and enormous budget size and, in the minds of many, concerns mainly high-energy physics. To my mind, however, gravitational physics cannot, and could not, escape being included in such big science, because any physical research must have a theoretical and experimental output (eventually an observational one, as in astrophysics, but even there the difference is slight). In the 1950s, my understanding of gravitational physics was such that it had to involve considerable means. After all, in the heroic times of Eddington, for instance, a worldwide expedition was undertaken to make possible the necessary observations of an eclipse at the right place on the globe. If, according to my conjecture, gravitational physics was to become a main consideration in research about the constitution of the material world, this would certainly imply that it would be involved in big science, whether understood merely in terms of costs or as that which is concerned with the very foundation of science, which inevitably could only develop within a framework of very large, international collaboration and organization. The Jubilee Conference was also thought of as a start in that direction.

I believe that the Jubilee Conference helped to reach these goals, for it fostered what later became known as the international GRG organization, with its committee, society, and conferences, as well as its journal (the conception of which, it should be mentioned, was Leopold Infeld's before he died).

In regard to these conferences, there was a controversy over the question of which one was the first GR 1. I propose calling the Berne Conference in 1955 Gr Zero (cf. the title of this section). Then, Chapel Hill would be Gr 1; Royaumont, Gr 2; and so on. As is well known, the present way to count the years in the West (and in practically the whole world) is the A.D. (anno domini) system, following B.C. (before Christ) with a zero in between. When this was decided (in the early Middle Ages, more precisely under the Merovingian King Childeric III in the middle of the 8th century), a mistake of at least 4 or 5 years, if not 6 or 7, was made in dating the birth of Jesus, and the zero was still unknown, as were negative numbers. Perhaps, counting the GR conferences has suffered a similar uncertainty! Today, of course, counting time is a straightforward business and allows for time between $-\infty$ and $+\infty$. This will be discussed further in connection with the myth of creation.

On one point, Pauli had some doubts, and he was right. In spite of my endeavor, we did not succeed in persuading Neils Bohr to attend the Jubilee Conference. At that time, I thought that the controversy between the two great men could be the reason for Bohr's reluctance. Today, I would be more careful and would submit that perhaps Bohr felt he had not thought enough about the possible future of a generally relativistic cosmology that would have given him premonitions of a development to come (I vaguely remember him putting forward in a conversation on the telephone some arguments suggesting this interpretation). Moreover, he was tremendously involved in the application of nuclear energy (the term was "atomic" energy in those days) to peaceful purposes. On the whole, I suspect that Bohr would have attended only if Einstein had been present, which was not the case, as it is well known that he died before the Jubilee Conference took place.

2. A Testimony About the Premonitions of the Possible Development of GRG After GR Zero

To summarize and arrive at a preliminary conclusion, we have reached the time when the physics of gravitation, identified with cosmology at large, had to take the lead in the search for an all-encompassing theory of macrophysical, if not also microphysical, reality. Yet, where is the boundary between macro- and microphysics? At human dimensions? Surely this would seem preposterous, even though some scientists have argued in favor of asserting what is currently called the anthropic principle. Moreover, observations have shown that even baryons (at least neutrons) are heavy, since they describe a gravity parabola when projected onto the surface of the earth. This means that either the Einsteinian dream about a unitary theory (in earlier days only concerned with gravitation and electromagnetism) would have to include gravitation and all further interactions, both those already known and those still to be discovered by physicists, or that the old difficulties, apparently insuperable, of unifying electromagnetism and gravitation were not of a technical nature, but were due to the fact that gravitation is different from whatever interaction might be discovered between particles endowed with a charge, or "perfume," mass not being such a charge or perfume. By the way, the first of these reasons could be the consequence of the second. Lichnerowicz is still hesitant about the possibility of "quantizing" gravitation. His reasons are of a geometric nature in relation to the light cone. To me, whether there are gravitons is a question that sounds very academic (as it did in the old days). My reason for hesitating in treating gravitation the way one treats electromagnetic, weak, or strong interactions is that whatever exists in the universe contributes to the structure of that universe locally as well as at large, and this is the picture that has replaced the Newtonian explanation of Keplerian motion. It yields the kind of (multidimensional) continuum, manifold, or variety in which to study physics. In the Minkowskian picture of a so-called space-time, even generalized to general relativity, it yields the independent, yet four-dimensional, variable, upon which everything else depends, just as in Newtonian physics where all properties of matter (bodies including particles) depend on time. This time was Newtonian time t, $-\infty < t < +\infty$, endowed with the same mathematical properties assigned by suitable axiomatics to real, negative, and positive numbers. Space-time is therefore not a generalization of space (to Newton, the sensorium of God; to Kant, the inner sensorium—"Form a priori der menschlichen Anschauung"—of man), but a generalization of time. In later publications, I have called four-time (in French, *quadritemps*) a continuum to be defined by a suitable axiomatic, taking the light-cone condition into account, locally if it is a Riemannian-like or otherwise structured (e.g., Finslerian, as I once attempted to calculate) manifold and more than four-dimensional, then called supertime, in unitary theories using certain additional conditions (of cylindricity, etc.).

So far (i.e., 1988), the development of particle physics, including the consideration of gravitation, does not seem to be in favor of my views. Yet, I still believe—this is nothing but a strong feeling—that any grand unification including gravitation will be a failure unless one considers the following circumstance.

In the Einsteinian picture, any general relativistic model of the world allows for the definition of a cosmic time that takes over the role of a linearly conceived independent variable capable of having lasted forever and going on forever. But, it could also have a big-bang beginning from a Lemaîtrian *atom primitif*. Whether this latter eventuality is identical to the creation of the world is questionable. It is, at most, the description of the first tiny fraction of the evolution out of the said primitive atom that God, the demiurge, would have to fetch in a preexisting foundry or melting pot, like Newton shooting a shell, also cast in a melting pot, from a gun with a given amount of powder, also prepared in a powder magazine, at a time $t = 0$ and calculating the trajectory for $t > 0$. Otherwise the foundry would have had to exist before the world and eventually even before God existed, unless God, the foundry, and the world are altogether one same capital thing, the pantheos. Rather, God has actually created the world by separating from it but this makes sense only if time and the world have a common beginning (with no past), and this implies that time and the world are one, or the same. Only because humans have as a result of evolution, been endowed with both the feeling, or consciousness, of a difference between the material and the temporal aspect of this same and the capacity, or intelligence, to construe geometry (geo-: first for the earth, then for the world, and finally for nothing but an abstract entity called space) has it been possible to elaborate a physics in the form of a formal (i.e., geometrically or more generally, mathematically describable) dependency of the material upon the temporal aspect. Thus, humans came to believe (1) that there is space in which (2) material properties as attached to the world depend upon time. But, actually, there does not exist such a space (can we buy and sell it?), whether three or more dimensional, whether Euclidean or not. The correct interpretation consists in saying that time, whether Newtonian, four-time, or eventually supertime, and world, identical at their common beginning, cannot remain identical for all future time, i.e., space-time or any higher manifold is and remains the material world. Such a manifold with all its geometrical peculiarities (curvature, affine connection, torsion) is, if not identical to the substitute for what in the pre-Einsteinian picture was implied by the concept of gravitation. Therefore, the theory that uses it is cosmology, and its accompanying cosmogony—in case the model is unstable—can take the form it has finally taken in our days with the appearance in the first fraction of the big bang of successive particles, quarks, leptons, baryons, nuclei, in an evolution that has led some to imagine tendencies so strange as to assert principles of an anthropic nature.

3. Points of Discussion

Whether one wishes to adhere to such grandiose views, if this is not big science, then I do not know what to call it. Of course, it is never so much the purely theoretical aspect as the observational and experimental one, that makes science big, especially when considerable technology is unavoidable. This is the case and has been since the 1950s, when observational astrophysics at large came into play. As explained the heroic times suggested it, and it seems pretty clear that from the moment the means of observation for cosmological conjectures were at hand the process had to start. It took less than nine years after GR Zero to convene people for the First Dallas Conference, which was more particularly devoted to its topic, *Gravitational Collapse and Other Topics in Relativistic Astrophysics*, and black holes were soon no more a mere conjecture. Those present had a taste not only of big science, but also of the huge development in astrophysics, especially relativistic astrophysics, undergone during the considered decade.

What further showed that more means of a technical and financial nature were necessary for the sound development of research was the help given by the US Aerospace Research Laboratories, where J. Goldberg played a leading role. Of course, a problem of conscience is posed: should scientific research be supported by agencies that are involved in the military business? Its origin is perhaps as ancient as humankind, for humans have always looked for ways to make more efficient weapons than those at the disposal of the enemy. This has required a respectable knowledge of some sort of engineering based on the knowledge of physics and, to a high degree, mathematics (although the early use of horses by the Indo-Europeans was neither exactly a physical invention nor a mathematical device). Descartes, for instance, the founder of analytical geometry, was an officer by profession, and Sadi Carnot, the founder of thermodynamics, was an officer, as was his father the politician, who wrote books on machines and geometry. A clear connection is illustrated by the French École Polytechnique, which from its very beginning was a creation of the French Revolution and until recently under military rule.

Traditions like this spread. At the same time, institutions, which are not physical persons endowed with a soul but juridical persons, lack both consciousness and conscience. It became a natural thing, besides being a technical necessity, to combine military preparation with the promotion of mathematics, physics, chemistry, etc. That this was still the case after World War II is not at all astonishing. If it was so in the United States, it was also the case in Switzerland, where subsidies for research in these fields were given by the military authorities, especially because neither higher education nor research were organized at the level of the federal authorities (with the exception of the Federal Institute of Technology), and the same applies to the United States, but this was a matter for the various states, which were not so very rich. This procedure was abandoned in Switzerland some years later, when the National Foundation for Scientific Research came into existence. The same would have applied to Germany, if that country had not been divided into two parts after World War II, West Germany not being allowed for several years to build anything comparable.

During the last three decades or so, things have changed considerably, owing somewhat to the rise of pacifism, especially among the intellectuals, but much more to the military use of nuclear power than to general relativity or for reasons of general scientific ethics. The Federal Republic of Germany has followed another way with its expansion and multiplication of the so-called Max Planck Institutes, which replaced the earlier Kaiser Wilhelm Institutes. Presently, France has probably the oldest and wisest system, the National Council of Scientific Research.

There is no doubt that both Einstein and Bohr were simultaneously, though totally independently, not only the earliest but also and by far the most outstanding protagonists of pacifist ideas about the benefit of science for humanity in the perspective of the great discoveries in relativistic and quantum physics. I have written about this in connection with Bohr's endeavors elsewhere (e.g., as early as *De la science à l'art et la morale*, Neuchâtel 1950).

To my knowledge, possible connections between World War II and the tremendous evolution of scientific knowledge have not been evaluated by historians. It is not evident that the same evolution including all the moral issues would have taken place without the war, but since history cannot be changed, it does not make much sense to speculate about the possible absence of what has actually been. Still, the field of relativity theory could have been a haven of peace for many during the war (which was not the case for those who worked, e.g., at Los Alamos, where even an engaged physicist decided to become a priest). Yet, at the same time, was such a haven then the right place for conscience?

There may be an explanation of an historical nature of the situation that led some physicists to say, as I indicated at the opening, that relativity theory had become a business for mathematicians and that physics proper should not bother with it. I mean the circumstance that caused the study of mechanics, from soon after Newton's time, and continuing for centuries after, to come under the care of savants considered to be mathematicians but not physicists proper; this was the case above all in France and in other Latin countries but also to a certain degree in England and elsewhere. This goes back to the period of the Enlightenment, with people such as d'Alembert, even perhaps back to Leibniz. It is well known that the teaching of so-called rational mechanics and analytical dynamics is attached to mathematical departments rather than to physical ones, and this is the case not only in universities of Latin countries. Newton occupied a chair of mathematics, the same which much later belonged to Dirac and, finally, to Hawking. Theoretical physics is a relatively recent concept. It was for a long time in England called applied mathematics, often dealing with elasticity, hydrodynamics, and similar subjects. A change in that tradition was due perhaps to Lord Kelvin (cf. the famous book by Tait and Thomson; the latter became Lord Kelvin). In countries of Germanic tradition, rational mechanics has long been taught by (theoretical) physicists (e.g., the very chair that I have occupied, the oldest of that kind in Switzerland, goes back to the turn of the century, so that when Einstein received, after overcoming some difficulties, the title of Privatdozent at Berne University, he was, so to speak, under the formal control of my predecessor). Now, Einstein's theory is, from the point of view of the evolution of physics, a generalization of classical mechanics, even though its roots were to be found in electromagnetism with its Lorentz invariance. Thus, (general) relativity quite naturally became an object for the attention of mathematicians more or less interested in physical problematics, going back to Poincaré, yet later entering not only the workshops of mathematicians such as Cartan and Weyl (not to speak of Cartan's pupil Lichnerowicz) but reaching far away geographically, as it did in India, for example. Of course, highly trained mathematical minds are likely to make considerable achievements in such a field. This was already well known in rational mechanics and analytical dynamics. However, one must possess a knowledge of how theoretical physics is elaborated if one is to progress and qualify as a theoretical physicist. Scientists of the stature of a Newton or Einstein, physicists and mathematicians at the same time, are discovered only once a century or so. The shift in research in relativity from mathematics back to physics probably had to take place sooner or later, since predictions of effects in nature could not but call the attention of observers keen to verify or falsify them. For example, think of black holes. They were originally mathematical singularities and then became something like hell for all material bodies that fell into their neighborhood. It may be that there is still a more or less purely mathematical trend in doing research in GRG. However, GRG has definitely found the way back to its home, which is physics. At the same time, it has definitely made astronomy a main field of physics. We call it astrophysics, but it is much more than just that. It is cosmology, meaning that any medieval-like conception of cosmology has to yield to the new approach, including the very question about the creation of the world. This does not mean that religion is made obsolete or become wrong. Religion is not a science. Perhaps there is no real incompatibility between the two.

I think that Eddington and his school have considerable influence in that respect. Eddington was religiously minded; he became a Quaker. Why not? Einstein was also religiously minded, but totally unorthodox from the point of view of his Jewish origin; he became a pantheist, like Spinoza who was forbidden to enter the synagogue. It may be a bad thing to mix religion with scientific thought; it certainly is, in case one of the two be judged by criteria borrowed from the other or conversely, but it is not when one will observe the necessary mutual respect.

These things are mentioned here because at the time of the turning point marked by GR Zero, the problematics thus envisaged would have to be confronted if relativity theory was to become cosmology, meaning the theory of the world and of its origin and history. Now, it has become precisely that, accompanied of course by the extremely subtle and elaborate views on particles, subparticles, and their disintegration and recombination. However, these views must be included in an encompassing frame necessarily borrowed from any generalization of general relativity theory. This development began in 1923 when Lemaître went to Cambridge for the first time. Of course, it can also be dated to De Sitter's model, published as early as 1917, a time during which Einstein believed that the universe must be static. There is no point in summing up the whole story about instability, expansion versus contraction, Hubble's observations, etc. This aspect of the problematics involved was not, perhaps, given due consideration. It was not accompanied by a big uproar, which happened when the name big bang was invented. Of course, one did not know in those days what later became known about the transmutation of matter, in order to complete the vision into something formidable. Yet, there is no big bang if not from a primitive atom; hence, the big bang is not the creation of the world out of nothing.

In that respect, the Jubilee Conference was again a turning point. For there too, the problematics of a combination of relativistic gravitation theory with quantum theory was asserted with steadfastness that did not leave much to be desired. Surely, this was one of Pauli's merits, though it was going to be carried out by others, and in that respect, thinking back to those days, let me mention the name of Peter Bergmann.

There is a final point that to my mind must be mentioned when asserting that research after World War II coincided with a turning point, especially with reference to general relativity. It concerns philosophical thought.

Of course, shortly after 1905, philosophers in the whole world were very much abashed by the publication of special relativity. In those days, the two leading philosophers in the world were Bergson and Husserl, both born in 1859, i.e., they were 20 years older than Einstein. B. Russell, three years younger, can hardly be said to have completed a comparable revolution in philosophical thought. He was not interested in general relativity, by the way (whereas Whitehead was a mathematician). Bergson, on the one hand, broke off with Cartesianism and more especially destroyed scientism, which had become the fashion, grounding the spiritual way of philosophizing in an intuitionistic approach to life and more specifically to time; Husserl, on the other hand, based his phenomenology on the Cartesian method, which he claimed to continue but also to supersede by searching for final, apodictic insights. In those days, a lot of would-be philosophers wrote for and, mostly, against relativity. Even Bergson argued against it. However, if he did so, it was because he had a biological-like conception rather than a physical-like conception of science. Today, this appears to have been a pioneering achievement. Later, after due consideration and having learned a lot more than is usually realized, Bergson abandoned his earlier critique and even forbade the republication of the book in which he had attacked relativity. It would be completely out of place to dispose of Bergson's intervention in the field of relativity theory by saying that, not being a professional physicist, what he wrote was in any case incompetent nonsense. Husserl, on his own side, did not start discussions on relativity proper, nor did he contribute much to the philosophical problematics of time, but his follower, Heidegger (born 10 years later than Einstein and 4 years after Bohr), did. In spite of all the criticism that can be addressed against him (see, for instance, my own appreciation of his work in my most recent book, *Roseau pensant*, Berne-Frankfurt/M., New York and Paris, 1988), Heidegger has doubtlessly been the most outstanding and influential philosopher among the generation following that of Husserl and Bergson. For all this, the problem of time has brought about one of the central philosophical concerns of the 20th century. Surely, relativity theory and quantum theory have to be taken into account in any comprehensive approach of the problematics. Some scientists of rank have done so, either in a very learned way (e.g., in most recent times, J.A. Wheeler) or in a way that may be called popular in the best sense, whereas much of the current popular work published is of poor quality; one outstanding example of the first case for the time before World War II is H. Weyl (born in 1885; yet, he was not concerned with quantum theory). Two outstanding examples of the second case are Eddington (born in 1882, with writings published between 1914 and the 1930s), who was rather inclined to neglect quantum theory, and Louis de Broglie (born in 1892, with writings published during the 1930s and 1940s), who was rather inclined to emphasize quantum theory. Einstein (born in 1879) and Bohr (born in 1885) should be put, I think, in a special category. They have sometimes, very wrongly, been assimilated as representatives of neopositivism. Nothing is more erroneous. Insofar as he also was a philosopher, Bohr was an outstanding representative of what can be called the philosophy of existence (and he has been correctly considered by one of his pupils, Oscar Klein, who also worked with Kaluza, as ranging in the Pascalian line), whereas Einstein ranks in the Spinozistic tradition, itself derived from Cartesianism. Therefore, the famous controversy between Einstein and Bohr is a sort of renewal of a dissension that can be traced back to the times of Descartes and Pascal, even though apparently the said controversy runs on a specifically physical plane.

Nevertheless, during the 1930s and the 1940s, professional philosophical considerations of science were mostly concerned with the so-called foundations of mathematics; they coincided with a tremendous evolution of formal logic. Relativity did not quite fall into oblivion but after the war it was not any more the point of attraction for philosophers. This has since changed again, and the middle 1950s mark a kind of turning point in this respect as well. However, from then on the questions are no longer attached to the problematics of "what is space" but to that of "what is time." Both the Cartesian and the Kantian approaches are obsolete, whereas the Bergsonian one is not (at least not yet).

Of course, only those who are professionally trained in theoretical physics can judge with final competence the correctness of arguments about matters of physics, but that training alone does not make a scientist a philosopher. Most physicists are not philosophers, because they do not care, they do not have the time, and they are not also professionally trained philosophers although they sometimes (wrongly) think that they can successfully philosophize. It needs a genius to spontaneously be a scientist and a philosopher; both Einstein and Bohr can be said to have been such, the latter perhaps a more profound one, though less clear in his sayings than the former. Pauli, also a genius, did not indulge in philosophizing publicly, i.e., in publications; however, in private conversations, I discovered in him a timid yet rather profound philosopher. Many physicists today, and also historians of the physics of the past decades, may think that philosophical problematics apparently raised by research in relativity are at most of secondary importance. Actually, this is not so, because what finally remains of any piece of acquired knowledge is always of philosophical relevance, even perhaps of philosophical relevance only, but no more of scientific relevance. The history of science is, for sure, instrumental to that achievement. Perhaps it is not more than instrumental, unless history (of science or anything else) is seen as having a value in itself. To me, however, the latter opinion is insufficient: if logic and mathematics lead to the clear and evident establishment of universals and if history leads to the ordered lineages of particulars, universals or particulars in their own right and in splendid isolation drift in a kind of vacuum. The relationship between universals and particulars per contra brings forth insights to consciousness. Therefore, the dialectics between the history and the philosophy of science cannot be eluded, and this chapter is perhaps, after all, nothing but such dialectical argumentation.

PART II

Conceptual Issues in General Relativity

Einstein and Mach's Principle

Julian B. Barbour

1.Introduction

There are few topics in the history of physics that are more tangled and confused than the question of Mach's principle. Looking at this problem from a broadly historical perspective, I should like to establish first what Einstein actually believed he was doing at the various stages in the creation of general relativity and then to identify the reasons for the confusion that surrounds the subject. I see the prime cause of confusion in the fact that at no stage in his work did Einstein make a serious attempt to analyze at a fundamental level the implications of the relativity of position and motion that were the core of Mach's criticism of Newtonian mechanics. However, my overall conclusion is somewhat paradoxical: namely, that although Einstein never made such an analysis, the theory he ultimately created implemented Mach's ideas as reasonably as one may expect.

Various aspects of Einstein's work on general relativity were already seen by his contemporaries at the time he was doing his work as entirely noncontroversial. In particular, all were soon agreed on the need to create a field theory of gravitation consistent with the basic requirements of special relativity. Other ideas were more revolutionary, but were generally accepted quite quickly. Moreover, there was little controversy about their significance, even if their need was perhaps questioned. This applies above all to the notion of a curved space-time and the representation of the gravitational field by a tensor. Almost all the argument surrounded, and still surrounds, the need for and the significance of Einstein's extension of the relativity principle and its relation to Mach's ideas. The connection in Einstein's mind between the extension of the relativity principle and the equivalence principle also needs to be clarified. These therefore are topics on which I concentrate.

2. What Mach Said

It will be helpful to begin with a brief review of some of the most important things that Mach said about the relativity of motion. In *History and Root of the Principle of the Conservation of Energy* (Mach 1872) and then again in *The Science of Mechanics* (Mach 1883), Mach (1837–1916) made some extremely trenchant criticisms of Newton's use of the concepts of absolute space, time, and motion in *Principia* (Newton 1687). Mach argued forcefully that motion is relative, conjectured that it would be possible to construct a form of mechanics in which only relative distances occur, and speculated that such a theory would lead to potentially observable effects in disagreement with Newtonian theory. For example, in his celebrated discussion of Newton's bucket experiment, Mach said:

> Newton's experiment with the rotating vessel of water simply informs us, that the relative rotation of the water with respect to the sides of the vessel produces *no* noticeable centrifugal forces, but that such forces *are* produced by its relative rotation with respect to the mass of the earth and other celestial bodies. No one is competent to say how the experiment would turn out if the sides of the vessel increased in thickness and mass until they were ultimately several leagues thick. (Mach 1883, p. 284)

Mach's proposal created quite a stir, and even before Einstein began to think seriously about his theory of gravitation, actual experiments to test for the putative "frame dragging" effect implicit in the above quotation were made by the Friedländer brothers (Friedländer and Friedländer 1896) and Föppl (1904a, 1904b). Mach really made the problem of the *relativity of motion* topical and acute. I used italics here because when we come to Einstein, we shall encounter a significant shift.

First, however, it will be necessary to say something about *inertial mass* and then the meaning of the word *inertia*. Mach made a much quoted criticism of Newton's definition of mass ("the quantity of matter is the nature of the same, arising from its density and bulk conjointly"), saying, "With regard to the concept of 'mass,' it is to be observed that the formulation of Newton... is unfortunate. As we can only define density as the mass of unit volume, the circle is manifest" (Mach 1883, p. 237). He replaced Newton's vicious circle with a beautiful operational definition that showed the mass concept cannot be separated from the content of Newton's third law, pointing out that, even if the concepts of absolute space and time are allowed, Newton did not quite succeed in clearly formulating the essence of dynamics. The point is that in the absence of interaction, bodies move in straight lines at uniform speed, but in an interaction (between two bodies, for example), they can deflect each other mutually from their inertial paths. Now, for any two given bodies, it is always found that the *accelerations* that they impart to each other in such interactions bear a constant ratio, and as Mach showed, it is precisely this fact that permits the introduction of the mass concept, i.e., numbers associated with various bodies in the universe that characterize the accelerations they impart to each other. Mach regarded this insight—the existence of mass—as the very heart of dynamics. Commenting on the discovery of dynamics, Mach said

> In reality only *one* great fact was established. Different pairs of bodies determine, independently of each other, and mutually, in themselves, pairs of accelerations, whose terms exhibit a constant ratio, the criterion and characteristic of each pair. (Mach 1883, p. 306)

Mach was completely satisfied that he had clarified the empirical basis of the concept of inertial mass:

> As soon therefore as we... have *perceived* in bodies the existence of a special property determinative of acceleration, our task with regard to it ends with the recognition and unequivocal designation of this *fact*. We shall not get beyond the recognition of this fact, and every venture beyond it will only be productive of obscurity. (Mach 1883, p. 270)

I have quoted the last sentence because, as we shall see, it proved to be prophetic.

Now, we need to be clear about the meaning of inertia. As originally used by Newton in *Principia* it meant *resistance to acceleration* (change of motion) and is therefore the *inertial mass* we have been discussing. However, by Mach's time, inertia had acquired a second meaning, being used as a shorthand for the content of Newton's first law, namely, the fact that bodies continue in a state of rest or uniform motion in a straight line unless acted upon by external forces. Indeed, by the 19th century Newton's first law was also known as the *law of inertia*. Mach was concerned with inertia in both its meanings, but his concern was qualitatively different in the two cases. As we have seen, his concern with inertial mass was solely that it should be properly defined. Having given what he regarded as its proper definition, he did not feel that anything more needed to be done. But with regard to the law of inertia, his position was quite different. Here, he believed that not only was epistemological clarification needed (how can one possibly say a body moves in a straight line through space if that space is invisible?), but also that Newton's theory could be fundamentally wrong and that the phenomenon described by Newton's law of inertia could have its origin not in motion with respect to absolute space, but in some kind of interaction with other matter in the universe. This comes out especially clearly in the earlier quotation about the bucket experiment and even more clearly in a passage from 1872:

> Now what share has every mass in the determination of direction and velocity in the *law of inertia?* No definite answer can be given to this by our experiences. We only know that the share of the nearest masses vanishes in comparison with that of the farthest. (Mach 1872)

Note the words that I have italicized. When Mach spoke of the problem of inertia, he was quite clearly referring to the law of inertia, not inertial mass. He advanced a conjecture, we may call it *Mach's conjecture* [to distinguish it from *Mach's principle*, which was Einstein's coining (Einstein 1918a) and to which we shall come later], and he advanced it because he believed in relativity of motion as an ontological fact.

3. Einstein and Extension of the Relativity Principle

Now what about Einstein? Let us note first that neither Mach nor any other physicist created a theory that in any sense realized Mach's conjecture prior to 1907 (when Einstein started serious work on his general relativity). In the first decade of this century, the conjecture was still a vague hunch. In addition, physics had been transformed when compared with the situation that existed in 1872 (when the conjecture was first put forward). One important development had actually been stimulated in large part by Mach's criticism of Newton's concepts of absolute space and time, which had led Ludwig Lange to attempt to retain the essence of Newton's law of inertia but to formulate its content in an epistemologically satisfactory manner, taking account of the fact that any observed motion is a relative motion. This led to his very beautiful dynamic definition of motion (in which the position of a studied body is measured relative to the positions of initially moving reference bodies) and the concepts of an inertial mass system (Lange 1885, 1886; see also Barbour 1989a, §12.3). This almost certainly had a strong influence on the manner in which Einstein approached this question; for it shifted attention from the law of inertia itself and the attempt, implicit in Mach's original writings, to recover it in some sort of limit from a theory formulated solely in terms of relative distances [which can be done, as Bertotti and I have shown (Barbour and Bertotti 1977, 1982)] to the existence of the distinguished inertial systems. This was especially understandable in view of the crucial role played by such systems in Einstein's own special theory of relativity, created in 1905. The problem that Einstein confronted was therefore rather more general than that of the origin of the law of inertia; it was that of the origin of distinguished frames of reference (inertial systems) in which the laws of nature took a particularly simple form. This was also the consequence of another very major change that had taken place between 1872 and 1907—the triumphant replacement of the Newtonian paradigm of mass particles acting through a distance instantaneously by the Faraday–Maxwell field concept and the associated notion of local laws of nature.

I think the shift in Einstein's perspective was decisive for the general approach that he adopted to the points raised by Mach. It should perhaps be said that, so far as I know, Einstein did not write an extensive discussion of how he saw the Machian problem. However, numerous comments scattered throughout his writings from 1907 to 1949 relate implicitly or explicitly to the Machian problem, and it is quite clear that the desire to solve it was an extremely strong stimulus for Einstein (Barbour 1989b). What is rather striking is that Einstein's Machian comments crystalize around two more or less distinct questions, between which, so far as I know, Einstein did not establish a really clear connection. Moreover, in neither case did Einstein explicitly address these questions at the truly fundamental level with which Mach was concerned—the definition of position and motion. Einstein operated throughout at a higher conceptual level.

The first Machian question was already raised in Einstein's 1907 paper, in which the equivalence principle (not yet given that name) made its first appearance (Einstein 1907). At the end of his review of the special theory of relativity, Einstein commented that the relativity principle as hitherto used required the laws of nature to be independent of the state of motion only for *unaccelerated* frames of reference. He immediately followed this with the key question that already adumbrated the line of argument that would be successively strengthened right through the successful completion of the general theory of relativity in 1915: "Could one suppose that the principle of relativity is also satisfied for systems moving relatively to each other with acceleration?" It was, of course, the equivalence principle that enabled Einstein to pose this seemingly absurd question. For, since "in a gravitational field all bodies are accelerated equally," he was able to argue that "at the present state of our knowledge" there were no grounds for believing that there were any respects in which a frame of reference accelerated uniformly in a region free of a gravitational field differed from one at rest in a homogeneous gravitational field.

Neither in the 1907 paper, nor the next paper dealing with gravitation (Einstein 1911) did Einstein mention Mach explicitly, but his concern with the Machian problem is revealed by a comment in the later paper: "In such an approach one cannot speak of the absolute acceleration of the coordinate system any more than in the special theory of relativity can one speak of the *absolute velocity* of the system" (Einstein's italics). If the Machian stimulus to this comment is more than evident, the direction of Einstein's thought is still somewhat surprising—and certainly bold. Whereas one might have expected Einstein, following a close reading of Mach's *Mechanics*, to have attempted to set up a relational theory, i.e., a theory involving only relative distances and relative velocities, and then from it recover Newton's first law (or, in his updated form of the problem, the family of distinguished inertial systems), the whole thrust of Einstein's work was in almost the diametrically opposed direction. Namely, he set out to show that the laws of nature must have such a form that they can be expressed in an identical manner in absolutely any frame of reference. If he could show this, then the Machian problem would have been solved in a very brilliant and surprising manner—the problem of distinguished frames of reference would not after all exist, since he would have shown that no such frames in reality exist! This whole direction of Einstein is very characteristic; instead of becoming bogged down in trying to explain a particular result by constructing a theory of specific form from which the desired result would follow, he sought to achieve his aim by the imposition of a broad and very powerful general principle. As I have written elsewhere (Barbour 1989) in more detail about the way in which this approach helped to guide Einstein to his final theory and as I am currently working on an extended account of the matter (to be given in Vol. 2 of Barbour 1989), I shall merely concentrate in this chapter on the evidence that establishes that this indeed was Einstein's basic strategy and the primary stimulus to his work.

The seeming absolute nature of rotation was, of course, the prime piece of evidence that Newton advanced for absolute space, and the problem of rotation was discussed very prominently and vividly by Mach. If Einstein was to make any progress in his attempt to extend the relativity principle (and thereby show that, in truth, there were no distinguished frames of reference at all), it was clearly necessary for him to extend the arguments he had used in 1907 from uniform rectilinear acceleration to rotation. In September 1909, he wrote to Sommerfeld along precisely such lines:

> The treatment of the uniformly rotating rigid body seems to me to be of great importance on account of an extension of the relativity principle to uniformly rotating systems along analogous lines of thought to those that I tried to carry out for uniformly accelerated translation in the last section of my paper published in *Zeitschrift für Radioaktivität.* (See Stachel 1980)

In the period following the first attempt (with Grossman) at a general theory of relativity (Einstein and Grossman 1913) and before the definitive formulation of the general theory of relativity in November 1915, Einstein expressed opinions such as this on several occasions in a very emphatic and strengthened form. Moreover, in these cases, he mentioned Mach explicitly. For example, in a particularly informative article published in *Scientia* (Bologna) in 1914 (Einstein 1914b), he opened a discussion of his work on the general theory of relativity by pointing out that both classical mechanics and the special theory of relativity suffer from a "fundamental shortcoming that no one open to epistemological arguments can deny." He went on to say that the weaknesses that he was going to describe "were laid bare in all clarity by E. Mach in his deeply penetrating investigations into the foundations of Newtonian mechanics." Einstein then gave what he called a very elementary example in order to illustrate the essential point. As the passage that follows comes closer to the spirit of Mach's writing than any other that I have been able to find in Einstein's papers and books and since the *Scientia* article is not readily accessible, I gave a translation:

> Two masses hover in space at a great distance from all celestial bodies. They are close enough to each other to have an influence on each other. Now, suppose an observer follows the motion of the two bodies by continually looking along the direction of the line joining the two masses to the sphere of the fixed stars. He will establish that this line traces out on the observable sphere of the fixed stars does not change. If the observer is possessed of natural understanding, but has learned neither geometry nor mechanics, he will draw the following conclusion: "My masses execute a motion that at least in part is causally determined by the system of fixed stars. The laws according to which masses in my neighborhood move are in part determined by the fixed stars." A man who has passed through the school of science will smile over the simplicity of the observer and say to him: "The motion of your masses has nothing to do with the heaven of the fixed stars; much rather the motion is determined by the laws of mechanics quite independently

> of the other masses. There is a space R, in which these laws hold. These laws are such that your masses remain fixed in a plane in this space. However, the system of fixed stars cannot rotate in this space, since otherwise it would be torn apart by powerful centrifugal forces. Therefore, if it is to be able to exist at all, it must necessarily be at rest (or at least nearly so!); this therefore is the reason why the plane in which your masses move always passes through the same fixed stars." But, our intrepid observer will say: "You are of course very learned. But, I believe no more in that giant thing of which you speak and call space than I could be brought to believe in ghosts. I can neither see such a thing nor form any conception of it; or should I think of your space R as a very fine material net with respect to which the other things are referred? But then, besides R, I can imagine a second such net R' that moves relative to R in an arbitrary manner (for example, rotates). Do your equations then also hold relative to R'?" The educated man denies this with certainty. Then, our simple-minded friend remarks: "But how do the masses know with respect to which of the 'spaces' R, R', etc., they should move in accordance with your laws; through what do they recognize the space, or the spaces, with respect to which they must move?" Our educated man is now in the greatest difficulty. He emphasizes, it is true, that such privileged spaces must exist, but he is unable to give any reason why those spaces could be distinguished with respect to others. To which the response of our simple-minded friend is: "Then, for the time being I shall continue to regard your distinguished spaces as an idle invention and hold fast to my belief that the sphere of the fixed stars plays a part in determining the mechanical behavior of my test masses."

Shortly after this discussion, on which I shall comment later, Einstein summarized his position as follows. "Since not only classical mechanics but also the theory of relativity in the restricted sense suffer from the fundamental defect that I have presented above, *I have set myself the aim of generalizing the theory of relativity in such a way that this imperfection is avoided*" (my italics). There followed immediately a comment that revealed the connection in Einstein's mind between the equivalence principle and the requirement of general relativity, i.e., general covariance. This is what Einstein said:

> I first recognized that in such a theory universal gravitation must be accorded a quite fundamental role. For it is already clear from what has been said earlier that every physical process must, since energy quantities correspond to it [Einstein is referring here to the energy-momentum tensor associated with every physical process], generate a gravitational field. On the other hand, the fact of experience that all bodies fall equally fast in a gravitational field suggests that *in a*

> *gravitational field physical processes unfold in exactly the same way as relative to an accelerated frame of reference* (equivalence hypothesis).

The italics are again mine and show that the supreme importance of the equivalence principle for Einstein was precisely that it opened up a possibility of extending the relativity principle to more frames of reference than simply those moving uniformly and rectilinearly relative to each other. This is precisely the logic that underlies the 1907 paper and shows how consistently and single-mindedly Einstein followed that logic right through to the completion of his theory in 1915.

This is a point worth making, since in many accounts of general relativity [including some by Einstein, for example, his first joint paper with Grossman (Einstein and Grossman 1913)], the equivalence principle is advanced as Einstein's prime motive for constructing general relativity. But, this confuses the tool with the motive. The above quotations and the whole logic of Einstein's approach show that what above all excited Einstein about the equivalence principle (and indeed dictated the name he chose for his theory) was the prospect it opened up for solving the Machian problem through the dramatic extension of the relativity principle.

This conclusion is confirmed by Einstein's famous summary paper "The Foundation of the General Theory of Relativity," published in early 1916 just after the completion of his theory in November 1915 (Einstein 1916). The crucial second section, which contains the wonderful thought experiment with the two fluid globes that hover freely in space, is entitled "The Need for an Extension of the Postulate of Relativity" and opens with the words "In classical mechanics, and no less in the special theory of relativity, there is an inherent epistemological defect that was, perhaps for the first time, clearly pointed out by Ernst Mach." The discussion of the thought experiment ended as follows (the italics are Einstein's):

> Of all imaginable spaces R_1, R_2, etc., in any kind of motion relatively to one another, there is none that we may look upon as privileged a priori without reviving the above-mentioned epistemological objection. *The laws of physics must be of such a nature that they apply to systems of reference in any kind of motion.* Along this road we arrive at an extension of the postulate of relativity.

It is only after "this weighty argument from the theory of knowledge" that Einstein mentioned the equivalence principle and the support it gave for his main argument.

4. Einstein and Inertial Mass

I said earlier that Einstein's Machian comments crystallize around two more or less distinct questions. The first of these, the need for an extension of the relativity principle, has now been adequately covered, and it is time to turn to the second question. This first appeared in a paper that Einstein wrote while still at Prague (Einstein 1912). I translate the title as "Is There a Gravitational Analogue of the Electromagnetic Induction?" In it, Einstein began to develop an idea that often appeared in discussions of general relativity and Mach's ideas and showed what a powerful influence Mach exerted by concluding his discussion of Newton's bucket experiment with the suggestive comment we have already quoted: "No one is competent to say how the experiment would turn out if the sides of the vessel increased in thickness and mass until they were ultimately several leagues thick."

This mere hint of Mach (in which a possible role of gravity is not mentioned) was developed by Einstein into the idea that the centrifugal forces that appear to arise from rotation pure and simple must really be thought of as gravitational forces of induction generated by the accelerated motion of distant masses in the frame of reference rotating with the bucket. So far as I know, this idea was first clearly and fully expressed by Einstein in a paper (Einstein 1914c) slightly later, from which it will be worth quoting. Having argued on the now familiar epistemological grounds for an extension of the relativity principle, Einstein mentioned Newton's counterargument based on the apparently absolute nature of rotation and the appearance in K' (a frame of reference rotating with the bucket) of centrifugal forces not present in a nonrotating frame K. Einstein continued:

> However, this argument is not conclusive, as E. Mach in particular has shown. Namely, we do not necessarily have to trace the origin of the centrifugal forces to a motion of K'; rather we can just as well attribute them to the average rotational motion of ponderable distant masses of the environment relative to K', which we regard as being "at rest."

Once again, the plausibility of this primary Machian argument is supported by the secondary argument from the equality of the inertial and gravitational masses of all bodies for "we have no means to distinguish a 'centrifugal field' from a 'gravitational field'."

I do not want to enter into a discussion about this idea at this point and have merely mentioned it to give the context in which Einstein introduced his second Machian aspect. Following up the induction idea in his 1912 paper (for the case of linearly accelerated motion), Einstein was led to examine the structure of the equations of motion that followed from his first theory of gravitation, in which c, the velocity of light, played the role of scalar gravitational potential. From the equation of motion of a particle moving at low velocity in an external gravitational field, Einstein found as a formal result that the particle would have an effective inertial mass m' given by

$$m' = m + \frac{kmM}{Rc_0^2},$$

where m is the mass of the particle, M is the mass of a much heavier body at distance R from the particle, and c_0 is the velocity of light at a great distance from both bodies. Einstein's comment on this was "In itself, this result is of great interest. It shows that the presence of the inertial shell K increases the inertial mass of the material point P within it. This makes it plausible for the *entire* inertia of a mass point is the effect of the presence of all other masses, resulting form a kind of interaction with them." In a footnote, Einstein said, "This is exactly the standpoint for which E. Mach has argued persuasively in his penetrating investigations of this question."

Two comments are appropriate here: (1) when Einstein spoke of "the *entire* inertia of a mass point" he was quite definitely referring to its *inertial mass*, not the fact that, in accordance with the *law of inertia*, it would, if free from external forces, follow a straight line with uniform speed; and (2) Einstein was quite wrong to attribute such an idea to Mach, who never put forward any such proposal. Einstein actually appears to be the victim of semantic confusion of two quite different meanings of the word *inertia*.

In the period immediately after the Prague paper, Einstein referred several times to Mach in connection with the idea that the inertia of a body is due to some kind of interaction with other bodies in the universe (in all such cases he was clearly referring to inertial mass). For example, in his first joint paper with Grossmann, the first section after the introduction was devoted to a review of Einstein's earlier scalar theory and ended with the derivation of the 1912 result that we have just described. Einstein commented that it "fits with Mach's bold idea that inertia has its origin in an interaction of the considered mass point with all others; for if we pile up masses in the neighborhood of the considered mass point, we thereby reduce the gravitational potential c and thus increase the quantity m/c, which determines the inertia." (Einstein and Grossman 1913, p. 228).

In a talk given in Frauenfeld, Switzerland, on September 9, 1913, Einstein gave a brief summary of the Einstein–Grossmann theory (in which, as in the Einstein–Grossmann paper, he used the equivalence principle as the main argument for the new theory). At the end of his talk, he again referred to what I have called the second Machian aspect, though there was now a significant amplification. This is what Einstein said:

> The theory that has been outlined eliminates an epistemological shortcoming that is present in not only the original theory of relativity but also in Galilean mechanics and has been emphasized by E. Mach in particular. It is also clear that the concept of acceleration of a material point can no more be given an absolute significance than can that of velocity. Acceleration can only be defined as relative acceleration of a point relative to other bodies. This circumstance indicates that it is meaningless to ascribe to a body a resistance relative to acceleration as such (inertial resistance of bodies in the sense of classical mechanics); much rather, it must be required that the appearance of an inertial resistance be tied to the relative acceleration of the considered body relative to other bodies. It must be required that the inertial resistance of a body can be increased by bringing unaccelerated ponderable masses into the neighborhood of the body; and this increase in the inertial resistance must again disappear if these masses also participate in the acceleration of the body. (Einstein 1913, p. 290)

Einstein concluded by claiming that the required behavior of the inertial resistance did indeed follow from the equations of motion of the Einstein–Grossmann theory and said that the effect could be characterized as *relativity of inertia*. He called it "one of the most important supports" of the theory.

It is noteworthy that in subsequent writings, above all the famous cosmological paper (Einstein 1917), Einstein often used the expression *relativity of inertia* in striking contrast to the expression *relativity of motion*, which virtually never occurred in his writings.

In the papers of 1914, the relativity of acceleration and inertia reappeared at least twice (Einstein 1914a, in a talk given on February 9, 1914, and in the *Scientia* paper already discussed). However, as the year passed, the second Machian aspect seems to have receded somewhat from the forefront of Einstein's thought. In the *Scientia* paper, only the relativity of acceleration was mentioned, and it was given relatively little prominence compared with the argument from the need for an extension of the relativity principle. The topic was not mentioned at all in the long review paper that Einstein presented into the Prussian Academy of Sciences on October 29, 1914 (Einstein 1914c). Like the *Scientia* account of the Einstein–Grossmann theory, this paper opened with an extended argument for extension of the relativity principle in order to eliminate the curious distinguished frames of reference of Newtonian mechanics and special relativity, after which came the discussion of the origin of centrifugal force that we have already considered. As we have seen, Einstein's argument followed very much the same direction in the introduction in the 1916 survey.

It was only in the 1917 cosmological paper (in which Mach is not actually mentioned) that the relativity of inertia came again to the fore in Einstein's arguments. Following discussions with De Sitter above all (see Barbour 1989b), Einstein was forced into the recognition that his principle of general covariance, in which he had set such store, believing that it would automatically solve the Machian problem, did not after all achieve the result he expected. In particular, it followed from the general theory of relativity that inertia (in both its meanings) could not be attributed exclusively to the effect of distant matter in the universe. This led to the rather tortuous reappraisal of his theory that Einstein, opening the second section of his paper ("The Boundary Conditions According to the General Theory of Relativity"), characterized with the words "I shall conduct the reader over the road that I have myself traveled, rather a rough and winding road, because otherwise I cannot hope that he will make much interest in the result at the end of the journey." Einstein then raised the question of the boundary conditions needed for his theory and said:

> The opinion that I entertained until recently, as to the limiting conditions to be laid down in spatial infinity, took its stand in the following conditions. In a consistent theory of relativity there can be no inertia *relatively to "space,"* but only an inertia of masses *relatively to one another*. If, therefore, I have a mass at a sufficient distance from all other masses in the universe, its inertia must fall to zero.

I have set off this passage for emphasis, since, on the basis of a rough survey of the post-Einstein literature on the Machian question, I have to come to the conclusion that about 60% of the people who have written on the subject seemed to have formed their idea of what Mach's principle (to the formulation of which we shall shortly come) should achieve on the basis of these three sentences. At this point, let me merely note that once again Einstein was quite clearly using the word *inertia* in the sense of the inertial mass. At the very least, this is most unfortunate, since it represents a very serious distortion of what Mach himself said. Many authors have been content to take Einstein at his word and have not attempted to understand Mach's position. This has been one major source of confusion, though not, I think, the most serious one of all, to which we shall come.

Einstein reported that he was not able to formulate conditions of his theory in a satisfactory manner yielding the result he wanted, and this led him, as the reader probably knows, to the idea that the universe should not require any boundary conditions at all, i.e., that it should be closed. This, in turn, led Einstein to introduce his famous cosmological constant and associated modifications of his field equations, the purpose of which was to construct a cosmological model that was both static (in accordance with the belief at the time) and contained matter. The subsequent fate of this proposal is discussed in my 1990 article (Barbour 1990).

The last important development in Einstein's thinking about the Machian problem appeared in a brief paper in 1918 (Einstein 1918a) in which Einstein responded to criticism by Kretschmann (Kretschmann 1917) of the conceptual basis of general relativity (Einstein 1918a). It was in this paper that Einstein first coined the expression *Mach's principle*. Einstein commented that he felt a need to return once more to the foundations of the general theory of relativity and said that "the theory, as it now appears to me, rests on three main points of view, which, however, are by no means independent of each other." He listed the main viewpoints:

> a) *Relativity principle*: The laws of nature are merely statements about time-space coincidences; they therefore find their only natural expression in generally covariant equations.
>
> b) *Einstein principle*: Inertia and weight are identical in nature (*wesensgleich*). It follows necessarily from this and from the result of the special theory of relativity that the symmetric "fundamental tensor" ($G_{\mu\nu}$) determines the metrical properties of space, the inertial behavior of bodies in it, as well as gravitational effects. We shall denote the state of space described by the fundamental tensor as the G-field.
>
> c) *Mach's principle*: The G-field is *completely* determined by the masses of the bodies. Since mass and energy are identical in accordance with the results of the special theory of relativity and the energy is described formally by means of the symmetric energy tensor ($T_{\mu\nu}$), the G-field is conditioned and determined (*bedingt und bestimmt*) by the energy tensor of the matter. (Einstein 1918a, pp. 241–242)

In a footnote Einstein commented that "Hitherto I have not distinguished between principles (a) and (c), and this was confusing. I have chosen the name 'Mach's principle' because this principle has the significance of a generalization of Mach's requirement that inertia should be derived from an interaction of bodies."

I do not propose at this stage to make any comments about Einstein's formulations except in connection with (a), from which a truly remarkable shift in viewpoint is evident. This came about in a two-stage process, of which the first was closely associated with Einstein's breakthrough to the definitive formulation of his field equations in generally covariant form and recognition of the error in his "hole argument," which had led him to believe incorrectly for two years that the gravitational field equations could not be generally covariant. It was the insight that all observations ultimately represent space-time coincidences that led him to see the flaw in the hole argument; at the same time, he saw this as a further powerful argument for requiring general covariance, and such an argument figures prominently in his 1916 summary paper (Einstein 1916, the two final paragraphs of §3). However, Kretschmann (Kretschmann 1917) succeeded in turning this argument against Einstein, who felt obliged to concede that the principle of general relativity "made no statement about physical reality, i.e., about the *content* of the laws of nature" and was "only a requirement about the mathematical *formulation*." Einstein accepted this on the ground that "because all of physical experience refers solely to coincidences, it must always be possible to represent experiences about the lawful connection of coincidences through generally covariant equations." In Einstein's eyes, this second stage of the process therefore deprived the relativity principle of pretty well all its mandatory power, and henceforth, he accorded it only heuristic value. But, he still had great faith in the Machian idea and therefore attempted to formulate it separately as (c) (commenting however that "it is by no means accepted by all experts in the field, though I feel that its fulfillment is absolutely necessary").

Einstein's concession of Kretschmann's point has spawned a huge amount of literature on the significance of general covariance. Earman and Glymour rightly commented that one would need to write a monograph to do the topic justice (Earman and Glymour 1978). Although I believe the question is very intimately related to the Machian problem and does need to be discussed, that would require much more space than can be allowed here, and I will merely restrict myself to a comment at the end of the chapter. Readers interested in the learning about Einstein's "hole argument" can consult the papers of Stachel and Norton (Stachel 1986, Norton 1984).

5. Why There Is So Much Confusion Surrounding Mach's Principle

The distortion of Mach's ideas about the relativity of motion by Einstein's idea that some sort of cosmological derivation of inertial mass was called for has already been mentioned. It was in truth a "red herring" that does not even seem to have much connection with the basic structure for general relativity. However, I believe that the root cause of the confusion relating to this topic is that when Einstein commenced the work in 1907 that eventually led in late 1915 to the general theory of relativity, he did not stop and ask two fundamental questions: (1) If, as is universally granted, all observation is relative, how do we define position and motion relationally, i.e., relative to observable matter? (2) How are we to construct a dynamical theory of motion that takes into account adequately and naturally this fact of the relativity of motion? It was these questions that Mach repeatedly posed with great urgency and vividness in all his writings on the subject. But, we search in vain through Einstein's writings for anything similar at this really basic level. The only place at which Einstein read at all like Mach was in the *Scientia* article and in his references to the relativity of acceleration, from which a need for the relativity of inertial mass was deduced (a conclusion that Mach most certainly did not draw). As I said, Einstein made virtually no references to the relativity of motion and none at all, I think, to the relativity position. But, acceleration is a concept that requires a considerable degree of sophistication, and a relational acceleration (for which Einstein called) would require prior clarification of the concept of relational position and relational motion, which would also involve definition of a relational time. Einstein did not address any of these problems; there is simply no kinematic analysis of the implications of relationalism. Instead, Einstein worked from the very beginning with frames of reference and these reduced, after he had adopted his definitive space-time concept, to mere labelings of the points of space-time. Had Einstein worked with concepts that had a more directly relational significance, it would have been much easier to see what the connection was between what he actually did and what Mach had advocated.

In this connection, it is especially revealing to see what each said about the use of coordinate systems. In his first published comments on the question of the definition of position and motion, Mach wrote

> In ordinary terrestrial cases, it will answer our purposes quite well to reckon the direction and velocity with respect to the top of a tower or a corner of a room; in ordinary astronomical cases, one or other of

> the stars will suffice. But, because we can also choose other corners of rooms, another pinnacle, or other stars, the view may easily arise that we do not need a point at all from which to reckon. But, this is a mistake; such a system of coordinates has a value only if it can be determined by means of bodies. . . . In fact, any one of the above points of origin of coordinates answers our purposes as long as a sufficient number of bodies keep fixed positions with respect to one another. But, if we wish to apply the law of inertia in an earthquake, the terrestrial points of reference would leave us in the lurch, and convinced of their uselessness, we would grope after celestial ones. But, with these better ones, the same thing would happen as soon as the stars showed movements that were noticeable. When the variations of the positions of the fixed stars with respect to one another cannot be disregarded, the laying down of a system of coordinates has reached an end. (Mach 1872)

The final sentence suggested that Mach regarded the use of coordinate systems as invalid or at least inappropriate. Several passages in his writings implied that one should formulate the theory of motion directly in relative terms, above all in his discussion of Newton's bucket experiment; Mach rejected the idea that the case of an earth rotating with stars at rest was different from that of an earth at rest and stars rotating about it. He said "The world is not *twice* given with an earth at rest and an earth motion; but only *once*, with its *relative* motions, alone determinable" (Mach 1883, p. 284).

Now, as we have seen, Einstein's response to the bucket experiment and the problem of centrifugal force was very different from what Mach advocated. Einstein said that there were *two* equally valid ways of looking at the bucket experiment: in the rest frame of the stars, in which there are no centrifugal forces, and in the frame of the bucket, in which centrifugal forces are generated by the rotation (accelerated motion) of the distant stars. Before commenting on this strategy of Einstein, let me quote a very interesting passage from a paper Einstein wrote after the general theory of relativity had been completed. He was addressing a problem that became extremely familiar, that of the description of nature by means of what are called *gauge variables*, i.e., quantities that are neither directly observable nor uniquely determined by observable quantities but extremely useful in the formulation of dynamical theories because they allow the physical content of the theory to be expressed by means of very simple equations. This is what Einstein said:

> We want to distinguish more clearly between quantities that belong to a physical system as such (are independent of the choice of the coordinate system) and quantities that depend on the coordinate system.

> One's initial reaction would be to require that physics should introduce in its laws only the quantities of the first kind. However, it has been found that this approach cannot be realized in practice, as the development of classical mechanics has already clearly shown. One could, for example, think—and this was actually done—of introducing in the laws of classical mechanics only the distance of material points from each other instead of coordinates; a priori one could expect that in this manner the aim of the theory of relativity should be most readily achieved. However, the scientific development has not confirmed this conjecture. It cannot dispense with coordinate systems and must therefore make use in the coordinates of quantities that cannot be regarded as the results of definable measurements. (Einstein 1918b)

I see this passage as the key to establishing the link between what Einstein actually did in his theory and the sort of thing Mach would have expected him to do. The point is, as we see, Einstein willingly accepted the use within his theory of what are now called gauge-dependent quantities. These are epitomized by the representation of the gravitational field by the Christoffel symbols, the use of which corresponds to the entirely arbitrary coordinate systems allowed in the theory. Therefore, to establish whether the theory that Einstein created does or does not implement Mach's ideas, we must establish the connection between the gauge-dependent quantities and operations that Einstein actually used and the gauge-independent quantities and operations that Mach would have liked him to use. If we can show that what Einstein did in gauge-dependent terms was in fact a faithful implementation of what Mach would have expected him to do in gauge-independent terms, then it will follow that Einstein did realize Mach's idea. So far as I know, Einstein never made any serious attempt to carry out an analysis of the kind that I am suggesting. Moreover, I rather doubt whether he saw the need for any such analysis, because, as we have seen, there were initially two more or less parallel strands in Einstein's Machian thinking—to show the absence of distinguished frames of reference and to establish the relativity of inertial mass. But, distinguished frames and inertial mass are high-level concepts corresponding to accelerations (distinguished frames are ones in which force-free particles move without acceleration, and inertial mass is measured by acceleration). Einstein never explicitly spelled out their connection to the much more primitive, lower-level concepts with which Mach was concerned: relational position and relational motion. My contention is that this can be done and that when the analysis has been made general relativity will be seen to be almost perfectly Machian for at least the case of a closed universe. In the final section of this paper, I shall try to outline the argument.

6. The Fundamental Problem of Motion

If it is accepted as an ontological fact that the position of a given body has no meaning except as a position relative to other bodies (which was Mach's standpoint), the really severe problem is not that of defining position but of defining *motion*. This can be seen as follows. Suppose a universe consists of n material points in Euclidean space (which for simplicity of visualization we take to be two dimensional) and nothing else. Suppose we take a "snapshot" of the positions of these particles relative to each other, i.e., the relative distances between them, and nothing else. Suppose we take a second snapshot a little later, when the relative positions of all the bodies have changed somewhat. We then ask: what motions have the particles made between the two snapshots? Because space is invisible, it is quite impossible to define the motions relative to space. All we have at our disposal is the relative disposition of the particles.

Suppose that in desperation we simply lay the second snapshot on the first. This will then show the displacements of the particles' images in the second snapshot relative to their positions in the first. We could define these displacements as the motion of the particles. However, this would clearly be arbitrary—each different placing of the second snapshot relative to the first would generate a new set of displacements. Obviously, we face a severe dilemma—there seems to be no way of arriving at an unambiguous definition of motion. This is what I call *the fundamental problem of motion*. The problem arises because of the fact that all the bodies in the universe are assumed to be in relative motion. This severs all connection between the successive time slices and makes the ambiguous definition of motion impossible. Note that even if we know the masses of the individual particles the problem is only partially eased. This information enables us to calculate the position of the center of mass of the complete system of particles in each of the two snapshots. When placing the second snapshot on top of the first, we can at least then ensure that the two centers of mass coincide. However, this still does not prevent us from rotating the second snapshot relative to the first about the fixed position of the center of mass. Serious ambiguity still remains.

This fact was the basis of an acute remark made by Poincaré in *La science et l'hypothèse* (Poincaré 1902). He posed the following. Suppose a system of material particles of known masses interacting in accordance with Newton's law of universal gravitation. Suppose further we are given at a certain instant the relative distances between these bodies and the rates of change of these relative distances, i.e., the *observable* initial data of the many-body problem of celestial mechanics. Can we predict the future evolution of the system uniquely? Poincaré pointed out that we cannot because we lack one crucial piece of information. We cannot determine the (intrinsic) angular momentum of the complete system, and this has drastic consequences for its future evolution. The observable initial data are not quite sufficient for unique predictions of the future. As Poincaré emphasized, the initial-value problem of the Newtonian dynamics is well posed and leads to unique evolution if the initial positions and initial velocities of the particles are specified in absolute space (or, in modern terms, in an inertial frame of reference). But, the essential data cannot all be extracted from the relative (and directly observable) data.

How, under these circumstances, can one set up a dynamical theory that operates with only relative distances and relative velocities and permits unique prediction of the future evolution from observable initial data? Now, in fact, contrary to Einstein's assertion noted above that it is not possible to set up theories of motion directly on the basis of relative distances, one can construct purely relational theories in which the basic dynamical variables are the relative distances between the various bodies in the universe (which, very importantly, is taken to be finite in extent). In particular, in the framework of the standard variational approach to dynamics, one can postulate a Lagrangian function $\mathcal{L}$ of a relational theory that contains only the relative distances r_{ij} between the particles and their rates of change $\dot{r}_{ij}$, i.e.,

$$\mathcal{L} = \mathcal{L}(r_{ij}, \dot{r}_{ij}). \tag{1}$$

(I am assuming for the purposes of this discussion the existence of an external, or absolute, time. However, one can also operate with a relational time.) Bertotti and I have worked out in some detail the consequences of postulating a Lagrange function of the form in (1) (Barbour and Bertotti 1977). The most interesting result is that, in such a framework, one can recover Newton's laws of motion, including the law of inertia, to a very good approximation with, however, one very important difference—the frames of reference in which the laws hold are explicitly determined by the average distribution of the matter of the universe.

However, such theories are not without difficulty. In particular, most theories based on Lagrange functions of the form in (1) predict a position-dependent and (most seriously) anisotropic effective inertial mass of the particles. Anisotropy of the effective inertial mass is ruled out observationally (see Barbour and Bertotti 1977). Bertotti and I therefore looked for alternative ways of constructing relational theories and finally hit upon a relatively simple device (Barbour and Bertotti 1982), which I should like to describe briefly because we eventually discovered that the variational principle of general relativity is actually a more sophisticated version of exactly the same device. Since the device was explicitly invented to overcome the fundamental Machian problem of motion described previously, I base my belief that general relativity is Machian on the close parallel between the two variational principles.

The basic idea is as follows. If, in the snapshot language used above, we place the second snapshot in some quite arbitrary manner on top of the first, then we shall find that particle i will have some displacement $d^{\,r}_{\mathbf{w}_i}$ relative to its position in the first snapshot. Let particle i have mass m_i ($i = 1, 2, \ldots, n$) and consider

$$\sum_{i=1}^{n} m_i \, d^{\,r^2}_{\mathbf{w}_i}. \tag{2}$$

The numerical value of (2) will clearly depend on the position of the second snapshot relative to the first. Now, move the second snapshot around the top of the first until the quantity (2) is minimized. Apart from exceptional symmetric configurations, which we can ignore (since we are interested in the generic situation), the required of minimization of (2) will lead to a unique position of the second snapshot relative to the first. One can think of this position as determining the minimal intrinsic variation that has taken place in the system between the taking of the two snapshots.

This device therefore leads to a definition of intrinsic motion. Such motion is defined for the system as a whole (which is why the considered system needs to be finite in some sense) and clearly overcomes what I called the fundamental problem of motion. Having obtained in this manner an unambiguous definition of the individual motions [the $d^{\,r}_{\mathbf{w}_i}$ corresponding to minimization of (2) in the manner described], we can now proceed to set up a variational principle of the conventional form with the Lagrangian function

$$\mathcal{L}_{\text{intr}} = \mathcal{L}_{\text{intr}}(\overset{r}{\mathbf{w}}_i, \overset{\dot r}{\mathbf{w}}_i), \tag{3}$$

where $\overset{r}{\mathbf{w}}_i$ and $\overset{\dot r}{\mathbf{w}}_i$ are constructed as follows. Imagine the system starts in one given relative configuration of the particles and ends in another, which must be intrinsically different. Suppose the system as a whole passes from the initial to the final configuration along some arbitrary virtual path, which we can assume is recorded by a succession of snapshots, each showing the successive relative configuration of the bodies.

Start with the initial snapshot, place that one that follows it on top of it, and then minimize the quantity (2) by moving the second snapshot with respect to the first. This *stacks* the second snapshot relative to the first. Then, put the third snapshot on the top and stack it relative to the second snapshot in the same way. Continue this procedure through the snapshot representing the final configuration. This stacking procedure defines a frame of reference in which the position $\underset{\mathbf{w}}{r}_i$ and velocity $\underset{\mathbf{w}}{\dot{r}}_i$ of each particle are well defined. These are the *intrinsic* variables that appear in (3). The physically realized history among all the putative virtual histories is assumed to be the one for which

$$\int \mathcal{L}_{\text{intr}}\, dt$$

from the initial to the final configuration is extremal. (It is worth noting that by introducing certain auxiliary gauge variables one can combine the two procedures just described—the finding of the intrinsic variables and the determination of the dynamically realized history—into a single variational problem; see Barbour and Bertotti 1982.)

Bertotti and I (Barbour and Bertotti 1982) showed that the physical histories that are in fact obtained in this way [for Lagrange function as in (3) of conventional form apart from the use of intrinsic variables] are identical to the histories that one would obtain in conventional Newtonian dynamics with, however, one important restriction, which reflects the manifestly Machian manner in which they were obtained, namely, only solutions corresponding to a zero value of the total intrinsic angular momentum of the system are obtained. (If the time is treated in a relational manner, too, only one fixed value of the total energy of the system—it can also be taken equal to zero—can be realized.) It is worth noting that in this class of Machian theories [which Bertotti and I called *intrinsic* to distinguish them from other theories of the *relative distance* type based on Lagrange functions in the form in (1)] the particles do not acquire an effective inertial mass that is position dependent and anisotropic. The properties of the inertial mass are exactly as in Newtonian theory. Thus, I think Einstein was wrong to claim that relativity of inertial mass is a necessary consequence of Mach's ideas. It is a possible consequence but depends on the particular structure of the chosen theory; for a natural class of Machian theories, it does not occur at all.

I must now show the connection between the basic idea of intrinsic dynamics and the structure of the variational principle of general relativity. To do this, it will be simpler to consider a slightly easier problem that could very easily have been posed in the 1870s and does not have to deal with the considerable complexities introduced by the problem of defining simultaneity. Let us suppose that soon after he had made his first trenchant criticisms of Newtonian dynamics in 1872, Mach had in fact solved his Machian problem in accordance with the principles of intrinsic dynamics outlined earlier. Mach might have met Clifford and learned about the revolutionary proposal that he (inspired by Riemann's work on the foundations of geometry) had made to the Cambridge Philosophical Society on February 21, 1870. Clifford said

> I hold in fact (1) that small portions of space *are* in fact of a nature analogous to little hills on a surface which is on the average flat; namely, that the ordinary laws of geometry are not valid in them. (2) That this property of being curved or distorted is continually being passed on from one portion of space to another after the manner of a wave. (3) That this variation of the curvature of space is what *really* happens in that phenomenon which we call the *motion of matter*, whether ponderable or ethereal. (4) That in the physical world nothing else takes place but this variation, subject (possibly) to the law of continuity.[1]

To this, Mach (he had not been such a reluctant theorizer) might have said the following. "If your proposal is correct, we must somehow be able to set up laws that show how little knots of curvature propagate and evolve in time. To do this, we shall need to define the rate of change with respect to the time of the curvature at a particular position in space. But, this means that we must somehow identify which point of space at the latter instant of time corresponds to which point of space at the earlier time. Now, how are you going to do that? The various points of your space at the first instant are individuated solely by the curvature relationships that at the instant hold around them (the points of space do not carry labels saying 'I am point x = 1, 4, 1'), and the points at the second instant are individuated solely by the curvature relationships around them that hold in the new space that exists in the later instant. But, apart from the fact that these two spaces differ only infinitesimally, all connection between them is severed by the fact that, in principle, the curvature relationships that are obtained in the two instants are quite different. How are you going to match up the points of the first space with those of the second in order to establish how the curvature at a given point is changing?"

To this, Clifford might have answered "I see your point. I think the only satisfactory solution will be in some really drastic generalization of the little trick you used in your intrinsic dynamics to overcome the analogous difficulty you faced in particle dynamics. However, I must admit that in my case the problems are greatly multiplied. In your case the relationships of Euclidean geometry are at least assumed to persist from one snapshot to another. This is why you can establish the intrinsic difference between the two snapshots by the simple device of sliding one snapshot around on top of the other. You are in fact using only rigid-body translations and rotations. But, I cannot do that because my geometry is changing. How can one match up the points of one curved space to those of another? We shall have to go over from rigid-body transformations to diffeomorphisms, i.e., we must start with an arbitrary mapping that associates the points of the first space in a one-to-one manner with the points in the second space, calculate on that basis the amount by which the curvature has varied between each pair of points as identified by that particular mapping, and construct some quantity analogous to your quantity in (2) by integration over the complete spaces. Then, we must change our original arbitrary mapping that establishes the connection between two spaces, doing this in a continuous manner so that bit by bit we work through all the infinitely many different ways in which the points of the first space can be identified with those of the second. In this way, we can find the tie-up between the two spaces that yields the minimal intrinsic difference between the two spaces. Then, we can define an intrinsic rate of change of the curvature and at long last get on with the dynamics."

Of course, the dialogue never took place, and general relativity was created without any paradigm of Machian dynamics for either point particles in Euclidean space or the much more ambitious Clifford–Mach dynamics of curved geometry that has just been outlined. However, several decades after Einstein had created this theory, the attempts to bring general relativity into line with quantum mechanics had forced investigators to examine very closely the basic dynamical structure for Einstein's theory. In particular, because the structure of quantum mechanics is so closely related to evolution in time of dynamical systems, it was necessary to "unpack" the Hilbert–Einstein variational principle of general relativity, which is formulated in four-dimensional terms, and establish precisely the rule by which the evolution of three-dimensional geometry occurs in that theory. This revealed that the theory in fact overcomes two very basic problems. One relates to the treatment of time and will not be considered here, since in this chapter we have concentrated almost exclusively on problems relating to position. However, the second problem turns out to be precisely the Clifford–Mach problem I have just outlined. The solution, too, is broadly in line with the proposal outlined above, as the reader can confirm by reading Wheeler's paper on the subject (Wheeler 1964). This, therefore, is the basis of my contention that general relativity does address the Machian problem in a very different way and solves it as well as one may reasonably expect.[2] However, this was long obscured by the fact that Einstein never analyzed the problem of defining motion in relational terms at a basic level. Nevertheless, although Einstein never thought of the dynamical problem in general relativity in the (3+1) terms described above, i.e., when one considers how a Riemannian three-geometry evolves in time (had he done so he would certainly have been forced to confront at a basic level the Machian problems outlined here), it does appear that this approach requiring general covariance in four-dimensional space-time achieved the desired result.[3]

In conclusion, let us look again at Einstein's final formulation of Mach's principle in 1918: "The G-field is *completely* determined by the masses of the bodies. Since mass and energy are identical in accordance with the results of the special theory of relativity and energy is described formally by means of the symmetric energy tensor ($T_{\mu\nu}$), the G-field is conditioned and determined by the energy tensor of the matter."

There are two problems with this formulation. The first is that Einstein is extremely vague about what he means by saying that the G-field is conditioned and determined by the energy tensor of the matter. He seems to imply that one can somehow specify an energy-momentum tensor (presumably in the whole space-time) and then from it, via the field equations, determine the space-time geometry. But, specification of an energy-momentum tensor without simultaneous specification of the space-time geometry is in fact impossible (and actually meaningless) in the framework of relativity. This therefore is a first difficulty with Einstein's 1918 formulation of Mach's principle. Attempts have been made to make Einstein's formulation more precise by a Green's function formulation of the theory. This work has been reviewed by Raine, who gives the necessary references (Raine 1981). Apart from the very considerable mathematical difficulties that arise, I am rather skeptical about such an approach because it seems to want to deny an integral part of general relativity, namely, that dynamical role accorded to space. Let us recall that Mach's original criticism was of Newtonian mechanics, which assumed a fixed structure of space. A priori it is difficult to see why space should not be "unfrozen" and have dynamical degrees of freedom like ordinary matter. Indeed, we have seen how Clifford wanted to abolish all matter and use only dynamical space (actually making his proposal before Mach published his first criticism of Newton). A point that should be made here is that the fundamental problem of motion (as formulated in the text) arises in any dynamical theory whenever everything that constitutes the universe—be it material particles in Euclidean space, fields in space (flat and nondynamical or curved and dynamical), or simply pure space as conjectured by Clifford—varies between successive time slices. This, therefore, is the second difficulty I see with Einstein's 1918 formulation: it rather unnaturally puts the entire responsibility for the determination of local inertial properties (represented by values of the metric tensor realized at and around a given space-time point) on the shoulders of matter alone, whereas the theory is actually about dynamical interaction and space and matter, which have their own independent dynamical degrees of freedom. This is particularly evident when one looks at the structure of general relativity from the point of view of the initial-value problem.

The importance of the initial-value problem in dynamical theory has been stressed several times in this chapter. The lack of agreement about the significance of general covariance in the framework of general relativity has also been mentioned. My impression is that most of the discussion on this last topic has been conducted at the level of general epistemological arguments. However, Poincaré's analysis of the initial-value problem in Newtonian dynamics showed that the important distinction between rival theories may hinge on comparatively subtle questions relating to the ability of any given theory to predict the future uniquely *from observable initial data.* I believe that the significance of general covariance as actually used by Einstein in the general theory of relativity should be examined from this point of view (see Barbour 1986).

Notes

[1] Quoted from Misner, Thorne, and Wheeler 1973, p. 1202.

[2] It should be said that the problem of establishing precisely what does happen in the framework of the Hilbert–Einstein variational principle is far from easy, and Wheeler, working with Isenberg, has somewhat modified his position compared with his initial analysis (see Isenberg and Wheeler 1979). However, I believe that in broad terms my assertion is correct. The qualification "as well a one may reasonably expect" is added to take account of the problems associated with the existence and uniqueness of the original "thin-sandwich" conjecture of Wheeler (see Wheeler 1964).

[3] We should note that hitherto in this part of the discussion we have considered only pure geometrodynamics, i.e., the case of an evolving three-geometry when no matter at all is present. In the more general case when matter fields are present (for the sake of simplicity let us take the case of a single scalar field Φ), the snapshots in the Clifford–Einstein scenario will show not only a curved three-geometry but also the intensity of the field Φ at each point on it. A procedure is now required to define the intrinsic variation of both Φ and the geometry; in

this case, both Φ and the geometry contribute to the quantity that generalizes (2) for this case.

REFERENCES

Barbour, Julian B. (1986). "Leibnizian Time, Machian Dynamics and Quantum Gravity." In *Quantum Concepts in Space and Time.* Roger Penrose and Christopher J. Isham, eds. London and New York: Oxford University Press, pp. 236–246.

——— (1989). *Absolute or Relative Motion?* Vol. 1. *The Discovery of Dynamics.* Vol. 2. *The Frame of the World* (in preparation) London and New York: Cambridge University Press.

——— (1990). "The Part Played by Mach's Principle in the Genesis of Relativistic Cosmology." In *Modern Cosmology in Retrospect.* Silvio Bergia and Bruno Bertotti, eds. London and New York: Cambridge University Press, pp. 47–66.

Barbour, Julian B., and Bertotti, Bruno (1977). "Gravity and Inertia in a Machian Framework." *Il Nuovo Cimento B* 38: 1–27.

——— (1982). "Mach's Principle and the Structure of Dynamical Theories." *Royal Society of London. Proceedings A* 382: 295–306.

Earman, John, and Glymour, Clark (1978). "Lost in the Tensors: Einstein's Struggles with Covariance Principles 1912–1916." *Studies in History and Philosophy of Science* 9: 251–278.

Einstein, Albert (1907). "Über das Relativitätsprinzip und die aus demselben gezogenen Folgerungen." *Jahrbuch der Radioaktivität und Elektronik* 4: 411–462.

——— (1911). "Über den Einfluss der Schwerkraft auf die Ausbreitung des Lichtes." *Annalen der Physik* 35: 898–908.

——— (1912). "Gibt es eine Gravitationswirkung die der elektrodynamischen Induktionswirkung analog ist?" *Vierteljahrsschrift für gerichtliche Medizin und öffentliches Sanitätswesen* 44: 37–40.

——— (1913). "Physikalische Grundlagen einer Gravitationstheorie." *Naturforschende Gesellschaft Zürich. Vierteljahrsschrift* 58: 284–290.

——— (1914a). "Zur Theorie der Gravitation." *Naturforschende Gesellschaft Zürich. Sitzungsberichte*: iv–vi.

——— (1914b). "Zum Relativitätsproblem." *Scientia* 15: 337–348.

——— (1914c). "Die formale Grundlage der allgemeinen Relativitätstheorie." *Königlich Preussische Akademie der Wissenschaften* (Berlin). *Sitzungsberichte*: 1030–1085.

——— (1916). "Die Grundlage der allgemeinen Relativitätstheorie." *Annalen der Physik* 49: 769–822. Quotations are from the English translation: "The Foundation of the General Theory of Relativity." In Lorentz et al. 1923, pp. 109–164.

——— (1917). "Kosmologische Betrachtungen zur allgemeinen Relativitätstheorie." *Königlich Preussische Akademie der Wissenschaften* (Berlin). *Sitzungsberichte*: 142–152. Quotations are from the English translation: "Cosmological Considerations on the General Theory of Relativity." In Lorentz, et al. 1923, pp. 175–188.

——— (1918a). "Prinzipielles zur allgemeinen Relativitätstheorie." *Annalen der Physik* 55: 241–244.

——— (1918b). "Dialog über Einwande gegen die Relativitätstheorie." *Die Naturwissenschaften* 6: 697–702.

Einstein, Albert, and Grossmann, Marcel (1913). "Entwurf einer verallgemeinerten Relativitätstheorie und einer Theorie der Gravitation." *Zeitschrift fur Mathematik und Physik* 62: 225–261.

Föppl, August (1904a). "Über absolute und relative Bewegung." *Bayerische Akademie der Wissenschaften. Mathematisch-physikalische Klasse. Sitzungsberichte* 34: 383–395.

——— (1904b). "Über einen Kreiselversuch zur Messung der Umdrehungsgeschwindigkeit der Erde." *Bayerische Akademie der Wissenschaften.Mathematisch-physikalische Klasse. Sitzungsberichte* 34: 5–28.

Friedländer, B., and Friedländer J. (1896). *Absolute und relative Bewegung.* Berlin.

Isenberg, James A., and Wheeler, John A. (1979). "Inertia Here is Fixed by Mass-Energy There in Every *W* Model-Universe." In *Relativity, Quanta, and Cosmology in the Developpment of the Scientific Thought of Albert Einstein.* Vol. 1. M. Pantaleo and F. de Finis, eds. New York: Johnson Reprint Corporation, pp. 267–293.

Kretschmann, Erich (1917). "Über den physikalischen Sinn der Relativitätspostulate, A. Einsteins neue und seine ursprüngliche Relativitätstheorie." *Annalen der Physik* 53: 575–614.

Lange, Ludwig (1885). "Über das Beharrungsgesetz." *Königlich Sächsischen Gesellschaft der Wissenschaften zu Leipzig. Mathematisch-physische Klasse. Berichte über die Verhandlungen* 37: 333–351.

——— (1886). *Die geschichtliche Entwicklung des Bewegungsbegriffs und ihr voraussichtliches Endergebnis.* Leipzig: W. Engelmann.

Lorentz, Hendrik Antoon, et al. (1923). *The Principle of Relativity: A Collection of Original Memoirs on the Special and General Theory of Relativity.* Arnold Sommerfeld, ed. W. Perrett and G. B. Jeffery, trans. London: Methuen; reprint New York: Dover, 1952.

Mach, Ernst (1872). *Die Geschichte und die Wurzel des Satzes von der Erhaltung der Arbeit.* Prague: Calve. The quoted passage are from the English translation: *History and Root of the Principle of the Conservation of Energy.* Philip E. B. Jourdain, trans. Chicago: Open Court, 1911.

——— (1883). *Die Mechanik in ihrer Entwickelung, historisch-kritisch dargestellt* Leipzig: Brockhaus. Page numbers cited from the English translation *The Science of Mechanics: A Critical and Historical Account of Its Develop-*

ment, 6th ed. T. J. MacCormack, trans. La Salle, Illinois: Open Court, 1960.

Misner, Charles W., Thorne, Kip S., and Wheeler, John Archibald (1973). *Gravitation.* San Fransisco, California: Freeman.

Newton, Isaac (1687). *Philosophiae Naturalis Principia Mathematica.* London.

Norton, John (1984). "How Einstein Found His Field Equations: 1912–1915." *Historical Studies in the Physical Sciences* 14: 253–316.

Poincaré, Henri (1902). *La science et l'hypothèse.* Paris: Flammarion.

Raine, Derek (1981). "Mach's Principle and Space-Time Structure." *Report on Progress in Physics* 44: 1151–1195.

Stachel, John (1980). "Einstein and the Rigidly Rotating Disk." In *General Relativity and Gravitation: A Hundred Years after the Birth of Einstein.* Alan Held, ed. New York: Plenum, pp. 1–15.

——— (1986). "What a Physicist Can Learn from the History of Einstein's Discovery of General Relativity." In *Proceedings of the Fourth Marcel Grossmann Meeting on General Relativity.* Remo Ruffini, ed. Amsterdam: Elsevier, pp. 1857–1862.

Wheeler, John A. (1964). "Geometrodynamics and the Issue of the Final State." In *Relativity, Groups and Topology.* Lectures Delivered at les Houches during the 1963 Session of the Summer School of Theoretical Physics, University of Grenoble. Cecile DeWitt, and Bryce S. DeWitt, eds. New York and London: Gordon and Breach, pp. 314–520.

Einstein and *Eindeutigkeit*: A Neglected Theme in the Philosophical Background to General Relativity

Don Howard

1. Introduction

Over the past ten years, historians and philosophers of science interested in the general theory of relativity have devoted considerable attention to Einstein's struggle with the "hole argument" or "*Lochbetrachtung.*"[1] The basic story is now well enough known not to require repeating in detail here: Sometime late in 1913, Einstein stumbled upon the hole argument, relieved to find what he then believed to be a fundamental argument legitimating his reluctant acceptance of less than generally-covariant field equations for gravitation at the time of his "Entwurf" theory (Einstein and Grossmann 1913).[2] But when he returned to generally-covariant field equations in November 1915, for reasons having little to do with the hole argument, he was faced with the embarrassment of the hole argument's presenting still a seemingly valid objection to the admissibility of such equations. This embarrassment was relieved by Einstein's development, also apparently in November 1915, of what Stachel (1980) has dubbed the "point-coincidence argument," in which the central mistake in the hole argument is rectified.

Many important physical and mathematical questions raised by the hole and point-coincidence arguments have been extensively investigated (which is not to say that all of these questions have been answered), and will no doubt continue to be studied.[3] Likewise, some of the metaphysical implications for our view of space-time are now being investigated.[4]

But relatively far less attention has been paid to the equally important methodological and epistemological issues at stake in both arguments. My aim in this paper is to begin the exploration of these issues by sketching the background, and to some small extent the later ramifications of what has long struck me as the most interesting methodological issue arising in the two arguments, namely, Einstein's assumption—an assumption essential to both arguments—that a theory of gravitation is acceptable only if its field equations determine "completely" (*vollständig*) or "uniquely" (*eindeutig*) a solution for the gravitational field in all regions of the space-time manifold, most importantly, in "holes," regions of space-time devoid of matter and energy, or what Einstein called "material processes" (see, for example, Einstein 1914, p. 1067). This assumption may be regarded as a special instance of a more general methodological principle to the effect that a theory is acceptable only if it is complete in the sense of determining for itself a unique model of the phenomena it aims to represent, that is, only if it is—permit me a deliberate, and pointed anachronism, and a momentary inaccuracy—a categorical theory. Let us call this, provisionally, the *Eindeutigkeit* principle.

If one speaks thus anachronistically, then one is led immediately to assume that the relevant background is to be found in the turn-of-the-century literature on the foundations of geometry, primarily in the work of Hilbert (1899), Huntington (1902), and Veblen (1904), in which the modern concepts of the completeness and categoricity of a formal system first began to find expression. One thinks as well about other developments in turn-of-the-century mathematics, such as the theory of invariants and Klein's Erlanger program, as well as the first assaults on the Cauchy problem. But without denying the relevance of this literature, I want to suggest that we should look elsewhere for the background pertinent to Einstein's invocation of the *Eindeutigkeit* principle.

The main reason for looking elsewhere is that, the anticipations of Hilbert, Huntington, and Veblen notwithstanding, the concept of categoricity that Einstein seems to be invoking was not to be articulated clearly in the foundations of mathematics literature until Rudolf Carnap (1927), Adolf Fraenkel (1928), H.G. Forder (1927), Hermann Weyl (1927), and others began exploring the topic in the 1920s,[5] which is to say that Einstein could not have found the concept of categoricity ready-formed in the mathematics literature by the years 1913–1915, when he was wrangling with the hole and point-coincidence arguments. And, in any case, the *Eindeutigkeit* idiom in which Einstein writes was not the idiom employed at this time in the foundations of geometry literature. Moreover, there is tantalizing circumstantial evidence suggesting that the order of influence was the opposite of what one might suspect, that is to say that Einstein's hole and point-coincidence arguments may have promoted the articulation of the categoricity concept, in the work of Carnap, Fraenkel, Weyl, and others, by way of its being disentangled from other, related conceptions of completeness.[6] Similarly, some of the most important work on the Cauchy problem was actually stimulated by the way in which the hole and point-coincidence arguments focused attention on the question of whether or not the field equations plus boundary conditions determine unique solutions, or a unique class of isomorphic solutions.[7]

Where then should we look for the pertinent background to the *Eindeutigkeit* principle? I think that one must look to the work of several perhaps unlikely individuals, including Joseph Petzoldt, Ernst Mach, Paul Volkmann, Paul Natorp, Ernst Cassirer, and Moritz Schlick. As we shall see, by 1913 Petzoldt, Natorp, Cassirer, and Schlick had all published important works defending points of view strongly suggestive of the kind of principle that Einstein invokes, Petzoldt (1895) going so far as to dub it "the law of *Eindeutigkeit*" ["das Gesetz der Eindeutigkeit"]. Mach had endorsed several of Petzoldt's ideas in his writings after 1897, and Volkmann had published a series of works in which the demand for *Eindeutigkeit* is criticized.

In other words, at the time Einstein introduced the hole and point-coincidence arguments, featuring prominently the *Eindeutigkeit* principle, he was working in the context of a rich and long-running *philosophical* debate about the methodological and epistemological significance and justification of that principle. Far from the principle being merely a mathematical common-place, as it may appear to modern readers, it was in the 1910s a controversial topic, the subject of deep and probing discussion.

There is at least circumstantial evidence for Einstein's acquaintance with this philosophical discussion at the time when he was struggling with the hole and point-coincidence arguments, evidence that is strongest in the case of Petzoldt's writings. And the vocabulary that Einstein employs in discussing the *Eindeutigkeit* principle recalls that employed in the philosophical literature. But even if Einstein had not previously been acquainted with the philosophical debate over *Eindeutigkeit*, even if he formulated the *Eindeutigkeit* principle independently of that earlier discussion, the convergence of opinion, at least as between Einstein and Schlick, was such that when Einstein did read Schlick's first major paper on relativity theory in December 1915, within a few weeks of completing work on the final formulation of general relativity, he would have found developed there a point of view on the relation between theory and reality that coincided nicely with that implicit in the hole and point-coincidence arguments. This circumstance helps to explain both Einstein's subsequent extremely high regard for Schlick's interpretation of the philosophical implications of relativity (through at least the mid-1920s) and the degree to which Schlick's philosophy of science influenced Einstein's thinking.

Another dividend to be derived from recognition of the relevance of this background to Einstein's thinking is an explanation of the puzzling fact that Einstein and others often referred to the problem raised in the hole argument as a "causality" problem, meaning that the hole argument implied the failure of generally-covariant theories to satisfy some kind of causality principle. This fact is puzzling because, on the face of it, the hole argument seems to have nothing to do with what is customarily designated a "causality" principle. As we shall see, however, the concept of *Eindeutigkeit* was originally introduced by Petzoldt in the course of an attempt to refine Mach's conception of causality and, thereby, to defend it from criticisms advanced by Wilhelm Wundt. It thus becomes clear how discussions of *Eindeutigkeit* became entangled with discussions of causality.

There is also an interesting history of the *Eindeutigkeit* principle after 1915. It seems that Einstein's invocation of the principle stimulated a new round of discussions, focusing in large part on the problem of *Eindeutigkeit* in relativity, discussions that drew in philosophers as diverse as Petzoldt (1921b, 1921c, 1923), Ernst Cassirer (1921), Rudolf Carnap (1922), and Max Faerber (1923). Among other consequences, this post-1915 discussion of *Eindeutigkeit* is what led ultimately, as suggested above, to the clarification of the modern concept of categoricity, especially considering Carnap's role in the story. A brief outline of this later history will be given in the concluding section of this paper, but for reasons of space a fuller discussion will have to be postponed.

An important historiographical lesson should also be drawn from the study of the history of the *Eindeutigkeit* principle, a lesson having little to do with Einstein's work itself. The *Eindeutigkeit* principle turns out to have been the focus of considerable debate among physicists and philosophers of science during the last decade of the nineteenth century and the first three decades of the twentieth. It excited the interest of thinkers as diverse as Mach and Cassirer, and it apparently had a considerable influence on the work of a thinker of Einstein's caliber. And yet the *Eindeutigkeit* principle and the discussions it provoked are now virtually forgotten by contemporary philosophers of science as well as by the few contemporary historians of the philosophy of science.[8] The lesson is a simple one, but no less valuable for its simplicity: it is that there is more to the history of the philosophy of science in the nineteenth and twentieth centuries than the history of the Vienna Circle. We have tended to read the history of the philosophy of science from the point of view of the problems and interests that typified the Vienna Circle and its allies in Berlin, Prague, and elsewhere; and we have done this because of the historical accident that it was the views of this group that largely shaped mid-twentieth century philosophy of science. Thus, our histories emphasize thinkers like Wittgenstein, Russell, and Frege, when in fact many of those thinkers were in their day on the periphery. Much of the excitement lay elsewhere, certainly in the eyes of thinkers like Einstein, and an adequate history of the philosophy of science should reflect that fact.

2. The Many Faces of *Eindeutigkeit*

Before we proceed, it will be helpful to distinguish in a preliminary way the different senses in which the concept of *Eindeutigkeit* will appear in the following discussion, all of them rather more technical than the mundane *Eindeutigkeit* of words that are unambiguous (not *mehrdeutig* or *vieldeutig*).

The principle that Einstein invokes in his hole and point-coincidence arguments, the principle that looks very much like a requirement that an acceptable theory be categorical, may be termed, for want of a better name, the "model-theoretic *Eindeutigkeit* principle." It is, again to dress the point up in more modern guise, the demand that a theory uniquely (*eindeutig*) determine its model. Or as Einstein and his contemporaries would more likely have said, since they did not speak the language of modern formal semantics, the demand that a theory determine uniquely that which it aims to describe, whether this be conceived of as some presumably real state of affairs, an aspect of an external physical reality, or as merely the face that the world presents to us in experience.

A second sense of *Eindeutigkeit* appears when one seeks a justification for the model-theoretic *Eindeutigkeit* principle, for the most tempting argument for the latter is that acceptable theories should determine their own models uniquely (*eindeutig*) because the world they aim to describe—whether this be the realist's real external world or the phenomenalist's world as it is presented in experience—is unique. The claim that the world described by our theories is thus unique may be termed the "metaphysical *Eindeutigkeit* principle." I do not regard this as a good argument for the model-theoretic *Eindeutigkeit* principle; it is, after all, much like the silly argument that simple theories are preferable to complex ones because reality is simple, but I suspect that it may have been the argument in the back of Einstein's mind when he deployed the model-theoretic *Eindeutigkeit* principle. A somewhat better argument for the metaphysical *Eindeutigkeit* principle (better if only because it is not obviously a silly argument) is based upon the Kantian (or neo-Kantian) thesis that *Eindeutigkeit* is a necessary feature of empirical concepts as concepts and of the grounds for the possibility of experience (the experience itself being possibly *mehrdeutig*). But I find this argument no more persuasive.

A third sense of *Eindeutigkeit* is involved in the question—in a sense complementary to the categoricity question—whether or not our choice of one among a set of competing theories is determined uniquely (*eindeutig*), either by reality itself, or by experience, in the form of experimental evidence (which one may or may not regard as the window through which an external reality manifests itself to us), or by some other constraints, like the Kantian categories, a biological imperative, or a nonempirical criterion of theory choice like simplicity. The claim that our choice of a theory is or should be determined uniquely by one or a combination of these factors—a claim questioned by the Duhemian, underdeterminationist version of the conventionalist thesis—may be termed the "epistemological *Eindeutigkeit* principle." We will encounter this principle below, but it will not be our primary focus.

At least the first and third senses of *Eindeutigkeit* play an important role in Einstein's thinking on questions of scientific method, Einstein endorsing the first and repudiating (in a very interesting way) the third. And, as noted, I suspect that the second sense is also there implicitly.

One final clarification is in order. It is important to recognize that when Einstein invokes the model-theoretic *Eindeutigkeit* principle in his work on gravitation, he fails to distinguish clearly between the demand that a theory determine exactly one model for itself and the demand that a theory determine at most a class of isomorphic models. But his *failure* to do this should not be seen as a *failing*, because the distinction was not clear even to mathematicians of the status of Hilbert at this time. Indeed, it was only in 1916 and in direct response to the challenge of the hole and point-coincidence arguments that Hilbert first reformulated the Cauchy problem for the gravitational field equations as the question whether or not those equations (plus boundary conditions) determine a unique class of isomorphic solutions, rather than a unique solution.[9] And it was again only in the 1920s that our modern understanding of categoricity as a matter of a theory's determining a class of isomorphic models was developed clearly in the work of Rudolf Carnap, Adolf Fraenkel, Hermann Weyl, and others (it was Carnap [1927] who introduced the apt expression, "monomorphic theories"). Here too, as I noted above, there is evidence that Einstein's invocation of the model-theoretic *Eindeutigkeit* principle contributed materially to this clarification.

3. A Review of the Hole Argument from a Methodological Point of View

Before we turn to look at the relevant works of Petzoldt, Volkmann, Natorp, Cassirer, and Schlick, let us recall the essential features of the hole and point-coincidence arguments, emphasizing those methodological features—in particular the model-theoretic *Eindeutigkeit* principle—that will be of central concern in the subsequent discussion.

In Einstein's clearest formulation of it, the hole argument proceeds as follows.[10] Consider a region of space-time, Σ, devoid of matter and energy and a set of generally-covariant field equations valid for the entire space-time manifold, both inside and outside of Σ. Given a coordinatization of the manifold, K, then the physical events ("das physikalische Geschehen") within Σ will be completely determined ("vollständig bestimmt") by the field equations, according to Einstein, if there exists a solution of those equations giving the quantities, $g_{\mu\nu}$ (the components of the metric tensor), as functions of the coordinates x_ν. We let the totality of these functions be represented by the expression $G(x)$. Consider now a second coordinatization, K', that coincides with K everywhere outside of and on the boundary of Σ, but differs from K inside of Σ, and consider, finally, the *same* functions, $G(x)$, only expressed now in terms of the new coordinates x'_ν, which we represent by $G'(x')$. We assume that the $g_{\mu\nu}$ and $g'_{\mu\nu}$ are all continuously differentiable functions of their respective coordinates, x_ν and x'_ν. It is important to remember, as Einstein notes, that $G(x)$ and $G'(x')$ are just different representations of the same gravitational field ("das nämliche Gravitationsfeld"), the physical reality that we are aiming to describe.

Next comes the construction central to the argument. In each of the functions $g'_{\mu\nu}(x'_\nu)$, replace the x'_ν by the numerically equivalent values x_ν, the totality of the resulting functions being represented by the expression, $G'(x)$.[11] Einstein then argued—and here is where he made his mistake, as he himself later noted—that if the field equations are generally covariant, the $G'(x')$ will constitute a solution of those equations if the $G(x)$ did (they are, as noted above, different expressions of the same solution),[12] and that, consequently, the $G'(x)$ are equally well a solution of the field equations, even though, in general, they differ from the $G(x)$ within the "hole," Σ. But the two different solutions, $G'(x)$ and $G(x)$, correspond to the same "reality," which in this case means the same distribution of matter and energy as represented by the stress-energy tensor $T_{\mu\nu}$.[13]

Einstein's reasoning in drawing this mistaken conclusion need not detain us.[14] What is important for our purposes is the further implication that Einstein drew, which is that because generally-covariant field equations admit non-equivalent solutions for events within Σ, such equations are therefore unacceptable as a satisfactory physical theory of gravitation. He went on to argue that we must therefore adopt a theory with restricted covariance properties, so as to block the coordinate transformation leading to the construction of the "different" solution $G'(x)$.

But why would the existence of non-equivalent solutions to the field equations render those equations unacceptable? Here is how Einstein presented the argument in his last major published statement of it:

> *Events* [das Geschehen] *in the gravitational field cannot be determined uniquely* [eindeutig festgelegt] *by means of generally-covariant differential equations for the gravitational field.*
>
> If we demand, therefore, that the course of events [der Ablauf des Geschehens] in the gravitational field be completely determined [vollständig bestimmt] by means of the laws that are to be established, then we are obliged to restrict the choice of the coordinate system. (Einstein 1914, p. 1067)[15]

The crucial assumption—what I term the model-theoretic *Eindeutigkeit* principle—is similarly stated in all versions of the argument, in published papers as well as in correspondence, starting from the earliest mention of the argument in the fall of 1913, through the above-quoted version from late 1914. For example, writing to Paul Ehrenfest in what was probably the late fall of 1913, Einstein noted:

> The questions regarding the theory of gravitation that were still undecided in the summer have clarified themselves in the meantime. A unique determination [eindeutige Bestimmung] of the $g_{\mu\nu}$ out of the $T_{\mu\nu}$ is possible only with the choice of special coordinate systems (rigorously provable). (EA 09-344)[16]

And writing to Ludwig Hopf on 2 November 1913, he reported:

> I am now quite satisfied with the theory of gravitation. The fact that the equations of gravitation are not generally covariant, which troubled me inordinately a little while ago, has turned out to be unavoidable; it can easily be proven that a theory with generally-covariant equations cannot exist in case it is demanded that the field be mathematically completely determined [vollständig bestimmt] by the matter. (EA 13-290)[17]

More or less the same formulation is found in other discussions of the argument.[18]

Thus the hole argument. The point-coincidence argument was introduced by Einstein late in 1915, when, after having returned to generally-covariant field equations for other reasons, he then had to find a way around the embarrassment of the hole argument, which he had presented for two years as a fundamental argument *against* the possibility of generally-covariant field equations. His strategy was to block the central move in the hole argument, the construction of the "different" solution $G'(x)$, by denying that the substitution of numerically equivalent values of the x_ν for the x'_ν in the functions $g'_{\mu\nu}(x'_\nu)$ results in the assignment of the *same* values g' to *different* points of the space-time manifold (and hence, in general, the assignment of different values g and g' to the same point). A crucial assumption behind the hole argument had been that a mere coordinatization of the manifold yields an individuation of the points of the manifold, in such a way that the numerically equivalent x_ν and x'_ν label *different* points. The point-coincidence argument maintains, instead, that a coordinatization is itself not sufficient to determine an individuation of the points of the manifold.[19] Einstein now argues that the points of the manifold are implicitly defined and thus individuated only as points of intersection or coincidence of world lines, and that since the coincidences are themselves determined by the $g_{\mu\nu}$, it is impossible to have two *different* sets of values of the functions $g_{\mu\nu}$ assigned *to one and the same point* of the manifold. (It is, of course, possible to have "different" solutions of the field equations, $g_{\mu\nu}(x_\nu)$ and $g'_{\mu\nu}(x'_\nu)$, relative to different coordinatizations; but then these are still just different representations of one and the same gravitational field, as Einstein had pointed out already in 1914.) Why are points to be individuated only in this fashion? It is because, as Einstein repeatedly says: (a) only intersections of world lines—and certainly not coordinate values—are invariant under arbitrary, continuous coordinate transformations, the class of transformations under which the field equations themselves are also to be invariant; and (b) the observations by means of which we test the predictions of any physical theory, general relativity included, consist, in principle, of just such intersections or coincidences. The point-coincidence argument thus provides the needed correction to the hole argument, the needed license for generally-covariant field equations. But while it corrects the principal error in the hole argument, it does so by appealing implicitly to the very same model-theoretic *Eindeutigkeit* principle that lay behind the hole argument; for another way to formulate the central move in the point-coincidence argument is to say that the points of the manifold are to be individuated as the points of intersection of world lines because these are determined uniquely—*eindeutig*—by the requirement of general covariance, in the sense that they are invariant under arbitrary, continuous coordinate transformations (whereas points individuated merely by coordinatizations lack this property of the invariance of identity under coordinate transformations), and that, therefore, the solutions of generally-covariant field equations, even in holes, are necessarily unique (*eindeutig*). In other words, if the elements of the ontology out of which models of our theory of gravitation are constructed are themselves individuated in the manner naturally suggested by the requirement of general covariance, then a gravitational theory incorporating generally-covariant field equations will be *eindeutig*.[20]

It is perhaps in his correspondence with Ehrenfest from December of 1915 and January of 1916 that Einstein makes this point most clearly. In the first of a series of three letters on the subject, this dated 26 December 1915, Einstein writes (referring to Einstein 1914, §12):

> In §12 of my work of last year, everything is correct (in the first three paragraphs) up to that which is printed with emphasis at the end of the third paragraph. From the fact that the two systems *G(x)* and *G′(x)*, referred to the same reference system [the same *x*], satisfy the conditions of the grav. field, no contradiction follows with the uniqueness of events [Eindeutigkeit des Geschehens]. That which was apparently compelling in these reflections founders immediately, if one considers that 1) the reference system signifies nothing real, 2) that the (simultaneous) realization of two different *g*-systems (or better, two different fields) in the same region of the continuum is impossible according to the nature of the theory.
>
> In place of §12, the following reflections must appear. The physically real in the world of events [Weltgeschehen] (in contrast to that which is dependent upon the choice of a reference system) consists *in spatio-temporal coincidences.** Real are, e.g., the intersections of two different world lines, or the statement that they *do not* intersect. Those statements which refer to the physically real therefore do not founder on any univocal [eindeutige] coordinate transformation. If two systems of the $g_{\mu\nu}$ (or in general the variables employed in the description of the world) are so created that one can obtain the second from the first through mere space-time transformation, then they are completely equivalent [gleichbedeutend]. For they have all spatio-temporal point coincidences in common, i.e., everything that is observable.
>
> These reflections show at the same time how natural the demand for general covariance is.
>
> * and in nothing else! (EA 9-363)[21]

This passage is interesting in part because the vocabulary that Einstein here employs—“Eindeutigkeit des Geschehens”—turns out to be characteristic of the philosophical literature to be explored below, and not characteristic of the literature on the foundations of geometry where the concept of categoricity was first introduced.

From a physical and metaphysical point of view, however, what is important about remarks like this are the ontological implications that Einstein draws: What are real are all and only the points determined as intersections of world lines, because these and the $g_{\mu\nu}$ associated with them are invariant under arbitrary coordinate transformations and hence determined uniquely—*eindeutig*. In other places, Einstein made the same point in a different way, arguing that the principal lesson of the point-coincidence argument is that "time and space thereby lose the last vestige of physical reality" (Einstein to Schlick, 14 December 1915, EA 21-611),[22] the point being that, since spatio-temporal coordinates are themselves not invariant under coordinate transformations, they cannot correspond to anything real. Some commentators, like Schlick, understood clearly what Einstein meant, themselves presenting this as the principal lesson of general relativity with respect to traditional philosophical conceptions of space and time.[23] Others, not appreciating the relevance of the *Eindeutigkeit* principle, have glossed such remarks as implying merely that space and time lose their separate reality, space-time being the only reality (a point made already several years before by Minkowski about the ontological implications of special relativity), not realizing that Einstein is posing much more fundamental questions about the manner in which the points of the space-time manifold are individuated.

A failure to appreciate the role of the *Eindeutigkeit* principle has contributed to other misunderstandings of Einstein. Consider this statement of the point-coincidence argument one week after the letter to Ehrenfest, this in a letter to Michele Besso of 3 January 1916:

> In the *Lochbetrachtung*, everything was correct up to the final conclusion. There is no physical content in the existence of two different solutions *G(x)* and *G′(x)* with reference to the same coordinate system *K*. Attributing two different solutions to the same manifold is senseless, and the system *K* has, indeed, no physical reality. The following consideration takes the place of the *Lochbetrachtung*. Physically *real* is nothing but the totality of spatio-temporal point-coincidences. If, e.g., physical events [das physikalische Geschehen] were to be built up solely out of the movements of material points, then the meetings of the points, i.e., the points of intersection of their world lines, would be the only reality, i.e., observable in principle. These points of intersection are naturally preserved under all transformations (and no new ones are added), if only certain uniqueness conditions [Eindeutigkeitsbedingungen] are maintained. Thus, it is most natural to demand of the laws that they do not determine *more* than the totality of the spatio-temporal coincidences. According to what has been said, this is already

> achieved by generally-covariant equations. (EA 7-272; Speziali 1972, pp. 63-64)[24]

Those who want to find in Einstein anticipations of ideas later to become important in Vienna positivism read remarks about the points of intersection of world lines being "the only reality, i.e., observable in principle" as implying some kind of reductionist phenomenalism, or at least anticipations of a verificationist theory of meaning.[25] But this is to get the order wrong. It is not that observability becomes Einstein's criterion of reality, with coincidences being real because, being invariant, they are observable. It is instead the *Eindeutigkeit* principle that yields, in effect, a criterion of ontic commitment for a theory: what is real for the theory is what is determined unambiguously as part of a model for the theory. A necessary condition for something's being observable may still be that it have the appropriate invariance properties. But that is not to say that a theory's ontology consists exclusively of observable elements.

To summarize then, the main point to remember for now concerning the hole argument and the point-coincidence argument is that in both the most important methodological principle that Einstein invokes is the model-theoretic *Eindeutigkeit* principle. In 1912, at the time of the hole argument, he faulted a gravitational theory built upon generally-covariant field equations for failing to satisfy this requirement; in 1915, at the time of the point-coincidence argument, he commends essentially the same theory because, as he then understood, it does satisfy the requirement. But in both cases the requirement is the same: an acceptable theory must determine uniquely—*eindeutig*—its own model.

The question then is: Where did Einstein get the *Eindeutigkeit* idea? As I noted in the introduction (and I will have more to say about this below), it would be a mistake to look where our instincts might tell us to look, namely, to the literature on formal axiomatics and the foundations of geometry, for the concept of categoricity was not prominent in that literature in the early 1910s, not being itself clearly defined and not being clearly distinguished from other completeness concepts until the 1920s. I want to argue instead that the relevant background to Einstein's invocation of the model-theoretic *Eindeutigkeit* principle is a philosophical tradition largely independent of the related contemporary work in geometry and axiomatics, a philosophical tradition that emerges in the late-nineteenth century and becomes by the second decade of our century a prominent theme in the philosophy of science literature, being represented in the writings of thinkers as various as the positivist Joseph Petzoldt, the theoretical physicist Paul Volkmann, the neo-Kantians Paul

Natorp and Ernst Cassirer, and the then realist Moritz Schlick. It is from this source, I believe, that Einstein drew his ideas and his vocabulary.

4. Petzoldt and the Beginning of the *Eindeutigkeit* Discussion

Let us begin with the metaphysical *Eindeutigkeit* principle. As far as I have been able to determine, explicit discussion of this principle in the turn-of-the-century philosophy of science literature seems to originate with Joseph Petzoldt's seminal 1895 paper, "Das Gesetz der Eindeutigkeit" (Petzoldt 1895), which was published in the *Vierteljahrsschrift für wissenschaftliche Philosophie und Soziologie*, the journal founded in 1877 by Petzoldt's mentor, the leader of Zurich positivism, Richard Avenarius. Though his work is little known today, Petzoldt was an important, if not major, figure in the emergent philosophy of science of the late-nineteenth and early twentieth centuries, then chiefly known for his allegiance with Mach and Avenarius (who were both personal friends of his),[26] for his defense of his own variation on their doctrines, which he termed "relativistic positivism,"[27] and for a long series of writings exploring the philosophical implications of relativity theory (not always with complete comprehension of the technical details).[28] Though he was the founder in 1912 of the Gesellsschaft für positivistische Philosophie,[29] Petzoldt never enjoyed a high professional standing, teaching for many years at a Gymnasium in Berlin-Spandau; it was only in 1904 that he acquired a university position, as Privatdozent at the Technische Hochschule Berlin-Charlottenburg, being promoted to Ausserordentlicher Professor in 1922.[30]

Petzoldt's "Das Gesetz der Eindeutigkeit" develops a theme first introduced in his often-cited Inaugural Dissertation, prepared under the psychologist Georg Elias Müller at Göttingen, "Maxima, Minima und Ökonomie" (Petzoldt 1890), in which he had criticized the biological-evolutionary economism of Mach and Avenarius, proposing a principle of "uniqueness" ["Eindeutigkeit"] and a tendency to "stability," as replacements for the Mach-Avenarius principles of economy. The Inaugural Dissertation concluded with these words: "*Not maxima, minima, and economy, but rather uniqueness* [Eindeutigkeit] *and stability highlight the aspects of reality that must be the focus of our interest*" (Petzoldt 1890, p. 78).[31] Petzoldt's immediate aim in his 1895 paper was to propose a friendly reinterpretation or refinement of the neo-Humean conception of causality first put forward by Mach in his early essay on the history of the conservation of energy (Mach 1872a) and then developed further in the first two editions of his *Mechanik* (1883, 1888), and simultaneously to defend this conception of causality against Wilhelm Wundt's criticisms (Wundt 1894). Petzoldt argued that every instance of a causal relationship could be reduced to one core conception, that being the quasi-mathematical idea of a (simultaneous) functional dependence (Petzoldt 1895, pp. 161–162), or, in what for our purposes is the more telling locution, the idea of an "eindeutige Funktion" (p. 171 and elsewhere), which translates poorly in English as a "single-valued function," or perhaps better, if liberally, as a one-to-one or many-to-one mapping.[32] It is this concept that Petzoldt elaborates into his "law of *Eindeutigkeit*." For wherever in the sciences one sees causal relationships or causal laws—which is to say everywhere, in the eyes of a Petzoldt or a Mach—one is now to see instead *eindeutige* functional dependence. Here is how Petzoldt himself formulates this conclusion:

> To the question raised at the outset, "What should take the place of the concepts cause and effect?" we may indeed answer: the concepts of uniquely [eindeutig] determining and determined elements or element complexes; and to the second question about the "meaning" or "essence" of processes in nature: they are of such a kind that we can reduce them to the *continuous* and *univocal* [*einsinnige*] change of a small number of ever recurring elements—i.e., *determining* elements, *means* of determination— that mutually *determine* one another *uniquely* [*eindeutig*]. (1895, p. 180)[33]

One of many reasons why the doctrine developed here is interesting as part of the background to Einstein's thoughts on scientific method is that it establishes a link, which we will encounter again in other thinkers, between *Eindeutigkeit* and determinism.

So important and pervasive are such determinate causal relationships or *eindeutige* functional dependences, that Petzoldt identifies here a biologically grounded methodological imperative. Implicitly invoking what one is tempted to call a "methodological anthropic principle," Petzoldt argues thus:

> All inorganic and organic systems that undergo development maintain themselves for a certain time as closed wholes and they change in the direction of certain stationary states.... In particular, the human central nervous system has the capacity to persist for a relatively long time as a particular organism, and all of the processes that occur within the human being—and specifically those that run "parallel" to "psychical" life—can most easily be understood by considering that tendency to stability. We must conceive of this conservation and development in connection with a certain general character of natural processes, with-

> out which they would not be possible. We must at the same time bring to bear on nature a certain general presupposition, without whose confirmation we ourselves could not live, either mentally or bodily. Such a presupposition lies at the base of all scientific research, something of which we may be more or less consciously aware, and we may be of the firm conviction that it will hold up everywhere, since we could not conceive of ourselves, with our particular mental nature, if we once imagine it being given up. Both our individual constitution and that postulate, as we may designate the relevant presupposition, belong inseparably together. The latter consists in nothing other than the assumption of the *thoroughgoing complete determination*, or—as we want to say in order to emphasize the most important side of the matter—in the assumption of the *uniqueness of all processes* [*Eindeutigkeit aller Vorgänge*]. (1895, pp. 167–168)[34]

And as if to dispel any doubt there might be about his meaning, Petzoldt adds on the next page: "Our highest mental existence, the most highly developed parts of the central nervous system are not at all conceivable without the uniqueness of all being and events" (1895, p. 169).[35]

Petzoldt then proceeds stunningly, if not always persuasively, to identify a number of other manifestations of the law of *Eindeutigkeit*, starting with the principle of continuity:

> Thus we cannot find an indefiniteness in the *succession* of appearances. We must only observe that uniqueness [Eindeutigkeit] shows here other sides than in the case of *simultaneous* dependence. It finds its particular expression here above all in the *continuity* of the change. Discontinuity would be saltatory change, and, thereby, in general the opposite of uniqueness. To this extent there is, in addition to the *simultaneous* dependence of the determining elements of facts, also a *successive* dependence in the numerical measures of these determining elements. Before a means of determination that has a specific value reaches another value differing from the first by a finite amount, all of the values lying in between must have been passed, supposing that, in the meantime, the series of changes has not, perhaps, been interrupted and then under other conditions resumed anew with a different relationship of the relevant determining element to other determining elements.... If nature permitted such saltatory changes, then there could be no talk of uniqueness, not even if in these cases the *simultaneous* dependence, i.e., the continued existence of the physical equations were completely assured, thus even if all of the other determining elements also made the corresponding jump prescribed for them by the equations. (1895, pp. 177–178)[36]

The list of manifestations of *Eindeutigkeit* continues, with Petzoldt mentioning: (1) least action or variational principles (pp. 185–188);[37] (2) the law of inertia (pp. 188–194); (3) Ostwald's "principle of the singular case" ["Prinzip des ausgezeichneten Falles"] (pp. 194–196);[38] and (4) the logical principles of identity and contradiction (pp. 199–200).[39] Students of Einstein may find especially interesting Petzoldt's answer, in connection with the law of inertia (2), to the question why we conceive the motion of a body not subject to external forces to be straight-line motion:

> *With the exclusion of all other means of determination, aside from a one-time impulse*, it is precisely only the straight-line path that is uniquely determined.... Indeed, we can define the straight line itself no better than as follows: in our three-dimensional space it is that line between two points that is uniquely determined—in the sense that for every deviation from it we can conceive innumerably many others that are equally well determined. For the straight line alone can we think of no other "equally justified" ["gleichberechtigte"] line. (1895, p. 190)[40]

One may well question whether or not precisely the same kind of *Eindeutigkeit* is manifested in all of the cases cited by Petzoldt, whether or not, for example, the *Eindeutigkeit* inherent in the functional relationship between a past state and future state in a causal process is the same in kind as that attaching to a path in phase space that minimizes a Hamiltonian. Whatever one's decision on that question, however, all of the mentioned cases (with the possible exception of the principles of identity and contradiction) would represent different aspects of what I am calling the metaphysical *Eindeutigkeit* principle, since they concern not so much the relation between theories and the world as the properties of that world itself. They represent different ways in which nature is itself uniquely determined, calling to mind Mach's remark in his *Mechanik*, a remark quoted approvingly by Petzoldt, "Die Natur ist nur *einmal* da" [roughly, "Nature is given only *once*"] (Mach 1883, p. 455; Petzoldt 1890, p. 57, 1895, p. 171).

But Petzoldt also introduces—indirectly and without emphasizing it—what I called above the epistemological *Eindeutigkeit* principle, this in the course of his discussion of the law of inertia, specifically in connection with the question why inertial motion must be uniform. Petzoldt argues that the uniformity of inertial motion is something that cannot be settled by a priori means, since we could just as well *imagine* unforced straight-line motion in the form of regular longitudinal oscillations or as alternately and regularly accelerated and retarded motion. "A priori," he says, "*everything* is conceivable" ["A priori ist *alles* denkbar"] (1895, p. 191, n. 2). He concludes:

> Here, therefore, thought is not "a priori" uniquely [eindeutig] determined, here we can demand only a periodical or stationary type of motion within the straight line. Thus, in this case, individual experience or experiment must step in, in order to determine thought uniquely. (1895, p. 191)[41]

We will find that Einstein, like Petzoldt, repudiates the Kantian position for good empiricist reasons. But we will also find him explicitly repudiating the claim that experience or experiment determines thought or our choice of a theory uniquely, when a priori considerations fail to do so, this denial being a crucial aspect of what I have elsewhere described as the (Duhemian) conventionalist component of Einstein's philosophy of science (see Howard 1990a).

Shortly, we will see how Petzoldt deployed the *Eindeutigkeit* principle in his many later writings on relativity. But let us pause for a moment to consider how it was first deployed by Petzoldt in a rather different context, this his defense of psycho-physical parallelism in his widely-read *Einführung in die Philosophie der reinen Erfahrung* (Petzoldt 1900, 1904).[42] Petzoldt's argument is simple. Unique determination [*eindeutige Bestimmung*] is a necessary condition on any adequate scientific explanation. But mental events are not uniquely determined by mental causes alone. Therefore, there cannot be an autonomous science of the mental. A psychology that seeks a scientific explanation of mental events must look to physical events in the brain and body for uniquely determining conditions, the causes, if you will, of those mental events.

Lest there be any misunderstanding of the principle upon which the argument is based, Petzoldt devotes the whole of chapter three in volume one to the articulation of "Das Gesetz der Eindeutigkeit" (Petzoldt 1900, pp. 34–56). The chapter repeats most of the essential content of the earlier paper by the same title (Petzoldt 1895), though with different examples, and no applications to technical physics. But it also contains a new, and in some ways clearer statement of the law itself:

> For every process means of determination may be found that determine it uniquely [eindeutig], in such a manner that for every variation in this process that one can conceive as being determined by the same means, one can find at least one other variation that would then be determined in the same way as, and thereby be equivalent to, the first, and thus would have, as it were, the same *right* to be realized as the first. We shall designate this principle as the law of *Eindeutigkeit*. (Petzoldt 1900, p. 39).[43]

The point is that what distinguishes real events and processes, what makes them truly *uniquely* determined is that, within a class of variations on a given type of process, the real event or process is that one for which no equivalent can be found. Petzoldt gives the example of a stationary ball on a flat, horizontal, frictionless surface being struck, straight on, by a moving ball. In what direction will the stationary ball recoil? The only possibility is for it to recoil in the same direction in which the incident ball was moving, because for any other conceivable direction, say ten degrees to one side of the incident ball's path, one can find another direction, thus ten degrees to the other side, that would be equally justified.

Petzoldt's arguments for the metaphysical *Eindeutigkeit* principle found an immediate and enthusiastic reception in certain quarters. Most importantly, in all subsequent editions of his *Mechanik*, starting with the third edition of 1897, Mach enthusiastically endorsed the application of the *Eindeutigkeit* principle, at least in the understanding of least action (1897, pp. 371–372; 1912, p. 256) and conservation principles (1897, p. 495), citing explicitly both Petzoldt 1890 (for example, 1897, pp. 372, 495) and 1895 (for example, 1912, pp. 256, 260), and thereby won for Petzoldt's introduction of the *Eindeutigkeit* principle a much larger audience than Petzoldt would have found on his own. Mach somewhat petulantly claimed priority, referring the reader to a few passages in his own earlier work on the conservation of energy (Mach 1872a),[44] but this did not prevent him from making the following generous acknowledgment: "But in reference to the dynamical cases, the meaning of *unique determination* has been portrayed better and *more perspicuously* by J. Petzoldt than I succeeded in doing" (1897, p. 372).[45]

It was not only in the *Mechanik*, however, that Mach discussed Petzoldt on *Eindeutigkeit*. Indeed, there is some mention of the issue in virtually every major work of Mach's after 1895. Thus, in the later chapters of the *Wärmelehre*, those concerned primarily with general methodological questions, Mach sought to defend aspects of his biological-evolutionary economism against Petzoldt's earlier criticisms in his "Maxima, Minima, Oekonomie" (Petzoldt 1890). And in the course of this discussion, there is a mention of the *Eindeutigkeit* principle that was to be picked up by other authors (see below, section 7, for Cassirer's citation of this passage). Mach wrote: "Just as with Petzoldt, I am convinced that in nature only that and only that much happens as can happen, and that this can happen in only *one* way" (Mach 1896, p. 393; see also pp. 324–327).[46]

Or consider the second (1900) and later editions of the *Analyse der Empfindungen*. Recall that the immediate foil for Petzoldt's first development of the *Eindeutigkeit* principle was Mach's conception of causality.

Petzoldt had read Mach as suggesting that in at least two different senses there can be less than complete determination in nature (Petzoldt 1895, pp. 164–167, 170–177); but the *Eindeutigkeit* principle implies complete, unique determination. In a later section of the *Analyse der Empfindungen* Mach replies to this criticism. The details need not detain us here. Let me just quote what Mach says, following an explicit reference to Petzoldt 1895: "This view is also quite compatible with the establishment of unique determinations, which always occur under the assumption of *given* circumstances and in *abstraction* from uncommon and unexpected changes" (Mach 1900b, p. 236).[47]

And, finally, the *Eindeutigkeit* principle is discussed again, with citations of Petzoldt 1895, in Mach's *Erkenntnis und Irrtum* (1905), where Mach wrote (hinting at the model-theoretic *Eindeutigkeit* principle as much as at the metaphysical one): "Only a *theory* which represents the facts of observation—always complicated and influenced by manifold accompanying circumstances—*more simply and more exactly* than these can actually be established by means of observation, only such a theory measures up to the *ideal of unique determination*" (1905, p. 449; see also pp. 278–280, 428–431, 448, and 450).[48]

Mach's citation of Petzoldt is important for our understanding of the background to Einstein's employment of *Eindeutigkeit* principles, because we know that Einstein read the *Mechanik* and the *Wärmelehre*, during his student days at the ETH in Zurich.[49] But far more important, for our purposes, is the fact that Petzoldt himself discussed the *Eindeutigkeit* principle extensively, and related it directly to relativity theory, in virtually all of the many books and papers that he wrote on relativity starting in 1912.[50]

Essentially the same argument was repeated in all of Petzoldt's publications on relativity, but nowhere more clearly or concisely than in his "Die Relativitätstheorie im erkenntnistheoretischen Zusammenhange des relativistischen Positivismus" (Petzoldt 1912), which was first delivered as a lecture to the Deutsche Physikalische Gesellschaft in Berlin on 8 November 1912, and published shortly thereafter in volume 14 of the *Verhandlungen*.

Petzoldt's main concern in all of his writings on relativity was to claim Einstein as an ally in the battle he saw himself waging, in concert with Mach and Avenarius, against the concept of substance, and for the point of view—Petzoldt's own variation on Mach and Avenarius—that he called "relativistic positivism."[51] According to Petzoldt, the essential insight of Mach and Avenarius was that all our knowledge of the world is perspectival, relative to a given observer, and that therefore there can be no talk of an independent reality; all that we can talk about is the world as it appears to one or another observer. It follows that the distinction between appearance and reality lapses, and that there is, therefore, no place left for the traditional philosophical conception of substance as the ground for that distinction.[52] It also follows, according to Petzoldt, that events can therefore be at most "described" and not "explained," as had been argued by Mach and Kirchhoff (1876, p. 1), theoretical concepts like "mass," "energy," "space," and "time" serving only as tools ("Hilfsmittel") in our descriptions.

Petzoldt believed that he saw a similar point of view expressed in Einstein's comments on the debate about whether length contraction and time dilation are "real" or "apparent" phenomena, and he was strengthened in this belief by the assumption that Einstein's relativization of the mathematical description of physical events to coordinate systems or frames of reference amounted to the same thing as Petzoldt's own relativization of knowledge to observers. The latter assumption, which we may characterize as the "subjectivist" misreading of relativity, was no doubt encouraged by Einstein's frequent talk of "observers" in different frames of reference; but however Petzoldt came by this view, it is a fundamental misunderstanding that he never overcame.

What is important to our story is not the history of Petzoldt's confusions and misunderstandings, which were many,[53] but the way in which he characterized the embodiment of perspectivalism in a relativistic physics. Petzoldt wrote:

> The task of physics becomes, thereby, the *eindeutige* general representation of events from different standpoints moving relative to one another with constant velocities, and the *eindeutige* setting-into-relationship of these representations. Every such representation of whatever totality of events must be *eindeutig* mappable onto every other one of these representations of the same[1)] events. The theory of relativity is one such mapping theory. What is essential is that *eindeutige* connection. Physical concepts must be bent to fit for its sake. We have theoretical and technical command only of that which is represented *eindeutig* by means of concepts.
>
> 1) Better: representations of events in arbitrarily many of those systems of reference that are *eindeutig* mappable onto one another are representations of '*the same*' event. Identity must be *defined*, since it is not given from the outset. (1912, p. 1059)[54]

This is at one and the same time a statement of a version of the metaphysical *Eindeutigkeit* principle, since it stipulates what it is that is unique (*eindeutig*) in what relativity theory aims to describe, and a statement of the model-theoretic *Eindeutigkeit* principle, since it asserts that we have adequate theoretical command only of that which our concepts (read "theories") represent uniquely. Petzoldt's footnote is clearly suggestive of the key insight in Einstein's point-coincidence argument, and, like the entire passage, suggestive of the Minkowskian invariant-theoretical formulation of special relativity (Minkowski 1909). It should be noted, however, that Petzoldt was not entirely comfortable with the Minkowskian approach—so typical of the Göttingen tradition inspired by Felix Klein (see Klein 1910)—because he rightly saw behind it a yearning for something absolute, such as the invariant metrical interval, and thus a yearning for a return to the kind of substantivalism that had been overcome, in his opinion, in the work of Mach.[55]

To what extent was Einstein familiar with Petzoldt's writings on relativity theory in the crucial years of his work on general relativity, 1913–1915? The modest correspondence between Einstein and Petzoldt that survives from the years 1919 to 1927 suggests that by that time there was between them a cordial, though certainly not intimate, personal relationship. Einstein patiently corrects some of Petzoldt's technical confusions concerning the rotating disk problem.[56] There are references to proposed personal meetings in which the philosophical implications of relativity theory would be a focus of discussion. And there is evidence of Einstein's having recommended Petzoldt for a position at the Technische Hochschule Dresden (where the energeticist Georg Helm had worked during his entire professional career) in the summer of 1919.[58] There is less evidence bearing upon their relationship before 1919, but this evidence still shows clearly that Einstein had studied some, if not all, of Petzoldt's writings on relativity perhaps as early as the fall of 1914.

There is, first, the fact that in the summer of 1912 Einstein allowed his name to be added to the list of founding members of the newly-formed Gesellschaft für positivistische Philosophie. Petzoldt was one of the organizers of the Gesellschaft and an editor of its new journal, the *Zeitschrift für positivistische Philosophie*, where his second major essay on relativity (Petzoldt 1914) appeared as the lead article in volume two.[59] It was just a few months after the founding of the Gesellschaft für positivistische Philosophie that Petzoldt's first major essay on relativity theory (Petzoldt 1912a), from which I quoted above, appeared in the *Verhandlungen* of the Deutsche Physikalische Gesellschaft, having first been delivered as a lecture to that body on 8 November 1912. Secondly, there is evidence that by 1915 Petzoldt was attending Einstein's lectures on relativity theory in Berlin.[60] And there is, thirdly, Petzoldt's own report in his "Anhang" to the eighth edition of Mach's *Mechanik* (Mach

1921)—entitled "Das Verhältnis der Machschen Gedankenwelt zur Relativitätstheorie" (Petzoldt 1921a)—that Einstein had read Petzoldt's 1914 essay, "Die Relativitätstheorie der Physik," and had expressed to Petzoldt his "agreement" ["Zustimmung"] with it (Petzoldt 1921a, p. 494); unfortunately, however, Petzoldt does not say *when* Einstein read this essay.

But the most important evidence for Einstein's study of Petzoldt is provided by Einstein himself. The Einstein Archive contains a copy of an undated postcard from Einstein to Petzoldt in which Einstein says the following:

> Today I have with great interest read your book in its entirety, and I happily infer from it that I have for a long time been your companion in your way of thinking. I have told a gentleman who is ill, a friend of mine, about your paper on relativity theory. He has shown a great interest in it. You would give him great pleasure if you wanted to send him a reprint of it. . . .
>
> I am sending you my new paper on the covariance of the grav. equations. (EA 19-067)[61]

Einstein's secretary, Helen Dukas, dated this card to May or June of 1919, but her reasons for doing so are unclear.[62] Thiele, who first published Einstein's letters to Petzoldt, dates it to 1918, but his only reason seems to be his misidentification of the paper on covariance as Einstein 1918 (Thiele 1971, p. 71). Given the interests that Petzoldt and Einstein shared, and given the way Einstein characterizes the paper, "meine neue Arbeit über Kovarianz der Grav. Gleichungen," it is far more likely that Einstein is referring here to his second paper with Grossmann, "Kovarianzeigenschaften der Feldgleichungen der auf die verallgemeinerte Relativitätstheorie gegründeten Gravitationstheorie" (Einstein and Grossmann 1914b). In that case, since he refers to the paper as his "new paper," the postcard should probably be dated to late 1914 or early 1915. This dating also places the posting of the card closer in time to the publication of those of Petzoldt's works that Einstein mentions, the book being most likely Petzoldt 1912b,[63] and the paper probably either Petzoldt 1912a or, more likely, Petzoldt 1914. Moreover, the offhand manner in which the paper is mentioned suggests that Einstein had been familiar with it for sometime before the card was written.

When I put all of this evidence together; when I recall, furthermore, that Petzoldt was one of only three *major* philosophical figures to have written on the philosophical implications of relativity by the end of 1915 (the other two being Natorp and Schlick, more on which below); and when I consider, finally, that Einstein was, from his university days forward, keenly interested in the philosophical issues raised by his own and other work in physics, especially as those issues were regarded by Mach and his followers—remembering that Mach cited Petzoldt on the topic of *Eindeutigkeit* in the *Mechanik*, the *Wärmelehre*, and elsewhere; when I add all of this up, it seems highly likely to me that Einstein was familiar with Petzoldt's writings on relativity from shortly after the time Petzoldt began to publish on the subject late in 1912, and certainly by late 1914 or early 1915. When I add to this the fact that the metaphysical and model-theoretic *Eindeutigkeit* principles are central to Petzoldt's analysis of the philosophical implications of relativity, I conclude that there is good reason to think that Petzoldt may well have been one of the sources which inspired Einstein's invocation of *Eindeutigkeit* principles in the hole and point-coincidence arguments. But Petzoldt is just one possible source. Let me now turn to another, the physicist, Paul Volkmann.

5. Paul Volkmann on *Eindeutigkeit*, Superposition, and Completeness

Almost simultaneously with the appearance of Petzoldt's "Das Gesetz der Eindeutigkeit" (1895), some of the same questions posed by Petzoldt were taken up, quite independently it would appear, and from a rather different, indeed critical standpoint, by the Königsberg theoretical physicist, Paul Volkmann, in his widely-read *Erkenntnistheoretische Grundzüge der Naturwissenschaften* (1896a). During the early part of his career, Volkmann worked extensively on problems in the theory of light and capillarity (see Volkmann 1891). But after about 1900, when he published his also widely-read *Einführung in das Studium der theoretischen Physik*, he concentrated his attention on epistemology and the philosophy of science, about which he had been writing since the mid-1880s. Though he was not an adherent of any philosophical school, his general orientation was in the direction of a neo-Kantian critical realism, heavily influenced by Helmholtz.[64]

The immediate occasion for Volkmann's remarks on *Eindeutigkeit* was a critique of Ernst Haeckel's monism (see Haeckel 1892, 1899), a topic to which Volkmann returned several times in later years (see, for example, Volkmann 1909a, 1909b, 1910b). But the attitude expressed in this early critique of Haeckel has interesting roots in Volkmann's more general views on the epistemology and methodology of the natural sciences. Volkmann says the following about monism:

> One doubt that I still want to raise against monism concerns the tacit assumption of the *Eindeutigkeit* of all questions and problems that natural science poses to itself. Such an assumption will so easily be accepted as self-evident that it may well be appropriate to return to it here with a few words.
>
> Credit for having first pointed out the importance of a proof of the *Eindeutigkeit* of problems may well be claimed by mathematics.[65] In order to answer the question to what extent problems in the natural sciences are *eindeutig*, one best calls upon the history of science.
>
> In the third and fourth lecture, it was explained how, e.g., different views about the nature of light, the emission view and the wave view, could be maintained side-by-side, and how then, when further experience had decided in favor of the wave view, the elastic and electromagnetic theories of light could for a long time play a role side-by-side.
>
> This example shows how, to be sure, advancing induction restricts the *Vieldeutigkeit* of the problems, but never entirely puts an end to it. The assertion of the complete *Eindeutigkeit* of the problems of natural science would, after all, come to the same thing as the abandonment of inductive natural scientific method.
>
> But the question of the *Eindeutigkeit* of the problems of physics is much clearer than in other natural scientific disciplines simply for the reason that physical problems are susceptible to treatment by mathematical means. In many cases, however, in spite of any *Eindeutigkeit* in the formal mathematical treatment, there still persists a *Vieldeutigkeit* in the physical interpretation of the mathematically *eindeutigen* formulas. (Volkmann 1896a, pp. 171–172).[66]

In this, his earliest remark about the *Eindeutigkeit* problem, Volkmann appears to conflate the model-theoretic and epistemological *Eindeutigkeit* theses, inasmuch as his comment about induction and the theory of light suggests a denial of the latter, and hence a kind of Duhemian, underdeterminationist view of scientific method, whereas his remarks about the *Eindeutigkeit* of mathematical problems suggests a denial of the model-theoretic *Eindeutigkeit* principle in the case of many physical problems. Such confusion is probably to be expected at a time when the basic distinction between theory and interpretation still had not been articulated. But the appearance of confusion on Volkmann's part is lessened when one turns back to the mentioned earlier discussion of the history of theories of light and the more general views on the epistemology and methodology of the sciences that lie behind both it and the present critique of monism.

Volkmann makes a fundamental systematic distinction between laws of nature and hypotheses (1896a, pp. 57–62; see also Volkmann 1900, pp. 12–20). Among the former, he counts such things as the conservation of energy, Newton's laws of motion, and the law of universal gravitation. Among the latter, he counts the hypotheses about the nature of light and the atomic hypothesis. Both laws of nature and hypotheses are the product of inductive inferences from empirical evidence and, hence, are subject to inductive uncertainties. Both serve, on the other hand, as bases for deduction, for the derivation of predictions about hitherto unobserved phenomena. But they differ in one crucial respect. Laws of nature claim universal validity for phenomena within the realm of appearances, and while afflicted with the unavoidable inductive uncertainty, nevertheless claim a near certainty within their intended domain; moreover, laws of nature are essentially conceptual in character.[67] Hypotheses, by contrast, are akin to sensual representations and intuitions, except that, by definition, they carry us beyond the realm of appearance and sense experience. Volkmann calls them "trans-sensual representations and intuitions" ["übersinnliche Vorstellungen und Anschauungen"] (1896a, p. 61).[68] To the extent that hypotheses must conform to laws of nature, and not the other way around, they partake more of the character of models, in the sense of modern formal semantics, whereas the laws are more like the theories our models are supposed to satisfy.[69] Thus, when Volkmann cites the (temporary) coexistence of the emission and wave hypotheses of light as an example of *Vieldeutigkeit*, he is, in effect, citing an example of *model-theoretic Vieldeutigkeit*. But again, since the basic distinction between model and theory had not been clarified in Volkmann's day, it is not good to press this interpretation of his views too far. It is better to survey some of his other pertinent remarks in order to appreciate the implications of his point of view.

A cluster of other concepts and related distinctions is closely connected in Volkmann's thinking with his views on the difference between laws and hypotheses. Consider, to begin with, his emphasis upon the essential role of analogies and models in physics. Very much in the tradition of Maxwell, Hertz, and Boltzmann, all of whom he cites in this connection,[70] Volkmann argued that models [*Bilder*] and analogies [*Analogien*] are a necessary part of all scientific knowledge, inasmuch as all (internal) representation or intuition [*innere Anschauung*] relies inevitably upon them.[71] The history of physics for Volkmann is, in part, a history of the gradual replacement of analogies grounded upon sensual, primarily visual representations, such as the wave model in optics and electrodynamics, by more abstract analogies, culminating at the end of the nineteenth century in the reign of the "mechanical analogy" (1896a, pp. 65–68, 145-152; 1900, pp. 30–33).[72] But however near- or far-removed these analogies may be from sensation, their function is the same. They are the tools by means of which we implement what Volkmann calls the "epistemologically so important *principle of comparison*" ["erkenntnistheoretisch so bedeutsame[s] *Princip der Vergleichung*"] (1896a, p. 29), the principle that he elsewhere characterizes as "an essential guide on the path of knowledge, of research" ["ein wesentlicher Führer auf dem Wege der Erkenntnis, der Forschung"] (1896a, pp. 64–65).[73] Just how this principle functions is suggested in the following passage:

> The newer presentations of mechanics by Hertz and Boltzmann appear to me... to be a step on the best way to drawing in the essential, legitimate, objective and subjective sides of research, this when they conceive of the task of mechanics from the point of view of a picturing, modeling, or mapping of reality. The opposition of subject and object can in fact be very well illustrated by the relationship between a model and the object modeled. (1900, p. vi)[74]

It is the hypothetical side of scientific knowledge that relies thus upon models and analogies to inform us about systems and events beyond the range of immediate experience by comparing them to systems and events within the range of our experience. The essential content of a hypothesis, on Volkmann's analysis, is precisely a representation in the form of a model or an analogy, and what knowledge it yields comes by way of comparison.

Given the central role that Volkmann assigns to hypotheses, to knowledge by analogy and by model, the obvious question is whether or not hypothetical representations in the form of models and analogies are capable of yielding a *complete* representation of that which they picture, and, related to this, whether or not there is in every situation only one correct model. Volkmann's answer is "no." And this answer rests upon his quite original doctrine of "isolation and superposition."[75]

According to Volkmann, all concrete reality, everything that presents itself to us in the world of experience, is complex [*zusammengesetzt*] and multifaceted [*vielfach*], rather than simple and univocal (1896a, p. 70; 1900, pp. 28–29). The task of science, he says, is to "isolate" by conceptual analysis the abstract constituent elements composing the concrete objects and phenomena we encounter in experience, and then to reconstruct these concrete wholes out of their abstract constituents by a process of synthesis or "superposition." A central role in the epistemology of science is thus assigned to the complementary processes of *isolation* and *superposition*; this is what Volkmann dubs the "principle of superposition" (1896a, p. 177).[76] Volkmann draws an explicit analogy to vector decomposition (1896a, pp. 80–87; 1900, pp. 29–30). For just as one can more easily do calculations on the "isolated" components of a vector, which are later "superposed" to recover the original quantity, so too we can better understand complex concrete objects and phenomena if science focuses its attention on the simple laws of nature governing their "isolated," abstract, conceptual components. These abstract "centers of isolation" ["Isolationscentren"], as Volkmann calls them (1896a, p. 88), rather than concrete objects and phenomena, are the proper concern of scientific laws. Thus "energy" is an abstraction, governed by the law of the conservation of energy; "inertia" is an abstraction governed by Galileo's law of inertia; and "gravity" is an abstraction governed by Newton's law of universal gravitation. A given phenomenon may incorporate all three components, and thus be governed, indirectly, by all three laws, as in the case of projectile motion. Still it is the components themselves that are the proper objects of scientific description by means of natural law.

The relations among the conceptual components of a concrete object or phenomenon are not in all respects the same as those among the components of a vector. For, in a given domain, there is a hierarchy among the former that is not found among the latter. Thus, the energy conservation principle governs all inertial phenomena, and the Galilean principle of inertia (Newton's first law) governs all gravitational phenomena, but not vice versa. And the relatively higher-order laws may hold across different domains, the lower-order laws in each domain being quite different. Thus, the energy principle and the inertia principle apply both to gravitational phenomena and to electromagnetic phenomena.

But in another respect, the analogy to vector decomposition is a helpful one. For just as in any application of vector methods there is no uniquely correct way of carrying out the decomposition, there being, as Volkmann notes, an "uncountable" ["unzählige"] number of possible decompositions, depending upon the set of basis vectors that we choose (1896a, p. 82); so too it is logically possible for a given concrete object or phenomenon, governed by one or more relatively higher-order laws, to be represented at a lower level by various different lower-order laws. Thus, electromagnetic radiation is governed by the energy principle, but it may equally well be represented at a lower level of description by the laws governing the motion of waves or those governing the motion of projectiles. In the same way, the choice of one basis vector for vector decomposition does not by itself determine the other bases. Of course, in vector analysis the different decompositions are equivalent, in the sense that any one can be translated into any other one by a purely mathematical transformation of the bases, whereas such equivalence does not obviously obtain in the examples that Volkmann gives of isolation and superposition in science. The wave and corpuscular theories of radiation may be empirically equivalent, up to a point, but they are clearly different models, not obviously translatable into one another.

Volkmann goes out of his way to stress that no one law of nature can alone describe concrete objects and phenomena completely:

> If the law of nature is an abstractum, and thus a center of isolation, but nature itself a concretum, something superposed, then we may not expect that there is a law of nature that completely comprehends the universe in its entirety. Every law of nature aims to comprehend nature or reality only from one direction, from one point of view, without it being the case that other points of view should thereby be excluded. (1896a, p. 88)[77]

Some laws of nature, like the energy principle, are universal in the sense that they govern one aspect of all phenomena (in this case, the energy); but none are universal in the sense of determining everything that there is to be said about a given object or phenomenon, let alone all objects and phenomena (1896a, p. 89).

Nor will the addition of further laws achieve for us the hoped-for completeness. In Volkmann's opinion, nature is too rich to be comprehended completely by any number of laws:

> Kirchhoff is well known for having defined the task of mechanics as being "to describe motions that occur in nature completely and in the simplest way"—a formulation that can easily be extended to all of physics, indeed, to all of natural science, if one replaces the word "motions" with the word "phenomena." This formulation has been characterized as too narrow in view of its stress upon description; I want to characterize it here as still too broad, in view of its stress on completeness. I assert: It is impossible to describe completely the phenomena that occur in nature, we must be content to describe them only in their essentials.
>
> One can object to the assertion of the impossibility of a complete description that our theories and representations should indeed help us to determine what actually occurs beyond the threshold of appearance—I recall the atomistic view of the constitution of matter. Processes that play themselves out beneath the threshold of consciousness certainly determine phenomena that appear in consciousness, but this still only partially; they also determine processes that play themselves out below the threshold value of consciousness but whose knowledge may still be necessary for completeness of description. The limits of our knowledge of nature appear to be determined thereby. (1896a, pp. 110–111)[78]

It is thus a fact about the world and about human knowledge that completeness remains a will-o'-the-wisp. About completeness and other methodological ideals (such as unity—a single, unified description of reality), Volkmann says:

> It is just as easy as it is worthless to indicate and establish such ideal goals. For them to be of value, ideals and goals must have some connection with reality, they may lie neither too far nor too near. Establishing such ideals is difficult, but epistemologically valuable. (1896a, p. 140)[79]

To return for a moment to Volkmann's critique of monism for its tacit assumption of the *Eindeutigkeit* of all problems—it is Volkmann's denial of the possibility of completeness that lies behind that critique. If a complete description of a given set of phenomena or a complete solution of a given problem were possible, there would be no *Vieldeutigkeit*. The limits on a complete theoretical description of nature are simultaneously the limits on an *eindeutige* description.

All of these themes come together in a remarkable, but now largely forgotten paper that Volkmann published in 1902 in the first volume of Ostwald's *Annalen der Naturphilosophie*: "Ueber die Fragen der Existenz, Eindeutigkeit und Vieldeutigkeit der Probleme und ihre mannigfaltige Bedeutung und Rolle für naturwissenschaftliche Auffassung und Erkenntnis" ["On the Questions of the Existence, Eindeutigkeit and Vieldeutigkeit of Problems and Their Manifold Significance for and Role in Natural Scientific Comprehension and Knowledge"].[80] The paper recapitulates with but minor modifications and refinements most of the essential ideas discussed above from the *Grundzüge* (1896a) and the *Einführung* (1900), but the problem of *Eindeutigkeit* is highlighted in a way that it had not been in the earlier works.

Volkmann begins with a more thorough discussion of the *Eindeutigkeit* problem in mathematics, noting its roots in potential theory, where Gauss and Dirichlet explored the question of the general circumstances under which a problem would be guaranteed from the outset to have a *unique [eindeutige*] solution.[81] What he emphasizes is that, even if one can assume the existence of a solution, one cannot always assume its uniqueness [*Eindeutigkeit*]:

> *The question concerning the Eindeutigkeit* of a solution was also of importance for mathematical and physical disciplines. It is not permitted, mathematically, to assume from the outset that a question—presupposing a proof of its [an answer's] existence—must always admit of only one answer; there can indeed even be *several* answers. (1902, p. 109)[82]

He remarks that the problems with multiple solutions are "of special interest" ["von besonderem Interesse"], and that it will be his aim to investigate not only the mathematical, but also the physical basis for such multiplicity [*Mehrdeutigkeit*] (1902, p. 110).

Especially interesting to Volkmann are those cases in physics where mathematical *Eindeutigkeit* and physical *Vieldeutigkeit* appear simultaneously, that is to say, those cases where different physical models are compatible with one and the same mathematical description of a group of phenomena. He cites two examples of this general state of affairs: (1) action-at-distance theories versus local-action theories in gravitation and electrostatics, where, in spite of the obvious differences between the two views, both are governed—so we learn from potential theory—by "one and the same differential equation, which thus can be derived just as well on the basis of the one view as the other" (1902, p. 115);[83] and (2) atomistic versus continuum theories of matter, which Volkmann had argued in an earlier article (Volkmann 1897) yield the same results for a broad range of phenomena.[84]

The important question is whether or not such states of affairs are always just temporary stops on science's path to a complete representation of the phenomena in question, the *Vieldeutigkeit* being a symptom of this temporary incompleteness, or whether there may perhaps be cases in which such *Vieldeutigkeit* will be an abiding feature of our theories. He notes in passing that the neglect of this question is the "principal failing" ["Hauptmangel"] of the earlier theories of scientific method (he has in mind Hertz and Boltzmann) that view scientific theories as models [Bilder], a view to which Volkmann is otherwise quite sympathetic (Volkmann 1902, p. 116).

Volkmann divides the various cases of physical *Vieldeutigkeit* into three categories: (1) those in which the *Vieldeutigkeit* is temporary, the mark of an incompleteness that will be overcome with the advance of inquiry; (2) those in which the *Vieldeutigkeit* may be enduring, but is inessential and scientifically insignificant; and (3) those in which the *Vieldeutigkeit* is enduring and essential. *Vieldeutigkeit* of the first kind is exemplified by the history of the theory of light, where first the emission and wave theories coexisted, and then, after the triumph of the wave theory, the electromagnetic and elastic wave theories coexisted for a considerable time, the electromagnetic theory eventually winning out (Volkmann 1902, pp. 117–119). But these instances of temporary *Vieldeutigkeit* are the least interesting from an epistemological point of view, if only because they accord with the traditional assumption that completeness and *Eindeutigkeit* should be the normal outcome of scientific inquiry.

More interesting are the second and third varieties of *Vieldeutigkeit*.

The second—enduring but inessential—kind is exemplified by the aforementioned case of the atomistic and continuum theories of matter (Volkmann 1902, pp. 120–122). This example is typical of the whole class in that one model is relatively more detailed than the other, but not manifestly inconsistent with it. Over a broad range of phenomena, the two models can be made to yield the same results, the only difference being that one accomplishes this end by simpler means; and so, for such phenomena, the more detailed model is inessential. Thus, the atomistic model is more detailed than the continuum model, and both can explain the phenomena associated with elasticity, though the continuum model does so with fewer special assumptions, which shows that one need not assume an atomistic theory of matter to explain all physical phenomena. Volkmann's principal aim with examples such as this is to combat the prejudice that the more detailed model is metaphysically more basic, somehow closer to reality, simply because of its being more detailed. He speaks in this connection of the "exaggerated estimation of the value of atomism for physics" [übertriebene Werthschätzung der Atomistik für die Physik"] (Volkmann 1902, p. 122). The spirit of his argument is to advocate a more liberal attitude toward questions of ontology and to caution against premature acceptance of a single model.[85]

The third and most interesting variety of *Vieldeutigkeit* is enduring and essential. What distinguishes such cases is the presence of two or more models each of which has its own independent justification and significance. And the explanation of all such cases is rooted in Volkmann's earlier elaborated doctrine of "isolation and superposition." For enduring and essential *Vieldeutigkeit* will arise, according to Volkmann, only in cases of "composite" ["zusammengesetzte"] phenomena (Volkmann 1902, pp. 122–126).

Volkmann once again adduces the example of vector decomposition. But to this he adds a new, more intriguing example from the theory of light, namely, that provided by the controversy that prevailed throughout much of the nineteenth century over the position of the plane of oscillation relative to the plane of polarization. Fresnel had argued that the former is orthogonal to the latter; Franz Neumann had argued that they coincide. Maxwell's electromagnetic theory of light finally resolved the controversy by showing that, in a sense, both were right: Fresnel's point of view is correct as regards the electrical oscillations, and Neumann's is correct as regards the magnetic oscillations.

With the benefit of hindsight, we could easily criticize Volkmann's example as not representing a case of enduring *Vieldeutigkeit*. And one might also argue that the relationship between the electrical and magnetic planes of oscillation in Maxwell's electrodynamics has more in common with the state of affairs that Bohr designated "complementarity" than with the kind of *Vieldeutigkeit* that is evinced in genuine cases of non-categoricity. But to argue thus would be to miss the point of Volkmann's discussion. His examples of physical *Vieldeutigkeit* may be poorly chosen to illustrate the logical and epistemological point that he was attempting to make, a point better grasped perhaps by keeping in mind the purely mathematical illustrations with which he began. For this situation—the existence of multiple solutions to a single problem—is exactly the kind of situation that forces itself upon our attention in Einstein's wrangling with the hole argument; and Volkmann deserves credit for being the first to argue that the demand for a single solution, the demand for *Eindeutigkeit* may be nothing more than a metaphysical or methodological prejudice.

There is no argument to be made for Volkmann's having had a direct or significant influence on Einstein, since, after all, Einstein's preference for the metaphysical *Eindeutigkeit* principle runs counter to the thrust of Volkmann's position. But was Einstein nevertheless familiar with Volkmann's writings? Did they form part of the background to Einstein's own thinking about the *Eindeutigkeit* principle?

I have found no mention of Volkmann in Einstein's own writings or in his correspondence. It is highly likely, however, that Einstein was at least aware of Volkmann and of some of his better-known views, and this not merely because, as mentioned above, Volkmann's *Grundzüge* and *Einführung* were rather widely-read during the period 1896–1910, nor because of the fact that they were the kinds of writings that would have piqued the curiosity of the young Einstein.

Far more important is the fact that in all editions of his *Mechanik*, starting with the 1897 third edition, Mach included a long quotation from the very section of Volkmann's *Grundzüge* most relevant to the *Eindeutigkeit* question, this being the section on "isolation and superposition." After a discussion of Galileo's analysis of projectile motion, Mach wrote: "P. Volkmann has characterized the decomposition of processes into mutually independent parts in a telling way as *isolation* and the composition of a process out of such parts as *superposition*" (Mach 1897, p. 147).[86] And then he quotes the following passage from the *Grundzüge*:

> Only in the *rarest* of cases does nature, with her profusion of appearances, present herself to us in a unified manner; on the contrary, in the majority of cases the world of appearance carries a thoroughly *composite* character... then it must be one of the tasks of our understanding to conceive of the phenomena as they present themselves as being com-

> posed out of a series of partial phenomena and at first to study these partial phenomena in their purity. Only when we know which portion each circumstance sustains individually in the total phenomenon, only then do we command the whole. (Mach 1897, pp. 147–148; Volkmann 1896a, p. 70)[87]

Mach follows this quotation immediately with three cross-references to his *Wärmelehre* (1896). Two of them direct the reader to what are presumably intended to be other illustrations of the principles of isolation and superposition: Fourier analysis (p. 123), and Prévost's method of conceiving thermal radiation as a process in which the "emitter" and the "absorber" both simultaneously emit and absorb (p. 151).[88] The third directs the reader to a page containing a discussion of *eindeutige* determination and the following footnote: "Petzoldt has brought forth objections against my older assertions concerning *eindeutige* determination that have given me to reflection; but I must reserve discussion of these for a later occasion" (p. 452).[89] In the fourth (1901) and later editions of the *Mechanik*, Mach added remarks critical of Volkmann's preference for Newton as a philosopher of science (see, for example, Mach 1912, p. 260) and of Volkmann's reactionary preference for an absolute conception of space (Mach 1912, p. 270). But he also added complimentary remarks about some of Volkmann's other views, such as his endorsement of the "principle of comparison" ["Prinzip der Vergleichung"] (Mach 1912, p. 259), and he retained the quoted remark about isolation and superposition, along with the associated cross-references to the *Wärmelehre* (Mach 1912, p. 145).[90]

Another source from which Einstein could have learned about Volkmann is Ferdinand Lindemann's notes to the 1904 German edition of Poincaré's *La Science et l'hypothèse* (1902). In a note on vector decomposition, Lindemann refers the reader to Volkmann's discussion of isolation and superposition in the *Grundzüge* (Poincaré 1904, pp. 310–311). One finds there, in addition, references to a number of other writings by Volkmann, including his 1891 (pp. 320, 323), 1896b (p. 317), 1898a (p. 294), 1900 (p. 294), and 1901 (p. 295).

Since we know that Einstein had read Mach 1897 during his student years and that he had probably read Poincaré 1904 by no later than 1904,[91] we may assume that he was at least familiar with Volkmann's name. But whether or not Einstein had actually read Volkmann's epistemological writings, the latter's focus upon the problem of *Eindeutigkeit*, combined with Petzoldt's championing of the principle of *Eindeutigkeit*, is evidence of the interest in such questions among philosophically sophisticated physicists and scientifically literate philosophers in the first decade of this century.

6. Paul Natorp and *Eindeutigkeit*

After Hermann Cohen, Paul Natorp was the leading representative of Marburg neo-Kantianism, or critical idealism, as it is often called.[92] The members and affiliates of the Marburg school—including Einstein's philosophy teacher at the ETH, August Stadler[93]—were well known for their critical attention to physics and mathematics as the epitome of human knowledge and, therefore, the proper focus of an investigation into the conditions for the possibility of knowledge. But in the first two decades of the twentieth century, Natorp was probably the best-known interpreter of the implications for the sciences of Marburg neo-Kantianism, this on the strength of his monumental *Die logischen Grundlagen der exakten Wissenschaften* (Natorp 1910).

One of the many distinguishing features of the book is that it was the first major philosophical work to treat at length the philosophical implications of relativity theory, this in its concluding chapter 7, "Die zeit-räumliche Ordnung der Erscheinungen und die mathematischen Prinzipien der Naturwissenschaft," especially §11, "Das Relativitätsprinzip von Lorentz, Einstein, Minkowski," and §12, "Kritische Beleuchtung des Relativitätsprinzips und Bestätigung des Idealismus." Natorp's principal aim in these sections is to show that, far from refuting Kant, the theory of relativity offers a stunning confirmation of the fundamental Kantian thesis of the ideality of space and time, since it provides a scientific confirmation (superior to the philosophical arguments of Mach) of the associated Kantian thesis—which Natorp thinks Kant took over from Newton—that *empirical* time and space are necessarily relative. This is, of course, a non sequitur, but Natorp's logical failings need not delay us.[94]

What is important for our story is Natorp's frequent and crucial employment in these sections of the metaphysical *Eindeutigkeit* principle. It appears, for example, in Natorp's criticism of Mach's denial of absolute space and time. According to Natorp, Mach is willing to concede the existence of at best an *eindeutige* functional connection (a one-to-one mapping?) between the "times" or "spaces" of different observers, while denying the existence of a single, absolute time or space behind these relative times and spaces. But this functional connection may be enough to satisfy Natorp, who writes:

> Really, therefore, the question can meaningfully be put only so: is there a single, a uniquely determined ordering [eindeutig bestimmte Stellenordnung] of that which exists, namely, those series of changes that we conceive of as being functionally connected with one another, the totality of which we call "Nature"? But this question reduces with-

> out remainder to this other one: Does precisely that which we call Nature really exist? Does this single functional connection of events exist? Or most concisely: *Does existence exist?* (1910, p. 335)[95]

According to Natorp, such existence would mean, to the empiricist

> nothing less than that a functional connection of "the" events merely would be or at least could be given in a uniquely determined fashion, in a single way. For existence means nothing else to him, and nothing else in general, than: complete, in no respect uncompleted determination of being, and indeed, since it is a question here of spatio-temporal being, completed determination in regard to time and space. (1910, pp. 335–336)[96]

And even if we deny that an absolute time is given in experience,

> it does not follow from that that the *requirement* of a unique determination of the ordering of that which exists is to be repudiated as meaningless. That would mean giving up the concept of existence itself; it would mean giving up the *project* of knowledge precisely as experience. (1910, p. 336)[97]

Mach certainly does not take this step, according to Natorp, and so he and Kant are really in agreement, except that Mach neglects the fact upon which Kant insists, that even as a mere *idea*, absolute time and absolute space are necessary as that to which all relative temporal and spatial determinations are reduced, as that which secures the possibility of a unique functional connection of events (1910, pp. 336–337). Other empiricist philosophers, says Natorp, have already been driven to this conclusion, and here he cites Petzoldt's "Das Gesetz der Eindeutigkeit" (Petzoldt 1895); but Natorp's fundamental misunderstanding of Petzoldt need delay us no longer than did his earlier non sequitur.

Natorp's interpretation and employment of the metaphysical *Eindeutigkeit* principle is deeply rooted in his reading of Kant, *Eindeutigkeit* or *Einzigkeit* (singularity—Natorp uses the two terms interchangeably) being on this reading the chief characteristic whereby that which is real—an object—in our experience is distinguished from mere concepts, and the *Eindeutigkeit* or *Einzigkeit* of time and space being therefore necessary grounds for the possibility of such experience:

> Reality means a determination, such that nothing remains undetermined. Indeterminacy is just mere possibility. In other words: possibility is multifaceted, it always permits a choice; reality is absolutely singular, it is conceived as being determined in a singular fashion, excluding all choice. This singularity plays a great role in Kant's theory of experience; upon its being required rests in particular his distinction of the

> intuition from the concept; intuition means to him "the representation that can only be given through a singular object." Time and space are in this sense "essentially singular" representations, and therefore intuitions; there is only one time, only one space, just as there is only one experience "in which all perceptions are represented as being in thoroughgoing and law-governed connection." Clearly the singularity of time and of space counts necessarily for him as the condition for the singularity of experience; how can the connection of experience be a singular one, without the singular time and the singular space being its foundation? (1910, p. 92)[98]

Here is the foundation upon which Natorp builds his argument for the existence of an absolute space and time, because empirically determined spatial and temporal relationships, being relative, do not have the requisite *Eindeutigkeit*:

> It is certain from the outset... that no really achievable foundation for temporal or spatial determination based upon empirical means of determination may claim any absolute security. Nevertheless— indeed, precisely thereby—one makes the fundamental presupposition that there is an absolute time, an absolute space for events in themselves [des Geschehens an sich], by means of which, if they were determinable, events themselves would be absolutely determined; by comparison with these every empirically possible determination could claim only the significance of a useful approximation. Upon what basis do we make this presupposition? It lies originally in the demand, in the final, fundamental supposition that experience is directed toward knowledge of that which is or proceeds "really." *Existence* can be conceived only in a single way, because it really means nothing other than determination in a single way. It is thus an "analytic proposition" that, insofar as existence should be spatio-temporally determined, this determination itself is required to be simply unique [eindeutig], although it is never given as such. Only a simply unique [eindeutig] (and that means: absolute) temporal and spatial determination *would be* the determination of existence itself; that the empirically possible determination can never be simply unique [eindeutig] in no way changes this conditional maxim. (1910, p. 328)[99]

Natorp goes on to infer more specific properties of space and time from the requirement of *Eindeutigkeit*, starting with the a priori necessity of the Euclidean character of space: "Euclidean space... [is] a 'necessary' presupposition in the specific sense that it is a condition for 'possible experience,' more specifically for the unique [eindeutige] law-governed determinability of *existence* in experience" (1910, p. 312).[100] He attempts, similarly, to ground Newton's three laws and Hamilton's principle—as principles of rational mechanics, not empirical physics—in the *Eindeutigkeit* requirement (pp. 370, 372, 382).

Given that Natorp sees absolute space and time, along with the principles of classical mechanics, as being grounded in the necessary assumption of the *Eindeutigkeit* of objects as given in experience and of the conditions for the possibility of such experience, how can he accommodate the (special) theory of relativity, which seems to be inconsistent with most of the central tenets of this classical physics? How can he accommodate relativity theory without compromising the *Eindeutigkeit* principle?

The less satisfactory part of his answer was suggested above, where I noted that Natorp regarded the principal achievement of Lorentz, Einstein, and Minkowski as having confirmed, in a surprising way, the Kantian thesis of the ideality of (absolute) space and time. More interesting, and more insightful, however, is a second answer that concerns not space and time, which, as empirical space and time, are not determined *eindeutig* according to special relativity, but rather the laws of nature:

> Otherwise, the fundamental law of unique [eindeutigen] determination loses thereby none of its logical validity; only the possibility of its empirical fulfillment is, in a certain way, restricted. The freedom in the choice of the time parameter, as in that of the spatial coordinates, does not, after all, threaten the unity of our knowledge of nature, because it does not touch the uniqueness [Eindeutigkeit] of the *laws of nature*. Precisely this is perhaps the most important result of Minkowski's investigations: the "invariance" of the laws of nature under all "Lorentz transformations." . . . A uniqueness [Eindeutigkeit] of empirical temporal and spatial determination more thoroughgoing than that which suffices for the establishment of laws of nature, is indeed not necessary. (1910, pp. 403–404).[101]

Though Natorp's neo-Kantian framework differs wholesale from Petzoldt's empiricist framework, the conclusion to which both come is nevertheless strikingly like that which underlies Einstein's point-coincidence argument: *Eindeutigkeit* is not compromised if we assume that what is *eindeutig* in nature, and that what therefore must be determined *eindeutig* by our theories, is all and only that which remains invariant under the class of transformations characterizing our theory.

Natorp was the first major philosophical figure to have discussed the relativity theory at any length, and that fact alone leads one to suspect that Einstein—always interested in good philosophy from which he could learn something of benefit to his physics—was aware of *Die logischen Grundlagen der exakten Wissenschaften.* But is there any real evidence of Einstein's possible acquaintance with the book and its discussion of *Eindeutigkeit* and relativity?

There is, indeed, as far as I can determine, no direct evidence of Einstein's having read Natorp. Still, there is circumstantial evidence suggesting that Einstein would have been familiar at least with Natorp's general position. A small correspondence between Einstein and Natorp survives, but the earliest item is a letter from Einstein to Natorp dated May 11, 1919, which concerns, anyway, only a political initiative in which both were involved, and the correspondence permits no inference, either positive or negative, regarding earlier acquaintance.[102] Far more important is some evidence bearing directly not upon the relationship between Einstein and Natorp, but upon Einstein's relationship with one of Cohen and Natorp's student's, Otto Buek.

Buek was, along with other students like Ernst Cassirer and Dimitry Gawronsky, a member of the inner circle around Cohen and Natorp.[103] He took a degree at Marburg in 1904, with a dissertation entitled, "Die Atomistik und Faraday's Begriff der Materie. Eine logische Untersuchung" (Buek 1905; see also Buek 1912), and later became a contributing editor for both the Cassirer (1912–1923) and Vorländer (1920–1940) editions of Kant's works. Most important for our purposes is the fact that by late 1914 he had developed a fairly close and regular relationship with Einstein, who had moved to Berlin, where Buek was then working, in April of that year. In a letter to Natorp on November 28, 1914, Cohen writes:

> It is very interesting that Buek is attending Einstein's lectures and often meets with him and is in the habit of having profound conversations with him. Yet he finds him unclear in philosophical matters, and still has no real opinion about the whole matter, with the sole exception of the difficult mathematics. (Holzhey 1986, vol. 2, p. 436; Universitäts-Bibliothek Marburg, Ms. 831/52)[104]

Some years later, Buek himself gave an account of these early meetings, lacking, sadly, the sharp comments on Einstein's philosophical competence, but adding details such as that many of these meetings took place at the home of the medical doctor and pacifist Georg Friedrich Nicolai (see below), and that he and Einstein often played music together.[105] In the winter semester of 1914, Einstein was indeed lecturing on relativity in Berlin,[106] a subject that Buek had declared an intention to work on already three years earlier.[107] And at least slim evidence for the conjecture that Einstein and Buek discussed philosophical issues is provided by the fact that a copy of the 1909 second edition of a collection of Kant's writings on natural philosophy (Buek 1909) was in Einstein's Princeton Library.

But Einstein's relationship with Buek was based as much on shared political interests as on scientific, philosophical, or musical ones. In late 1914, Buek—who reportedly tended toward anarchism in his politics (Holzhey 1986, vol. 2, p. 343)—co-authored with Einstein and the pacifist Georg Friedrich Nicolai a "Manifesto to Europeans" that was intended as a critical reply—defending the idea of European unity and international intellectual cooperation—to the infamous "Manifesto to the Civilized World," signed by ninety-three prominent German scientists and intellectuals (hence also commonly known as the "Manifesto of the 93"), and intended to justify to the international intellectual community the German invasion of neutral Belgium at the outset of the First World War.[108]

Given the nature of Einstein's relationship with Buek, and given Buek's central place within the circle around Natorp and Cohen (who had himself returned to Berlin upon his retirement from Marburg in 1912), it seems highly likely that Einstein would have learned about Natorp's view of relativity, as expressed in *Die logische Grundlagen der exakten Wissenschaften*, by middle or late 1914 at the latest, if, indeed, he had not already read Natorp earlier. Thus we have here another possible source from which Einstein could have learned about the metaphysical *Eindeutigkeit* principle.

7. Cassirer on *Eindeutigkeit* and Reality

Paul Natorp was not the only representative of Marburg neo-Kantianism whose ideas may have found their way to Einstein through the agency of Otto Buek. Indeed, Buek was part of what can only be described as a Marburg colony active in Berlin at the time of Einstein's arrival in 1914. Hermann Cohen himself returned to Berlin upon his retirement from Marburg in 1912 and worked there continuously until his death in 1918. And Ernst Cassirer, who was the preeminent student of Cohen and Natorp, having taken a degree with them in 1899 for a dissertation on "Descartes Kritik der mathematischen und naturwissenschaftlichen Erkenntnis," was a Privatdozent at the University of Berlin from 1906 to 1919, when he finally received a regular academic appointment as Ordinarius at the University of Hamburg. It is quite conceivable, if not likely, that Einstein would have gotten to know both Cassirer and Cohen through his relationship with Buek or other mutual acquaintances; but, failing that, it is more than likely that he at least would have been exposed to their works and ideas through conversation with Buek. And

Cassirer is another thinker from whom Einstein could have learned about contemporary discussions of the *Eindeutigkeit* principle.

In 1910, the same year in which Natorp's *Die logischen Grundlagen der exakten Wissenschaften* was published, Cassirer published his first systematic work on epistemology—it is really better described as a treatise on the philosophy of science—that being his *Substanzbegriff und Funktionsbegriff. Untersuchungen über die Grundfragen der Erkenntniskritik* (Cassirer 1910). Given the fact that Cassirer was later to publish one of the classic works on the philosophical implications of relativity theory (Cassirer 1921), it is more than a little surprising that there is no mention of Einstein or the relativity theory in this earlier work. Cassirer does discuss the question of the relativity of space and time, but he does so entirely within the framework of classical, Newtonian physics. Still, the general epistemological point of view developed there is rich with implications for the principle of *Eindeutigkeit*.

Cassirer's principal aim in *Substanzbegriff und Funktionsbegriff* is to argue that knowledge in general, especially the kind of knowledge epitomized by modern mathematical and scientific theories, is properly understood as concerning itself not with individual substances and their attributes, but instead with patterns of functional relationships. Such a view of knowledge was anticipated by Leibniz, according to Cassirer (see especially Cassirer 1902), but he believed that it had come ever more self-consciously to the fore in the profound foundational investigations in mathematics and the sciences that began to appear in the late-nineteenth and early-twentieth centuries. Cassirer's view has much in common with that later to be developed by Schlick in his *Allgemeine Erkenntnislehre* (1918), where Schlick argued that all knowledge is knowledge of form, rather than content. But, of course, Schlick was a more-or-less thoroughgoing empiricist, believing in the corrigibility of all knowledge claims, whereas the Marburg neo-Kantian influence survives in Cassirer's insistence that some, at least, of the relations constitutive of human knowledge are a priori—and hence immune to empirically motivated calls for their replacement—in the sense of being necessary and universal features of all knowledge claims.

In Cassirer's words, "the concept of function constitutes the general schema and model according to which the modern concept of nature has been molded in its progressive historical development" (Cassirer 1910, p. 27 [21]),[109] but this circumstance finds its clearest exemplification in the science of number, where "consciousness of the meaning and value of the formation of concepts first develops. In the thought of number, all the power of knowledge seems contained, all possibility of the logical determination of the sensuous" (p. 35 [27]).[110] As Cassirer puts it, "the concept of number is fulfilled in the general concept of function, which permeates it" (p. 96 [74]);[111] and "the relational structure as such, not the absolute property of the elements, constitutes the real object of mathematical investigation" (p. 122 [93]).[112]

More to the point for the present investigation, however, is the manner in which the relational character of knowledge is manifested in geometry, where Hilbert's *Grundlagen der Geometrie* (1899) and the associated doctrine of implicit definition best expresses the tendency of thought Cassirer wants to highlight: "The nature of the original geometrical objects is here exclusively defined by the conditions to which they are subordinated. . . . The point and the straight line signify nothing but structures which stand in certain relations with others of their kind, as these relations are defined by certain groups of axioms" (p. 123 [93–94]).[113]

But how do we choose among the various logically possible systems of geometrical relations? Like Kant, Cassirer believes that experience does not present to us bare givens, uninformed by those "conceptual interpretations" that are necessary for the very constitution of the objects presented in experience; and so he believes that experience alone cannot determine our choice of a geometry. Not all conceivable systems are compatible with the objects we encounter in experience, logical consistency being merely a "negative condition," but many are, and our choice among them is guided ultimately by simplicity, a criterion that the Cassirer of 1910 thinks will favor Euclidean geometry. Still, he recognizes that new experiences might force a reevaluation. What then is left of the a priori? "The pure concept on its side is prepared and fitted for all conceivable changes in the empirical character of perceptions; the universal serial form is the means by which every order of the empirical is to be understood and logically mastered" (p. 147 [111]).[114]

Quite separate from the question of how we choose among alternative systems of geometry is the question of the relation between any one of those systems and the "reality" it purports to describe. But what Cassirer has to say about this latter question will shed important light on the manner in which he understands the relational character of knowledge.

Recall Natorp's reading of Kant, upon which what distinguishes the real object in experience or intuition from the concept is the *Eindeutigkeit* of the former as compared with the inherent *Vieldeutigkeit* of the latter. Cassirer starts from a similar assumption, but he supplements Natorp by asking, in effect, how inherently *vieldeutige* concepts can be employed to construct a complete representation of an *eindeutige* reality.

Cassirer's starting point for this discussion is, quite interestingly, Volkmann's discussion of "*Isolation*" and "*Superposition.*" But in what can only be characterized, charitably, as an act of creative and perhaps willful misreading, Cassirer interprets superposition in a manner significantly different from Volkmann. The essential, if ineptly employed, metaphor for Volkmann was that of vector addition and decomposition: there is always a multiplicity of different possible ways of decomposing a vectorial quantity, depending upon one's choice of basis vectors, and hence, for Volkmann, the reality represented by the superposition of vector components is not something *eindeutig*. For Cassirer, by contrast, the whole point of constructing a representation by superposition is to combine sufficiently many relations as to determine up to the point of isomorphism the reality one is seeking to represent. And it is precisely to the extent that one can hope thus to succeed in specifying a domain of objects unambiguously (*eindeutig*) by the relations they satisfy that one can claim to have eliminated concepts of substance, or "thing" concepts, from science in favor of relations.

Richard Dedekind's definition of "number" by means of the concept of a "progression" (Dedekind 1893) is Cassirer's model for this way of understanding how a science represents its reality. Cassirer quotes Dedekind to the effect that mapping (*Abbildung*) is the basic mental capacity upon which counting and thus the concept of number is based: "Upon this one, but absolutely inevitable, foundation the whole science of number must be erected" (Cassirer 1910, p. 46 [36]: Dedekind 1893, p. viii). Cassirer then comments:

> The starting point here seems to be the traditional logical doctrine of a plurality of *things* and the power of the mind to copy [abzubilden] them; nevertheless, it becomes evident with deeper understanding that the old terms have here gained a new import and meaning. The "things," which are spoken of in the further deduction, are not assumed as independent existences present anterior to any relation, but they gain their whole being, so far as it comes within the scope of the arithmetician, first in and with the relations which are predicated of them. Such "things" are terms of relations, and as such can never be "given" in isolation, but only in ideal community with each other. (1910, pp. 46–47 [36])[115]

After sketching the particular system of relations constituting the kind of serial order or "progression" that Dedekind regards as the essence of number, Cassirer explains the significance of what Dedekind has accomplished:

> All the propositions of arithmetic, all the operations that it defines, are related solely to the general properties of a progression; hence

> they are never directed primarily upon "things" but upon the ordinal relation, which prevails between the elements of certain systematic wholes.... The function of "number" is, in its meaning, independent of the factual diversity of the objects which are enumerated; this diversity must therefore be disregarded when we are concerned merely to develop the determinate character of this function. Here abstraction has, in fact, the character of a liberation; it means logical concentration on the relational connection as such with rejection of all psychological circumstances, that may force themselves into the subjective stream of presentations, but which form no actual constitutive aspect of this connection.... What is here expressed is just this: that there is a system of ideal objects whose whole content is exhausted in their mutual relations. The "essence" of the numbers is completely expressed in their positions. (1910, pp. 49–51 [38-39])[116]

Of course, Dedekind's way of regarding the numbers can be generalized to other sciences. In Cassirer's words:

> Whenever a *system of conditions* is given that can be realized in different contents, there we can hold to the form of the system itself as an *invariant*, undisturbed by the difference of the contents, and develop its laws deductively. In this way, we produce a new "objective" form, whose structure is independent of all arbitrariness; but it would be uncritical *naïveté* to confuse the object which thus arises with sensuously real and effective things. We cannot read off its "properties" empirically, nor do we need to, for its form is revealed in all determinateness as soon as we have grasped in its purity the relation from which it develops. (1910, pp. 52–53 [40–41])[117]

And Cassirer sees at work here once more the legacy of Leibniz:

> Again we are led to the Leibnizian conception of mathematics. According to this conception, mathematics is not the general science of magnitude but of form, not the science of quantity but of quality.... Wherever a definite form of connection is given, which we can express in certain rules and axioms, there an identical "object" is defined in the mathematical sense. (1910, p. 121–122 [92–93])[118]

In the late-nineteenth century, Felix Klein's Erlanger program sought to characterize the whole of mathematics in such terms, this by defining the various areas of investigation group theoretically, in terms of a set of transformations and their associated invariants (see Cassirer 1910, p. 116–118). But it is again Hilbert's way of regarding geometry that represents the culmination of this tendency. Earlier I quoted Cassirer's characterization of Hilbert's view, according to which the concepts of point and line are defined implicitly by the geometrical axioms; that quotation continues: "Only this systematic 'complexion' of the elements, and not their particular characters, is taken here as the expression of their essence. In this sense, Hilbert's geometry has been correctly called a pure theory of relations" (1910, p. 123 [94]).[119]

What is the general lesson to be drawn from an understanding of the way arithmetic and geometry are related to the "realities" they pretend to represent?

> The determination of the individuality of the elements is not the beginning but the end of the conceptual development; it is the logical goal, which we approach by the progressive connection of universal relations. The procedure of mathematics here points to the analogous procedure of theoretical natural science, for which it contains the key and the justification. (Cf. Ch. V.) (1910, p. 124 [94]) [120]

The forward reference to "Ch. V." is to the first chapter of part II, entitled "The System of Relational Concepts and the Problem of Reality," where Cassirer takes up Volkmann's doctrine of "isolation" and "superposition."

Permit me to quote at length the central paragraphs of chapter V, where Cassirer advances the most important conclusions of this point of view for our understanding of the relation between theory and reality and for the general methodology of science. Cassirer writes:

> While reality presents us a mixture of heterogeneous circumstances seemingly inseparably interwoven and confused, thought demands the separate consideration of each particular moment and the exact determination of the part it plays in the structure of the whole. . . . The complete picture of the total process arises when we connect the partial system[s] and place them, as it were, upon each other. This picture of the total process appears not merely as a unitary intuition, but as a differentiated whole, in which the type of dependency between the individual elements is exactly defined. . . . Thus when the principle of "isolation" and "superposition" is explained by the fact that all reality represents only the sum of the manifestations of particular laws of nature, and is to be conceived as having issued from these laws, the real epistemological meaning of the thought is concealed. We cannot be concerned here with the origin of things, but only with the origin and character of our insight into things. The "real," as it is grasped in the sensuous impression, is not in and for itself a "sum" of various elements, but first appears to us as an absolutely simple and unanalysed whole. This original "simplicity" ["Einfalt"] of intuition is only transformed into a manifold [inneren Vielgestaltigkeit] by the logically analytic work of the concept. The concept is here just as much a source of plurality as it elsewhere appears to be a source of unity. Since we

> conceive a particular process as successively introduced into different systems, whose general structure can be mathematically deduced, we thereby increase its determinateness, in so far as we define its position in the general plan of our thought with increasing exactness. (1910, pp. 339–342 [256–258])[121]

Cassirer's remark about the inherent "plurality" ["Vielheit"] of concepts echoes Natorp's claim that, for Kant, *Eindeutigkeit* is the mark distinguishing intuitions from concepts. But Cassirer goes beyond Natorp in arguing that by accumulating conceptual relations, one can achieve an ever greater determinateness through concepts alone, "the determination of the individuality of the elements" being, as he wrote earlier, "not the beginning but the end of the conceptual development."

That complete, unambiguous [*eindeutige*] determination of the goal is confirmed, not so ironically, as it turns out, in a comment on Mach. Once again, I quote Cassirer at length:

> If we represent mathematically a given totality of observations by the "superposition" of several series, we do not indeed thereby increase our knowledge of the absolute and transcendent causes of the process; but we have raised ourselves to a new *type of knowledge*.... To describe a group of phenomena, then, means not merely to record receptively the sensuous impressions received from it, but it means to transform them intellectually. From among the theoretically known and developed forms of mathematical connection... a selection and combination must be made such that the elements given here and now appear as constructively deduced elements in the system, which arises. The logical moment given here cannot be denied even in the theories of empiricism, or under whatever names it may be concealed. The "adjustment of ideas to reality" presupposes the very concept of this reality, and thus a system of intellectual postulates. It is in the principle of the unambiguity of natural processes [*Eindeutigkeit des Geschehens*—emphasized with Sperrdruck in the German original] above all, that all these postulates are ultimately combined. "I am convinced," says Mach himself, "that in nature only so much happens as can happen, and that this can happen only in one way." All physical process is thus completely determined by the momentarily effective circumstances, and can thus take place in only one way. However, if we analyse the grounds of this conviction, we are implicitly led back to all those fundamental conceptions, which the sensationalistic explanation explicitly denies. The conception of the unambiguity [Eindeutigkeit] and "stability" of being obviously is not in the content of perception as given in our first immediate experience, but it indicates the goal, which the intellectual labor of science seeks to have this content approximate. This goal can only be reached, if we are able to establish certain permanent relations in the flux of sensations,

> which differ and have their truth limited to a single moment of time; the rules of these permanent relations we can call to mind independently of change of the momentary material. To the extent that this takes place, the scientific concept of nature develops and is confirmed. (1910, pp. 349–351 [263–264])[122]

Of course, the passage that Cassirer cites from Mach comes from the very pages of the *Wärmelehre* (Mach 1896, 1900, p. 393) where, in the course of defending his biological-evolutionary economism against earlier criticisms by Petzoldt (1890), Mach expresses his qualified endorsement of the *Eindeutigkeit* principle. And while Cassirer himself nowhere cites Petzoldt, his citation of this passage from Mach makes it clear that his invocation of the *Eindeutigkeit* principle nevertheless places him squarely in the middle of the discussion of *Eindeutigkeit* inaugurated by Petzoldt and by this time carried forward by, at the very least, Mach, Volkmann, and Natorp.

Cassirer, however, is the only one of these thinkers, other than Volkmann, to have advanced what is arguably an assertion of the model-theoretic *Eindeutigkeit* principle, as well as the more standard metaphysical version of the principle. That is to say that Cassirer not only asserts the *Eindeutigkeit* of the real, but also the *Eindeutigkeit* of those objects implicitly constructed or defined within a mathematical theory by the accumulation of sufficiently many axioms or principles intended to characterize the essential properties of the objects supposed to fall within the domain of the theory, in effect, the objects constituting a model for the theory. Moreover, profiting especially from the work of Hilbert, Cassirer's treatment of these issues is superior to Volkmann's by virtue of its sophistication about formal issues. To be sure, the metaphysical and model-theoretic *Eindeutigkeit* principles are not distinguished as clearly as they might be by Cassirer, this no doubt owing to his Marburg neo-Kantian orientation, which eschews any knowledge of things in themselves, regarding the reality described by science as consisting of those "real" objects that are constituted by the operations of the mind embodied in scientific theories, thus blurring the distinction between model and reality. Still, the tendency of Cassirer's thinking is to emphasize in an especially clear manner the *Eindeutigkeit* (or lack thereof) of the objects constituting models for our theories, such as the numbers described by Dedekind's number theory and the points and lines defined by Hilbert's geometry.

I can find no positive evidence of Einstein's having read Cassirer's *Substanzbegriff und Funktionsbegriff* prior to 1915, or at any later time for that matter. There is abundant evidence of Einstein's having later read carefully, while it was still in manuscript, Cassirer's *Zur Einsteinschen Relativitätstheorie* (1921),[123] where the *Eindeutigkeit* principle is taken up explicitly in connection with relativity, and there is an interesting correspondence between Einstein and Cassirer starting at about this same time, namely, after Cassirer's move to Hamburg in 1919;[124] but that is too late for this chapter in my story of Einstein and *Eindeutigkeit*. On the other hand, as I suggested at the beginning of this section, the same close, personal relationship between Einstein and Buek that could have served as a vehicle for Einstein's introduction to Natorp's writings could have served equally well as a vehicle for his introduction to Cassirer's writings, if not to Cassirer himself, who was, one will recall, living and working in Berlin in 1914–1915, and lecturing as Dozent at the same University of Berlin where Einstein was then just starting to lecture.

Whatever the evidence might say about Einstein's having read or not read *Substanzbegriff und Funktionsbegriff* by the fall of 1915, one is especially struck by the occurrence in the latter work of the very same phrase, "Eindeutigkeit des Geschehens," that we earlier encountered in Einstein's December 26, 1915 letter to Ehrenfest (EA 9-363) setting out the point-coincidence argument, and before that in Petzoldt's "Das Gesetz der Eindeutigkeit" (1895). The phrase is so distinctive that one cannot help but ask where Einstein picked it up. Cassirer is clearly a possible source.

8. Moritz Schlick and *Eindeutigkeit* as the Essence of Truth

There is one final possible source for Einstein's acquaintance with the *Eindeutigkeit* principle, and this specifically the model-theoretic version of the principle. It is Moritz Schlick's 1910 essay, "Das Wesen der Wahrheit nach der modernen Logik" (Schlick 1910), in which he first asserted his subsequently very influential doctrine that the essence of truth consists in an "eindeutige" coordination of judgments or propositions to facts.

The starting point for Schlick's definition of truth is an analysis of the nature of judgments,[125] which he says are nothing more than symbols [Zeichen] representing facts [Tatsachen]. Insofar as they function only as symbols, the most we can demand of them, as of any symbol, is that they be correlated one-to-one with that which they are intended to designate. Truth, according to Schlick, consists precisely in this relationship. In his own words:

> Judgments [are]. . . symbols for states of affairs. What conditions do we impose on a symbol? Only a single one, namely, that it be unam-

> biguous (*eindeutig*), or better still, one-to-one in character, i.e., every symbol must correspond to one and only one significatum and every significatum to just one symbol. Since, by its very nature, a symbol can have no other meaning than this, it follows immediately:
>
> *A judgment is true if it uniquely* (eindeutig) *designates a definite state of affairs.* (1910, p. 466)[126]

Elsewhere he makes it clear that the relationship he intends here between judgment and fact, between symbol and signified, is essentially the same as what the mathematician calls an "eineindeutige Zuordnung," a one-to-one correlation (see 1910, pp. 469, 474, etc.).

The same conception of truth is repeated, in a more concise form, in virtually all of Schlick's later writings on relativity (Schlick 1915, 1917a, 1917b, 1919, 1921, 1922a, and 1922b) and in his *Allgemeine Erkenntnislehre* (Schlick 1918, §§8, 10; 1925). Through the latter, especially, it entered the philosophy of science mainstream, and became the topic of regular discussion, as in Hans Reichenbach's *Relativitätstheorie und Erkenntnis Apriori* (Reichenbach 1920).[127]

But for all its influence, Schlick's definition of truth is not ultimately very satisfactory. One of the more serious problems with it is evident already in the fact that in the very paper where he first introduced it, Schlick wavered between characterizing truth as an "eineindeutige" correlation between judgments and states of affairs and a merely "eindeutige" correlation. While it is hard to express the difference between these two conceptions in fluid English, the difference is nevertheless a clear one. To advert to the mathematical vocabulary, it is the difference between a one-to-one correlation, and a correlation that may be many-to-one. The reason for Schlick's ambivalence is obvious. It is that, as Schlick himself noted (1910, p. 472), there will naturally be cases where two or more perfectly true judgments or propositions are correlated with one and the same fact, one and the same state of affairs; and while we may be tempted, in such cases, to think that for this reason these propositions "mean the same thing" or are, in some sense, the same proposition, it is not clear how one might go about proving that claim without begging the question.

The problem that arises here is, in fact, the very problem raised above under the heading of the "epistemological" *Eindeutigkeit* principle: the problem of whether or not a proposition, set of propositions, or theory is uniquely determined by a fact or set of facts. From another point of view, it is the problem posed by Duhem in his critique of the concept of the *experimentum crucis* (and by Poincaré, in yet another form), for Duhem argued, in effect, that for any given body of empirical evidence, any given set of facts, there will always be a multiplicity of alternative theories accounting equally—by empirical standards—for those facts, the choice among these alternatives being a matter of convention. In this guise, the present problem became the focus of a long-running discussion, involving Schlick, Einstein, and many others, about whether or not general relativity is uniquely favored by the empirical evidence, with Einstein and Schlick (at least in the years before about 1925) favoring the Duhemian (Poincaréan) point of view.[128] Given the subsequent direction of Schlick's thinking on these questions, it is no surprise that he hesitated already in 1910 to assert, without qualification, that truth is a matter of a one-to-one correlation between judgment and fact.

A second, and still more revealing problem concerns false judgments. If a true judgment is one correlated *eindeutig* to a given state of affairs, then a false judgment must be one not so correlated. That would mean either that it is correlated with no state of affairs or that it is correlated with more than one. According to Schlick, the latter is the case, a false judgment is, he says "zwei- oder mehrdeutig" ["doubly or multiply ambiguous"] (1910, p. 471). He adds:

> And it is indeed immediately clear that the only significance of the distinction between true and false statements is, in fact, to preserve the *Eindeutigkeit* of linguistic and conceptual expression, which is the necessary precondition for all understanding, without which every signification and every expression would be altogether meaningless. (1910, p. 471)[129]

One may agree with Schlick on the desirability of unambiguous means of expression, without going so far as to assert that it is a necessary precondition for all understanding (since a complete lack of ambiguity, of *Vieldeutigkeit*, is not achievable for any interesting proposition or theory), and certainly without following him to the curious conclusion that ambiguity is tantamount to falsity. That Schlick's conception of falsity is unworkable is shown already by such simple examples of true, but ambiguous propositions as that "John is twenty-one years old" (there having been many people named John born in 1970).[130]

But while the unworkability of Schlick's conception of falsity brings out, thus, the unworkability of the underlying definition of truth, it does not follow that the basic idea from which Schlick started—that of a one-to-one or many-to-one correlation between propositions and states of affairs—is without interest. The potential significance of the idea is evident from that fact that in the hands of thinkers equipped with a broader array of formal tools than Schlick had at his disposal, including, especially, a better understanding of the relation between syntax and semantics, it was transformed into Tarski's semantic definition of truth as satisfaction (Tarski 1931).

More important for our purposes, however, is the fact that the idea of an *eindeutige* correlation between propositions and states of affairs is a version of what we are calling the model-theoretic *Eindeutigkeit* principle, the principle that an acceptable theory should determine uniquely its own model, that it should, in this sense be correlated uniquely with a single state of affairs. Consider again what Schlick says about "false" propositions, namely, that they are *vieldeutig*, correlated with more than one real state of affairs. We would now characterize such propositions not as false, but as non-categorical. That something like categoricity was the property of a "good" theory or proposition that Schlick was really aiming to capture in his theory of "truth," whatever name he may have given to the property, is suggested even more clearly by the following remark from near the end of the paper in question:

> Under all circumstances, the world of facts must permit a system of symbols to be uniquely correlated with it and there is always only *one* truth, and it is for us unalterable and eternal, because every fact is only *one*, not alterable by means of any interpretation or any human manner of conception, even if it is there *before* any conception. (1910, p. 472)[131]

If reality or the world of facts is one, and if a "true" proposition is to be correlated uniquely with this one reality, then that proposition cannot admit several different models, it must admit only one model, a model corresponding in every relevant detail to the reality, the world of facts with which the proposition is to be correlated. Or, so as not to import subtle distinctions developed only later, let us just say that the one reality, the one world of facts, is the one model permitted to the proposition. (Notice that the "one truth" to which Schlick refers is the world of facts, not the set of propositions or symbols correlated with it.) To put it all in a nutshell then, the property of a good theory that Schlick labels "truth" is exactly what we are calling the model-theoretic *Eindeutigkeit* principle.

As noted above, Schlick repeated the essentials of this doctrine in all of his major writings on relativity theory. Consider what he said in the first of these, "Die philosophische Bedeutung des Relativitätsprinzips," (Schlick 1915). (Notice by the way that the Duhemian conventionalism is now more pronounced, the earlier error about whether or not the correlation must be one-to-one or may be many-to-one having been corrected):

> The totality of our scientific propositions, in word and formula, is in fact nothing else but a system of symbols *correlated* to the facts of reality; and that is equally certain, whether we declare the real to be

> a transcendent being or merely the totality and interconnection of the immediately "given." The system of symbols is called "true," however, if the correlation is completely *eindeutig*. Certain features of this symbol-system are left to our arbitrary choice; we can select them in this way or that without damaging the *Eindeutigkeit* of the correlation. It is therefore no contradiction, but lies, rather, in the nature of the matter, that under certain circumstances, several theories may be true at the same time, in that they achieve indeed a different, but each for itself completely *eindeutige* designation of the facts. (1915, p. 149)[132]

Schlick's deployment of the theory of truth as *eindeutige* correlation in his 1915 paper on relativity is especially interesting for understanding Einstein's attitude toward the *Eindeutigkeit* principle. Einstein read the essay within two weeks of completing work on the general theory of relativity, in mid-December 1915, and he responded immediately and enthusiastically, writing:

> Yesterday I received your essay and I have already studied it through completely. It is among the best which have until now been written about relativity. From the philosophical side, nothing appears to have been written on the subject which is nearly so clear. At the same time, you really have complete command of the subject. There is nothing in your exposition with which I can find fault. (Einstein to Schlick, December 14, 1915; EA 21-610)[133]

Elsewhere I have shown how a close (though not intimate) and respectful relationship grew up between Schlick and Einstein after 1915, a relationship in which each was heavily influenced by the other (Howard 1984; see also Hentschel 1984). For now, let us recall merely that Einstein is reacting thus enthusiastically to Schlick's essay, an essay featuring prominently the doctrine of truth as *eindeutige* correlation, at precisely the time when he is developing the point-coincidence argument in his above-quoted letters to Ehrenfest and Besso (see above, section three). It is also important to note that the point-coincidence argument figures prominently in Schlick's subsequent writings, most importantly his widely-read *Raum und Zeit in der gegenwärtigen Physik* (1917a, 1917b, 1919, 1920, 1922a), whose principal aim Schlick himself described thus: "The essay [is] less a representation of the general theory of relativity itself as it is an exhaustive explanation of the thesis that, in physics, space and time have now forfeited all objectivity" (Schlick to Einstein, February 4, 1917; EA 21-568).[134] One should recognize this "thesis"—it is *exactly* the thesis that Einstein himself repeatedly presented as the principal conclusion of the point-coincidence argument, including in his December 14, 1915 letter to Schlick (again, see above, section three).

So the presumption is strong that Einstein saw a deep connection between Schlick's theory of truth and his own deployment of the metaphysical and model-theoretic *Eindeutigkeit* principles in the point-coincidence argument. This could, of course, be wholly a matter of an after-the-fact convergence of views. But one wonders, was Einstein familiar with Schlick's earlier statement of his theory of truth?

There is no conclusive evidence of Einstein's having had any contact with Schlick or of his having read any of Schlick's writings before December of 1915. But there is some tantalizing circumstantial evidence pointing at least to the possibility of earlier contact. Most important in this connection is the fact that Schlick spent three years, from 1907 through 1910, studying philosophy at the University of Zurich with Gustav Störring and Friedrich Schumann.[135] Einstein's first regular academic appointment was at the University of Zurich, starting in the fall of 1909. That means that Schlick's time in Zurich overlapped with Einstein's by one year, that being exactly the time when Schlick was preparing his "Das Wesen der Wahrheit nach der modernen Logik" (Schlick 1910). Given the nature of Einstein's circle of acquaintances in Zurich at that time, including the philosopher Friedrich Adler as one of his common companions (see Howard 1990a); given the fact that Schlick was one of Max Planck's favorite students, having completed a Ph.D. under Planck in Berlin in 1906; and given that Schlick was a close friend of Max Laue, a fellow student of Planck's who had been in contact with Einstein for several years; given all of these considerations, I think it likely that the paths of Einstein and Schlick would have crossed in Zurich in 1909 and 1910. One should also remember that the philosophically *engagé* Laue himself taught at the University of Zurich for two years, 1912 to 1914, a period coinciding with Einstein's return to Zurich from Prague to teach at the ETH. Laue's correspondence with Schlick from this period makes it clear that Laue was in regular contact with Einstein, this at precisely the time when Einstein was working on general relativity and when Schlick was beginning his own study of the relativity theory;[136] so that even if there had been no contact between Schlick and Einstein prior to this time, Laue would have been in a position to introduce Einstein to Schlick's writings.

9. The Subsequent History of the *Eindeutigkeit* Principle and Relativity

Many strands in the history of late-nineteenth and early-twentieth century physics, mathematics, and philosophy come together in the hole and point-coincidence arguments. The philosophical tradition centering around discussions of the *Eindeutigkeit* principle is just one of these. But equally important for the history of physics, mathematics, and philosophy in this century are the lines of development that issue from the hole and point-coincidence arguments, as the significance of these arguments came to be more widely appreciated in the years following 1915. And the *Eindeutigkeit* principle plays, if anything, an even more important role in this later history than it did in the philosophical developments prior to 1915 sketched in this paper. Let me indicate briefly the more important of these later developments.

One of these later lines of development in mathematics has yet to be investigated at all, so far as I can tell. It is the influence of general relativity, and specifically the point-coincidence argument, on topology and differential geometry. I strongly suspect that the modern formulation of the concept of an abstract topological manifold owes much to the insights into the manner in which the points of the manifold are individuated that derive from the point-coincidence argument. Here is a topic deserving of the attention of historians of mathematics who are prepared to examine in detail the way the manifold concept evolves after 1915 in the work of thinkers like Weyl, Schouten, and Struik.

Another one of these later lines of development in mathematics, namely, the Cauchy problem in general relativity, has recently been investigated, in a preliminary fashion, by John Stachel (1988). The crucial step in this history, from the point of view of the present study, was taken in the second of Hilbert's classic notes on general relativity (Hilbert 1917). For Hilbert modified what he called "the principle of causality" in such a way as to stipulate the physical equivalence of all solutions of the field equations related to one another by mere coordinate transformations, this specifically in response to the point-coincidence argument (Hilbert 1917, p. 61). From another point of view, what Hilbert was saying here is that the Cauchy problem is solvable for the field equations plus boundary conditions in the sense that a solution unique (*eindeutig*) up to the point of isomorphism is determined.

On the side of philosophy, some of the fallout from the point-coincidence argument and Einstein's use of the *Eindeutigkeit* principle has been hinted at above. Thus Cassirer, for example, who in 1910 argued that the aim of all physical theory is to accumulate sufficiently many relations governing a given domain of phenomena so as to be able to characterize that reality in an *eindeutige* fashion, takes up in 1921 the question of whether or not and, if so, in what sense general relativity satisfies the *Eindeutigkeit* requirement, arguing that the topological properties of space can still be regarded as being given a priori because these, unlike the metrical properties of space, are determined *eindeutig*. And it was suggested that most of Schlick's writings on relativity after 1915 had it as their express aim to explicate the central thesis of the point-coincidence argument and to extract from it its chief philosophical consequences. Foremost among these was a conception of the role of conventions in science that, with the benefit of further refinement at the hands of Reichenbach, was to become a hallmark of Vienna positivism.[137]

There are, however, other philosophical implications of the point-coincidence argument and the *Eindeutigkeit* principle whose importance almost no one has yet recognized. I strongly suspect—and I can only state this as a conjecture for now, a topic for further research—that the modern concept of categoricity in formal semantics, far from having been presupposed prior to 1915 by the philosophers and physicists discussed here, was instead first clearly formulated as a direct result of later reflections on the significance of the point-concidence argument and the *Eindeutigkeit* principle.

My evidence for this conjecture is so far just circumstantial. The first adequate formulation of the modern concept of categoricity, namely, that a categorical theory is one that determines the class of its possible models up to the point of isomorphism, was given, curiously, by Rudolf Carnap in 1927; his term for such a theory was "monomorphic" (Carnap 1927). In understanding what led Carnap to this formulation, one must recall that in Carnap's Ph.D. dissertation on the problem of space in contemporary physics, the point-coincidence argument and the *Eindeutigkeit* principle are central to Carnap's argument that only the topological properties of space are determined *eindeutig* by experience, and that topological space therefore constitutes, conversely, the necessary ground for the possibility of experienced objects (Carnap 1922, pp. 38–39, 66).[138] One must remember as well, that Reichenbach, with whom Carnap was in regular contact by 1927, was himself at this time working out the implications of the *Eindeutigkeit* principle, as it appears in Schlick's definition of truth as *eindeutige* correlation, in his writings on the epistemology of relativity theory (Reichenbach 1920, 1924, 1928).

Almost simultaneously, essentially the same conception of categoricity made its appearance in the literature on set theory (see Fraenkel 1928, where Carnap is cited, pp. 349–350) and the foundations of geometry (Forder 1927). Also, in that same year, one finds one of the most interesting discussions of the philosophical significance of categoricity, or as he called it, the "isomorphism of material interpretations," in Hermann Weyl's classic *Philosophie der Mathematik und Naturwissenschaft*, where Weyl writes: "The idea of isomorphism represents the self-evident, unsurpassable boundary of knowledge" ["Der Isomorphiegedanke bezeichnet die selbstverständliche unübersteigbare Schranke des Wissens"] (Weyl 1927, p. 22). It is no accident, in my opinion, that Weyl is, of course, one of the most important contributors to the foundations of topology and general relativity. Lastly, it should be noted that the individual who first introduced the term "categoricity" in 1904, Oswald Veblen, had by the late 1920s turned his attention to differential geometry and general relativity.

A full history of the categoricity concept remains to be written.[139] But I am convinced that when it is written, it will accord a prominent place to the point-coincidence argument and the *Eindeutigkeit* principle.

Let me conclude with one really outrageous suggestion. Many people have puzzled over the curious conception of "completeness" that seems to be assumed in the famous Einstein–Podolsky–Rosen argument for the "incompleteness" of quantum mechanics. Recent scholarship has led to a major revision in our understanding of both Einstein's reasons for thinking quantum mechanics incomplete and the background to this worry in his earlier writings on the quantum hypothesis.[140] One result of these investigations has been to show that the operative criterion of completeness in Einstein's thinking was this: a theory is complete if and only if it assigns a unique theoretical state, such as a psi-function, to every unique real physical state. But if one understands a theoretical state as, in effect, a model for a set of equations plus boundary conditions (much as the metric tensor in general relativity is simultaneously a solution of the field equations plus boundary conditions and a representation of the reality described by the theory), then Einstein's conception of a completeness requirement should really be understood as a categoricity requirement. In other words, Einstein is saying that a "complete" (read "categorical") theory is one that determines a unique (*eindeutige*) model for the reality it aims to represent. Or to put it yet differently, the very same methodological principle that underlies the hole and point-coincidence arguments also underlies the far more famous argument for the "incompleteness" (read "noncategoricity") of quantum mechanics.

Acknowledgments: Among the many individuals to whom special thanks are owed both for helpful comments on earlier versions of this paper and for help in investigating various details of Einstein's life and work are John Stachel, John Norton, Robert Schulmann, David Cassidy, Jürgen Renn, A.J. Kox, and Thomas P. Ryckman. Thanks to John Stachel as well, and to Robert S. Cohen, for extending the hospitality of, respectively, the Center for Einstein Studies and the Center for the Philosophy and History of Science, both at Boston University, where much of the research for this paper was conducted. At various times, essential research support has been provided by the National Science Foundation (Research Grant No. SES-8421040), the American Philosophical Association, the Deutscher Akademischer Austauschdienst, and the University of Kentucky Research Foundation. Finally, I thank the Hebrew University of Jerusalem, which holds the copyright, for permission to quote from Einstein's unpublished letters and papers; items from the Einstein Archive are cited by their numbers in the Control Index.

NOTES

[1] See especially Earman and Glymour 1978a, 1978b, Stachel 1980, Pais 1982. Norton 1984, 1987, 1990, Earman and Norton 1987.

[2] For further details, including a dating of the discovery of the hole argument, see Stachel 1980.

[3] See, for example, Stachel 1986a, 1986b, where the point-coincidence argument is used to criticize the employment of a space-time manifold as the framework within which to construct space-time theories.

[4] Earman and Norton 1987, for example, have turned Einstein's point-coincidence argument into an argument against space-time substantivalism; see also Norton 1988.

[5] See below, section 9.

[6] Again, see below, section 9.

[7] For more on the connection between the Cauchy problem and general relativity, see Stachel 1988.

[8] The history of the *Eindeutigkeit* principle may have been forgotten, but the questions it raises survive in the work of at least two contemporary philosophers of science, W.V.O. Quine and Nelson Goodman. In different ways, each has developed a position entailing explicit denial of the *Eindeutigkeit* principle, Quine with his doctrines of referential inscrutability and ontological relativity (see, for example, Quine 1968), Goodman with his argument that the world is not one way, but many ways (see Goodman 1960); each evinces a thorough understanding of the issues at stake.

[9] See Stachel 1988.

[10] This is the version found in Einstein 1914, p. 1067; for a discussion of the evolution of the argument, see Stachel 1980.

[11] In effect, Einstein is assuming here an active, point transformation—a homeomorphic mapping of the manifold onto itself—in which the $g_{\mu\nu}$ are dragged along; for further details, see again, Stachel 1980, pp. 78–81.

[12] The $G'(x')$ constitute the image of the $G(x)$ under the passive coordinate transformation $x_\nu \to x'_\nu$.

[13] The $T_{\mu\nu}(x_\nu)$ and the $T'_{\mu\nu}(x_\nu)$ are everywhere identical, because outside of the hole, $x_\nu = x'_\nu$, and inside of the hole, $T_{\mu\nu} = c$, for any constant c, inside the region Σ; the designation "hole argument" (*Lochbetrachtung*), is thus somewhat misleading.

[14] For more detail, see again Stachel 1980; see also Norton 1984.

[15] "*Durch allgemein kovariante Differentialgleichungen für das Gravitationsfeld kann das Geschehen in demselben nicht eindeutig festgelegt werden.*

"Verlangen wir daher, dass der Ablauf des Geschehens im Gravitationsfelde durch die aufzustellenden Gesetze vollständig bestimmt sei, so sind wir genötigt, die Wahl des Koordinatensystems . . . einzuschränken."

[16] "Die im Sommer noch schwebenden Fragen betreffend die Gravitationstheorie haben sich unterdessen geklärt. Eine eindeutige Bestimmung der $g_{\mu\nu}$ aus den $T_{\mu\nu}$ ist nur bei Wahl spezieller Koordinatensysteme möglich (streng beweisbar)."

[17] "Mit der Gravitationstheorie bin ich nun sehr zufrieden. Die Thatsache, dass die Gravitationsgleichungen nicht allgemein kovariant sind, welche mich vor einiger Zeit noch so ungemein störte, hat sich als unumgänglich herausgestellt; es lässt sich einfach beweisen, das eine Theorie mit allgemein kovarianten Gleichungen nicht existieren kann, falls verlangt wird, dass das Feld durch die Materie mathematisch vollständig bestimmt wird."

[18] See also Einstein to Ehrenfest, fall 1913 (EA 09-342), as well as Einstein and Grossmann 1914a, p. 260, and 1914b, pp. 216–219.

[19] It is this claim that lay behind Einstein's common characterization of the main implication of the point-coincidence argument as being that "time and space thereby lose the last vestige of physical reality" ["dadurch verlieren Zeit und Raum den letzten Rest von physikalischer Realität"] (Einstein to Schlick, December 14, 1915, EA 21-611), the point being that, since spatio-temporal coordinates are themselves not invariant under coordinate transformations, they cannot correspond to anything real; see also Einstein 1915, p. 831, Einstein 1916, p. 776, and numerous other similar remarks.

[20] In a recent, careful reconstruction of the point-coincidence argument employing the full apparatus of differentiable manifolds (something unavailable to Einstein in 1915, lack of which contributes to various obscurities in Einstein's account of these matters), John Norton contends that, in addition to the principle of (active) general covariance, the point-coincidence argument tacitly presupposes a principle of (active) "Leibniz equivalence," according to which any two diffeomorphic models of a space-time theory represent the same physically possible

space-time (Norton 1989, pp. 1226–1231).

21 "In §12 meiner Arbeit vom letzten Jahre ist alles richtig (in den ersten 3 Absätzen) bis auf das am Ende des dritten Absatzes gesperrt Gedruckte. Daraus, dass die beiden Systeme $G(x)$ und $G'(x)$, auf das gleiche Bezugssystem bezogen, den Bedingungen des Grav. Feldes genügen, folgt noch gar kein Widerspruch gegen die Eindeutigkeit des Geschehens. Das scheinbar Zwingende dieser Überlegung geht sofort verloren, wenn man bedenkt, dass

1) das Bezugssystem nichts Reales bedeutet
2) dass die (gleichzeitige) Realisierung zweier verschiedener g-Systeme (besser gesagt zweier verschiedener Grav. Felder) in demselben Bereiche des Kontinuums der Natur der Theorie nach unmöglich ist.

"An die Stelle des §12 hat folgende Überlegung zu treten. Das physikalisch Reale an dem Weltgeschehen (im Gegensatz zu dem von der Wahl des Bezugssystem Abhängigen) besteht *in raumzeitlichen Koinzidenzen.** Real sind z.B. die Schnittpunkte zweier verschiedener Weltlinien, bezw. die Aussage, dass sie einander *nicht* schneiden. Diejenigen Aussagen, welche sich auf das physikalisch-Reale beziehen, gehen daher durch keine (eindeutige) Koordinatentransformation verloren. Wenn zwei Systeme der $g_{\mu\nu}$ (bezw. allg. der zur Beschreibung der Welt verwandten Variabeln) so beschaffen sind, dass man das zweite aus dem ersten durch blosse Raum-Zeit-Transformation erhalten kann, so sind sie völlig gleichbedeutend. Denn sie haben alle zeiträumlichen Punktkoinzidenzen gemeinsam, d.h. alles Beobachtbare.

"Diese Überlegung zeigt zugleich wie natürlich die Forderung der allgemeinen Kovarianz ist.

* und in nichts anderem!"

22 "Dadurch verlieren Zeit und Raum den letzten Rest von physikalischer Realität." See also Einstein 1915, p. 831, Einstein 1916, p. 776, and numerous other similar remarks.

23 See especially Schlick 1917b, pp. 50–63. In a letter to Einstein of February 4,1917 (EA 21-568), Schlick said about this monograph: "And so the essay is less an account of the general theory of relativity itself, than it is a thorough elucidation of the proposition that space and time in physics have now forfeited all objectivity." ["Und so ist der Aufsatz weniger eine Darstellung der allgemeinen Relativitätstheorie selbst als eine eingehande Erläuterung des Satzes, dass Raum und Zeit nun in der Physik alle Gegendständlichkeit eingebüsst haben."] And what was Einstein's judgment of Schlick's handling of this theme? In a letter to Schlick on May 21, 1917 (EA 21-618) he writes: "Again and again I look at your little book and am delighted by the splendidly clear expositions. Likewise, the last section, "Relations to Philosophy," seems to me splendid." ["Immer wieder sehe ich mir Ihr Büchlein an und freue mich der vortrefflich klaren Ausführungen. Auch der letzte Abschnitt 'Bezeihungen zur Philosophie' scheint mir vortrefflich."]

24 "An der Lochbetrachtung war alles richtig bis auf den letzten Schluss. Es hat keinen physikalischen Inhalt, wenn inbezug auf *dasselbe* Koordinatesys-

tem K zwei verschiedene Lösungen $G(x)$ und $G'(x)$ existieren. Gleichzeitig zwei Lösungen in dieselbe Mannigfaltigkeit hineinzudenken, hat keinen Sinn und das System K hat ja keine physikalische Realität. Anstelle der Lochbetrachtung tritt folgende Überlegung. *Real* ist physikalisch nichts als die Gesamtheit der raum-zeitlichen Punktkoinzidenzen. Wäre z.B. das physikalische Geschehen aufzubauen aus Bewegungen materieller Punkte allein, so wären die Begegnungen der Punkte, d.h. die Schnittpunkte ihrer Weltlinien das einzig Reale, d.h. prinzipiell beobachtbare. Diese Schnittpunkte bleiben natürlich bei allen Transformationen erhalten (und es kommen keine neue hinzu), wenn nur gewisse Eindeutigkeitsbedingungen gewahrt bleiben. Es ist also das natürlichste, von den Gesetzen zu verlangen, dass sie nicht *mehr* bestimmen als die Gesamtheit der zeiträumlichen Koinzidenzen. Dies wird nach dem Gesagten durch allgemein kovariante Gleichungen erreicht."

[25] This is the reading that one finds, for example, in Friedman 1983, pp. 22–25. Thus, Friedman (p. 24) quotes the following passage from Einstein 1916: "All our space-time verifications [Konstatierungen] invariably amount to a determination of space-time coincidences. If, for example, events consisted merely in the motion of material points, then ultimately nothing would be observable but the meeting // of the material points of our measuring instruments with other material points, coincidences between the hands of the clock and points on the clock dial, and observed point-events happening at the same place at the same time" (Einstein 1916, p. 776; Friedman quotes from p. 117 of the translation in Lorentz et al. 1923.) And Friedman comments: "I think it fair to say that this passage represents the birth of the modern observational/theoretical distinction. It also contains the beginnings of the empiricist and verificationist interpretations of science characteristic of later positivism" (p. 24). But this is a forced reading. For one thing, Einstein's German term, "*Konstatierungen*," need not carry the verificationist connotations suggested by the philosophically unsophisticated translation by Perrett and Jeffery. It later acquired such connotations when, in the 1930s in Schlick's hands, it became a technical term—Schlick's alternative to "protocol sentence"—meaning something like those basic statements reporting the contents of immediate experience that ground observation statements (see Schlick 1934). In Einstein's usage, however, it carries the more ordinary German meaning of "*Feststellungen*" or "determinations." Moreover, at the place marked above by the slashes "//," Friedman omits two whole lines that subtly change Einstein's meaning. The affected passage reads, in full, with the omitted material highlighted: "If, for example, events consisted merely in the motion of material points, then ultimately nothing would be observable but the meetings *of two or more of these points. Moreover, the results of our measurings are nothing but verifications of such meetings* of the material points of our measuring instruments with other material points, coincidences. . ." One must remember that the coincidences of primary interest to Einstein are infinitesimal points of intersection of world lines, the points corresponding to the space-time "events" that constitute the space-time manifold. Surely such points are in no way the kinds of direct

"observables" later to be so important in the scientific epistemology of the Vienna positivists.

Friedman is wrong as well in seeing Schlick as the medium through which these alleged Einsteinian anticipations of verificationism were carried to Vienna. The Schlick of the late 1910s, the author of *Raum und Zeit in der gegenwärtigen Physik* (1917b) and the first edition of the *Allgemeine Erkenntnislehre* (1918), was a thoroughgoing realist. Here is what he said about space-time coincidences in general relativity, this in the section of Schlick 1917b that Einstein adjudged "splendid" (see note 23):

"Physics introduces, as its ultimate indefinable conceptions, the coincidences of two *events*: on the other hand, the psycho-genetic analysis of the idea of objective space ends in the conception of the space-time coincidence of two *elements of perception.* Are they to be regarded simply as one and the same thing?

"Rigorous positivism, such as that of Mach, affirms them to be so. According to him, the directly experienced elements such as colours, tones, pressures, warmths, etc., are the sole reality, and there are no other actual events beyond the coming and going of these elements. Wherever else in physics other coincidences are mentioned, there are only abbreviated modes of speech, economical working-hypotheses, not realities as perceptions are. Looked at from this point of view, the conception of the physical world in its objective four-dimensional scheme would merely be an abridged statement of the correspondence of the subjective time-space experiences in the realms of the various senses, and *nothing more*.

"This view is, however, not the only possible interpretation of scientific facts. If distinguished investigators in the domain of the exact sciences do not cease to urge that the picture of the world as offered by Mach fails to satisfy them, the ground for it is doubtless to be sought in this, that the quantities which occur in physical laws do not all indicate 'elements' in Mach's sense. The coincidences which are expressed by the differential equations of physics are not immediately accessible to experience. They do not directly signify a coincidence of sense-data; they denote non-sensory qualities, such as electric and magnetic field strengths and similar quantities. There is no argument whatsoever to force us to state that only the intuitional elements, colours, times, etc., exist in the world. We might just as well assume that elements or quantities which cannot be directly experienced also exist. These can likewise be termed 'real,' whether they be comparable with intuitional ones or not.... The picture of the world, as presented by physics, would then be a system of symbols arranged in a four-dimensional scheme, by means of which we get our knowledge of reality; that is *more* than a merely auxiliary conception, allowing us to find our way through given intuitional elements."

"The two views stand in opposition; and I believe that there is no rigorous proof of the correctness of the one and the falseness of the other. The reasons which induce me to declare myself in favour of the second—which may, in contrast to the strictly positivist view, be called realistic—are as follows:..."

(Schlick 1917b, pp. 58–59: quoting from the translation in Schlick 1979, vol. 1, pp. 264–265.)

For more on the relationship between Schlick and Einstein as it bears upon the *Eindeutigkeit* principle, see below, section 8.

26 See the correspondence between Petzoldt and Mach in Blackmore and Hentschel 1985; see also Müller 1966. Another helpful source of information, especially concerning the relationship between Mach and Petzoldt, is Wolters 1987.

27 See, for example, Petzoldt 1912a, 1912b.

28 See Petzoldt 1912a, 1914, 1918, 1920a, 1920b, 1921a, 1921b, 1923.

29 See note 59.

30 For further biographical details on Petzoldt, see Müller 1966 and Hentschel 1989, 401–405.

31 "*Nicht Maxima, Minima und Oekonomie, sondern Eindeutigkeit und Stabilität heben die Seiten der Wirklichkeit hervor, die für uns im Vordergrunde des Interesses stehen müssen.*" Petzoldt cites Mach 1872a, 1872b, 1882, 1883, and 1886, as well as Avenarius 1876 as sources of the view he criticizes. Mach replied to Petzoldt's criticisms in his *Wärmelehre* (1896, pp. 391–395).

32 See below, section 7, for Cassirer's deployment of the concept of a many-to-one ("eindeutige") or one-to-one ("ein-eindeutige") mapping ("Zuordnung").

33 "Wir dürfen also wohl auf die Eingangs aufgeworfene Frage: was soll an die Stelle der Begriffe Ursache und Wirkung treten? antworten: die Begriffe eindeutig bestimmende und bestimmte Elemente, bez. Elementencomplexe: und auf die zweite Frage nach dem 'Sinn' order dem 'Wesen' der Naturvorgänge: sie sind von der Art, dass wir sie auf die *stetige* und *einsinnige* Aenderung einer kleinen Zahl von immer wiederkehrenden elementen—d.h. *Bestimmungs*elementen, Bestimmungs*mitteln*—zurückführen können, die sich gegenseitig *eindeutig bestimmen.*"

34 Alle anorganischen und organischen Systeme, die sich entwickeln, erhalten sich eine Zeit lang als abgeschlossene Ganze und ändern sich in der Richtung auf gewisse stationäre Zustände.... Im Besonderen vermag das Centralnervensystem des Menschen eine längere Zeit als besonderer Organismus zu bestehen und sind alle in ihm stattfindenden Vorgänge—namentlich auch die, denen das 'psychische' Leben 'parallel' verläuft—am leichtesten durch die Beachtung jener Tendenz zur Stabilität zu begreifen. Diese Erhaltung und Entwicklung müssen wir nun im Zusammenhang mit einer gewissen allgemeinen Beschaffenheit der Naturvorgänge denken, ohne die sie nicht möglich wäre. Wir müssen gleichsam an die Natur eine gewisse allgemeine Voraussetzung herantragen, ohne deren Bestätigung wir selbst weder geistig noch körperlich leben könnten. Eine solche Voraussetzung liegt mehr oder weniger bewusst aller wissenschaftlichen Forschung zu Grunde, und wir dürfen die feste Ueberzeugung haben, dass sie sich überall bewähren wird, da wir uns selbst in unserer geistigen Eigenart nicht denken können, wenn wir sie einmal als aufgehoben vorstellen. Beides, unser individueller Bestand und jenes Postulat, wie wir die betreffenden Vorausset-

zung auch bezeichnen dürfen, gehören untrennbar zusammen. Letztere besteht in nichts anderem als in der Annahme der *durchgängigen vollkommenen Bestimmtheit* oder—wie wir, um die wichtigste Seite der Sache hervorzuheben, sagen wollen—in der Annahme der *Eindeutigkeit aller Vorgänge*."

[35] "Unsere höchste geistige Existenz, die höchstentwickelten Theile des Centralnervensystems sind ohne die Eindeutigkeit alles Seins und Geschehens gar nicht zu denken."

[36] "Wir können also in der *Succession* der Erscheinungen eine Unbestimmtheit nicht finden. Wir müssen nur beachten, dass die Eindeutigkeit hier andere Seiten zeigt als bei der *simultanen* Abhängigkeit. Sie findet da zunächst ihren besonderen Ausdruck in der *Stetigkeit* der Aenderung. Unstetigkeit wäre sprungweise Aenderung und damit im Allgemeinen das Gegentheil der Eindeutigkeit. Insofern findet ausser der *simultanen* Abhängigkeit der Bestimmungselemente der Thatsachen auch eine *succedane* Abhängigkeit der Masszahlen dieser Bestimmungselemente statt. Ehe ein Bestimmungsmittel, das einen bestimmten Werth hat, einen anderen um eine endliche Grösse von ihm verschiedenen Werth erreicht, muss es alle dazwischen liegenden Werthe passirt haben, falls inzwischen nicht etwa die Aenderungsreihe unterbrochen und dann unter anderen Bedingungen, in anderem Zusammenhange des betreffenden Bestimmungselementes mit anderen Bestimmungselementen von Neuem aufgenommen wurde.... Gestattete die Natur solche sprungweisen Aenderungen, so könnte von Eindeutigkeit keine Rede sein, auch nicht wenn in diesen Fällen die *simultane* Abhängigkeit, d.h. der Bestand der physikalischen Gleichungen, vollkommen gewarht bliebe, wenn also auch alle übrigen Bestimmungselemente den entsprechenden, ihnen dann durch die Gleichungen vorgeschriebenen Sprung mitmachten."

[37] Petzoldt had first made this point with regard to Hamilton's and related principles five years earlier; these are, he said: "nothing other than *analytic expressions for the fact of experience that natural processes are uniquely determined*" ["nichts anderes als *analytische Ausdrücke für die Erfahrungsthatsache, dass die Naturvorgänge eindeutig bestimmte sind*" (Petzoldt 1890, p. 211).

[38] See Ostwald 1893. Petzoldt characterizes the principle thus: "real processes appeared here as unique, singular cases among the infinitely many conceivable cases" ["Die wirklichen Vorgänge zeigten sich hier als einzigartige, als singuläre Fälle unter unendlich vielen denkbaren" (p. 194); and he remarks that Ostwald regarded this principle as essential to the energeticist program, illustrating this point by Ostwald's use of the principle of virtual displacements, which Petzoldt characterizes as a special instance of the principle of the singular case (p. 195).

[39] These logical principles are considered in the course of Petzoldt's discussion of the *Eindeutigkeit* of mental processes. He argues that, contrary to Wundt's position (Wundt 1894), there can be no purely psychical causality, because no *Eindeutigkeit* is to be found among mental processes alone. But since all natural processes are subsumed under the law of *Eindeutigkeit*, mental processes included, we must, he says, look elsewhere for the conditions of the *eindeutige*

determination of the latter, namely to the physical realm. This, according to Petzoldt, is an argument for psycho-physical parallelism: "There remains no other way out than to assume a complete, *eindeutige* relationship between psychical phenomena and some kind of physical processes. . . .*if we are to understand psychical life completely, then we must associate it* eindeutig, *in all of its phases, with processes in the central nervous system*" ["Es bleibt gar kein anderer Ausweg als eine vollkommene eindeutige Beziehung zwischen den psychischen Erscheinungen und irgendwelchen physischen anzunehmen. . . .*wenn wir das psychische Leben voll verstehen wollen, so müssen wir es in allen seinen Phasen eindeutig Vorgängen des Centralnervensystems zuordnen*"] (1895, p. 201). He concludes by asserting that Avenarius's entire program of a "critique of pure experience" ["Kritik der reinen Erfarhung"] (Avenarius 1888, 1890) can be seen as an effort to establish the *eindeutige* determination of psychical processes.

[40] "Eindeutig bestimmt ist eben *bei Ausschluss aller weiteren Bestimmungsmittel ausser eines einmaligen Anstosses nur die geradlinige Bahn*. . . . Die gerade Linie selbst können wir wohl nicht besser definiren, als: sie ist die in unserem dreidimensionalen Raume durch zwei Punkte eindeutig bestimmte—in dem Sinne, dass wir zu jeder Abweichung von ihr noch unzählig viele andere denken können, die ebensogut bestimmt wären. Zu der Geraden allein lässt sich nicht noch eine 'gleichberechtigte' andere denken." Petzoldt credits Friedrich Poske (1884) with first having promoted this way of viewing the law of inertia.

[41] "Hier is also des Denken 'a priori' nicht eindeutig bestimmt, hier können wir nur eine periodische, bez. stationäre Bewegungsart innerhalb der Geraden fordern. In diesem Falle muss also die Einzelerfahrung, bez. das Experiment eintreten, um das Denken eindeutig zu bestimmen."

[42] While ostensibly an "introduction" to his mentor Avenarius's *Kritik der reinen Erfahrung* (Avenarius 1888, 1890), Petzoldt's *Einführung* deserves to be read as significant philosophical work in its own right, as it was read by many contemporaries; see, for example, Boyce Gibson's review of volume one (Petzoldt 1900) in *Mind* (Boyce Gibson 1900). It should be noted that Boyce Gibson devotes four of the twelve and one-half pages of his review to a discussion of the consequences Petzoldt extracts from the *Eindeutigkeit* principle. Thanks to Tom Ryckman for a reference that led to his review.

[43] "Für jeden Vorgang lassen sich Bestimmungsmittel auffinden, durch die er eindeutig bestimmt ist, derart, dass man zu jeder Variation dieses Vorgangs, die man durch dieselben Mittel bestimmt denken wollte, mindestens noch eine finden könnte, die dann in gleicher Weise bestimmt, ihr somit gleichwertig wäre und also gleichsam dasselbe *Recht* auf Verwirklichung hätte wie jene. Wir wollen diesen Satz als das Gesetz der Eindeutigkeit bezeichnen."

44 Mach 1897, p. 495, cites Mach 1872a, p. 42. There we read: "If the appearances α, β, τ, ... depend on the appearances x, y, z, ..., then certain equations

$$\alpha = f_1(x, y, z, \ldots),$$
$$\beta = f_2(x, y, z, \ldots),$$
$$\tau = f_3(x, y, z, \ldots),$$

obtain, from which the α, β, τ, ... are uniquely determined when the x, y, z, ... are given." ["Wenn nun die Erscheinungen $\alpha\,\beta\,\tau$ ... von den Erscheinungen $x\,y\,z$ ... abhängen, so bestehen gewisse Gleichungen ... aus welchen sich die $\alpha\,\beta\,\tau$ eindeutig bestimmen, wenn die $x\,y\,z$ gegeben sind."] And a few pages earlier (p. 37), Mach writes: "Let us call the totality of the appearances upon which an appearance α may be viewed as being dependent the cause of α. If this totality is given, then α is determined, and indeed uniquely determined." ["Nennen wir die Gesammtheit der Erscheinungen, von denen eine Erscheinung α als abhängig betrachtet werden kann, die Ursache von α. Wenn diese Gesammtheit gegeben ist, so ist α bestimmt und zwar eindeutig bestimmt."] But it it is not clear that these remarks suffice to ground Mach's claim to priority in enunciating the *Eindeutigkeit* principle. For when he states a definition of causality, Mach gives only these loose formulations: "*The law of causality is thus sufficiently characterized if one says that it presupposes a dependence of appearances upon one another*" ["*Das Causalgesetz ist also hinreichend charakterisiert, wenn man sagt, es setzte eine Abhängigkeit der Erscheinungen von einander voraus*"] (p. 35, Mach's emphasis): "One can therefore also express the law of causality in this form: 'the effect is determined by the cause'" ["Mann kann also das Causalgesetz auch in der Form ausdrücken: 'Die Wirkung ist durch die Ursache bestimmt'"] (p. 37). What is lacking is an emphasis upon the crucial property of *Eindeutigkeit*, which Mach seems to take for granted, but which Petzoldt elevates to the status of a principle in its own right. Mach's English translator, Philip E.B. Jourdain, is generous, however, in crediting Mach with priority. In a note to the earlier of Mach's two quoted uses of the "Eindeutigkeit" idiom, Jourdain writes: "The principle of the unique determination of natural events by others has been developed by Joseph Petzoldt, starting from Mach's considerations of 1872" (Mach 1911, p. 102). I thank Tom Ryckman for drawing my attention to this note.

45 "In bezug auf die dynamischen Fälle ist aber die Bedeutung der *eindeutigen Bestimmtheit* besser und *durchsichtiger*, als es mir gelungen war, von J. Petzoldt dargestellt worden."

46 "Ich bin ebenso wie Petzoldt überzeugt, dass in der Natur nur das und so viel geschieht, als geschehen kann, und dass dies nur auf *eine* Weise geschehen kann."

47 "Diese Ansicht ist auch sehr wohl verträglich mit der Aufstellung eindeutiger Bestimmtheiten, welche immer unter Voraussetzung *gegebener* Umstände und unter *Abstraction* von ungewöhnlichen und unerwarteten Aenderungen stattfindet."

48 "Nur eine *Theorie*, welche die immer komplizierten und durch mannigfache Nebenumstände beeinflussten Tatsachen der Beobachtung *einfacher und genauer* darstellt, als dies durch die Beobachtung eigentlich verbürgt werden kann, entspricht dem *Ideal der eindeutigen Bestimmtheit*."

49 See Einstein to Mileva Marić, September 10, 1899 (Stachel et al. 1987, pp. 229–231, especially n. 8) and February 8?,1902 (Stachel et al. 1987, pp. 334–335). For more on Einstein's reading of Mach, see Stachel et al. 1989, p. xxiv.

50 See Petzoldt 1912a, 1912b, 1914, 1918, 1920a, 1920b, 1921a, 1921b, 1921c, 1923. Petzoldt's formal study of relativity theory (or at least of Einstein's own papers) seems to have begun in 1910; see the letters from Petzoldt to Mach of September 22, 1910 and June 1, 1911 (Blackmore and Hentschel 1985, pp. 84–85, 91–92).

51 For more on both of these battles, see Petzoldt's 1906, 1912b, and 1921c.

52 Cassirer, another proponent of the *Eindeutigkeit* principle, also argued for the elimination of the concept of substance, but for different reasons; he saw it being replaced in modern science by the concept of functional relationships. See below, section 7.

53 For more on Petzoldt's interpretation of relativity, see Hentschel 1989, pp. 405–420.

54 "Die Aufgabe der Physik wird damit die eindeutige allgemeine Darstellung Vorgänge von verschiedenen relativ mit konstanter Geschwindigkeit gegeneinander bewegten Standpunkten aus und eindeutige In-Beziehung-Setzung dieser Darstellungen. Jede solche Darstellung irgendwelcher Gesamtheit von Vorgängen muß eindeutig abbildbar auf jede andere dieser Darstellungen derselben[1)] Vorgänge sein. Die Relativitätstheorie ist eine solche Abbildungstheorie. Das Wesentliche ist jener eindeutige Zusammenhang. Ihm zuliebe müssen sich die physikalischen Begriffe Biegung gefallen lassen. Erst das eindeutig durch Begriffe Dargestellte beherrschen wir theoretisch und technisch.

"1) Besser: eindeutig aufeinander abbildbare Darstellungen von Vorgängen beliebig vieler jener Bezugssysteme sind Darstellungen '*desselben*' Vorganges. Die Identität muss *definiert* werden, da sie nicht von vornherein gegeben ist."

55 For the critique of the Klein-Minkowski point of view, see Petzoldt 1912, pp. 1063–1064, and 1914, pp. 36–37; see also Hentschel 1989, pp. 405–420 and Wolters 1987, pp. 182–187.

56 See Petzoldt to Einstein, July 26, 1919 (EA 19-055) and Einstein to Petzoldt, 19 August, 1919 (EA 19-069).

57 See Einstein to Petzoldt, August 23, 1919 (EA 19-072), Petzoldt to Einstein, July 6 , 1920 (EA 19-057), and Einstein to Petzoldt, July 21, 1920 (EA 19-058).

58 See, again, Petzoldt to Einstein, July 26, 1919 (EA 19-055) and Einstein to Petzoldt, August 19, 1919 (EA 19-069).

59 See the announcements on the founding of the Gesellschaft in, for example, the *Naturwissenschaftliche Rundschau* 27 (1912): 336, the *Physikalische*

Zeitschrift 13 (1912): 735–736, and *The Journal of Philosophy* 9 (1912): 419–420. Among the other founding members were Eugen Dietzgen, Sigmund Freud, Georg Helm, David Hilbert, Felix Klein, Ernst Mach, and F.C.S. Schiller. In 1921 the Gesellschaft was absorbed in the Kantgesellschaft (Hentschel 1989, p. 404).

[60] See Thiele 1971, p. 70. Einstein had moved to Berlin in April 1914. He lectured on relativity in winter semester 1914/1915 (see Einstein's autograph notes on these lectures, EA 3-008), and in summer semester 1915 (see *Königliche Friedrich-Wilhelms-Universität zu Berlin. Verzeichnis der Vorlesungen Sommer Semester 1915*).

[61] "Ich habe heute mit grossem Interesse Ihr Buch zu Ende gelesen, aus dem ich mit Freude entnehme, schon längst Ihr Gesinnungsgenosse gewesen zu sein.—Ich habe einem kranken Herrn, mit dem ich befreundet bin, von Ihrer Relativitätsarbeit erzählt. Er hat sich sehr dafür interessiert. Sie würden ihm eine grosse Freude machen, wenn Sie ihm ein Separatum senden wollten. . . .

"Ich schicke Ihnen meine neue Arbeit über die Kovarianz der Grav. Gleichungen."

The "kranker Herr" was Einstein's friend, Dr. Ludwig Kraft.

[62] She supports her dating by references to two other items in the Einstein–Petzoldt correspondence, but there is no discernible connection between these two letters and the card in question.

[63] In this, the second edition of Petzoldt 1906, a new section was added detailing the relevance of relativity theory to Petzoldt's assault on the concept of substance, which is otherwise the main topic of the book, and the title was changed slightly, presumably to emphasize the link with relativity theory. Petzoldt did not publish another book on relativity until 1921 (Petzoldt 1921b, 1921c). Of course the book in question need not have concerned relativity, but it seems unlikely that Einstein would have been so interested in any of Petzoldt's other books: and, in any case, Petzoldt published no other books of any kind until 1921.

[64] Volkmann's entire professional career was lived at Königsberg. After taking his degree there in 1880, he became a Privatdozent in 1882, was promoted to Extraordinary Professor in 1886, and was named Ordinarius in 1894. He retired in 1924 and died in 1938 at the age of 82. The *Grundzüge* (1896a, 1910a) and the Einführung (1900, 1913) were both published by Teubner in Leipzig, and both went through two editions. As a philosopher of science, Volkmann was a thinker of striking originality and independence: though he had little lasting influence, his many writings deserve much more careful scrutiny by contemporary historians of the philosophy of science.

[65] As his own later writings made clear (see especially Volkmann 1902), what Volkmann has in mind here is the Cauchy or initial-value problem, the general question whether or not a given set of equations and boundary conditions (initial values) together determine a unique solution for those equations. From the mathematical point of view, this is exactly the problem raised in the hole argument. See below, section 9; see also Stachel 1988.

[66] "Ein Bedenken, welches ich hier noch gegen Monismus geltend machen möchte, liegt in der stillschweigenden Vorausetzung der *Eindeutigkeit* aller der Fragen und Probleme, welche sich die Naturwissenschaft stellt. Eine solche Voraussetzung wird so leicht als selbstverständlich hingenommen, dass es wohl angemessen sein möchte, hier mit einigen Worten darauf zurückzukommen.

"Das Verdienst auf die Wichtigkeit eines Nachweises der Eindeutigkeit der Probleme zuerst hingewiesen zu haben, darf wohl die Mathematik für sich in Anspruch nehmen. Zur Beantwortung der Frage, in wieweit die naturwissenschaftlichen Probleme eindeutig sind, wird man am besten die Geschichte der Wissenschaft heranziehen.

"In dem dritten und vierten Vortrag ist ausgeführt worden, wie z.B. die Anschauugen über die Natur des Lichtes, die Emanations- und die Undulations-Anschauung lange Zeit nebeneinander hergehen durften, wie dann, als das weitere Erfahrungsmaterial zu Gunsten der Undulations-Anschauung entschieden hatte, die elastische und elektomagnetische Lichttheorie lange Zeit nebeneinander eine Rolle spielen durften.

"Dieses Beispiel zeigt, wie allerdings die fortschreitende Induction die Vieldeutigkeit der Probleme einschränkt, aber niemals ganz aufhebt. Die Behauptung der vollständigen Eindeutigkeit der naturwissenschaftlichen Probleme käme so gut wie mit dem Aufgeben der inductiven naturwissenschaftlichen Methode überhaupt überein.

"Nun liegt die Frage nach der Eindeutigkeit der Probleme in der Physik einfach aus dem Grunde um vieles klarer, als in den anderen naturwissenschaftlichen Disciplinen, weil die physikalische Behandlung den mathematischen Hülfsmitteln zugänglich ist. In vielen Fällen besteht aber weiter bei aller Eindeutigkeit der mathematischen formalen Behandlung noch eine Vieldeutigkeit der physikalischen Deutung der mathematisch eindeutigen Formeln."

[67] In a supplement (1896a, pp. 162–169), and again in Volkmann 1900 (pp. 12–20), the term "law of nature" ["Naturgesetz"] is reserved for those high-level regularities, such as Newton's law of gravitation or Coulomb's law, that claim validity only within a narrow domain, whereas the terms "principle" ["Princip"] (1896a, p. 164) and "postulate" ["Postulat"] (1900, p. 12) are introduced for those high-level conceptual principles, such as the energy conservation principle, that claim validity for all phenomena investigated by the physicist. Laws of nature, on the one hand, and principles or postulates, on the other hand, thus differ only in their range of application, and both are to be distinguished from intuitive [*anschauliche*] hypotheses.

[68] More specifically, hypotheses partake of the character of "internal" representations or intuitions [*innere Anschauungen*], to use the Kantian idiom favored by Volkmann, since they are not themselves perceptions or "external" representations or intuitions [*äussere Anschauungen*] (1896a, p. 148: see below).

[69] The distinction that Newton drew between "certainties" (empirical generalizations, like laws of spectral decomposition) and "conjectures" (constructive models, like the corpuscular theory of light) in the course of his debates with

Hooke and others over the theory of light (see, for example, Rosenberger 1895) is as much a part of the background to Volkmann's distinction between laws of nature and hypotheses as Kant's distinction between concepts and intuitions. See, for example, Volkmann 1898a (which is reprinted in Volkmann 1910a, the second edition of his 1896a). The tradition stemming from Newton also may have informed Einstein's more famous distinction between theories of principle and constructive theories, with Volkmann as one of the possible intermediaries.

[70] In his 1896a, p. 150, Volkmann cites Maxwell 1856, or more specifically, the German translation (Maxwell 1898) in the Ostwald's Klassiker series. Hertz 1894 and Boltzmann 1892, 1896b, and 1897b are cited in Volkmann 1900, pp. vi and 45.

[71] "Just as in external intuition the senses admit of a sharpening by means of tools and instruments, so too analogy constitutes such a sharpening for internal intuition." ["Ebenso wie für die äussere Anschauung die Sinne eine Verschärfung durch Instrumente und Werkzeuge zulassen, so bildet für die innere Anschauung eine solche Verschärfung die Analogie."] (1896a, p. 148).

[72] According to Volkmann: "It is well known that from a mathematical and physical point of view mechanics is the most highly developed natural scientific discipline, and that is the reason why precisely the mechanical analogy plays a relatively large role in physics. ...[the] *principles and theorems of mechanics* have become *forms of intuition or representation of physical research.*" ["Die Mechanik ist bekanntlich die mathematisch und physikalisch durchgebildetste naturwissenschaftliche Disciplin, und das ist der Grund, warum gerade die mechanische Analogie in der Physik eine verhältnismässig grosse Rolle spielt. ...[die] *Principien und Sätze der Mechanik* [sind] *Anschauungsformen der physikalischen Forchung* geworden."] (1896a, pp. 147–148). A physics that relies too heavily upon the more sensual analogies and models preferred earlier, among which he includes models featuring the concept of "force" ["Kraft"], is condemned as "anthropomorphic" ["anthropomorph"] (1900, p. v). Volkmann's views should be compared to Planck's later argument in his famous "Leyden lecture" that the history of physics is a history of "a certain emancipation from the anthropomorphic elements" ["eine gewisse Emanzipation von den anthropomorphen Elementen"] (Planck 1908, p. 64).

[73] Volkmann (1896a, p. 29) cites Mach 1894 as his source for the principle of comparison.

[74] "Die neueren Darstellungen der Mechanik von Hertz und Boltzmann scheinen mir...auf dem besten Wege, die in der Natur der Sache liegenden, berechtigten objectiven und subjectiven Seiten der Forschung aufzunehmen, wenn sie die Aufgabe der Mechanik unter dem Gesichtspunkt einer Abbildung der Wirklichkeit erfassen. In der That kann der Gegensatz von Subject und Object in mancher Richtung sehr gut durch das Verhältniss von Bild und Gegenstand veranschaulicht werden." (I have embellished the translation in order to bring out the many relevant connotations of the terms "Bild" and "Abbildung" in German scientific and philosophical prose.) The most important source for the view

of scientific theories as models or pictures [*Bilder*] is the introduction to Hertz 1894. Boltzmann elaborated a similar position in several essays; see especially his 1897a, pp. 1–2. Neither Hertz nor Boltzmann, however, is careful to distinguish as clearly as does Volkmann between laws and postulates, on the one hand, and hypotheses, on the other; and as a result they do not restrict the role of models and analogies to the hypothetical side of science, as does Volkmann.

[75] The first discussion of this doctrine is found in Volkmann 1894.

[76] Volkmann (1896a, p. 177) credits Boltzmann (1896a, p. 45) with having first recognized the general significance of the principle of superposition.

[77] "Wenn das Naturgesetz ein Abstractum, also ein Isolationscentrum ist, die Natur aber ein Concretum, ein Supersitionum, dann werden wir nicht erwarten dürfen, dass es ein Naturgesetz giebt, welches das Universum in seiner Gesammtheit völlig begreift. Jedes Naturgesetz will die Natur, die Wirklichkeit nur nach einer Richtung, nach einem Gesichtspunkt begreifen, ohne dass damit andere Gesichtspunkte ausgeschlossen sein sollen."

[78] "Kirchhoff hat bekanntlich es als Aufgabe der Mechanik präcisirt: 'die in der Natur vor sich gehenden Bewegungen vollständig und auf die einfachtste Weise zu beschreiben'—eine Formulirung, die sich leicht auf die gesammte Physik, ja auf die gesammte Naturwissenschaft ausdehnen lässt, wenn man das Wort 'Bewegungen' durch das Wort 'Erscheinungen' ersetzt. Man hat diese Formulirung nach der Seite der Beschreibung als zu eng bezeichnet, ich möchte sie hier nach der Richtung der Vollständigkeit noch immer als zu weit bezeichnen. Ich sage: es ist unmöglich, die in der Natur vor sich gehenden Erscheinungen vollständig zu beschreiben, wir müssen uns dabei bescheiden, sie nur im Wesentlichen zu beschreiben.

"Man könnte gegen die Behauptung der Unmöglichkeit einer vollständigen Beschreibung einwenden, dass unsere Theorien und Vorstellungen ja mit dazu helfen sollen, das festzustellen, was jenseits der Schwelle der Erscheinungen thatsächlich vor sich geht—ich erinnere an die atomistische Anschauung von der Constitution der Materie. Gewiss bedingen Vorgänge, die sich unterhalb der Schwelle des Bewusstseins abspielen, Erscheinungen, die in unser Bewusstsein treten, aber dies doch nur theilweise: sie werden auch Vorgänge bedingen, die sich unter dem Schwellenwerth des Bewusstseins abspielen und deren Kenntnis doch für eine Vollständigkeit der Beschreibung nothwendig sein dürfte. Die Grenzen unseres Naturerkennens scheinen mit dadurch bedingt."

[79] "Es ist ebenso leicht wie werthlos, solche ideale Ziele anzugeben und aufzustellen. Ideale und Ziele, die Werth haben sollen, müssen mit der Wirklichkeit Fühlung haben, sie dürfen weder zu fern noch zu nah liegen. Solche Ideale aufzustellen ist schwierig, aber erkenntnistheoretisch werthvoll."

[80] This paper is reprinted with only minor changes as the "Zehnter Vortrag" in the second edition of Volkmann's *Grundzüge* (Volkmann 1910a, pp. 196–217).

[81] For more on the history of this, the Cauchy problem, see Stachel 1988.

[82] "Auch die *Frage nach der Eindeutigkeit* einer Lösung war für die mathematischen und physikalischen Disciplinen von Wichtigkeit. Es ist mathema-

tisch unerlaubt, von vornherein anzunehmen, dass eine Fragestellung—ihre Existenzberechtigung einmal vorausgesetzt—immer nur *eine* Antwort zulassen muß, es könnten ja auch der Antworten *mehrere* sein."

[83] "Und doch lehrt die Potentialtheorie, dass beide Anschauungen eine und dieselbe Differentialgleichung verbindet, welche daher ebenso auf Grund der einen wie der anderen Anschauung abgeleitet werden kann."

[84] This essay was written in response to Boltzmann's influential essay, "Über die Unentbehrlichkeit der Atomistik in der Naturwissenschaft" (Boltzmann 1896b). Neither hostile to atomism, nor the vigorous proponent that Boltzmann was, Volkmann argued that all physical phenomena can be divided into three groups: (i) those whose explanation requires the atomistic hypothesis, among which he counts electrolytic phenomena and the absorption and dispersion of light, along with the kinetic theory of gases; (ii) those for whose explanation the atomistic hypothesis is definitely in no way essential, among which he includes such "coarser" or "grosser" ["gröbere"] phenomena as elasticity: and (iii) those for which it is doubtful whether their explanation requires the atomistic hypothesis, and where therefore either model is admissible, among which he includes capillary phenomena. The atomistic hypothesis is necessary only in those cases where the scale of the phenomena is such that the mathematical model of infinitely divisible volume elements is not applicable. As he had argued earlier (Volkmann 1896a, pp. 108–114), a complete explanation is never achievable, so the question in every domain is not whether or not one has given a complete explanation, but whether or not one has explained all that is essential. His point about atomism is that it is not essential for the explanation of many phenomena, and he concludes the paper by noting that his larger aim is to oppose a "false monism" ["falsche Monismus"] (Volkmann 1902, p. 203), advocating which is presumably the sin that Boltzmann is supposed to have committed. Boltzmann replied to Volkmann later that year (Boltzmann 1897c), arguing that in defending the universal validity of atomism—by which he means merely the assumption of systems with a finite number of degrees of freedom—he does not mean to deny the obvious usefulness of infinitesimal methods assuming infinitely divisible volume elements. He concedes that the atomistic hypothesis may be inessential in such cases, but he argues that it is not, for that reason, incorrect, and that it has the advantage of being conceptually a clearer model of the phenomena. Finally, he notes that in those cases—he mentions explicitly the work of Navier and Poisson on theory of elasticity—where the atomistic hypothesis had originally served as the basis for the development of what proved to be the correct mathematical description, we may be deceiving ourselves if we think that we can actually free ourselves of the atomistic model.

[85] It is no accident, of course, that this paper appeared in the first volume of a journal founded and edited by *Ostwald.* Volkmann concludes his discussion of atomism thus: "The antipathy of physical chemists toward an exaggerated estimation of the value of atomism is only to be comprehended from this point of view and it finds its epistemological explanation on the basis of the view

developed here regarding the necessity of a distinction between those features that are essential and inessential for natural scientific understanding" ["Die Abneigung physikalischer Chemiker gegen eine übertriebene Werthschätzung der Atomistik wird lediglich unter diesem Gesichtspunkt aufzufassen und findet auf Grund der hier entwickelten Anschauung von der Nothwendigkeit einer Unterscheidung wesentlicher und unwesentlicher Momente für naturwissenschaftliche Auffassungen ihre erkenntnissmässige Erklärung"] (Volkmann 1902, p. 122).

[86] "Die Zerlegung der Vorgänge in voneinander unabhängige Theile hat P. Volkmann in treffender Weise als *Isolation*, die Zusammensetzung eines Vorganges aus solchen Theilen als *Superposition* bezeichnet."

[87] "Nur in den *seltensten* Fällen tritt uns die Natur mit ihrer Fülle der Erscheinungen einheitlich gegenüber, in der Mehrzahl der Fälle trägt die Erscheinungswelt im Gegentheil einen durchaus *zusammengesetzten* Charakter. ... dann wird es eine der Aufgaben unserer Erkenntniss sein müssen, die Erscheinungen, wie sie sich bieten, aus einer Reihe von Theilerscheinungen zusammengesetzt aufzufassen und zunächst diese Theilerscheinungen in ihrer Reinheit zu studieren. Erst wenn wir wissen, welchen Antheil jeder Umstand einzeln an der Gesammterscheinung trägt, dann beherrschen wir das Ganze."

[88] Both of these examples share with the example of vector decomposition the essential feature that there is no uniquely correct way of carrying out the analysis of decomposition.

[89] "Gegen einige meiner älteren Aufstellungen über eindeutige Bestimmung hat Petzoldt Einwendungen vorgebracht, die mir zu denken gegeben haben; ich muss die Erörterung derselben einer späteren Gelegenheit vorbehalten."

[90] Mach's criticisms elicited a reply from Volkmann in a letter to Mach of June 3, 1901, in which Volkmann located the basis of the disagreement in his belief that Mach's phenomenalism provides an insufficient foundation for science "because of the limits of our senses" ["wegen der Schranken unserer Sinne"] (Thiele 1978, p. 131). For more on Volkmann's views on Mach, see Volkmann 1910a, pp. 30–31.

[91] For Einstein's reading of Mach, see above, note 49. His reading of Poincaré is reported to have taken place with the other members of the "Akademie Olympia" in Bern; see, for example, Solovine 1956, p. viii, and Stachel et al. 1989, p. xxv. Solovine does not say explicitly that the members of the academy read the 1904 German edition of Poincaré 1902.

[92] For more on Cohen and Natorp, see Holzhey 1986.

[93] Stadler was born in 1850 to a prominent Zurich family, and he began his studies there at the University of Zurich under Friedrich Albert Lange. In 1872 he went to Berlin, where he studied with Helmholtz and Cohen, the relationship with Cohen, evolving into a close, life-long personal friendship. He returned to Zurich in 1877, first as a Privatdozent at the ETH, and then as Professor of Philosophy and Pedagogy from 1892 until his death in 1910. Like other representatives of the Marburg school, he showed a special interest in the implications of Kant's work for the sciences; see especially his three best-known books, Stadler 1874,

1876, and 1883. Einstein took two courses from him, one an introduction to Kant in summer semester 1897, the other an introduction to the "Theorie des wissenschaftlichen Denkens," essentially a course in logic, in winter semester 1897/1898 (see Stachel et al. 1987, pp. 364–365). For further information, see Cohen's *Nachruf* for Stadler, Cohen 1910.

[94] Petzoldt (1914, pp. 21–23) develops an interesting criticism of Natorp's argument, pointing out that if an ideal, absolute spatio-temporal order were to be postulated standing behind the various relative phenomenal (empirical) spatio-temporal orderings, it would of necessity comprise an infinity of inherently contradictory spatio-temporal relationships (e.g., the temporal order of space-like separated events may be reversed for different observers), and be thus, quite literally, inconceivable. One is not advised to dwell, however, upon the dispute between Petzoldt and Natorp, since Natorp, no less than Petzoldt, is guilty of what I called above the "subjectivist" mistake of identifying frames of reference with the phenomenal worlds of different observers. For a better critical discussion from the empiricist point of view, see Moritz Schlick's review of Natorp 1910 (Schlick 1911), and the discussion in Schlick's first major work on relativity, Schlick 1915, pp. 155–163. See also Hentschel 1989, pp. 212–217.

[95] "Die Frage kann also mit Sinn überhaupt nur so gestellt werden: Gibt es eine einzige, eine eindeutig bestimmte Stellenordnung des Existierenden, nämlich der in funktionalem Zusammenhang miteinander gedachten Veränderungsreihen, deren Inbegriff wir 'Natur' nennen? Diese Frage aber reduziert sich restlos auf die andere: Existiert überhaupt eben dies, was wir Natur nennen? Existiert dieser einzige Funktionalzusammenhang des Geschehens? Oder ganz kurz: *Existiert die Existenz?*"

[96] "nichts Geringeres. . . also dass ein Funktionalzusammenhang 'des' Geschehens, schlechthin eindeutig, in einziger Weise bestimmt, 'gegeben' wäre, oder doch gegeben werden könnte. Denn nichts anderes bedeutet ihm und bedeutet überhaupt Existenz als: vollständige, in keiner Hinsicht unvollendete Bestimmtheit des Seins, und zwar, da es sich hier um zeit–räumliches Sein handelt, vollendete Betimmtheit in Bezug auf Zeit und Raum."

[97] "so folgt daraus nicht etwa, dass die *Forderung* der eindeutigen Bestimmtheit der Stellenordnung des Existierenden als sinnlos abzuweisen sei. Das hiesse den Begriff der Existenz selbst, es hiesse, die *Aufgabe* der Erkenntnis gerade als Erfahrung preisgeben."

[98] "Wirklichkeit bedeutet eine Bestimmtheit, so dass nichts unbestimmt bleibt. Unbestimmtheit ist eben blosse Möglichkeit. In anderer Wendung: die Möglichkeit ist vielfach, sie lässt stets eine Wahl, Wirklichkeit ist schlechthin einzig, sie wird gedacht als auf einzige, jede Wahl ausschliessende Art bestimmt. Diese Einzigkeit spielt eine grosse Rolle in Kants Erfahrungslehre: auf ihrer Forderung beruht besonders seine Unterscheidung der Anschauung vom Begriff: Anschauung heisst ihm 'die Vorstellung, die nur durch einen einzigen Gegenstand gegeben werden kann.' Zeit und Raum sind in solchem Sinne 'wesentlich einzige' Vorstellungen, darum Anschauungen; es gibt nur eine Zeit, nur einen

Raum, so wie es nur eine Erfahrung gibt, 'in welcher alle Wahrnehmungen als in durchgängigem und gesetzmässigem Zusammenhange vorgestellt werden.' Offenbar gilt ihm die Einzigkeit der Zeit und des Raumes für notwendig als Bedingung der Einzigkeit der Erfahrung; wie könnte der Zusammenhang der Erfahrung ein einziger sein, ohne dass die einzige Zeit und der einzige Raum ihm zugrunde läge?"

99 "Voraus... steht fest, dass keine für empirische Bestimmungsmittel überhaupt erreichbare Grundlage der Zeit- und Ortsbestimmung je absolute Festigkeit wird beanspruchen dürfen. Man macht dennoch, ja eben damit die grundsätzliche Voraussetzung, dass es eine absolute Zeit, einen absoluten Ort des Geschehens an sich gebe, mit welchen erst, wenn sie bestimmbar wären, das Geschehen selbst absolut bestimmt wäre, denen gegenüber also jede empirisch mögliche Bestimmung nur die Bedeutung einer brauchbaren Annährerung beanspruchen kann. Woraufhin macht man diese Voraussetzung? Sie liegt ursprünglich in der Forderung, in der letzten, grundsätzlichen Supposition der Erfahrung als gerichtet auf Erkenntnis dessen, was 'wirklich' sei oder vorgehe. *Existenz* kann nur auf einzige Weise bestimmt gedacht werden, weil sie überhaupt nichts anderes besagt als Bestimmtheit auf einzige Weise. Es ist also ein 'analytischer Satz', dass, sofern die Existenz zeit-räumlich bestimmt sein soll, diese Bestimmung selbst als schlechthin eindeutige gefordert, obzwar nie gegeben ist. Nur eine schlechthin eindeutige (und das heisst: absolute) Zeit- und Raumbestimmung *würde* die der Existenz selbst sein; an diesem konditionalen Ausspruch ändert es nichts, dass die empirisch mögliche Bestimmung schlechthin eindeutig niemals sein kann."

100 "Euklidische Raum... [ist] eine 'notwendige' Voraussetzung in dem bestimmten Sinne, dass er bedingend ist für 'mögliche Erfahrung': bestimmter: für die eindeutige gesetzmässige Bestimmbarkeit von *Existenz* in der Erfahrung."

101 "Übrigens verliert das Grundgesetz der eindeutigen Bestimmung damit nichts von seiner logischen Geltung: nur die Möglichkeit seiner empirischen Erfüllung wird in bestimmter Weise eingeschränkt. Die Freiheit der Wahl des Zeitparameters wie der Raumkoordinaten ist für die Einheit des Naturerkennens schon darum nicht bedrohlich, weil sie die Eindeutigkeit der *Naturgesetze* überhaupt nicht berührt. Gerade dies ist vielleicht das wichtigste Ergebnis der Minkowskischen Untersuchung: die 'Invarianz' der Naturgesetze gegenüber allen 'Lorentz-Transformationen.'... Eine weitergehende Eindeutigkeit der empirischen Zeit- und Raumbestimmung, als die für die Aufstellung von Naturgesetzen zureicht, ist ja nicht erforderlich."

102 Einstein to Natorp, May 11, 1919 (Universitäts-Bibliothek Marburg, Ms. 831/694): see also Natorp to Einstein, June 26, 1919 (EA 44-519). The initiative was an "Aufruf des deutschen Geistes zum Sozialismus" (draft in Ms. 831 (Nachlass Natorp), Universitäts-Bibliothek Marburg).

103 See the frequent mentions of Buek in the correspondence of Cohen and Natorp published in Holzhey 1986, vol. 2.

104 "Sehr interessant ist es, dass Buek die Vorlesung von Einstein hört & mit diesem öfter zusammenkommt & gründliche Aussprache pflegt. Er findet ihn

im Philosophischen doch unklar, & hat noch immer kein rechtes Urteil über die ganze Sache, bei der nur die schwierige Mathematik ausser Frage steht."

105 Buek reports that he often provided piano accompaniment for Einstein's violin. See Buek's unpublished essay, "The Einstein I Knew," a copy of which is in the Einstein Archive (EA 59-353).

106 See Einstein's autograph notes for these lectures (EA 3-008).

107 See Ernst Cassirer to Natorp, May 16, 1911 (Universitäts-Bibliothek Marburg, Ms. 831/643; cited in Holzhey 1986, vol. 2, p. 437).

108 The only other signer of the Nicolai manifesto was the Berlin astronomer Wilhelm Förster. For more on this episode, see Nathan and Norden 1968, p. 3–7. There was at least sporadic contact between Einstein and Buek in later years. Thus, in 1918, Einstein recommended Buek, who was then returning to Zurich, to his friend Michele Besso; see Einstein to Besso, July 9, 1918 and July 29, 1918 (EA 7-321 and 7-322; Speziali 1972, pp. 128–130—note that Speziali has mistranscribed the name "Buek" as "Buck"; I thank Robert Schulmann for confirming the correct transcription). They met again in the spring of 1925 during Einstein's visit to Argentina, where Buek was then a correspondent for *La Nación* (Buek, "The Einstein I Knew," p. 5, EA 59-353). And in the summer of 1953, Einstein helped arrange emergency financial assistance for Buck, who was then living in poverty in Paris (see, for example, Einstein to Buek, August 4, 1953, EA 59-351, and Buek to Einstein, August 28, 1953, EA 59-352).

109 "Der Funktionsbegriff enthält in sich zugleich das allgemeine Schema und das Vorbild, nach welchem der moderne Naturbegriff in seiner fortschreitenden geschichtlichen Entwicklung sich gestaltet hat." Unless otherwise noted, English translations are taken from Cassirer 1923, page number references to which are given in square brackets.

110 "An ihm entwickelt sich zuerst das Bewusstsein von dem Wert und der Bedeutung der Begriffsbildung überhaupt. Im Gedanken der Zahl scheint alle Kraft des Wissens, alle Möglichkeit der logischen Bestimmung des Sinnlichen beschlossen."

111 "Der Zahlbegriff erfüllt und durchdringt sich mit dem allgemeinen Funktionsbegriff."

112 "Die Relationsstruktur als solche, nicht die absolute Beschaffenheit der Elemente macht den eigentlichen Gegenstand der mathematischen Betrachtungs- und Untersuchungsweise aus."

113 "Der Bestand der ursprünglichen geometrischen Objekte [wird hier] ausschliesslich durch die Bedingungen bestimmt, denen sie gehorchen. ... Der Punkt, die Gerade bedeutet uns nichts anderes, als ein Gebilde, das mit anderen seinesgleichen in Beziehungen steht, wie sie durch gewisse Axiomgruppen definiert sind."

114 "Es zeigt, dass der reine *Begriff* für alle nur erdenklichen Änderungen in der empirischen Beschaffenheit der Wahrnehmungen seinerseits vorbereitet und gerüstet ist: die unversellen Reihenformen bieten die Handhabe, jegliche Ordnung des Empirischen zu verstehen und logisch zu beherrschen."

After the general theory of relativity mounted its more formidable challenge in the years following 1915, Cassirer refined his views on the relation between geometry and experience and on the a priori component in geometry, arguing that it is, more specifically, the fundamental, local topological relations common to Euclidean, Riemannian, and Lobachevskian geometry (essentially Hilbert's axioms of order and incidence) that retain an a priori status and, hence, are immune to revision on empirical grounds. For Einstein's rather critical reaction to this view, see his letter to Cassirer on 5 June, 1920, EA 8-386); see also Schlick 1921. For further discussion, see Hentschel 1989, pp. 224–231.

115 "Hier scheint ganz im Sinne der traditionellen logischen Doktrin von einer Mehrheit von *Dingen* und von dem Vermögen des Geistes, sie abzubilden, ausgegangen zu werden—; aber dennoch zeigt es sich bei tieferem Eindringen sogleich, dass die überlieferten Bezeichnungen selbst einen neuen Gehalt und eine neue Bedeutung gewonnen haben. Die 'Dinge,' von denen in der weiteren Ableitung die Rede ist, werden nicht als selbständige Existenzen vor jeder Beziehung als vorhanden gesetzt, sondern sie erhalten ihren gesamten Bestand, soweit er für den Arithmetiker in Betracht kommt, erst in und mit den Beziehungen, die von ihnen ausgesagt werden. Sie sind *Relationstermen*, die niemals losgelöst, sondern nur in idealer Gemeinschaft miteinander 'gegeben' sein können."

Dedekind himself writes: "A thing is completely determined by all that can be affirmed or thought concerning it. A thing *a* is the same as *b* (identical with *b*), and *b* the same as *a*, when all that can be thought concerning *a* can also be thought concerning *b*, and when all that is true of *b* can also be thought of *a*." ["Ein Ding ist vollständig bestimmt durch alles Das, was von ihm ausgesagt oder gedacht werden kann. Ein Ding *a* ist dasselbe wie *b* (identisch mit *b*), und *b* dasselbe wie *a*, wenn Alles, was von *a* gedacht werden kann, auch von *b*, und wenn Alles, was von *b*, auch von *a* gedacht werden kann."] (Dedekind 1893, p. 1; English from the translation in Dedekind 1901, p. 44.)

116 "Alle Sätze der Arithmetik, alle Operationen, die sie definiert, beziehen sich lediglich auf die allgemeinen Eigenschaften der Progressionen: sie gehen daher niemals unmittelbar auf 'Dinge', sondern auf die ordinalen Beziehungen, die zwischen den Elementen bestimmter Inbegriffe obwalten.... Die Funktion der '*Zahl*' ist ihrer Bedeutung nach unabhängig von der inhaltlichen Verschiedenheit der *Gegenstände, die gezählt werden* können; diese Verschiedenheit kann und muss daher ausser acht bleiben, wenn es sich darum handelt, lediglich die Bestimmtheit dieser Funktion zu entwickeln. Hier wirkt daher die Abstraktion in der Tat als eine Befreiung: sie bezeichnet die logische *Konzentration* auf den Relationszusammenhang als solchen unter Abweisung aller psychologischen Nebenumstände, die sich im subjektiven Vorstellungsverlauf herandrängen mögen, die aber kein sachlich-konstitutives Moment dieses Zusammenhangs bilden.... Was hier zum Ausdruck kommt, ist eben dies, dass es ein Gefüge idealer Gegenstände gibt, deren gesamter Inhalt in ihren gegenseitigen Beziehungen erschöpft ist. Die 'Essenz' der Zahlen geht in ihrem Stellenwert auf."

On the subject of abstraction as liberation, Cassirer quotes Dedekind again:

"With reference to this liberation, of the elements from every other content (abstraction), we can correctly call numbers the free creation of the human mind" ["In Rücksicht auf diese Befreiung der Elemente von jedem anderen Inhalt (Abstraktion) kann man die Zahlen mit Recht eine freie Schöpfung des menschlichen Geistes nennen"] (1910, p. 50 [38]: Dedekind 1893, p. 21). Dedekind first advances the view of the numbers as "free creations of the human mind" in the Foreword to the first edition; see Dedekind 1893, p. vii. Given Einstein's frequent employment of the "free creations" image, it should be noted that Einstein reportedly read Dedekind 1893 sometime between 1903 and 1905 along with his fellow members of the Akademie Olympia in Bern; see Stachel et al. 1989, pp. xxiv–xxv.

117 "Wo immer ein *System von Bedingungen* gegeben ist, das sich in verschiedenen Inhalten erfüllen kann, da können wir, unbekümmert um die Veränderlichkeit dieser Inhalte, die Systemform selbst als *Invariante* festhalten und ihre Gesetze deduktiv entwickeln. Wir erschaffen dadurch ein neues 'objektives' Gebilde, das in seiner Struktur von aller Willkür unabhängig ist: aber unkritische Naivität wäre es, den *Gegenstand*, der auf diese Weise entsteht, mit den sinnlich wirklichen und wirksamen Dingen zu verwechseln. Diesem Gegenstand können wir nicht empirisch seine 'Eigenschaften' ablesen; und wir bedürfen dessen nicht, da er in all seiner Bestimmtheit vor uns steht, sobald wir einmal die Relation, aus der er erwächst, in ihrer Reinheit ergriffen haben."

118 "Wieder ist es die Leibniz'sche Grundkonzeption der Mathematik, zu der uns wir hierbei zurückgeführt sehen. Die Mathematik ist danach nicht die allgemeine Wissenschaft der *Grösse*, sondern der *Form*, nicht der *Quantität*, sondern der *Qualität*.... Wo immer eine bestimmte Weise der Verknüpfung gegeben ist, die wir in gewissen Grundregeln und Axiomen aussprechen können, da ist im mathematischen Sinne ein identisches 'Objekt' fixiert."

119 "Lediglich diese systematische 'Komplexion' der Elemente, nicht ihre Einzelbestimmtheit, wird hier als Ausdruck ihrer Wesenheit gebraucht und festgehalten. In disem Sinne hat man mit Recht die Hilbertsche Geometrie eine *reine Beziehungslehre* genannt."

120 "Die Bestimmung der *Individualität* der Elemente steht daher nicht am Anfang, sondern am Ende der Begriffsentwicklung; sie ist das logische Ziel, dem wir uns durch die fortschreitende Verknüpfung allgemein gültiger Beziehungen annähern. Das Verfahren der Mathematik weist hier auf ein analoges Verfahren der *theoretischen Naturwissenschaft* voraus, für welches es den Schlüssel und die Rechtfertigung enthält. (S. Kap. V.)"

121 "Während die Wirklichkeit uns eine vielfältige Mischung heterogener Umstände zeigt, die untrennbar ineinander verwoben und verwirrt scheinen, fordert der Gedanke die gesonderte Betrachtung jedes einzelnen Moments und die genaue Bestimmung des Anteils, der ihm in der Struktur des Ganzen zukommt. ... Indem wir die Teilsysteme miteinander verknüpfen und sie gleichsam sich übereinander lagern lassen, entsteht uns damit wiederum das vollständige Bild des Gesamtvorgangs, das jedoch nunmehr nicht nur in der Art einer einheitlichen

Totalanschauung, sondern als ein differenziertes Begriffsganzes erscheint, in welchem die Art der Abhängigkeit zwischen den Einzelmomenten fest bestimmt ist.... Wenn daher das Prinzip der 'Isolation' und 'Superposition' gelegentlich dadurch erklärt und begründet werden soll, dass alles Wirkliche nur die Summe von Äusserungen einzelner Naturgesetze darstellt und aus diesen als *hervorgegangen* zu denken ist, so verhüllt diese Wendung bereits den eigentlichen erkenntnistheoretischen Sinn des Gedankens. Nicht um den Ursprung der Dinge, sondern um den Ursprung und die Beschaffenheit unserer Einsicht in die Dinge kann es sich hier allein handeln. Das 'Wirkliche', wie es im sinnlichen Eindruck erfasst wird, ist nicht an und für sich bereits eine 'Summe' verschiedenartiger Elemente, sondern steht zunächst als schlechthin einfaches und unzerlegtes Ganzes vor uns. Diese ursprüngliche 'Einfalt' der Anschauung wandelt sich erst unter der logisch zergliedernden Arbeit des Begriffs zu einer inneren Vielgestaltigkeit um. Der Begriff ist somit hier ebenso der Quell der Vielheit, wie er sonst allenthalben als der Quell der Einheit erscheint. Indem wir einen einzelnen Vorgang successiv verschiedenen Systemen einordnen, deren allgemeine Struktur sich mathematisch-deduktiv ableiten lässt, geben wir ihm damit eine immer weitergehende Bestimmtheit, sofern damit seine Stellung in dem allgemeinen Orientierungsplan unseres Denkens immer genauer bezeichnet wird."

122 "Haben wir einen gegebenen Inbegriff von Beobachtungen durch 'Superposition' mehrerer Grundreihen mathematisch dargestellt, so haben wir damit freilich unsere Kenntnis von den absoluten und transzendenten Ursachen des Geschehens nicht vermehrt; wohl aber ist es eine neuer *Typus des Wissens*, zu dem wir uns damit erhoben haben.... Eine Gruppe von Phänomenen beschreiben heisst alsdann nicht lediglich rezeptiv die sinnlichen Eindrücke verzeichnen, die wir von ihr empfangen, sondern es heisst sie gedanklich umprägen. Unter den theoretisch bekannten und entwickelten Formen des mathematischen Zusammenhangs... soll eine derartige Auswahl getroffen und eine derartige Zusammensetzung gefunden werden, dass in dem Inbegriff, der auf diese Weise entsteht, die hier und jetzt *gegebenen* Elemente als konstruktiv abgeleitete Elemente erscheinen. Das logische Moment, das hiermit gegeben ist, verleugnet sich denn auch in den Theorien des Empirismus nicht: unter wie verschiedenen Namen es sich hier auch immer verbergen mag. Die 'Anpassung der Vorstellungen an die Wirklichkeit' setzt eben den Begriff dieser Wirklichkeit selbst und damit eine Ganzes intellektueller Forderungen voraus. Vor allem ist es das Prinzip der *Eindeutigkeit des Geschehens*, in welchem all diese Forderungen sich zuletzt zusammenfassen. 'Ich bin überzeugt,' so heisst es bei Mach selbst, 'dass in der Natur nur das und so viel geschieht, als geschehen kann, und dass dies nur auf *eine* Weise geschehen kann.' Alles physische Geschehen wird demnach durch die augenblicklich wirksamen Umstände vollständig bestimmt und daher nur in *einer* Weise vor sich gehen können. Analysiert man indes die *Gründe* eben dieser Überzeugung, so wird man damit implizit auf alle jene Grundgedanken zurückgeführt, die die sensualistische Grundlegung explizit verleugnet. Der Gedanke der Eindeutigkeit und der 'Stabilität' des Seins liegt of-

fenbar nicht in dem Inhalt der Wahrnehmungen selbst, so wie sie uns im ersten unmittelbaren Erleben gegeben sind, sondern er bezeichnet das Ziel, dem die wissenschaftliche Denkarbeit diesen Inhalt mehr und mehr anzunähern strebt. Dieses Ziel ist nur dann erreichbar, wenn es gelingt, in dem Wechsel der Empfindungen, deren jede von der anderen verschieden ist und deren jede in ihrer Bedeutung und Wahrheit zunächst nur auf einen einzigen Zeitmoment beschränkt ist, gewisse gleichbleibende Relationen der Verknüpfung festzuhalten, deren Regeln wir uns, unabhängig von der Veränderlichkeit des jeweiligen Materials, zum Bewusstsein bringen können. In dem Masse, als das geschieht, entsteht und festigt sich der wissenschaftliche Begriff der *Natur*."

[123] See the preface to Cassirer 1921, as well as Einstein's letter to Cassirer of June 5, 1920 (EA 8-386).

[124] For the correspondence, see items EA 8-384 through 8-396 in the Einstein Archive. See also the references to Einstein in Cassirer 1981.

[125] The title of Schlick 1910 notwithstanding, Schlick was the inheritor of an older logical tradition that regarded judgments as the bearers of meaning, truth, and hence logical relationships. But already in the present paper he was working his way toward the more modern conception, itself in part a product of Schlick's work, that sees linguistic entities—propositions or statements—as the bearers of these properties, for he was careful to stress that he is concerned with judgments only in their objective aspect, only as entities that can be the common property of different speakers and knowers; see Schlick 1910, p. 460.

[126] "Urteile [sind]. . . *Zeichen* für Tatbestände. Welche Anforderungen stellen wir an eine Zeichen? Nur eine einzige, nämlich die, dass es *eindeutig* sei, oder besser noch, eineindeutig, d. h. jedem Zeichen muss ein und nur ein Bezeichnetes und jedem Bezeichneten nur ein Zeichen entsprechen. Da einem Zeichen seinem Wesen nach keine andere Bedeutung zukommen kann als diese, so folgt ohne weiteres:

"*Ein Urteil is wahr, wenn es einen bestimmten Tatbestand eindeutig bezeichnet.*"

[127] The crucial chapter four of Reichenbach 1920, which is wholly devoted to Reichenbach's somewhat idiosyncratic interpretation of Schlick's doctrine of truth as "eindeutige Zuordnung" concludes: "Thus we may give the critical question this form: *With which principles does the correlation of equations to reality become unambiguous*?" ["Darum dürfen wir der kritischen Frage diese Form geben: *Mit welchen Prinzipien wird die Zuordnung von Gleuchungen zur Wirklichkeit eindeutig*?"] (p. 45). And in a later footnote, Reichenbach adds: "The characterization of knowledge as unambiguous correlation is Schlick's analysis of reason, and *Eindeutigkeit* is his synthetic judgment a priori" ["Die Charakterisierung der Erkenntnis als eindeutige Zuordnung ist Schlicks Analyse der Vernunft, und die Eindeutigkeit sein synthetisches Urteil a priori"] (p. 110).

We tend not to realize today just how influential Schlick's theory of truth was in its day. For example, it was crucial to the evolution of the distinction between empirical hypotheses and coordinating definitions, a distinction made famous in

Reichenbach's *Philosophie der Raum-Zeit-Lehre* (1928), and employed in one way or another by virtually all of the members of the Vienna Circle and their allies; for more on this, see Howard 1987.

[128] For more on the history of this discussion, see Howard 1984, 1987, 1990a.

[129] "Es ist ja auch ohne weiteres klar, dass die Unterscheidung von wahren und falschen Aussagen tatsächlich nur den Sinn hat, allem sprachlichen, und gedanklichen Ausdruck die Eindeutigkeit zu wahren, welche die notwendige Vorbedingung alles Verständnisses ist und ohne die jede Bezeichnung und jeder Ausdruck überhaupt sinnlos wäre."

[130] Of course, a defender of Schlick might argue that it is the term "John," not the proposition containing it that is ambiguous. But this defense leads either to the no less silly medieval doctrine of Adamic proper names, or to Russell's doctrine of definite descriptions, which is more a persuasive evasion than a solution.

[131] "Es muss sich der Welt der Tatsachen ein Zeichensystem unter allen Umständen eindeutig zuordnen lassen, und es gibt immer nur *eine* Wahrheit, und die ist für uns unveränderlich und ewig, weil jede Tatsache nur *eine* ist, durch kein Deuteln und keine menschliche Auffassungsweise veränderlich, ist sie doch *vor* allem Auffassen da."

[132] "Die Gesamtheit unserer naturwissenschaftlichen Sätze in Wort und Formel nämlich ist nichts als ein Zeichensystem, das den Tatsachen der Wirklichkeit *zugeordnet* ist; und das ist gleich sicher, mag man nun die Wirklichkeit für ein transzendentes Sein erklären oder nur für den Inbegriff und Zusammenhang des unmittelbar 'Gegebenen'. Das Zeichensystem heisst aber 'wahr', wenn die Zuordnung vollständig *eindeutig* ist. Gewisse Eigenschaften dieses Zeichensystems sind unserer Willkür überlassen, wir können sie so oder so wählen, ohne doch der Eindeutigkeit der Zuordnung zu schaden. Es ist also kein Widerspruch, sondern liegt vielmehr in der Natur der Sache, dass unter Umständen mehrere Theorien zugleich wahr sein können, indem sie eine zwar verschiedene, aber doch jede für sich völlig eindeutige Bezeichnung der Tatsachen leisten."

Compare the Duhemian, underdeterminationist thesis expressed here, with the rather similar remark in the "Einleitung" to Hertz 1894, p. 2: "The models that we make of things are still not uniquely determined by the requirement that the consequences of the models be again the models of the consequences. Different models of the same objects are possible, and these models can be distinguished from one another in several respects." ["Eindeutig sind die Bilder, welche wir uns von den Dingen machen wollen, noch nicht bestimmt durch die Forderung, dass die Folgen der Bilder wieder die Bilder der Folgen seien. Verschiedene Bilder derselben Gegenstände sind möglich und diese Bilder können sich nach verschiedenen Richtungen unterscheiden."] Notice that Hertz does not use the term "model" ["Bild"] in the sense that term has in modern formal semantics: for him it means something more like "image," and these "models" or "images" function epistemologically like theories.

[133] "Ich habe gestern Ihre Abhandlung erhalten und bereits vollkommen durchstudiert. Sie gehört zu dem Besten, was bisher über Relativität geschrieben worden ist. Von philosophischer Seite scheint überhaupt nichts annäherend so Klares über den Gegenstand geschrieben zu sein. Dabei beherrschen Sie den Gegenstand materiell vollkommen. Auszusetzen habe ich an Ihren Darlegungen nichts."

[134] "Der Aufsatz [ist] weniger eine Darstellung der allgemeinen Relativitätstheorie selbst als eine eingehende Erläuterung des Satzes, dass Raum und Zeit nun in der Physik alle Gegenständlichkeit eingebüsst haben."

[135] For most of my information about Schlick's stay in Zurich I am indebted to Dr. A.J. Kox, Assistant Director of the Vienna Circle Archive in Amsterdam and co-editor, along with Henk Mulder, of the forthcoming edition of Schlick's correspondence.

[136] See Laue to Schlick, January 5, 1912 (EA 71-35), April 2, 1913 (EA 71-20), and August 19, 1913 (EA 71-21).

[137] For details, see Howard 1984, 1987, and 1990a.

[138] Carnap cites at these points, among other works, Cassirer 1921, Natorp 1910. Petzoldt 1921b, and Schlick 1917b, 1918, and 1920.

[139] Though a helpful start has been made in Corcoran 1980, 1981.

[140] Among other things, it has been shown that Einstein did not write the famous EPR paper, did not like the argument it contained, and developed his own rather different argument for incompleteness within weeks of the publication of the EPR paper. For details, see Fine 1981, Howard 1985, 1990b.

References

Avenarius, Richard (1876). *Philosophie als Denken der Welt gemäss dem Princip des kleinsten Kraftmasses. Prolegomena zu einer Kritik der reinen Erfahrung*. Leipzig: Fues's Verlag (O. R. Reisland).

——— (1888). *Kritik der reinen Erfahrung*, vol. 1. Leipzig: Fues's Verlag (O. R. Reisland).

——— (1890). *Kritik der reinen Erfahrung*, vol. 2. Leipzig: O. R. Reisland.

Blackmore, John, and Hentschel, Klaus, eds. (1985). *Ernst Mach als Aussenseiter. Machs Briefwechsel über Philosophie und Relativitätstheorie mit Persönlichkeiten seiner Zeit. Auszug aus dem letzten Notizbuch (Faksimile) von Ernst Mach*. Vienna: Wilhelm Braumüller.

Boltzmann, Ludwig (1892). "Über die Methoden der theoretischen Physik." In *Katalog mathematischer und mathematisch-physikalischer Modelle, Apparate und Instrumente*. Walther Dyck, ed. Munich: Deutsche Mathematiker-Vereinigung (K. Hof u. Universitätsbuchdruckerei von Dr. C. Wolf & Sohn), pp. 89–98.

——— (1896a). "Ein Wort der Mathematik an die Energetik." *Annalen der Physik und Chemie* 57: 39–71.

——— (1896b). "Über die Unentbehrlichkeit der Atomistik in der Naturwissenschaft." *Kaiserliche Akademie der Wissenschaften* (Vienna). *Mathematisch-naturwissenschaftliche Klasse. Abtheilung IIa. Sitzungsberichte* 105: 907–922. Rpt. *Annalen der Physik und Chemie* 60 (1897): 231–247.

——— (1897a). *Vorlesungen über die Principe der Mechanik.* Part 1. Leipzig: Johann Ambrosius Barth.

——— (1897b). "Über die Frage nach der objektiven Existenz der Vorgänge in der unbelebten Natur." *Kaiserliche Akademie der Wissenschaften* (Vienna). *Mathematisch-naturwissenschaftliche Klasse. Abtheilung IIa. Sitzungsberichte* 106: 83–109.

——— (1897c). "Nochmals über die Atomistik." *Annalen der Physik und Chemie* 61: 790–793.

Boyce Gibson, W. R. (1900). Review of Petzoldt 1900. *Mind* 9: 389–401.

Buek, Otto (1905). "Die Atomistik und Faradays Begriff der Materie. Eine logische Untersuchung." *Archiv für Geschichte der Philosophie* 11 (1905): 65–110, 139–165.

——— (1912). "Michael Faradays System der Natur und seine begrifflichen Grundlagen." In *Philosophische Abhandlungen. Hermann Cohen zum 70sten Geburtstag (4. Juli 1912) dargebracht.* Berlin: Bruno Cassirer.

Buek, Otto, ed. (1909). *Immanuel Kants kleinere Schriften zur Naturphilosophie*, 2nd ed. Leipzig: Dürr.

Carnap, Rudolf (1922). *Der Raum. Ein Beitrag zur Wissenschaftslehre.* "Kant-Studien" Ergänzungshefte, no. 56. Berlin: Reuther & Reichard.

——— (1926). *Physikalische Begriffsbildung.* Wissen und Wirken. Einzelschriften zu den Grundfragen des Erkennens und Schaffens. Emil Ungerer, ed., vol. 39. Karlsruhe: G. Braun.

——— (1927). "Eigentliche und uneigentliche Begriffe." *Symposion. Philosophische Zeitschrift für Forschung und Aussprache* 1: 355–374.

Cassirer, Ernst (1902). *Leibniz' System in seinen wissenschaftlichen Grundlagen.* Marburg: N. G. Elwert.

——— (1910). *Substanzbegriff und Funktionsbegriff. Untersuchungen über die Grundfragen der Erkenntniskritik.* Berlin: Bruno Cassirer.

——— (1921). *Zur Einsteinschen Relativitätstheorie. Erkenntnistheoretische Betrachtungen.* Berlin: Bruno Cassirer.

——— (1923). *Substance and Function and Einstein's Theory of Relativity.* William Curtis Swabey and Marie Collins Swabey, trans. Chicago: Open Court; reprint New York: Dover, 1953.

Cassirer, Ernst, ed. (1912–1923). *Immanuel Kants Werke*, 11 vols. Berlin: Bruno Cassirer.

Cassirer, Toni (1981). *Mein Leben mit Ernst Cassirer.* Hildesheim: Gerstenberg.

Cohen, Hermann (1910). "August Stadler: geb. am 24. August 1850; gest. am 16 Mai 1910. Ein Nachruf." *Kant-Studien* 15: 403–420.

Corcoran, John (1980). "Categoricity." *History and Philosophy of Logic* 1: 187–207.

——— (1981). "From Categoricity to Completeness." *History and Philosophy of Logic* 2: 113–119.

Couturat, Louis (1905). *Les Principes des mathématiques. Avec un appendice sur la philosophie des mathématiques de Kant.* Paris: Félix Alcan.

——— (1908). *Die philosophischen Prinzipien der Mathematik.* Carl Siegel, trans. Philosophisch-soziologische Bücherei, vol. 7. Leipzig: Alfred Kröner.

Dedekind, Richard (1893). *Was sind und was sollen die Zahlen?* 2nd ed. Braunschweig: Friedrich Vieweg und Sohn.

——— (1901). *Essays on the Theory of Numbers.* Wooster Woodruff Beman, trans. Chicago: Open Court; reprint New York: Dover, 1963.

Earman, John and Glymour, Clark (1978a). "Lost in the Tensors: Einstein's Struggle with Covariance Principles 1912–1916." *Studies in History and Philosophy of Science* 9: 251–278.

——— (1978b). "Einstein and Hilbert: Two Months in the History of General Relativity." *Archive for History of Exact Sciences* 19: 291–308.

Earman, John and Norton, John (1987). "What Price Spacetime Substantivalism? The Hole Story." *British Journal for the Philosophy of Science* 38: 515–525.

Einstein, Albert (1914). "Die formale Grundlage der allgemeinen Relativitätstheorie." *Königlich Preussische Akademie der Wissenschaften* (Berlin). *Sitzungsberichte*: 1030–1085.

——— (1915). "Erklärung der Perihelbewegung des Merkur aus der allgemeinen Relativitätstheorie." *Königlich Preussische Akademie der Wissenschaften* (Berlin). *Sitzungsberichte*: 831–839.

——— (1916). "Die Grundlage der allgemeinen Relativitätstheorie." *Annalen der Physik* 49: 769–822.

——— (1918). "Bemerkung zu Herrn Schrödinger's Notiz: Über die Lösungssystem der allgemein kovarianten Gravitationsgleichungen." *Physikalische Zeitschrift* 19: 165–166.

Einstein, Albert and Grossmann, Marcel (1914a). "Entwurf einer verallgemeinerten Relativitätstheorie und einer Theorie der Gravitation." *Zeitschrift für Mathematik und Physik* 62: 225–261.

——— (1914b). "Kovarianzeigenschaften der Feldgleichungen der auf die verallgemeinerte Relativitätstheorie gegründeten Gravitationstheorie." *Zeitschrift für Mathematik und Physik* 63: 215–225.

Faerber, Max (1923). "Eindeutigkeit und Relativitätstheorie." *Kant-Studien* 28: 127–135.

Fine, Arthur (1981). "Einstein's Critique of Quantum Theory: The Roots and Significance of EPR." In *After Einstein: Proceedings of the Einstein Centennial Celebration at Memphis State University, 14–16 March 1979.* Peter Barker and Cecil G. Shugart, eds. Memphis, Tennessee: Memphis State University Press, pp. 147–158.

Forder, H.G. (1927). *The Foundations of Euclidean Geometry.* Cambridge: Cambridge University Press.

Fraenkel, Adolf (1928). *Einleitung in die Mengenlehre,* 3rd. rev. and exp. ed. Berlin: Julius Springer.

Friedman, Michael (1983). *Foundations of Space-time Theories: Relativistic Physics and Philosophy of Science.* Princeton, New Jersey: Princeton University Press.

Gawronsky, Dimitry (1949). "Ernst Cassirer: His Life and Work." In *The Philosophy of Ernst Cassirer.* Paul Arthur Schilpp, ed. The Library of Living Philosophers, vol. 6. Evanston, Illinois: The Library of Living Philosophers, pp. 3–37.

Gehrcke, Ernst (1911). "Die erkenntnistheoretischen Grundlagen der verschiedenen physikalischen Relativitätstheorien." *Kant-Studien* 19: 481–487.

Goodman, Nelson (1960). "The Way the World Is." *Review of Metaphysics* 14: 48–56. Reprinted in *Problems and Projects.* Indianapolis and New York: Bobbs-Merrill, 1972, pp. 24–32.

Haeckel, Ernst (1892). *Der Monismus als Band zwischen Religion und Wissenschaft. Glaubensbekenntnis eines Naturforschers, vorgetragen am 9. Oktober 1892 in Altenburg beim 75. jährigen Jubiläum der Naturforschenden Gesellschaft des Osterlandes.* Bonn: Emil Strauss.

—— (1899). *Die Welträthsel. Gemeinverständliche Studien über monistische Philosophie.* Bonn: Emil Strauss.

Hentschel, Klaus (1984). "Zum Verhältnis Philosophie-Physik anhand der Korrespondenz Schlick-Einstein und ergänzender Dokumente." Master's thesis, University of Hamburg.

—— (1989). "Interpretationen und Fehlinterpretationen der speziellen und der allgemeinen Relativitätstheorie durch Zeitgenossen Albert Einsteins." Ph.D. Dissertation, University of Hamburg.

Hertz, Heinrich (1894). *Gesammelte Werke.* Vol. 3, *Die Prinzipien der Mechanik in neuem Zusammenhange dargestellt.* Philipp Lenard, ed. Leipzig: Johann Ambrosius Barth (Arthur Meiner).

Hilbert, David (1899). "Grundlagen der Geometrie." In *Festschrift zur Feier der Enthüllung des Gauss-Weber-Denkmals.* Leipzig: B.G. Teubner, pp. 3–92.

—— (1917). "Die Grundlagen der Physik. (Zweite Mitteilung)." *Königliche Gesellschaft der Wissenschaften zu Göttingen. Mathematisch-physikalische Klasse. Nachrichten:* 53–76.

Holzhey, Helmut (1986). *Cohen und Natorp.* Vol. 1, *Ursprung und Einheit. Die Geschichte der "Marburger Schule" als Auseinandersetzung um die Logik des Denkens.* Vol. 2, *Der Marburger Neukantianismus in Quellen. Zeugnisse kritischer Lektüre. Briefe der Marburger. Dokumente zur Philosophiepolitik der Schule.* Basel and Stuttgart: Schwabe.

Howard, Don (1984). "Realism and Conventionalism in Einstein's Philosophy of Science: The Einstein-Schlick Correspondence." *Philosophia Naturalis* 21: 616–629.

—— (1985). "Einstein on Locality and Separability." *Studies in History and Philosophy of Science* 16: 171–201.

——— (1987). "Einstein's Conventionalism." Paper delivered to the Department of Philosophy, Johns Hopkins University, February 25, 1987. (Typescript.)

——— (1990a). "Einstein and Duhem." *Synthese* 83: 363–384.

——— (1990b). "'Nicht sein kann was nicht sein darf,' or the Prehistory of EPR, 1909–1935: Einstein's Early Worries about the Quantum Mechanics of Composite Systems." In *Sixty-two Years of Uncertainty: Historical, Philosophical, and Physical Inquiries into the Foundations of Quantum Mechanics.* Arthur I. Miller, ed. New York: Plenum Press, pp. 61–111.

Howard, Don, and Stachel, John, eds. (1989). *Einstein and the History of General Relativity: Based on the Proceedings of the Osgood Hill Conference, North Andover, Massachusetts, May 8–11, 1986.* Boston: Birkhäuser Boston, 1989.

Huntington, Edward V. (1902). "A Complete Set of Postulates for the Theory of Absolute Continuous Magnitude." *American Mathematical Society. Transactions* 3: 264–279.

Kirchhoff, Gustav (1876). *Vorlesungen über mathematische Physik.* Vol. 1, *Vorlesungen über Mechanik.* Leipzig: B.G. Teubner.

Klein, Felix (1910). "Ueber die geometrischen Grundlagen der Lorentzgruppe." *Deutsche Mathematiker-Vereinigung. Jahresbericht* 19: 281–300.

Lorentz, Hendrik A. et al. (1923). *The Principle of Relativity: A Collection of Original Memoirs on the Special and General Theory of Relativity.* W. Perrett and G. B. Jeffery, trans. London: Methuen, 1923; reprint New York: Dover, 1952.

Mach, Ernst (1872a). *Die Geschichte und die Wurzel des Satzes von der Erhaltung der Arbeit.* Prague: Calve.

——— (1872b). *Die Gestalten der Flüssigkeit. Die Symmetrie. Zwei populäre Vorträge.* Prague: Calve.

——— (1882). *Die ökonomische Natur der physikalischen Forschung. Vortrag.* Vienna: Gerold's Sohn.

——— (1883). *Die Mechanik in ihrer Entwickelung historisch-kritisch dargestellt.* Leipzig: Brockhaus.

——— (1886). *Beiträge zur Analyse der Empfindungen.* Jena: Gustav Fischer.

——— (1888). *Die Mechanik in ihrer Entwickelung historisch-kritisch dargestellt,* 2nd ed. Leipzig: Brockhaus.

——— (1894). "Über das Princip der Vergleichung in der Physik." In *Verhandlungen der Gesellschaft Deutscher Naturforscher und Ärzte.* Part 1. Leipzig, pp. 44–56.

——— (1896). *Die Principien der Wärmelehre. Historisch-kritisch entwickelt.* Leipzig: Johann Ambrosius Barth.

——— (1897). *Die Mechanik in ihrer Entwickelung historisch-kritisch dargestellt,* 3rd impr. and enl. ed. Leipzig: Brockhaus.

——— (1900). *Die Principien der Wärmelehre. Historisch-kritisch entwickelt,* 2nd ed. Leipzig: Johann Ambrosius Barth.

——— (1900b). *Die Analyse der Empfindungen und das Verhältniss des Physi-*

schen zum Psychischen, 2nd ed. Jena: Gustav Fischer. [2nd ed. of Mach 1886.]

——— (1901). *Die Mechanik in ihrer Entwickelung historisch-kritisch dargestellt*, 4th impr. and enl. ed. Leipzig: Brockhaus.

——— (1904). *Die Mechanik in ihrer Entwickelung historisch-kritisch dargestellt*, 5th impr. and enl. ed. Leipzig: Brockhaus.

——— (1905). *Erkenntnis und Irrtum. Skizzen zur Psychologie der Forschung*. Leipzig: Johann Ambrosius Barth.

——— (1908). *Die Mechanik in ihrer Entwickelung historisch-kritisch dargestellt*, 6th impr. and enl. ed. Leipzig: Brockhaus.

——— (1911). *History and Root of the Principle of the Conservation of Energy*. Philip E. B. Jourdain, trans. Chicago: Open Court; London: Kegan Paul, Trench, Trübner & Co. [Translation of 1872a.]

——— (1912). *Die Mechanik in ihrer Entwickelung historisch-kritisch dargestellt*, 7th impr. and enl. ed. Leipzig: Brockhaus.

——— (1921). *Die Mechanik in ihrer Entwickelung historisch-kritisch dargestellt*, 8th ed. Joseph Petzoldt, ed. Leipzig: Johann Ambrosius Barth.

Maxwell, James Clerk (1856). "On Faraday's Lines of Force." *Cambridge Philosophical Society. Transactions* 10 (1864): 27–82. [Delivered on December 10, 1855 and February 11, 1856.]

——— (1898). *Ueber Faraday's Kraftlinien*. Ludwig Boltzmann, ed. Ostwalds Klassiker der exakten Wissenschaften, no. 69. Leipzig: Wilhelm Engelmann.

Minkowski, Hermann (1909). "Raum und Zeit." *Physikalische Zeitschrift* 10: 104–111, 216.

Müller, Horst (1966). "Joseph Petzoldt." *Humanismus und Technik* (Technische Universität Berlin) 11, no. 1.

Nathan, Otto, and Norden, Heinz, eds. (1968). *Einstein on Peace*. New York: Schocken Books.

Natorp, Paul (1910). *Die logischen Grundlagen der exakten Wissenschaften*. Wissenschaft und Hypothese, vol. 12. Leipzig and Berlin: B. G. Teubner.

Norton, John (1984). "How Einstein Found His Field Equations." *Historical Studies in the Physical Sciences* 14: 253–316. Reprinted in Howard and Stachel 1989, pp. 101–159.

——— (1987). "Einstein, the Hole Argument and the Reality of Space." In *Measurement, Realism and Objectivity*. J. Forge, ed. Boston and Dordrecht: D. Reidel, pp. 153–188.

——— (1988). "The Hole Argument." In *PSA 1988*, vol. 2. Arthur Fine and Jarrett Leplin, eds. East Lansing, Michigan: Philosophy of Science Association, 1989, pp. 56–64.

——— (1989). "Coordinates and Covariance: Einstein's View and the Modern View." *Foundations of Physics* 19: 1215–1263.

Ostwald, Wilhelm (1893). "Ueber das Princip des ausgezeichneten Falles." *Königlich sächsische Gesellschaft der Wissenschaften. Mathematisch-physika-*

lische Classe. Berichte über die Verhandlungen: 599–603.

Pais, Abraham (1982). *'Subtle is the Lord...': The Science and the Life of Albert Einstein.* Oxford: Clarendon Press; New York: Oxford University Press.

Petzoldt, Joseph (1890). "Maxima, Minima und Oekonomie." *Vierteljahrsschrift für wissenschaftliche Philosophie und Soziologie* 14: 206–239, 354–366, 417–442. Page numbers are cited from the reprinting: *Maxima, Minima und Ökonomie.* Inaugural-Dissertation zur Erlangung der Doctorwürde der philosophischen Facultät der Georgs-August-Universität zu Göttingen. Altenburg, S.-A.: Schnuphase (Max Lippold), 1891.

—— (1895). "Das Gesetz der Eindeutigkeit." *Vierteljahrsschrift für wissenschaftliche Philosophie und Soziologie* 19: 146–203.

—— (1900). *Einführung in die Philosophie der reinen Erfahrung.* Vol. 1, *Die Bestimmtheit der Seele.* Leipzig: B.G. Teubner.

—— (1904). *Einführung in die Philosophie der reinen Erfahrung.* Vol. 2, *Auf dem Wege zum Dauernden.* Leipzig: B.G. Teubner.

—— (1906). *Das Weltproblem von positivistischem Standpunkte aus.* Aus Natur und Geisteswelt, no. 133. Leipzig: B. G. Teubner.

—— (1908). "Die Gebiete der absoluten und relativen Bewegung." *Annalen der Naturphilosophie* 7: 29–62.

—— (1912a). "Die Relativitätstheorie im erkenntnistheoretischer Zusammenhange des relativistischen Positivismus." *Deutsche Physikalische Gesellschaft. Verhandlungen* 14: 1055–1064.

—— (1912b). *Das Weltproblem vom Standpunkte des relativistischen Positivismus aus, historisch-kritisch dargestellt,* 2nd ed. Wissenschaft und Hypothese, vol. 14. Leipzig and Berlin: B. G. Teubner. [2nd ed. of Petzoldt 1906.]

—— (1914). "Die Relativitätstheorie der Physik." *Zeitschrift für positivistische Philosophie* 2: 1–53.

—— (1918). "Verbietet die Relativitätstheorie Raum und Zeit als etwas Wirkliches zu denken?" *Deutsche Physikalische Gesellschaft. Verhandlungen:* 189–201.

—— (1920a). "Kausalität und Relativitätstheorie." *Zeitschrift für Physik* 1: 467–474.

—— (1920b). "Mechanistische Naturauffassung und Relativitätstheorie." *Annalen der Philosophie und philosophische Kritik* 2: 447–462.

—— (1921a). "Das Verhältnis der Machschen Gedankenwelt zur Relativitätstheorie." Postscript to Mach 1921, pp. 490–517.

—— (1921b). *Die Stellung der Relativitätstheorie in der geistigen Entwicklung der Menschheit.* Dresden: Sibyllen-Verlag.

—— (1921c). *Das Weltproblem vom Standpunkte des relativistischen Positivismus aus, historisch-kritisch dargestellt,* 3rd rev. ed., "unter besonderer Berücksichtigung der Relativitätstheorie." Wissenschaft und Hypothese, vol. 14. Leipzig and Berlin: B. G. Teubner.

—— (1923). *Die Stellung der Relativitätstheorie in der geistigen Entwicklung der Menschheit*, 2nd impr. and enl. ed. Leipzig: Johann Ambrosius Barth.

Planck, Max (1908). "Die Einheit des physikalischen Weltbildes." *Physikalische Zeitschrift* 10 (1909): 62–75. (Lecture delivered 9 December 1908 at the University of Leyden.)

Poske, Friedrich (1884). "Der empirische Ursprung und die Allgemeingültigkeit des Beharrungsgesetzes." *Vierteljahrsschrift für wissenschaftliche Philosophie und Soziologie* 8: 385ff.

Poincaré, Henri (1902). *La science et l'hypothèse.* Paris: Ernest Flammarion.

—— (1904). *Wissenschaft und Hypothese.* Lisbeth Lindemann, trans. Notes by Ferdinand Lindemann. Leipzig: B. G. Teubner. [Translation of Poincaré 1902.]

Quine, W.V.O. (1968). "Ontological Relativity." *The Journal of Philosophy* 65: 185–212. Reprinted in *Ontological Relativity and Other Essays.* New York and London: Columbia University Press, 1969, pp. 26–68.

Reichenbach, Hans (1920). *Relativitätstheorie und Erkenntnis apriori.* Berlin: Julius Springer.

—— (1924). *Axiomatik der relativistischen Raum-Zeit-Lehre.* Die Wissenschaft, vol. 72. Braunschweig: Friedrich Vieweg und Sohn.

—— (1928). *Philosophie der Raum-Zeit-Lehre.* Berlin: Julius Springer.

Rosenberger, Ferdinand (1895). *Isaac Newton und seine physikalischen Principien. Ein Hauptstück aus der Entwickelungsgeschichte der modernen Physik.* Leipzig: Johann Ambrosius Barth (Arthur Meiner).

Schlick, Moritz (1910). "Das Wesen der Wahrheit nach der modernen Logik." *Vierteljahrsschrift für wissenschaftliche Philosophie und Soziologie* 34: 386–477.

—— (1911). Review of Natorp 1910. *Vierteljahrsschrift für wissenschaftliche Philosophie und Soziologie* 35: 255–260.

—— (1915). "Die philosophische Bedeutung des Relativitätsprinzips." *Zeitschrift für Philosophie und philosophische Kritik* 159: 129–175.

—— (1917a). "Raum und Zeit in der gegenwärtigen Physik. Zur Einführung in das Verständnis der allgemeinen Relativitätstheorie." *Die Naturwissenschaften* 5: 161–167, 177–186.

—— (1917b). *Raum und Zeit in der gegenwärtigen Physik. Zur Einführung in das Verständnis der allgemeinen Relativitätstheorie.* Berlin: Julius Springer.

—— (1918). *Allgemeine Erkenntnislehre.* Berlin: Julius Springer.

—— (1919). *Raum und Zeit in der gegenwärtigen Physik. Zur Einführung in das Verständnis der Relativitäts- und Gravitationstheorie*, 2nd ed. Berlin: Julius Springer.

—— (1920). *Raum und Zeit in der gegenwärtigen Physik. Zur Einführung in das Verständnis der Relativitäts- und Gravitationstheorie*, 3rd ed. Berlin: Julius Springer.

—— (1921). "Kritizistische oder empiristische Deutung der neuen Physik." *Kant-Studien* 26: 96–111.

—— (1922a). *Raum und Zeit in der gegenwärtigen Physik. Zur Einführung in das Verständnis der Relativitäts- und Gravitationstheorie,* 4th ed. Berlin: Julius Springer.

——— (1922b). "Die Relativitätstheorie in der Philosophie." In *Verhandlungen der Gesellschaft Deutscher Naturforscher und Ärzte. 87. Versammlung, Hundertjahrfeier,* pp. 58–69. Leipzig.

——— (1925). *Allgemeine Erkenntnislehre,* 2nd ed. Berlin: Julius Springer.

——— (1934). "Über das Fundament der Erkenntnis." *Erkenntnis* 4: 79–99.

——— (1979). *Philosophical Papers.* 2 vols. Henk L. Mulder and Barbara F. B. van de Velde-Schlick, eds. Peter Heath, trans. Vienna Circle Collection. Henk Mulder, ed. Vol. 11. Dordrecht, Boston, and London: D. Reidel.

Solovine, Maurice (1956). "Introduction." In Albert Einstein. *Lettres à Maurice Solovine.* Maurice Solovine, ed. Paris: Gauthier-Villars, pp. v–xiii.

Speziali, Pierre, ed.(1972)*Albert Einstein-Michele Besso. Correspondance1903–1955.* Paris: Hermann.

Stachel, John (1980). "Einstein's Search for General Covariance, 1912–1915." Paper delivered at the Ninth International Conference on General Relativity, Jena, German Democratic Republic, 17 July 1980. First published in Howard and Stachel 1989, pp. 63–100.

——— (1986a). "What a Physicist Can Learn from the History of Einstein's Discovery of General Relativity." In *Proceedings of the Fourth Marcel Grossmann Meeting on General Relativity.* Remo Ruffini, ed. Amsterdam: Elsevier, pp. 1857–1862.

——— (1986b). "How Einstein Discovered General Relativity: A Historical Tale with Some Contemporary Morals." In *General Relativity and Gravitation: Proceedings of the 11th International Conference on General Relativity and Gravitation, Stockholm, July 6–12, 1986.* M. A. H. MacCallum, ed. Cambridge: Cambridge University Press, 1987, pp. 200–208.

——— (1988). "The Early History of the Cauchy Problem in General Relativity, 1916–1937." See this volume, pp. 407–418.

Stachel, John, et al., eds. (1987). *The Collected Papers of Albert Einstein.* Vol. 1, *The Early Years, 1879–1902.* Princeton, New Jersey: Princeton University Press.

Stachel, John, et al., eds. (1989). *The Collected Papers of Albert Einstein.* Vol. 2, *The Swiss Years: Writings, 1900–1909.* Princeton, New Jersey: Princeton University Press.

Stadler, August (1874). *Kants Teleologie und ihre erkenntnisstheoretische Bedeutung. Eine Untersuchung.* Berlin: Ferdinand Dümmler, Harrwitz & Gossmann.

——— (1876). *Die Grundsätze der reinen Erkenntnistheorie in der Kantischen Philosophie. Kritische Darstellung.* Leipzig: S. Hirzel.

——— (1883). *Kants Theorie der Materie.* Leipzig: S. Hirzel.

Tarski, Alfred (1931). "Der Wahrheitsbegriff in den formalisierten Sprachen." *Studia Philosophica* 1 (1936): 261–405. Reprint of the paper first delivered in Polish to the Warsaw Scientific Society on March 21, 1931.

Thiele, Joachim (1971). "Briefe Albert Einstein's an Joseph Petzoldt." *NTM—*

Schriftenreihe zur Geschichte der Naturwissenschaften, Technik und Medezin 8: 70–74.

—— (1978). *Wissenschaftliche Kommunikation. Die Korrespondenz Ernst Machs*. Kastellaun: A. Henn.

Veblen, Oswald (1904). "A System of Axioms for Geometry." *American Mathematical Society. Transactions* 5: 343–384.

Volkmann, Paul (1891). *Vorlesungen über die Theorie des Lichtes. Unter Rücksicht auf die elastische und die elektromagnetische Anschauung*. Leipzig: B. G. Teubner.

—— (1894). "Hat die Physik Axiome? Erkenntnistheoretische Studien über die Grundlagen der Physik." *Schriften der physikalisch-ökonomische Gesellschaft zu Königsberg. Sitzungsberichte* 35: 13–22.

—— (1896a). *Erkenntnistheoretische Grundzüge der Naturwissenschaften und ihre Beziehungen zum Geistesleben der Gegenwart. Allgemein wissenschaftliche Vorträge*. Leipzig: B.G. Teubner.

—— (1896b). *Franz Neumann. Ein Beitrag zur Geschichte deutscher Wissenschaft*. Leipzig: B.G. Teubner.

—— (1897) "Ueber nothwendige und nicht nothwendige Verwerthung der Atomistik in der Naturwissenschaft." *Annalen der Physik und Chemie* 61: 196–203.

—— (1898a). "Über Newton's 'Philosophiae naturalis principia mathematica' und ihre Bedeutung für die Gegenwart." *Schriften der physikalisch-ökonomische Gesellschaft zu Königsberg. Abhandlungen* 39: 1–17; reprinted in 1910a, pp. 344–390.

—— (1898b). "Ueber das Princip von der Gleichheit der Actio und Reactio bei Newton." *Annalen der Physik und Chemie* 66: 781–784.

—— (1900). *Einführung in das Studium der theoretischen Physik, insbesondere in das der analytischen Mechanik. Mit einer Einleitung in die Theorie der physikalischen Erkenntnis. Vorlesungen*. Leipzig: B.G. Teubner.

—— (1901). "Die gewöhnliche Darstellung der Mechanik und ihre Kritik durch Hertz." *Zeitschrift für den physikalischen und chemischen Unterricht* 14: 266–283.

—— (1902). "Ueber die Fragen der Existenz, Eindeutigkeit und Vieldeutigkeit der Probleme und ihre mannigfaltige Bedeutung und Rolle für naturwissenschaftliche Auffassung und Erkenntnis." *Annalen der Naturphilosophie* 1: 105–132; reprinted, with revisions, in 1910a, pp. 196–217.

—— (1909a). *Die materialistische Epoche des neunzehnten Jahrhunderts und die phänomenologisch-monistische Bewegung der Gegenwart. Rede am Krönungstage, 15. Januar 1909 in der Aula der Königl. Albertus Universität zu Königsberg i. Pr.* Leipzig and Berlin: B.G. Teubner.

—— (1909b). *Fähigkeiten der Naturwissenschaften und Monismus der Gegenwart*. Leipzig and Berlin: B.G. Teubner.

—— (1910a). *Erkenntnistheoretische Grundzüge der Naturwissenschaften und ihre Beziehungen zum Geistesleben der Gegenwart. Allgemein wissen-*

schaftliche Vorträge, 2nd rev. ed. Wissenschaft und Hypothese, vol. 9. Leipzig and Berlin: B.G. Teubner.

—— (1910b). *Die Eigenart der Natur und der Eigensinn des Monismus.* Leipzig and Berlin: B.G. Teubner.

—— (1913). *Einführung in das Studium der theoretischen Physik, insbesondere in das der analytischen Mechanik. Mit einer Einleitung in die Theorie der physikalischen Erkenntnis. Vorlesungen,* 2nd rev. ed. Leipzig: B.G. Teubner.

Vorländer, Karl, ed. (1920–1940). *Immanuel Kant. Sämmtliche Werke,* 10 vols. Leipzig: Felix Meiner.

Weyl, Hermann (1927). *Philosophie der Mathematik und Naturwissenschaft.* Munich and Berlin: R. Oldenbourg.

Wolters, Gereon (1987). *Mach I, Mach II, Einstein und die Relativitätstheorie. Eine Fälschung und ihre Folgen.* Berlin and New York: Walter de Gruyter.

Wundt, Wilhelm (1894). "Ueber psychische Causalität und das Princip des psycho-physischen Parallelismus." *Philosophische Studien* 10: 1–124.

The Correspondence of Albert Einstein and Gustav Mie, 1917–1918

József Illy

1. Introduction

The years between 1915 and 1920 were an exciting period for the general theory of relativity. Einstein published the first major synthesis in 1916 (Einstein 1916a); others made often successful attempts to reconcile their ideas with Einstein's theory or even to contribute to it, and most of them (e.g. H.A. Lorentz, W. De Sitter, P. Ehrenfest, F. Klein, H. Weyl) carried on shorter or longer but always lively correspondences with Einstein.

Gustav Mie, at the time a professor at Halle University, was a physicist who was obsessed with the idea of unified field theory or ether monism, with all the peculiarities of an intermediate position between classical and relativistic physics.

In 1912 and 1913, he published his theory of matter (Mie 1912a, 1912b, 1913). He set himself the task of deriving the laws of electromagnetism and gravitation from a single world function by means of the calculus of variations. However, he succeeded only in deriving Maxwell's equations, and in making some cautious steps toward the derivation of the existence of the electron. In order to obtain the laws of gravitation, he had to supplement the set of fundamental electromagnetic quantities with those of gravitation, so that a unified electromagnetic theory was not achieved. Mie's draft theory was held in high esteem by Hilbert, Weyl, and others (Hilbert 1915, p. 395; Weyl 1918, section 25).

Mie's underlying world view was that there exists nothing but the ether. Matter is a peculiar state of it. "All the elementary particles of matter are nothing else but spots of accumulation of enormous quantities of energy in very small regions of the ether." In this way, he defined matter (Mie 1917a, p. 553). Mie's ether is not, however, a fluid, a peculiar solid body, or a special kind of dielectric. "One calls empty space conceived as a physical object 'ether' or also 'world ether'," he wrote in *Lehrbuch der Elektrizität* (Mie 1910, p. 6).

Let us look at the documentary sources. The original letters from Mie to Einstein are to be found in the Albert Einstein Archive of the Hebrew University, Jerusalem; those from Mie to Hilbert in the library of Göttingen University. It is appropriate to express my thanks to the Jewish National and University Library for having permitted me to quote from Einstein's unpublished letters.[1]

2. The Correspondence

On June 5–8, 1917, Mie delivered a series of lectures at Göttingen University (Mie 1917a,b,c). The lectures are called Wolfskehl lectures because the grant for them was given by the Wolfskehl-Stiftung on Hilbert's recommendation.

Einstein was prevented by illness from attending the lectures to which he had been invited by Hilbert and very probably by Mie as well, and his letter of excuse to Mie was the opening letter of their correspondence. This was on June 2, 1917 (EA 17-210).[2] He excused himself not only for being unable to participate but also for a debate with Mie four years earlier. In the summer of 1913, at the Vienna Naturforscherversammlung, Einstein gave a survey of his and Grossmann's novel theory of gravitation, together with a sketch of similar attempts by Nordström (in the published version also of those of Abraham) (Einstein 1913). In the discussion following the lecture, Mie reproached Einstein for not having mentioned his theory. Einstein apologized by saying that he had left Mie's theory out of consideration because the strict proportionality of inertial and gravitational masses does not hold in it, and he had taken into account only theories that satisfy the principle of equivalence. That was their first encounter.

Einstein's new gravitation theory, published in 1913 (Einstein and Grossmann 1913), was a challenge for Mie with which to compare his theory. This he did in three papers (Mie 1914a, 1914b, 1915) and in his Wolfskehl lectures in 1917.

In his first letter, Einstein commented on Mie's views, which must have been expounded in the nonextant inviting letter. Mie adhered to a covariance under linear transformations, while, as Einstein stressed, it was sufficient to require covariance under similarity transformations. Mie's conclusion that the general transformability of coordinates (this is general covariance in Mie's parlance) was incompatible with the requirements of a closed theory that had been shared by Einstein three years before but was no longer held by him now (an allusion to the "Lochbetrachtung"). At the end, he agreed with Mie that an a posteriori specialization of the coordinate system would make the theory far more beautiful, although the attempts in this direction had not yet satisfied him.

Mie's reply is unknown, but he attached to it his Wolfskehl lectures, probably in manuscript, because Einstein thanked him for them. As Einstein wrote on December 22, 1917 (EA 17-213), the main difference between their opinions was that he did not believe that the ground state of the universe was flat and that space was filled with matter in only finite regions. He could not accept that the metric tensor was determined by boundary conditions at infinity and only in part by matter.

Einstein closed the letter with the remark: "I agree with your view on matter." I am of the opinion that he agreed with the view that matter is energy, but not with Mie's statement that this energy is that of the ether, because, as is well known, in 1910, he said: "Mass and energy are equivalent quantities, like for example heat and mechanical work, and it is only one further step to consider mass as an enormous concentration of energy" (Einstein 1910, p. 144).

Enclosed in the letter Einstein sent Mie were the reprints of his recent papers, among them the "Kosmologische Betrachtungen," in which he introduced the cosmological term (Einstein 1917); the others are not mentioned by title in Mie's reply.

Mie's first known letter to Einstein dates from February 5, 1918 (EA 17-215). It initiates their discussions on the distribution of matter in the universe, the cosmological term, and the preferred system of reference.

2.1 The Global Curvature of Space and the Masses

In his "Kosmologische Betrachtungen" (Einstein 1917), Einstein arrived at an expression for the relation between the radius of curvature of the spherical universe and the average density of matter in it. In his first letter, Mie asserted that any predetermined value of the radius admits arbitrary densities. Let us consider a particular density for a definite radius. In this case, the corresponding gravitational potential prevailing in the vacuum between celestial bodies is Minkowskian. If we had another constant value of density, the metric tensor would become non-Minkowskian. We could change, however, the units of mass, length, and time in such a way as to restore its Minkowskian character.

In replying to Mie on February 8 (EA 17-217), Einstein emphasized that Mie's argument left untouched the fact that the relation between the radius of curvature and the density was independent of the choice of coordinates.

Mie admitted that Einstein was right (February 17–18, EA 17-220). However, new doubts arose in him about the interpretation of the field equations with the cosmological term. If they were to be valid in each element of space, then he had to suppose that these elements were pervaded by continuous matter of very low density in all cases where a real accumulation of matter (viz. bodies) was absent. The reason was that far from celestial bodies the geometry must be uniform, of constant curvature, zero not excluded.

Mie seemed to discover that Einstein's equations allowed an interpretation coinciding with his concept of a very low density space-filling ether with nodules of extreme density. A further problem for Mie was that the theory seemed to yield no means to determine the amount of discrete material particles necessary for shaping the universe.

In his reply (February 22, EA 17-221), Einstein pointed to the circumstance that, although with the assumption of a homogeneous distribution of matter the local gravitational field might not be fixed precisely, the geometry at large could be adequately outlined. Mie only answered (March 21, EA 17-223) that these arguments had not shaken his conviction that a constant density of mass must be assumed in regions void of stars.

2.2 The Cosmological Term

Mie welcomed this new term in Einstein's field equations for two reasons. The first, to be found in his letter of February 5 (EA 17-215), was that it proved to him that Euclid's axiom of parallels and general covariance were independent of each other. The following argumentation might lay behind the statement. The parameter λ may take on any value, zero included (which means Euclidean metric), without altering the general covariance of the field equations in which it figures. Einstein's motive for introducing λ (i.e., to avoid nonrelativistic boundary conditions at infinity) was not accepted by Mie. As he wrote, considerations on conditions at infinity may give rise to interesting ideas but not to decisions on how the events in finite regions are to be described. He held Euclid's axiom for a principle of physics like the principle of relativity or Hamilton's principle; it belonged to the mathematical foundations of the universe.

Mie expounded the second reason why he welcomed the cosmological term in his letter of March 21 (EA 17-223). The term was the missing link for him between Euclid's axiom and physics. If $\lambda = 0$, the universe was Euclidean; if not, it was non-Euclidean.

In Einstein's reply of March 24 (EA 17-224), there was no reference to this interpretation of λ, only the remark that one argument for his having an aversion to the quasi-Euclidean solution $\lambda = 0$, $g_{\mu\nu} = \text{const.}$ can be found in a note attached to the letter. The note must be his paper entitled "Prinzipielles zur allgemeinen Relativitätstheorie," which reached the editor of *Annalen der Physik* on March 6 (Einstein 1918a). In it, he pointed out that $g_{\mu\nu} = \text{const.}$ is a general solution to the field equations in the absence of matter, which contradicts a new principle called the Mach principle.[3]

The third and most significant subject of their correspondence was the existence of a preferred coordinate system, interwoven with the problem of the relationship between general covariance and general relativity.

2.3 The Preferred Frame

On February 5, Mie introduced the discussion by an analysis of how the real world was mapped onto a "spacetime scheme." He quoted Einstein's "Kosmologische Betrachtungen" in which Einstein approximated the real space-time continuum by a space-time of constant (three-dimensional) curvature. Einstein compared his procedure with that of geodetic surveyors who approximated the complicated surface of the earth by an ellipsoid of revolution (Einstein 1917, p. 148). Mie used this analogy to prove the necessity of an absolute frame. Just as surveyors cannot do without a simple geometric form when they want to label mountain peaks and the sea floor with numbers indicating their height, so physicists cannot do without an absolute space-time scheme "in which to arrange the world."

Maybe the approximation was not unique, he admitted, since conventions may play a certain role in selecting the approximating structure, but not to the extent that mathematicians used to think. Mie mentioned a convention that, in his view, lay at the foundation of geometry, namely, that the space-time scheme must be uniform in order to be able to receive the world in itself. What happens if one drops this convention? Mie reiterated Poincaré's model of the universe in which the velocity of light becomes a function of place, and at the same time, the laws of nature, the size of atoms, elementary quanta, and the like change in a way that compensates exactly for the change in the velocity of light. The new description of the world, relying on measurements carried out with distorted rods and clocks, will be the same as the old one except that it will establish a non-Euclidean geometry. However, from the standpoint of logic (this meant methodology for Mie), the assumption that differences in the velocity of light and in other physical quantities considered real may be elicited by mere mathematical stipulations contradicts physical causality. This was precisely the case with Einstein's fictitious fields.[4]

Mie's favorite paradox, which he deployed in his struggle for an absolute system of reference was that of the wriggling rod. Let us suppose, he wrote both in his lectures (Mie 1917c, p. 599) and in his letters (February 5 and May 6, 1918, EA 17–215, 17–225), that we hold a straight rod motionless. Its straightness is checked by running our eyes along it, i.e., by comparing it with a light ray supposed to be straight in a Euclidean space without gravitation. Now, let us assume a gravitational field along the rod, changing both in space and time. As a consequence, the velocity of light will also change and the rod will prove to wriggle. If a gravitational field is equivalent to a frame without gravity and if every frame is to be equivalent to each other, the rod will be straight and motionless or curved and wriggling, depending only on our arbitrary choice of frames. To solve the paradox, Mie suggested dropping the general equivalence of frames.

The paradox is not difficult to solve: as Einstein noted, (February 8, EA 17–217), the specter of absolute space haunts it. The rest and the wriggling take place in different frames.

On May 6 (EA 17-225), Mie found a place in Einstein's paper "Näherungsweise Integration" (Einstein 1916b) in which Einstein followed a similar train of thought. Indeed, Einstein derived six types of gravitational waves among which there were three that did not carry energy. These waves arose from the circumstance that a wave-like oscillating coordinate system was used. Given $\sqrt{(-g)} = 1$, i.e., after a particular choice of frame, these wave types disappeared, which supplied for this choice "a deep physical justification" (Einstein 1916b, p. 696). The correspondence having come to an end, Einstein's reflections remain unknown.

Mie's next and most serious objection to general relativity was the nonexistence of the relativity of rotation. In his letter of February 5 (EA 17-215), he considered Einstein's example of two fluid bodies uniformly rotating around a common axis with respect to each other at a distance sufficiently large for the bodies not to exert gravitational effects on each other and apart from other masses in the universe (Einstein 1916a, pp. 771–773). Their shapes as determined by local measurements were a sphere and an ellipsoid of revolution, respectively. Einstein asked why they differed in shape. The Newtonian answer is as follows: the laws of mechanics are valid in the space (frame) in which the sphere is at rest but not in the one in which the ellipsoid is at rest. This Galilean space, however, was an unobservable, "mere fictitious" cause of the different behavior of the bodies, Einstein maintained. None of the other "spaces" moving in an arbitrary way lacked this defect; consequently, the laws of physics must be immune to the frames, all frames being equivalent. Distant masses as the only observable causes must have taken over the role of absolute space.

Mie insisted that the frames fixed to the bodies could be considered equivalent only if a universe filled with a sourceless, "fictitious" gravitational field could be taken as equivalent to a universe having a "homogeneous space-time scheme" and filled with a gravitational field produced by material sources. He rejected this possibility without arguing. Obviously, the fictitious field was to be interpreted as a field "geometrized away" by changing the affine structure of the Euclidean or Minkowskian manifold, and Mie's aversion to the equivalence of frames was an aversion to the geometrization of "impressed forces."

Moreover, as far as I know, Einstein never stated explicitly that the effect of distant masses was a gravitational effect, though it was suggested by his comparing static gravitational effects to those of electrostatics and inertia to electromagnetic induction.[6] On February 8, Einstein replied only that the two bodies behave differently because their motions are different with respect to the distant masses (EA 17-217).

Mie's other case was the rotation of the earth. The statement that the earth rotated, he wrote on February 17–19 (EA 17-220), expressed in shorthand a number of phenomena: the Foucault experiment, the passate winds, etc. If we imagine the earth at rest, we must attribute to the "scheme of space-time setting" particular properties (that means a nonflat metric). This metric will not depend on the presence of real bodies. Undoubtedly, all phenomena will be described adequately, for the procedure is mathematically impeccable, but it will include "arbitrary factors" (Mie uses Schlick's expression), thereby violating the principle of parsimony.

Einstein objected (March 24, EA 17-224) that according to his theory of 1915 there is no ground to consider the earth actually revolving, because one cannot distinguish between gravitational and inertial fields. It is their sum, the total field, that governs the motion of Foucault's pendulum and satisfies the differential equations in a frame at rest with respect to the earth.

Mie's last example of the absurdity of the relativity of rotation was the case of the rotating electron (February 17–19, EA 17-220). The principle of special relativity, he began arguing, was a true principle of relativity because with its help the field of a uniformly moving electron could be deduced from that of an electron at rest. However, the principle of general covariance has so far proved unable to produce the field of an electron rotating around a central body from that of a motionless electron, which eloquently proves that the principle of general covariance is not identical with a principle of general relativity of motion. Should such a principle exist, it should have the form Kretschmann gave it.[7]

Einstein agreed with Mie in that the transformation of the electron into rest (Mie posed the inverse problem!) may formally simplify the solution in a number of cases, but not to the same extent as the Lorentz transformations in the case of the translating electron (an electron in zero gravitational field) (February 22, EA 17-221). He added that a peculiar feature of special relativity lay in the fact that its empirical character was guaranteed by the Lorentz transformations.

Mie returned to the problem of the rotating electron in 1920 at the Nauheim Naturforscherversammlung. He set forth in detail how to calculate the electromagnetic field of a charged particle rotating around a gravitating center (Mie 1920a). He called attention to the fact that to derive the field from that of an electron at rest is possible only if the electron moves in a gravitational field; Einstein's theory fails when the acceleration is due to nongravitational effects. As a consequence, the principle of general relativity does not merit its name: it is only a principle of relativity of gravitational effects. The problem posed is, however, worthy of consideration, he added, because according to that principle, a rotating electron must not radiate energy. He was fully aware of the quantum theoretical connotations.

How did Mie conceive the preferred frame to be constructed? By approximation, as he expounded in his letter of March 21 (EA 17-223). The process starts with terrestrial observations, e.g., made on the earth covered by thick clouds. Phenomena thus observed speak of the earth's revolution around its axis and of its center of mass being at rest. The frame fixed to it suffices for the description of observations. If the precision of our instruments goes higher, a precession of the earth may be revealed with the consequence that an additional motion must be attributed to it: a curvilinear motion around the sun. Switching over to a frame with its origin at the center of mass of the planetary system, the Newtonian theory could be reconstructed. As precision continues to grow, more and more bodies are to be taken into account in order to understand the phenomena, and a series of frames could be anchored to their centers of mass. In this way, a frame is approximated and may be called preferred or absolute. Because approximation depends solely on the precision of instruments available, in principle, there is no obstacle to reaching this frame. The only obstacles are practical. Einstein's answer was that the covariant equations are always satisfied, no matter how far space is stretched and how frames are chosen (March 24, EA 17-224).

For me the absolute frame of Mie does not contradict Einstein's assertion that it is only the relation to the masses of the universe that matters, because "relation" means a position with respect to a frame determined by them. Einstein's motive to reject Mie's frame might have been that in his eye this frame was a chart while his frame was a physical manifold, coordinatable at will and not only in a way to be preferred.

The necessity of a preferred or "natural" system of reference was argued also by Mie's claim that Einstein and Schwarzschild, with their apparently purely mathematical stipulations, also selected a frame, though implicitly (May 6, 1918, EA 17-225). Schwarzschild arrived at a solution of Einstein's field equations in the case of a sphere (in effect of a mass point!) with two constants of integration, α and ρ, and with the stipulation that $\sqrt{(-g)} = 1$; α was determined by the mass of the central body; ρ was left arbitrary. As Mie suspected, Schwarzschild placed the discontinuity of the solution in the origin of the frame out of respect for Newton's potential m/r. Mie, however, saw no reason why it could not be put at another place within the body. According to his calculations, the field of a gravitating sphere in the "natural" system of reference proposed by him in the Wolfskehl lectures turned out to be the same as Schwarzschild's with $\rho = 0$. Outside the sphere, $\sqrt{(-g)} = 1$ (i.e., Einstein's stipulation) in fact leads to postulating the "natural" coordinate system, at least in the case of a spherically symmetric field.

As a matter of fact, it would have been difficult for Schwarzschild not to put the singularity at the center of the body because he considered not a sphere, as Mie asserted, but a mass point.

Toward the end of the letter, Mie raised the question of what was the physical sense of the whole wonderful theory if in the end general covariance was canceled by the introduction of a preferred system of coordinates. He emphasized that just this problem was the subject of the Wolfskehl lectures in which he gave a solution by formulating the so-called extended principle of relativity of gravitational effect, which read like this: the physical laws preserve their form in non-Minkowskian regions, too, where it makes no sense to speak of transformations. The general validity of fundamental equations presupposes their general covariance. He closed the letter with the remark that he was toying with the idea of attacking a special problem with this principle and expressing in this way what he had in mind better than he could in words. Mie was true to his promise. In 1920, he published a paper on the subject (Mie 1920b) that, as Eisenstaedt has shown (Eisenstaedt 1982), is of importance for the history of the Schwarzschild solution.

The discussion on the preferred frame entered the field of philosophy, too. In his letter of February 8 (EA 17-217), Einstein promised Mie that a copy of Moritz Schlick's *Raum und Zeit* would be sent to him by Berliner, the editor of *Die Naturwissenschaften*. Schlick's booklet was also published in this journal, though with the omission of the final, philosophical chapter (Schlick 1917a, 1917b). Upon receiving the booklet, and after unreservedly identifying himself with Schlick, Mie quoted (February 17–19, EA 17-220) Schlick's statement that theories describing the same observations and proposing the same predictions are only various ways to express the same situation. Among them, however, one must be found that is the simplest, not because it is the most practical or most comfortable, but because it contains the smallest number of arbitrary factors (Schlick 1917b, p. 62). The principle of general covariance, Mie continued, affords a transformation of the earth into rest, but this would entail the introduction of a metric that would not depend on the presence of real bodies, i.e., it would represent an arbitrary moment in Schlick's sense.

What was a real body for Mie? He answered this question in his Wolfskehl lectures: "When we suppose that a body exists, with it we reveal our expectation that its existence will at some time be provable, even if it is unknown for the time being. A material body unobservable in itself would be a logical absurdity, like a four-angled triangle" (Mie 1917a, p. 553). The temporary impossibility of its observation does not weaken the in-principle possible existence of a physical body because, as Mie emphasized in the lectures and in a letter to Einstein (May 6, EA 17-225), the world exists objectively, independently of our perception. He called this statement a postulate lying at the foundation of scientific principles without which there is no experimental science.

Einstein agreed with what Schlick wrote but not with the conclusions that Mie drew from them. "I do not contest," he wrote, "that the description of the world turns out simpler if one uses a system of reference with respect to which the earth revolves in a certain way. I do contest, however, that a corresponding preference of a single frame of reference should have a primary importance" (February 22, EA 17-221). Einstein's reasoning against a preferred system of reference was this: Mie adhered to the flat ground state of the universe because a sufficiently small region of a constantly curved manifold could always be taken as flat; in it gravitating masses would appear as sources only of local curvature. This picture is analogous to the ancient idea of the surface of the earth being flat. An essentially flat and infinite universe demands nonrelativistic boundary conditions and, in order not to be curved by its masses, the assumption that for the most part it is empty. Nonrelativistic boundary conditions, however, must be dropped because a theory, covariant under coordinate transformations but not covariant under boundary conditions would be inconsistent. By dropping these boundary conditions, the argument for the preference of a frame loses its footing.

3. Epilogue

With Mie's letter of May 6, the correspondence came to end[8] not because any of the disputed problems had been settled, but rather because Einstein seems to have lost interest in a quarrel for which he had not had much interest even earlier. Although they assured each other several times that they verged on full harmony, in the last letter, Mie confessed that their exchange of views was a squabble from which the only escape would be repeated conversations. Einstein had been of the same opinion in his very first letter: "A conversation is usually much more useful than a mere correspondence" (June 2, 1917, EA 17-210). They succeeded in meeting in Einstein's Berlin home not long after March 24. Their discussion impressed Mie to the extent that, as he confessed to Einstein, for about half an hour his doubts about curved space were partially resolved but, as time went on, this impression lost its strength (May 6, 1918, EA 17-225).

Mie found a deeper reason why Einstein was unable to understand him fully in the difference between mathematical and physical frames of mind. The mathematician is allowed to make the most paradoxical assumptions and hypotheses and to accept the most arbitrary conventions. However, these must be revised by physical considerations. For example, a physicist has to distinguish between a frame in which apparent gravitational fields prevail and a frame in which there are only real fields. Mie had the impression that Einstein oscillated between the two positions, although he was essentially on the side of the physicist. When Einstein emphasized the fundamental equivalence of frames, he acted as a mathematician; when he distinguished between frames, he was a physicist. The pure mathematical procedure does not satisfy the physicist because it is far from the concept formation he needs in order to draw experimentally important conclusions from his theory. In Mie's opinion, Einstein achieved his wonderful results (those acknowledged as such by Mie!) precisely because he did not yield to the mathematical siren song of the Hilbertian world (May 6, EA 17-225).

Mie admitted a psychological motive, too. In a letter to Hilbert of December 22, 1917 (Library of Göttingen University), he wrote that terms such as "equal" and "infinite" might connote other meanings for him than for Hilbert and Einstein because he was a calculator physicist for whom "equal" meant "equal up to such and such decimals" and "infinite" never connoted "actually infinite."

Although Einstein had ended his correspondence with Mie, in 1918 he returned to the most important issues in a witty dialogue in *Die Naturwissenschaften* between a critic of relativity and a relativist. The critic posed the question of how can one accept that mere fictitious (i.e. unreal) gravitational fields can exert an influence on the running of clocks. The relativist answered: "The distinction between 'real' and 'unreal' does not further us much... Instead of drawing a distinction between 'real' and 'unreal' we will distinguish quantities that are due to the physical system as such (independently of the choice of the coordinate system) from quantities that depend on the coordinate system. Accordingly, no quantity, independent of the choice of coordinates, corresponds to the components of the gravitational field at a space-time point; nothing 'physically real' corresponds to the components of the gravitational field at a given point, but it does correspond to this gravitational field in conjunction with other data. For this reason, we cannot claim that the gravitational field at a given point should be something 'real,' nor that it should be something 'merely fictitious'" (Einstein 1918b, p. 699–700). The difference between the two distinctions, as I view it, is not one of kind but one of degree: the historically evolved, vague distinction gives way to a more precise definition.

The reason why scientists are not ready to accept or at least to understand the difference lies, as Einstein put it, "in the circumstance that according to the general theory of relativity the connection between quantities in the equations and measurable quantities is far more indirect than in customary theories" (Einstein 1918b, p. 700). The method of selecting the frame-independent quantities from all those figuring in the equations (i.e., singling out absolute objects) is even today one of the theory's cardinal problems.

4. Conclusions

To draw conclusions requires a meticulous physical and historical analysis of Mie's work. Both kinds of analysis wait to be done; my brief remarks scattered here and there only indicate their directions. One question must, however, be asked before I end my review. Did Einstein and Mie, two very pronounced personalities, influence each other in any respect?

I think they did. A minute sign of change in Mie's position was a shift in his naming of the preferred frame: he started with "absolute," then continued with " 'absolute'," then "natural," and in 1920, "rational." He also accepted non-Euclidean geometry, though only as a mathematical device.

It is not so easy to show instances of Einstein being influenced by Mie. I have the impression that he gathered an additional impetus from Mie, ranked behind that of De Sitter and Lorentz, in his progression from a definite negation of the existence of the ether toward its acceptance by 1920. While in 1916, he only felt that general relativity was nearer to an ether concept than special relativity, in 1918, in the "Dialog" already mentioned (Einstein 1918b), he asserted that space, void of ordinary matter and electromagnetic fields, might have physical properties. This is equivalent to stating that space is a physical reality in itself. The decisive impulse to vote for the ether was given to him by De Sitter's matter-free solution of the field equations in 1917 and 1918 (De Sitter 1917, 1918). At the end of 1919, Einstein confessed to Lorentz that "it would have been more correct had I limited myself in my earlier publications to insisting that it is the velocity of the ether that is not real, instead of having defended at all the nonexistence of the ether. For since then, I have realized that with the word ether nothing is said, but that space must be considered as the carrier of physical qualities" (November 15, EA 16-494).[9] I only remind you how Mie defined ether in 1910: "One calls the empty space conceived as a physical object 'ether'."

The case of Einstein and Mie underlines a rather sceptical thesis: there is no perfect understanding of each other. Theories change in the hands of scientists, taking part in a relay race called science with each member of the team leaving the fingerprints of his commitments and beliefs on it. These fingerprints are often nonrational and incomprehensible to others. But, this is the way science actually grows, and to quote Emile Meyerson, "les croyances scientifiques font partie intégrante de l'être intellectuel le plus intime de l'homme" (Meyerson 1925, p. ix).

NOTES

[1] It is a pity that I have been unable to get permission to quote Mie's letters. I may therefore only paraphrase their contents. Professor Helmut Spehl of Freiburg University kindly informed me of the regrettable fact that Gustav Mie's estate is out of the reach of historians. It consists of at least 1000 items. A 60-page manuscript on the Einstein–Mie correspondence had already been prepared by Professor H. Hönl, but it had to be left unpublished, because Mie's

heirs withheld their permission to quote or publish anything. The documents are now kept by a private solicitor.

[2] The numbers refer to the Control Index of the Einstein Archive at the Hebrew University of Jerusalem.

[3] At the time, Einstein did not accept De Sitter's matter-free solution; the day after he had submitted the paper, on March 7, he read a lecture before the Prussian Academy criticizing De Sitter (Einstein 1918c).

[4] Mie did not mention an important feature of Poincaré's universe: its finite size before introducing the force field. Supposing that geometricians in this universe construct their geometry with the help of (allegedly) rigid rods and (allegedly) uniformly running clocks, they will arrive at a non-Euclidean space without field and infinite in its dimensions. In "reality" the universe is finite and Euclidean, with a field of force prevailing in it (Poincaré 1902, Section 4). Poincaré concluded that none of geometries may have a real existence; they are mere conventions, while Mie adhered to the variant with the field of force.

[5] For a thorough analysis of Einstein's rotating spheres see Kerszberg 1987, pp. 56–62.

[6] That there must have been some confusion in what the effect of distant masses might consist of is witnessed by De Sitter's distinction, though for convenience, between ordinary matter producing gravity and world-matter producing inertia (De Sitter 1917).

[7] Mie's mentioning Kretschmann is not merely a reference to a paper. There was a closer contact between them. There are features in Kretschmann's paper (Kretschmann 1917) that suggest that he shared Mie's view in some respects: he, too, used the term "ether," and denied the general principle of relativity in Einstein's sense. Kretschmann remarked in a footnote to the paper that Mie had given him advice in a letter of February 1916.

[8] Strictly speaking, Mie wrote Einstein a letter on June 29, 1919, in which he asked Einstein's opinion of Hermann Weyl as a prospective professor of mathematics at Halle University, briefly outlined his investigation of the electric field of a charged particle rotating around a gravitating center (later published as Mie 1920a), and invoked Einstein's help in an experiment. Einstein's answer is unknown.

[9] For a discussion of Einstein's return to the ether, see Illy 1989.

References

De Sitter, Willem (1917). "On the Curvature of Space." *Koninklijke Akademie van Wetenschappen te Amsterdam. Section of Sciences. Proceedings* 20: 229–242.

——— (1918). "Further Remarks on the Solutions of the Field-Equations of Einstein's Theory of Gravitation." *Koninklijke Akademie van Wetenschappen te Amsterdam. Section of Sciences. Proceedings* 20: 1309–1312.

Einstein, Albert (1910). "Le principe de relativité et ses conséquences dans la physique moderne." *Archives des Sciences Physiques et Naturelles* 29: 5–28, 125–44.

—— (1913). "Zum gegenwärtigen Stande des Gravitationsproblems." *Physikalische Zeitschrift* 14: 1249–1266 (lecture of September 23, 1913).

—— (1916a). "Die Grundlage der allgemeinen Relativitätstheorie." *Annalen der Physik* 49: 769–822.

—— (1916b). "Näherungsweise Integration der Feldgleichungen der Gravitation." *Königlich Preussische Akademie der Wissenschaften* (Berlin). *Sitzungsberichte*: 688–696.

—— (1917). "Kosmologische Betrachtungen zur allgemeinen Relativitätstheorie." *Königlich Preussische Akademie der Wissenschaften* (Berlin). *Sitzungsberichte*: 142–152.

—— (1918a). "Prinzipielles zur allgemeinen Relativitätstheorie." *Annalen der Physik* 55: 241–244.

—— (1918b). "Dialog über Einwande gegen die Relativitätstheorie." *Die Naturwissenschaften* 6: 697–702.

—— (1918c). "Kritisches zu einer von Herrn De Sitter gegebenen Lösung der Gravitationsgleichungen." *Königlich Preussische Akademie der Wissenschaften* (Berlin). *Sitzungsberichte*: 270–272.

Einstein, Albert, and Grossmann, Marcel (1913). "Entwurf einer verallgemeinerten Relativitätstheorie und einer Theorie der Gravitation." *Zeitschrift für Mathematik und Physik* 62: 225–261.

Eisenstaedt, Jean (1982). "Histoire et singularités de la solution de Schwarzschild (1915–1923)." *Archive for History of Exact Sciences* 27: 157–198.

Hilbert, David (1915). "Die Grundlagen der Physik. (Erste Mitteilung)." *Königliche Gesellschaft der Wissenschaften zu Göttingen. Mathematisch-physikalische Klasse. Nachrichten:* 395–407.

Illy, József (1989). "Einstein Teaches Lorentz, Lorentz Teaches Einstein: Their Collaboration in General Relativity, 1913–1920." *Archive for History of Exact Sciences* 39: 247–289.

Kerszberg, Pierre (1987). "The Relativity of Rotation in the Early Foundations of General Relativity." *Studies in History and Philosophy of Science* 18: 53–79.

Kretschmann, Erich (1917). "Über den physikalischen Sinn der Relativitätspostulate, A. Einsteins neue und seine ursprüngliche Relativitätstheorie." *Annalen der Physik* 53: 575–614.

Meyerson, Emile (1925). *La déduction relativiste.* Paris: Payot.

Mie, Gustav (1910). *Lehrbuch der Elektrizität und des Magnetismus.* Stuttgart: Ferdinand Enke.

—— (1912a). "Grundlagen einer Theorie der Materie, I." *Annalen der Physik* 37: 511–534.

—— (1912b). "Grundlagen einer Theorie der Materie, II." *Annalen der Physik* 39: 1–40.

——— (1913). "Grundlagen einer Theorie der Materie, III." *Annalen der Physik* 40: 1–66.

——— (1914a). "Bemerkungen zu der Einsteinschen Gravitationstheorie, I." *Physikalische Zeitschrift* 15: 115–122.

——— (1914b)."Bemerkungen zu der Einsteinschen Gravitationstheorie, II." *Physikalische Zeitschrift* 15: 169–176.

——— (1915). "Das Prinzip von der Relativität des Gravitationspotentials." In *Arbeiten aus den Gebieten der Physik, Mathematik, Chemie. Festschrift Julius Elster und Hans Geitel zum sechzigsten Geburtstag gewidmet von Freunden und Schülern.* Braunschweig: Friedrich Vieweg und Sohn, pp. 251–268.

——— (1917a). "Die Einsteinsche Gravitationstheorie und das Problem der Materie, I." *Physikalische Zeitschrift* 18: 551–556.

——— (1917b). "Die Einsteinsche Gravitationstheorie und das Problem der Materie, II." *Physikalische Zeitschrift* 18: 574–580.

——— (1917c). "Die Einsteinsche Gravitationstheorie und das Problem der Materie, III." *Physikalische Zeitschrift* 18: 596–602.

——— (1920a). "Das elektrische Feld eines um ein Gravitationszentrum rotierenden geladenen Partikelchens." *Physikalische Zeitschrift* 21: 651–659.

——— (1920b). "Die Einführung eines vernunftgemässen Koordinatensystems in die Einsteinschen Gravitationstheorie und das Gravitationsfeld einer schweren Kugel." *Annalen der Physik* 62: 46–74.

Poincaré, Henri (1902). *La science et l'hypothèse.* Paris: Flammarion.

Schlick, Moritz (1917a). "Raum und Zeit in der gegenwärtigen Physik. Zur Einführung in das Verständnis der allgemeinen Relativitätstheorie." *Die Naturwissenschaften* 5: 161–167, 177–186.

——— (1917b). *Raum und Zeit in der gegenwärtigen Physik. Zur Einführung in das Verständnis der Relativitäts- und Gravitationstheorie.* Berlin: Springer.

Weyl, Hermann (1918). *Raum-Zeit-Materie.* Berlin: Springer.

An Outline of the History of Einstein's Relativistic Ether Concept

Ludwik Kostro

1. Introduction

Albert Einstein has acquired the reputation of having destroyed the concept of the ether. Today, this view is put forward in textbooks, encyclopedias, and scientific reviews. It is true that from 1905 until the end of his life Einstein denied the existence of an ether as it was conceived in 19th-century physics, in particular of Lorentz's ether, which was in the first place a privileged reference frame. He denied its existence because it violated his principle of relativity, according to which there is no privileged reference frame for the formulation of the laws of nature. Nevertheless, in 1916, after the definitive formulation of the general theory of relativity, Einstein, provoked by Lorentz and Lenard, proposed a completely new conception of the ether. In this conception, the new relativistic ether does not violate the principle of relativity because the space-time of the theory of relativity is conceived as a material medium sui generis that can in no way constitute a frame of reference.

Einstein's relativistic ether conception is today almost unknown among physicists, philosophers, and sometimes, also, historians of physics. The main goal of this paper is to present a historical outline of Einstein's ideas concerning the ether, especially the new ether (Section 2), and to discuss its physical meaning and properties (Section 3).

2. A Historical Outline of Einstein's Ideas Concerning the Ether

2.1 Einstein Believed in the Existence of the Ether in the Old Sense (1894–1899)

As a boy of 15 or 16 years, Einstein believed in the existence of a stationary elastic ether that transports light at different speeds. In 1894 or 1895, he wrote his first "scientific paper" entitled "On the Investigation of the State of the Ether in the Magnetic Field." It was first published many years after his death (Mehra 1971). In this paper, he wondered in particular how "the three components of elasticity affect the velocity of an ether wave" generated when a current is turned on. At the beginning of his studies at the Eidgenössische Technische Hochschule, Einstein had no doubts about the existence of the ether and even wanted to investigate experimentally the motion of the ether with respect to the earth. He put forward several proposals for instruments that would be able to measure accurately the motion of the earth through the ether. In a speech at Kyoto University in 1922 (Ono 1983), he talked about one of these proposals, in which two thermocouples have to be used. Another proposal, which is based on the usual interference phenomena, was mentioned by him in 1901 in a letter to his friend Marcel Grossmann.[1]

2.2 Einstein Had Doubts about the Existence of a Stationary Ether

Einstein's doubts about the existence of the ether began as early as the end of the 19th century. In a letter to his girl friend Mileva Marić, written in August 1899, he expressed his doubts about the electrodynamics of moving bodies of those days and the existence of the ether in the following way:

> I am more and more convinced that the electrodynamics of moving bodies, as it is presented today, does not correspond with reality, and that it will be possible to formulate it in a simpler way. The introduction of the term "ether" into the theories of electricity led to the notion of a medium of whose motion one can speak without being able, I believe, to attach a physical meaning to this statement.[2]

2.3 Einstein Denied the Existence of the Ether

Einstein began to deny the existence of the ether in 1905 in his paper on special relativity (Einstein 1905). He wanted in particular to deny the existence of Lorentz's ether, which he identified with absolute space at absolute rest. Einstein did not know exactly what Lorentz's ether was (Illy 1988). Lorentz did not consider his ether as being at absolute rest: he only maintained that the parts of his ether were at rest with respect to each other and he attributed a certain substantiality to his ether. According to him, the vacuum was "too empty" to qualify as an ether (Kox 1988).

Einstein identified Lorentz's ether with absolute space under the influence of Paul Drude's textbooks and papers (see, e.g., Drude 1894, 1900), which he read during his studies at the ETH in Zürich. Drude presented Lorentz's results in electromagnetism in an ether that was identified with physical space at rest, endowed with physical properties. Einstein proclaimed the ether to be superfluous because, according to his paper of 1905, based on the principle of relativity, an absolute space at absolute rest understood as a privileged reference frame does not exist.

Einstein denied the existence of the ether for 11 years, from 1905 to 1916. He did so in scientific papers (Einstein 1909, 1910), as well as in newspaper articles, for instance, in 1914 in the *Vossische Zeitung* (Einstein 1914).

2.4 Einstein Introduced the Concept of the Relativistic Ether

New ideas are sometimes born in a discussion or controversy between scientists. The new ether concept was born in the correspondence with Lorentz. Lorentz accepted Einstein's theory (Lorentz 1914) after having read the Einstein–Grossmann papers and, in particular, after the definitive formulation of general relativity, and he became convinced that Einstein's new theory of gravitation allowed the hypothesis of a stationary ether. On June 6, 1916, Lorentz wrote a long letter to Einstein in which he tried to convince Einstein of his views. In his reply, Einstein proposed a definition of a new relativistic ether that can be expressed in the following short form:

$$\text{field}\, g_{\mu\nu} = \text{ether}.$$

In Einstein's letter to Lorentz of June 17, 1916 we read:

> I admit that the general theory of relativity is closer to the ether hypothesis than the special theory. This new ether theory, however, would not violate the principle of relativity, because the state of this $g_{\mu\nu}$ = ether would not be that of a rigid body in an independent state of motion, but every state of motion would be a function of position, determined through the material processes.[3]

As we can see, physical space (intimately connected with time), the local state of which is described in relativity theory by the fundamental covariant metrical tensor $g_{\mu\nu}$, was regarded by Einstein as the new relativistic ether. Einstein did not publish this new conception in either 1916 or 1917. A controversy between Einstein and Lenard provoked the first appearance of the new view in print. In a lecture in 1917 and the next year in a published paper (Lenard 1918), Lenard raised the objection against Einstein's general theory of relativity that in this theory the disqualified ether came back under the new name "space." In reply, Einstein wrote an essay (Einstein 1918) in which he published for the first time his new conception. At the same time, he proclaimed once again that the "ether in the old sense does not exist" (Einstein 1918).

2.5 Three Models of the Relativistic Ether

After 1918, Einstein published several interpretative papers in which he interpreted his models of space-time as models of the new ether. In these papers, one can make a distinction between three models of the relativistic ether.

1) The first one is the ether of the special theory of relativity. In the mathematical description of this ether, the 10 components of the tensor $g_{\mu\nu}$ are constants. The related ether is flat and infinite. Its metric is pseudo-Euclidean. It can be called "inertial ether" because in this ether model the physical space accomplishes its active function by "determining the inertial behavior of a test body introduced in it" (Einstein 1954b). It also has, however, "the physical property of transmitting electromagnetic waves" (Einstein and Infeld 1938), and it "is also absolute because its influence on inertia and on the propagation of light is taken to be independent of every kind of physical influence" (Einstein 1924).

Einstein took the four-dimensional ether of the special theory of relativity to be to some extent as "rigid" and absolute as Newton's three-dimensional space and Lorentz's three-dimensional ether.

> Physical space and the ether are only different terms for the same thing.... The four-dimensional space of the special theory of relativity is just as rigid and absolute as Newton's space. (Einstein 1954c)
>
> The rigid four-dimensional space of the special theory of relativity is to some extent a four-dimensional analogue of H. A. Lorentz's rigid three-dimensional ether. (Einstein 1954b)

2) The second model of the ether is that of the general theory of relativity. In its mathematical description, the 10 components of the tensor $g_{\mu\nu}$ are no longer constants. This ether is no longer flat. Its metric is pseudo-Riemannian. It can be called the "inertio-gravitational" or "gravitational"

(Einstein 1920a) ether because it determines the inertio-gravitational behavior of material bodies. It is no longer rigid and absolute in the above-mentioned sense because it "not only conditions the behavior of inert masses, but it is also conditioned in its state by them" (Einstein 1922a).

The pseudo-Riemannian space-time with its curvature varying not only from place to place but also in time shows the nonrigidity and dynamic nature of the general relativistic ether. Its changing metrical structure also depends on the existence and motion of the physical bodies.

3) The third model of the relativistic ether is the one that appears in Einstein's attempts to construct a unified field theory. In the mathematical description of this ether, the symmetrical tensor $g_{\mu\nu}$ (where $g_{\mu\nu} = g_{\nu\mu}$), according to Einstein, no longer completely describes the ether because the geometrical structure of the relativistic ether is more than Riemannian. In Einstein's general theory of relativity, as well as in special relativity, the electromagnetic field still appears as something that "fills space" (Einstein 1954b), i.e., as something that does not belong to the structure of the space-time as described by the symmetrical tensor $g_{\mu\nu}$. Therefore, Einstein looked for

> a theory of the continuum in which a new structural element appears side by side with the metric such that it forms a single whole together with the metric. (Einstein 1954c)

In this way, he wanted to unify the electromagnetic and the gravitational field. Therefore, the third ether may be called the "total ether." Einstein presented several versions of the relativistic unified field theory, but none of them can today be considered satisfactory. At the end of his life, Einstein was even convinced that he had "found the most natural form" of the "total field" in the "theory of the unsymmetrical field" (Einstein 1955) (where $g_{\mu\nu} \neq g_{\nu\mu}$). But, this representation of the four-dimensional continuum is today considered unsatisfactory as well.

2.6 The New Ether Terminology and its Meaning

Beginning in 1916, Einstein's physics of space-time became step by step a physics of a new ether. In this physics, the three physical notions of physical space, ether, and field have become to some extent synonyms. We can even say that they have found their unification through subsequent identification, though with some variable ontological nuances that manifested themselves during the evolution of the new conception. In 1934, Einstein wrote:

> Physical space and the ether are only different terms for the same thing; fields are physical states of space. (Einstein 1954c)

Einstein's ether concept is a result of a gradual activation, dynamization, and materialization of physical space. This process of activation, dynamization, and materialization of physical space occurred step by step after the definitive formulation of the general theory of relativity. First, during the period 1905–1916 (in his research program) under the influence of epistemological positivism (Mach, Ostwald, Avenarius), Einstein wanted to remove the concept of space conceived as endowed with physical properties from physics. He considered it, as did positivism, as a metaphysical inclusion not allowed in physics. In 1913, he wrote to Mach:

> For me, it is absurd to ascribe physical properties to "space." The totality of masses generates a $g_{\mu\nu}$-field (gravitational field), which in its turn controls the course of all processes, including the propagation of light rays and the behavior of measuring rods and clocks.[4]

In his general relativity paper of 1916, he showed himself convinced that by introducing the principle of general covariance he had succeeded in taking "away from space and time the last remnant of physical objectivity" because "all our space-time verifications invariably amount to a determination of space-time coincidences" (Einstein 1916).

In Einstein's ether concept, four-dimensional space-time not only step by step regained its physical reality, but also became an active dynamic field determining inertio-gravitational, electromagnetic, and other processes, including the production of elementary particles. Physical space, as an active field of this kind, possesses energy and therefore mass as well, and that is why it is itself of a material nature. It constitutes an active matter sui generis for which the term ether is the most appropriate name. Thus, physical space, at first considered by Einstein as an inadmissible metaphysical inclusion, became a material field that constitutes the only bearer of reality. He called it the ether, physical space, or total field. It must be noted, however, that these three terms, which to some extent can be considered as synonyms, were used by Einstein in different periods with different frequencies and with different ontological values attributed to their meanings.

During the period 1916–1925 (see e.g., Einstein 1918, 1920a, 1920b, 1924), the term ether was frequently used by Einstein. It denoted a "physical space endowed with physical properties."[5] In Einstein's research program, the new ether constituted the primary reality, i.e., the ontologically most important reality: "in a consequent field theory all objects of physics must be embraced in the notion of the ether" (Einstein 1924). But, according to Einstein, one cannot say that in the current state of his theory this program has already been carried out.

During the period 1926–1935 (see e.g., Einstein1930a,1930b,1930c, 1930d, 1930e), the term "physical space" played the principal part. According to Einstein, physical space constitutes the primary reality endowed with the properties of the ether and the field and is able to generate the elementary particles. For example, he stated

> Space, brought to light by the corporeal object, made a physical reality by Newton, has in the last few decades swallowed ether and time and seems about to swallow also the field and the corpuscles, so that it remains as the sole medium of reality. (Einstein 1930a)

During the third period, 1936–1955, (see e.g., Einstein and Kaufman 1955), the term "total field" (or sometimes also "entire field") was the term preferred by Einstein. According to him, the total field constitutes the primary reality, possessing spatial properties and constituting a matter sui generis:

> space as opposed to "what fills space" has no separate existence.... There is no such thing as an empty space, i.e., a space without field. Space-time does not claim existence on its own, but only as a structural quality of the field. (Einstein 1954b)

After 1925, Einstein began to use the word ether less and less often, although in 1938, he still wrote:

> This word ether has changed its meaning many times in the development of science.... Its story, by no means finished, is continued by the relativity theory. (Einstein and Infeld 1938)

After 1938, i.e., in the last 17 years of his life, Einstein did not use the word ether to denote his models of space-time. In his technical papers of this period, he only used the terms total field and entire field. Nevertheless, it must be noted that he never rejected the interpretation of his space-time models as models of a new ether. Several times during 1938–1955 he republished or allowed the republication of two papers presenting his new view of the ether, a paper from 1920 (Einstein 1920a) and one from 1934 (Einstein 1934b). They were republished in the books that presented Einstein's outlook on life and the physical world. This fact has a particular meaning: during the period 1938–1955, Einstein admitted the interpretation of his models of space-time as models of a new relativistic ether.

In 1954, when *Mein Weltbild* (Einstein 1934a) was republished (1954a), Einstein asked Carl Seelig, the editor of this work, to omit the last part because it no longer corresponded with the ideas he considered true in 1954,[6] but he left unchanged the words "Physical space and the ether are only different terms for the same thing; fields are physical states of space." (Einstein 1954c). It means that in 1954, i.e., one year before his death, Einstein still considered the terms physical space, ether, and total field as terms that, to some extent, mean the same thing, although during the periods indicated, he gave a particular emphasis to the ontological importance of the meaning of one or another of these three terms.

The term ether is not important here; what is important is the conception that real space-time as a field possessing energy and mass constitutes a matter sui generis and therefore possesses physical properties. In 1924, Einstein indicated that "Instead of speaking about ether, one might as well speak about the physical qualities of space" (Einstein 1924).

2.7 Some Details of the History of the New Ether

Having proposed his new ether conception in 1916, Einstein began to consider his denial of the ether during the period 1905–1916 as overly radical: "in 1905, I was of the opinion that it was no longer allowed to speak about the ether in physics. This opinion, however, was too radical . . ." (Einstein 1920b). Einstein also began to regret his denial of the ether. In a letter to Lorentz of November 15, 1919, he wrote:

> It would have been more right if I had limited myself, in my earlier publications, to emphasizing only the nonexistence of an ether velocity, instead of arguing the total nonexistence of the ether, for I can see that with the word ether we say nothing else than that space has to be viewed as a carrier of physical qualities.[7]

In 1920, the anti-relativistic attitude of several German physicists became of a political and anti-Semitic nature. Lenard, Gehrcke, and others began to propagate a "German physics" against a "Jewish physics." Paul Weyland, who was not a physicist, created an organization called the "Arbeitsgemeinschaft deutscher Naturforscher zur Erhaltung reiner Wissenschaft" with the express purpose of attacking Einstein and his theory. On August 24, 1920, he organized a demonstration against Einstein in the auditorium of the Berlin Philharmonic. At the entrance, swastika badges were distributed, and Lenard's book *Über Relativitätsprinzip, Äther, Gravitation* (a reprint of Lenard 1918) was sold. Weyland gave a speech entitled "Betrachtungen über Einstein's Relativitätstheorie und die Art ihrer Einfürung" (Weyland 1920), and the physicist Gehrcke spoke on "wissenschaftliche Massensuggestion" (Gehrcke 1920).

Since Lenard and Gehrcke were fervent defenders of the ether hypothesis, the problem of the existence of the ether was one of the principal points of Gehrcke's lecture. Einstein attended the meeting, and on August 27, he published an article criticizing Lenard and Gehrcke in the newspaper *Berliner Tageblatt* (Einstein 1920c).

The same year, the 86th meeting of the "Gesellschaft Deutscher Natruforscher und Ärzte" was held in Bad Nauheim (September 19–25). The existence of the ether was one of the principal topics of the discussion; Einstein called his definition of the new ether "eine neuartige Definition für den Begriff Äther" (Einstein, Lenard, et al., 1920).

The new ether concept matured, however, not in the controversy with Lenard, but in the correspondence between Lorentz and Einstein. Lorentz asked Einstein in several letters for a comprehensive publication of the new ideas. Einstein replied that he would do it "when the first suitable opportunity arises."[8] Such an opportunity occurred when Lorentz invited Einstein to be visiting professor at the University of Leiden. Einstein devoted his inaugural lecture (delivered on October 27, 1920; see Pais 1982, p. 313) to the problem of the ether in the theory of relativity. The text of this lecture was translated in several languages and reprinted many times during the period 1938–1955, in which Einstein no longer used the term ether in his technical papers. The inaugural lecture contains the following statement in favor of the ether hypothesis.

> ...there is a weighty argument to be adduced in favor of the ether hypothesis. To deny the ether is ultimately to assume that empty space has no physical qualities whatsoever. (Einstein 1922a)

According to Einstein's lecture, both general relativity and special relativity admit an ether hypothesis.

> As to the mechanical nature of the Lorentzian ether, it may be said of it, in a somewhat playful spirit, that immobility is the only mechanical property of which it has not been deprived by H.A. Lorentz. It may be added that the whole change in the conception of the ether, which the special theory of relativity brought about, consisted of taking away from the ether its last mechanical quality, namely, its immobility. More careful reflection teaches us, however, that the special theory of relativity does not compel us to deny ether. We may assume the existence of an ether; only we must give up ascribing a definite state of motion to it, i.e., we must by abstraction take from it the last mechanical characteristic that Lorentz had still left it.... Generalizing we must say this: There may be supposed to be extended physical objects to which the idea of motion cannot be applied. They may not be thought of as consisting of particles that allow themselves to be separately tracked through time.

> In Minkowski's idiom, this is expressed as follows: Not every extended conformation in the four-dimensional world can be regarded as composed of world threads. The special theory of relativity forbids us to assume the ether to consist of particles observable through time, but the hypothesis of ether in itself is not in conflict with the special theory of relativity. (Einstein 1922a)

According to Einstein, the general theory in particular is incomprehensible without an ether.

> ... according to the general theory of relativity, space is endowed with physical qualities; in this sense, therefore, there exists and ether. According to the general theory of relativity, space without ether is unthinkable; for in such space, not only would there be no propagation of light, but also no possibility of existence for standards of space and time (measuring rods and clocks), nor therefore any space-time intervals in the physical sense. But, this ether may not be thought of as endowed with the quality characteristic of ponderable media, as consisting of parts that may be tracked through time. The idea of motion may not be applied to it. (Einstein 1922a)

> The ether of the general theory of relativity is a medium that is itself devoid of all mechanical and kinematical qualities, but it helps to determine mechanical (and electromagnetic) events. (Einstein 1922a)

As we see, Einstein's ether is not composed of particles, nor is the concept of motion applicable to it.

In 1919, the English were very excited about relativity because of Eddington's eclipse expedition. A special issue of the journal *Nature* was devoted to relativity. Einstein had been invited to contribute to this special issue. He wrote a 35-page essay on the development of the theory of relativity. The paper was, however, too long, so only a three-page summary was eventually published in 1921 (Einstein 1921). But, the longer manuscript is quite interesting because it also contains two sections on the new ether. The first section on the ether of special relativity is on page 15:

> (13) *Special relativity and the ether.* It is clear that in the theory of relativity there is no place for the notion of the ether at rest. If the reference systems K and K' are completely equivalent for the formulation of the laws of nature, it is inconsistent to base the theory on a conception that distinguishes one of these systems from the other ones. If one postulates an ether at rest with respect to K, it moves with respect to K', which is not in accord with the equivalence of the two systems.

> Therefore, in 1905, I was of the opinion that it was no longer allowed to speak about the ether in physics. This opinion, however, was too radical, as we will see later when we discuss the general theory of relativity. It does remain allowed, as always, to introduce a medium filling all space and to assume the electromagnetic fields (and matter as well) are its states. But, it is not allowed to attribute to this medium a state of motion in each point, in analogy to ponderable matter. This ether may not be conceived as consisting of particles that can be individually tracked in time. (Einstein 1920b)

The second section is on p. 34:

> (22) *General relativity and ether.* It is not difficult to incorporate the laws of nature, already known from special relativity, into the broader framework of general relativity. The mathematical methods were readily available in the "absolute differential calculus," based on the work of Gauss and Riemann and further developed by Ricci and Levi-Cevita in particular. It concerns a rather simple way of generalizing the equations from the special case of the constant $g_{\mu\nu}$ to the case of the spatiotemporary varying $g_{\mu\nu}$. In all laws generalized in this way, a role is played by the gravitational potentials $g_{\mu\nu}$, which, in a word, express the physical properties of empty space.
>
> Thus, once again "empty" space appears as endowed with physical properties, i.e., no longer as physically empty, as seemed to be the case according to special relativity. One can thus say that the ether is resurrected in the general theory of relativity, though in a more sublimated form. The ether of the general theory of relativity differs from the one of earlier optics by the fact that it is not matter in the sense of mechanics. Not even the concept of motion can be applied to it. It is furthermore not at all homogeneous, and its state has no autonomous existence but depends on the field-generating matter. Since in the new theory, metric facts can no longer be separated from "true" physical facts, the concepts of "space" and "ether" merge together. Since the properties of space appear as determined by matter, according to the new theory, space is no longer a precondition for matter; the theory of space (geometry) and of time can no longer be presupposed prior to actual physics and expounded independently of mechanics and gravitation. (Einstein 1920b)

In the first papers on the new ether (i.e., in the papers from the period 1916–1920), physical space, identified with the ether and the fields, was still conceived as something secondary to ponderable matter and actually generated by it. Step by step, however, Einstein changed his opinion. The first signs of such a change can be found in his paper "Spielen Gravitationsfelder im Aufbau der materiellen Elementarteilchen eine wesentliche Rolle?" (Einstein 1919).

The opinion, according to which the ether (identified with physical space and the fields) constitutes a primary reality with respect to ponderable matter composed of elementary particles, was expressed by Einstein in 1924 in his paper "Über den Äther" (Einstein 1924). There one reads:

> ... one can defend the view that this notion [the ether] includes all objects of physics, since according to a consistent field theory, ponderable matter and the elementary particles from which it is built also have to be regarded as "fields" of a particular kind or as particular "states" of space. (Einstein 1924)

Thus, in 1924, the view of physical space being primary with respect to ponderable matter had become the main element of Einstein's research program. He propagated it particularly in 1929 and 1930. In a lecture delivered at a meeting organized by the "Kaiser Wilhelm Gesellschaft zur Forderung der Wissenschaft," he said:

> According to the views here presented, the axiomatic foundation of physics appears as follows. The real is conceived as a four-dimensional continuum with unitary structure of a definite kind (metric and direction). The laws are differential equations, which the structure mentioned satisfies, namely, the fields that appear as gravitation and electromagnetism. The material particles are positions of high density without singularity.
>
> We may summarize in symbolic language. Space, brought into light by the corporeal object and made a physical reality by Newton, has in the last few decades swallowed ether and time and also seems about to swallow the field and the corpuscles, so that it remains as the sole medium of reality. (Einstein 1930a)

In a particularly emphatic way, the primacy of physical space was expressed by Einstein in a speech at the University of Nottingham:

> The strange conclusion to which we have come is this—it now appears that space will have to be regarded as a primary thing and that matter is derived from it, so to speak, as a secondary result. Space is now turning around and eating up matter. We have always regarded matter as a primary thing and space as a secondary result. Space is now having its revenge, so to speak, and is eating up matter. But, that is still a pious wish. (Einstein 1930d)

In 1929 and 1930, Einstein published several papers on the problem of space, ether, and the field. The above-mentioned lecture at the meeting organized by the Kaiser Wilhelm Gesellschaft was published on December 15, 1929, in the *Deutsche Bergwerks-Zeitung* in the form of an extensive summary by an anonymous author (Einstein 1929). The full text was published in German and English in 1930 in *Forum Philosophicum* (Einstein 1930a). An essay with some new ideas concerning the same problem was published in the review *Die Koralle* (Einstein 1930c).

It is interesting to note that the problem of space, ether, and the field was the subject of Einstein's lecture at the Second World Power Conference (Berlin, June 16–25, 1930), a conference devoted to the resources of energy. It means that Einstein was convinced that physical space, with the properties of an ether and conceived as the total field possessing energy and able to generate elementary particles, is the most fundamental resource of energy.

Since, according to Einstein's research program, elementary particles have to be conceived as generated by physical space, Einstein searched for solutions of the field equations of general relativity that might be "interpreted as representing corpuscles" (Einstein 1936). In 1935, he was convinced that he found such solutions. Together with Rosen, he published a paper (Einstein and Rosen 1935) in which they presented solutions of the centrally symmetric gravitational field equations that were interpreted by them as representing both neutral and electrically charged particles. Having found these, they stated with emphasis that "The neutral as well as the electrical particle is a portion of space" (Einstein and Rosen 1935).

In 1934, Einstein's book *Mein Weltbild* was published (Einstein 1934a). In this book, which presents Einstein's outlook on life and the physical world, we find a new paper regarding the problem of space, ether, and the field. It is in this paper that Einstein wrote the following.

> Physical space and the ether are only different terms for the same thing; fields are physical states of space. If no particular state of motion can be ascribed to the ether, there does not seem to be any ground for introducing it as an entity of a special sort alongside of space. (Einstein 1954c)

The article under consideration was Einstein's last paper presenting his new ether, but at the same time it was the one that was reprinted many times and translated into many languages during the next 20 years, i.e., until Einstein's death. It was reprinted in books presenting Einstein's outlook on life and the physical world. As mentioned, it was reprinted with some changes suggested by Einstein himself (in a letter to Carl Seelig, the editor of the book), but without any change in the principal ideas concerning the new ether. We find it in every new edition of the books *Mein Weltbilt* (1934a) and *Ideas and Opinions* (Einstein 1954a) and also in the books *Essays in Science* (Einstein 1934c, 1934d) and

Man and the Universe (Einstein 1947). It means that during the last 20 years of his life, Einstein did not reject the interpretation of the space-time models of relativity theory as models of a new ether, although he did not write new papers on this subject, and in his technical papers, he only used the term total field or entire field.

It should be noted that Einstein never used the term ether in the new sense in his technical papers. He used it only in his interpretative articles written and published in the period 1918–1938 and reprinted in the period 1938–1955.

3. Essential Features of Einstein's Relativistic Ether

3.1 Einstein's Ether Constituted an Ultrareferential Fundamental Reality

In identifying the ether with physical space, Einstein made a very clear distinction between physical space as such, which is one and unique, and reference spaces, the number of which is infinite. According to him, the ether cannot be identified with any of the reference spaces because that would mean that one of them is favored or privileged with respect to others. This contradicts the principle of relativity. According to Einstein, only physical space as such constitutes the relativistic ether because it is not all a reference space. Such a clear distinction between physical space as such and reference spaces does not exist in pre-Einsteinian physics. Physical space of pre-Einsteinian physics, conceived as connected or identified with the ether in the old sense, was at the same time a reference space. It was a privileged reference space that was often also taken to be at absolute rest.

According to Einstein, in the theory of relativity, the notion of motion can only be applied to reference spaces because only these move or are at rest with respect to each other. Physical space as such does not move and is not at rest. The notion of motion is not applicable to it. This is a completely new point of view.

According to the theory of relativity, if an observer is in physical space as such, he can never decide whether he moves or is at rest with respect to it. When the observer introduces an infinite number of points that are at rest with respect to him, he introduces, according to special relativity, his own reference space, but this space does not constitute physical space as such. In this way, the observer can introduce an infinite number of spaces that move or are at rest with respect to him, but none of these constitutes physical space as such. All are reference spaces.

Physical space as such, identified with the relativistic ether, constitutes an ultrareferential reality, a background reality. It means that, to a certain extent, physical space is over or behind all reference spaces. With respect to these, it constitutes a more fundamental background reality, which makes possible the existence and the motion of reference spaces, but is not in motion or at rest itself.

3.2 The Relativistic Ether is Not Composed of Points or Particles

According to Einstein, a reference space is usually connected with a body of reference and forms its artificial extension. We introduce a reference space through an infinite number of points that are at rest with respect to such a body. Together, all points introduced in this way constitute the body's own space and its artificial extension.

According to Einstein, the reference spaces of Newtonian physics and those of special relativity are idealizations of extended physical, practically rigid bodies, composed of particles that are considered immobile. Reference spaces constitute quasi-objects composed of points that do not move with respect to each other. According to the theory of relativity, such a universal quasi-object composed of motionless points does not exist.

Space as such cannot be composed of points because, according to the theory of relativity, a point of space constitutes a set of colocal events, and colocality is something totally relative and closely connected with a reference space. The events that are colocal in one reference space are not colocal in the other. Therefore the ultrareferential space, identified with the relativistic ether, cannot be composed of points.

3.3 The Relativistic Ether is Not Composed of Parts

Reference spaces that are quasi-objects can be divided into parts because they consist of points. The distance between two points of a reference space can be represented with the help of a segment of a straight line or a geodesic line. Connecting different points of a reference space, we can divide it into parts with the help of the mentioned segments. Physical space as such cannot be divided into parts because it is not composed of points, although it makes possible the existence of reference spaces that can be divided into parts. The indivisibility of physical space as such seems to be something strange because we are accustomed to dividing material objects and quasi-objects, i.e., reference spaces. Einstein's discovery of the indivisibility of physical space as such is very interesting. In order to clarify this discovery somewhat, let us consider a one-dimensional rectilinear space. In such a space, we can introduce an infinite number of straight lines that move with respect to each other. Every straight line introduced in this way can be divided into segments because it is composed of points. A segment constitutes a distance between two points of a straight line. But, the one-dimensional rectilinear space as such cannot be divided into segments because it is not composed of points.

In an analogous way, physical time is not divisible in time intervals because it is not composed of moments. Only reference times closely connected with reference spaces are composed of moments. Although time as such is not composed of moments and therefore not divisible into time intervals, it still constitutes an extended reality composed of past, present, and future. With respect to a freely chosen event considered as present, there always exists a set of events that are absolutely (i.e., ultrareferentially) future and a set of events that are absolutely past.

3.4 Dynamic and Static Images of Motion of the Reference Spaces

According to Einstein, in the description of the motion of the reference spaces in physical space as such, which makes possible their existence and their motion, we can use two images: a dynamic one and a static one (see Einstein and Infeld 1938). In the dynamic image, we try to investigate the motion of reference spaces in the time of our laboratory reference frame. In this image, the positions of the reference spaces change in time. In the static image, we take into consideration the intimate connection between space and time. We consider space and time as a four-dimensional continuum. Using such an image, we present the motion of reference spaces as something taking place in space-time.

Until now, we have only used the dynamic image. Let us now use the static one. In this image, we are dealing with a clear distinction between space-time as such, which is unique, and reference space-times, which are infinite in number. The new ether, of course, has to be identified with space-time as such and not with any reference space-time. In the static image, space-time as such constitutes the fundamental background ultrareferential reality that makes possible the existence of reference space-times. The latter are present in space-time as such and established particular orientations in it.

In the static image, to every point of a reference space, there corresponds a world line that constitutes a set of colocal events in this reference space, and to every moment of a reference time, there corresponds an instantaneous space that constitutes a set of simultaneous events. It is clear that in the static image reference space-times are composed of world lines and instantaneous spaces. But, space-time as such, identified with the relativistic ether, is not composed of world lines or instantaneous spaces. It constitutes merely a four-dimensional extended continuum composed of events only. "In Minkowski's idiom, this is expressed as follows: Not every extended conformation in the four-dimensional world can be regarded as composed of world threads." (Einstein 1922a)

3.5 The Relativistic Ether Constitutes a Medium That is Not Comparable with Any Ponderable Medium

According to Einstein, every material medium composed of particles, atoms, or molecules can serve as a convenient three-dimensional reference frame. He used material media as reference frames when he studied Brownian motion, for example. Sometimes, he also used a gas composed of photons as a reference frame (see, for instance, Einstein and Hopf 1910). As is well-known, an observer who moves with respect to a material medium composed of particles (a gas, liquid, etc.) feels a wind or a change of temperature on one side, for instance when we are dealing with motion with respect to a gas of photons. The cosmic background radiation discovered nine years after Einstein's death also constitutes a material medium composed of particles (photons) and can serve as a convenient reference frame in cosmology. Also, the supergas composed of galaxies can serve as a reference frame in cosmology. The theory of relativity admits such convenient reference frames and recognizes the influence of such material media on the metric structure of physical space as such, but at the same time, it emphasizes that none of them can be identified with physical space as such, because the latter in no respect constitutes a reference frame.

A gas of photons, also in the case when it constitutes a cosmic background radiation, cannot be considered an absolute reference frame. We can always imagine a great number of very large containers filled with photon gas that move with respect to each other, and there is no reason to consider one of them as an absolute reference frame. We can also imagine an empty space without any photon gas.

Physical space as such, identified with the relativistic ether, constitutes an ultrareferential fundamental background reality that lays behind every material medium and every reference space, and therefore, it cannot be compared with any material medium that is composed of particles and can serve as a reference frame.

3.6 The Relativistic Ether Plays an Active Part in Physical Phenomena

It seemed utterly absurd to the physicists of the 19th century to attribute physical functions or states to space itself. That is not the case in Einstein's theory of relativity. Physical space as such plays a truly active part in physical processes. When we speak about the new ether, we are dealing with "those things considered physically real, which alongside ponderable matter composed of elementary particles play a role in the causal connection of physics." (Einstein 1924).

According to Einstein's new conception, it is impossible to formulate a complete physical theory without an (at least latent) ether hypothesis, because every complete physical theory must take into consideration the real properties of physical space, i.e., the "Milieu-Einflüsse" (Einstein 1924). One might not use the word ether, but one still has to recognize that physical space has real properties that play an active role in physical happenings, and therefore, Einstein maintained that "the ether hypothesis was bound always to play some part in physical science, even if first only a latent part" (Einstein 1922a).

According to the new conception in its most developed form, physical space closely connected with time is not a passive and static container of events, nor is it a physically indifferent or neutral arena of physical phenomena, but it is an active and dynamic material field that determines the inertio-gravitational, electromagnetic, and other processes, including the production of elementary particles.

As we can see, Einstein considered a nonatomic and nonmechanical ether as the fundamental source of every physical activity, including the creation of particles. His presentation of the ether activity, however (except the inertio-gravitational one), cannot be considered satisfactory today. On this point, Einstein's research program cannot be regarded as having succeeded in a definitive way. Today, this program, as has been indicated by Fadeev (1979), is continued in hypotheses in which elementary particles are represented as solitons on an active relativistic field.

Acknowledgments: The author would like to express his sincere thanks to the Volkswagen-Stiftung for a scholarship that made this research possible and to Dr. Jürgen Teichmann and all others at the Deutsches Museum in Munich for their kind assistance. He would also like to thank the Hebrew University of Jerusalem for the kind permission to quote from Einstein's papers and letters and Professor Don Howard for his assistance in finding some of Einstein's papers and for very useful discussions, which have been expressed in this paper.

NOTES

The numbers in brackets refer to the control index of the Einstein Archive at the Hebrew University of Jerusalem.

[1] Einstein to Marcel Grossmann, September 6, 1901. Published in Stachel 1987, pp. 315–316.

[2] Einstein to Mileva Marić, August 10, 1899. Published in Stachel et al. 1987, pp. 225–227.

[3] Einstein to H.A. Lorentz, June 17, 1916 (EA 16-453).

[4] Einstein to Ernst Mach, December 1913 (EA 16-454).

[5] Einstein to H.A. Lorentz, November 15, 1919 (EA16-494).

[6] Einstein to Carl Seelig, quoted in Einstein's "Mein Weltbild," reedited in 1983.

[7] Einstein to H.A. Lorentz, November 15, 1919 (EA 16-494)

[8] Einstein to H.A. Lorentz, November 15, 1919 (EA 16-494).

REFERENCES

Drude, Paul (1894). *Physik des Aethers auf elektromagnetischer Grundlage.* Stuttgart: Ferdinand Enke.

—— (1900). *Lehrbuch der Optik.* Stuttgart: S. Hirzel.

Einstein, Albert (1905). "Zur Elektrodynamik bewegter Körper." *Annalen der Physik* 17: 891–921.

—— (1909). "Über die Entwicklung unserer Anschauungen über das Wesen und die Konstitution der Strahlung." *Physikalische Zeitschrift* 10: 817–825.

—— (1910). "Le principe de relativité et ses conséquences dans la physique moderne." *Archives des Sciences Physiques et Naturelles* 29: 5–28, 125–44.

—— (1914). "Vom Relativitätsprinzip." *Vossische Zeitung,* April 26, 1914 (Suppl. 8), pp. 33–34.

—— (1916). "Die Grundlage der allgemeinen Relativitätstheorie." *Annalen der Physik* 49: 769-822.

—— (1918). "Dialog über Einwande gegen die Relativitätstheorie." *Die Naturwissenschaften* 6: 697-702.

—— (1919). "Spielen Gravitationsfelder im Aufbau der materiellen Elementarteilchen eine wesentliche Rolle?" *Preussische Akademie der Wissenschaften* (Berlin). *Sitzungsberichte*: 349-356.

—— (1920a). *Äther und Relativitätstheorie. Rede Gehalten am 5. Mai 1920 an der Reichs-Universität zu Leiden.* Berlin: Springer.

—— (1920b). "Grundgedanken und Methoden der Relativitätstheorie in Ihrer Entwicklung Dargestellt." Unpublished manuscript (EA 2-070).

——— (1920c). "Meine Antwort über die antirelativitätstheoretische G.m.b.H." *Berliner Tageblatt und Handelszeitung,* August 27, 1920, pp. 1–2.
——— (1921). "A Brief Outline of the Development of the Theory of Relativity." *Nature* 106: 782–784.
——— (1922a). "Ether and the Theory of Relativity." In *Sidelights on Relativity.* G.B. Jeffery and W. Perrett, trans. London: Methuen; New York: E.P. Dutton, pp. 1–24. Translation of Einstein 1920a.
——— (1922b). "Geometry and Experience." In *Sidelights on Relativity.* G.B. Jeffery and W. Perrett, trans. London: Methuen; New York: E.P. Dutton.
——— (1924). "Über den Äther." *Schweizerische naturforschende Gesellschaft. Verhandlungen* 105:85–93.
——— (1929). "Professor Einstein spricht über das physikalische Raum- und Aether-Problem." *Deutsche Bergwerks-Zeitung,* December 15, 1929, p. 11.
——— (1930a). "Raum, Äther und Feld in der Physik." *Forum Philosophicum* 1: 173–180. English translation "Space, Ether and the Field in Physics." *Forum Philosophicum* 1: 180–184.
——— (1930b). "Raum-, Feld- und Äther-Problem in der Physik." In *Gesamtbericht, Zweite Weltkraftkonferenz, Berlin, 1930,* Vol. 19. Franz zur Neden and Carl Theodor Kromer, eds. Berlin: VDI-Verlag, pp. 1–5.
——— (1930c). "Das Raum-, Feld- und Äther-Problem in der Physik." *Die Koralle. Monatshefte für alle Freunde von Natur und Technik* 11: 486–487.
——— (1930d). "Address at the University of Nottingham." *Science* 71: 600–610.
——— (1934a). *Mein Weltbild.* Amsterdam: Querido Verlag.
——— (1934b). "Das Raum-, Äther- und Feld-Problem in der Physik." In *Einstein* 1934a, pp. 229–248.
——— (1934c). *Essays in Science.* Alan Harris, trans. New York: Philosophical Library. Abridged translation of Einstein 1934a.
——— (1934d). "The Problem of Space, Ether, and the Field in Physics." In *Einstein* 1934c, pp. 61–77.
——— (1936). "Physik und Realität." *Journal of the Franklin Institute* 221: 313–347.
——— (1947). "The Problem of Space, Ether, and the Field in Physics." In *Man and the Universe.* Saxe Commins and R. N. Linscot, eds. New York: Random House, pp. 471–482.
——— (1954a). *Ideas and Opinions.* Carl Seelig, ed. Sonja Bargmann, trans. New York: Crown, 1960.
——— (1954b). "Relativity and the Problem of Space." In *Einstein* 1954b, pp. 360–377.
——— (1954c). "The Problem of Space, Ether, and the Field in Physics." In *Einstein* 1954b, pp. 276–285.
——— (1955). "Appendix II." In *The Meaning of Relativity,* 5th. ed. Princeton, New Jersey: Princeton University Press.
——— (1983). "Mein Weltbild." Frankfurt/M, Berlin, Wien: Verlag Ullstein GmbH, p. 199.

Einstein, Albert, and Hopf, Ludwig (1910). "Statistische Untersuchungen der Bewegung eines Resonators in einem Strahlungsfeld." *Annalen der Physik* 33: 1105–1115.

Einstein, Albert, and Infeld, Leopold (1938). *The Evolution of Physics: The Growth of Ideas from Early Concepts to Relativity and Quanta.* New York: Simon & Schuster.

Einstein, Albert, and Kaufman, Bruria (1955). "A New Form of the Relativistic Field Equations." *Annals of Mathematics* 62: 128–138.

Einstein, Albert, and Lenard, Philipp (1920). "Algemeine Diskussion über Relativitätstheorie." *Physikalische Zeitschrift* 21: 666–668.

Einstein, Albert, and Rosen, Nathan (1935). "The Particle Problem in the General Theory of Relativity." *Physical Review* 48: 73–77.

Einstein, Albert (1920). "Allgemeine Diskussion über Relativitätstheorie." *Physikalische Zeitschrift* 21: 666–668.

Faddeev, L. D. (1979). "Le vedute di Einstein e talune vedute contemporanee sulle particelle elementari." In *Centenario di Einstein 1879–1979.* Florence: Giunti Barbèra, pp. 765–793.

Gehrcke, Ernst (1920). "Die Relativitatstheorie, eine wissenschaftliche Massensuggestion." *Schriften aus dem Verlage der Arbeitsgemeinschaft deutscher Naturforscher zur Erhaltung reiner Wissenschaft e.V.*, No. 1. , Berlin.

Illy, József (1988). "Einstein Teaches Lorentz, Lorentz Teaches Einstein. Their Collaboration in General Relativity, 1913–1920." *Archive for History of Exact Sciences* 39: 247–289.

Kox, A.J. (1988). "Hendrik Antoon Lorentz, the Ether, and the General Theory of Relativity." *Archive for History of Exact Sciences* 38: 67–78.

Lenard, Philipp (1918). "Über Relativitätsprinzip, Äther, Gravitation." *Jahrbuch der Radioaktivität und Elektronik* 15: 117–136.

Lorentz, Hendrik A. (1914). "La gravitation." *Scientia* 16: 28–59. Reprinted in *Collected Papers*, Vol. 7, The Hague: Nijhoff, 1934, pp. 116–146.

Mehra, Jagdish (1971). "Albert Einsteins erste wissenschaftliche Arbeit." *Physikalische Blätter* 27: 386–391.

Ono, Y. A. (1983). "Einstein's Speech at Kyoto University, December 14, 1922." *NTM-Schriftenreihe für Geschichte der Naturwissenschaften, Technik, und Medizin* 20: 25–28.

Pais, Abraham (1982). *"Subtle is the Lord ...": The Science and Life of Albert Einstein.* London and New York: Oxford University Press.

Weyland, Paul (1920). "Betrachtungen über Einstein's Relativitätstheorie und die Art ihrer Einführung." *Schriften aus dem Verlage der Arbeitsgemeinschaft deutscher Naturforscher zur Erhaltung reiner Wissenschaft e.V.*, No. 2., Berlin.

The Physical Content of General Covariance

John Norton

1. Introduction

"My wildest dreams have been fulfilled," wrote a jubilant Einstein in early December 1915 to his dear friend Michele Besso (Speziali, 1972, p. 60) "*General* covariance. Perihelion motion of Mercury wonderfully exact...," he continued, checking off the achievements that brought to a dramatic and successful close nearly three years of struggle with his general theory of relativity, during which time he had mistakenly come to believe that he must forgo general covariance. The general covariance of his theory was to be stressed by Einstein as one of its most distinctive properties and, in particular, the one that gave mathematical expression to the theory's extension of the principle of relativity to all states of motion. He explained this connection to the principle of relativity in lectures he gave at Princeton University:

> We shall be true to the principle of relativity in its broadest sense if we give such a form to the laws that they are valid in every such four-dimensional system of coordinates, that is, if the equations expressing the laws are covariant with respect to arbitrary transformations. (Einstein 1922, p. 60)

Such proclamations are common in Einstein's work.[1] Unfortunately, they have proved to be a problem for later commentators who seek to understand Einstein's views because it is now commonplace for any reasonably coherent space-time theory to have a generally-covariant formulation. One need only formulate it by the standard methods of modern differential geometry. Since these generally-covariant theories include the versions of Newtonian space-time theory that unequivocally violate the usual relativity principles, any interesting connection between general covariance and relativity principles seems extremely dubious. Thus, some modern texts on general relativity lay out the foundations of the theory without recourse to principles of relativity or covariance. See, for example, Hawking and Ellis (1973) and Sachs and Wu (1977), who do not even have entries for these terms in their indices![2]

The purpose of this chapter is not to offer yet another vindication of Einstein's claim that general relativity is a theory that extends the principle of relativity of the special theory. Rather, I explain why a literal reading of Einstein's claims should appear incoherent to modern readers in the first place. Further, I suggest that Einstein's covariance principles are principles with physical content and that they are the analogues of physically significant covariance principles in the modern context. However, I also argue that the relativity-principle-like character of Einstein's covariance principles is a peculiarity of his older and less adequate formulation of general relativity, a conclusion that provides *no* support to the view that general relativity, freed of the peculiarities of specific, known formulations, has effected a generalization of the principle of relativity. My explanation rests on the fact that we now use a far richer set of mathematical machinery in formulating general relativity than Einstein did in the 1910s. One need only scan a text such as Hawking and Ellis' (1973) to find a wealth of mathematical terms, distinctions, and symbols that are just not present in Einstein's work of the 1910s.

Modern readers routinely translate Einstein's claims both consciously and unconsciously into this richer language and are generally very successful. My thesis is that this translation has been routinely carried out incorrectly in one important area, and that this incorrect translation is responsible for much of the apparent incoherence of Einstein's claims about covariance. The root of the problem lies in the following:

> Where we now represent a space-time mathematically by a differentiable manifold with a point set of unspecified elements, Einstein simply used number manifolds, open sets of the R^4.

This representation relation is a *coordination* of an actual or physically possible space-time with a general differentiable manifold or, for Einstein, with a number manifold. Thus Einstein naturally gave this representation the name *coordinate system*, and we have:

> Einstein's coordinate systems are not the coordinate charts of a general differentiable manifold of the modern approach. Rather, they correspond to the modern representation of an actual or physically possible space-time by a general differentiable manifold.

I shall try to show that if we read Einstein in accord with the above proposal, then we shall find that his principle of general covariance is a principle with significant physical content, and that that content is of the character of a generalized relativity principle.

2. The Modern View of Space-Time Theories

We now usually take a space-time theory to be synonymous with its set of models. Thus, general relativity is the set of all triples $\langle M, g_{ab}, T_{ab}\rangle$, where M is a four-dimensional differentiable manifold, g_{ab} is a Lorentz signature metric, and T_{ab} is a symmetric stress-energy tensor, such that the g_{ab} and T_{ab} of each pair satisfy the field equation

$$G_{ab} = kT_{ab} \tag{1}$$

where G_{ab} is the Einstein tensor and k is a constant. In everyday practice, it is usually convenient to treat a single triple $\langle M, g_{ab}, T_{ab}\rangle$ as representing each actual or physically possible space-time and sometimes even to speak of the triple as simply being the space-time. But, in more careful presentations, for example, Hawking and Ellis (1973, p. 56), we are reminded that each physically possible space-time is actually represented by an equivalence class of diffeomorphic triples.

To be more precise, now let us define two requirements that may be satisfied by space-time theories with models of the form $\langle M, O_1, O_2, \ldots\rangle$, where M is a differentiable manifold and $O_1, O_2, \ldots$ are geometric object fields defined on M. The first simply provides for the existence of the members of the equivalence classes of diffeomorphic models previously mentioned. It is automatically satisfied by such theories as general relativity by virtue of the general covariance of its defining field equation Eq. (1).

> *(Active)*[3] General Covariance: If $\langle M, O_1, O_2, \ldots\rangle$ is any model of the space-time theory and h any diffeomorphism from M to hM, then the carried along tuple $\langle hM, h^*O_1, h^*O_2, \ldots\rangle$ is also a model of the theory.
>
> *(Active)* Leibniz Equivalence: If $\langle M, O_1, O_2, \ldots\rangle$ and $\langle hM, h^*O_1, h^*O_2, \ldots\rangle$ are diffeomorphic models of a space-time theory, then they represent the same physically possible space-time.

The following is crucial to my story:

> The requirements of (active) general covariance and Leibniz equivalence are not forced on us by mathematical necessity; they are physical principles that we can choose to accept or deny.

The justification for this claim depends on the fact that two diffeomorphic models are distinct mathematical objects, unless the diffeomorphism concerned is the identity map or a symmetry of the model. Prior to further assumptions, we must assume that each property of a theory's model represents some physical property of a physically possible space-time. General covariance, asserted without Leibniz equivalence, allows us to take a model of a theory and from it generate arbitrarily many new and distinct mathematical models, each of which asserts the physical possibility of a distinct space-time. Leibniz equivalence asserts that two diffeomorphic models represent the same physically possible space-time. This can only be the case if the properties that distinguish the two models have no physical significance, that is, represent nothing in the physically possible space-time. Finally, general covariance and Leibniz equivalence, asserted jointly, amount to the assertion that the physically significant properties of a theory's models are just those that are invariant under arbitrary diffeomorphism. They assert that a model and all possible diffeomorphic copies of it represent the same physically possible space-time. Thus, the physically significant parts of each model can only be those upon which they all agree, that is, their invariants under arbitrary diffeomorphism. In sum, since these principles take a stand on which space-times are physically possible and what their physical properties are, the principles are physical principles, not mathematical stipulations.

The properties that distinguish diffeomorphic models are hardly of great importance because of the sustained attempt to purge our formulations of space-time theories of properties without physical significance. But, they are there. The simplest case is the one in which the manifolds M and M' are the same and the diffeomorphism h is not an identity or symmetry of the geometric object fields. Then, we can always find a point p of M at which O_1 and h^*O_1 differ. If the point sets of M and M' are disjoint, the two models are still mathematically distinct entities distinguished by the property of set membership. If p is a member of the point set of M, then it cannot also be a member of the point set of M'. This difference—if noticed at all—is usually not taken to be physically significant. But, no incoherence would follow from deciding otherwise.[4]

There are two grounds for accepting Leibniz equivalence, and since Leibniz equivalence is a physical principle, these grounds depend on physical considerations and cannot be proofs:

1. Under canonical interpretation, the properties that distinguish two diffeomorphic models do not correspond to any observable physical properties. So, we admit Leibniz equivalence in order to minimize instances of distinct physical states of affairs that cannot be distinguished

> by any possible observation.

Of course, we would not be guilty of incoherence if we denied Leibniz equivalence and allowed the possibility of observationally indistinguishable but distinct space-times.

> 2. The generation of diffeomorphic models is a gauge freedom of theories with generally-covariant field equations. If we deny Leibniz equivalence, we force indeterminism in many space-time theories. For example, even general-relativistic space-times that admit Cauchy surfaces become indeterministic.

The forcing of indeterminism in cases such as these seems unwarranted. In particular, there are no observables corresponding to the properties that remain undetermined, that is, those properties that distinguish the diffeomorphic models. Notice again, however, that there is no incoherence in denying Leibniz equivalence and thereby forcing indeterminism.

These two considerations are essentially the "point-coincidence" and "hole" arguments presented in greater detail in Earman and Norton (1987) and Norton (1987). Both arguments find early expression in the work of Einstein in the 1910s, and I shall return to the first argument in Section 8.

3. Number Manifolds

The ability of space-time theories to provide mathematical representations of physically possible space-times depends on the availability of a continuous, finite-dimensional mathematical structure. The modern study of such structures was initiated by Riemann (1854) in his classic inaugural lecture in which he introduced the concept of the "n-fold extended manifold." Riemann's analysis of his n-fold extended manifolds was brief and imprecise. We can now turn to the theory of point set topology for a very detailed account of precisely what it is to have manifold structure. But, this resource was not available to those geometers who sought to develop Riemann's idea in the later part of the 19th century. Fortunately, this proved not to be a serious obstacle since these geometers had at their disposal one example of a class of differentiable manifolds that had properties that were well understood and could be used whenever their theories called for a differentiable manifold. That class of differentiable manifold was the number manifold R^n. Thus, Felix Klein made Riemann's concept precise by explaining:

> At the foundation of his research, Riemann laid n variables x_1, x_2, ..., x_n, each of which can take all real values. Riemann denoted the totality of their value systems as a manifold of n dimensions; by a fixed value system x'_1, x'_2, ..., x'_n, he meant a point in this manifold. (Klein 1928, p. 289)

This identification of a manifold of n-dimensions with the number manifold R^n was already well established in 1873, at which time Klein (1873, p. 315) could report that it was "in agreement with the usual terminology." The tradition survived into the 1920s and appears on the opening page of Levi-Civita's 1925 treatise on the "absolute differential calculus," as the relevant branch of mathematics was then called. (Levi-Civita 1925, p. 9; 1926, p. 1). While the point set topological approach was being developed in the 1910s through such works as Weyl (1913) and Hausdorff (1914), the identification of geometrical manifolds with number manifolds remained attractive since it promoted the application of geometrical methods to the problems of real analysis and vice versa.[5]

This number manifold tradition entered relativity theory in its infancy. Minkowski's famous "world" was none other than R^4. He wrote in his 1908 address[6] (Minkowski 1908, pp. 56–67):

> We will try to visualize the state of things by the graphic method. Let x, y, z be rectangular coordinates for space and let t denote time.... A point of space at a point of time, that is, a system of x, y, z, t, I will call a world point. The manifold of all thinkable x, y, z, t systems of values we will christen the world.

The following summarizes the crucial difference between the older approach of Klein and Minkowski and the modern approach.

> Modern approach: An actual or physically possible space-time is represented mathematically by a general differentiable manifold, that is, one with a point set of unspecified elements.
>
> Older approach: An actual or physically possible space-time is represented mathematically by a special case of the general differentiable manifold, a number manifold, R^n, or its open subsets. The representation, which may only be "patchwise," is a coordination of the space-time with points of R^n, so that the maps x_1, x_2, ..., x_n (Klein) or x, y, z, t (Minkowski) that effect this coordination are called "coordinate systems."

Einstein's work in space-time theories lies within the older approach. I refer here in particular to the three major expositions of general relativity that Einstein gave in the 1910s: Einstein and Grossmann (1913), Einstein (1914), and Einstein (1916a). These expositions set the pattern for

Einstein's and many others' later expositions. Each of these three expositions presented a review of the current state of the theory and included a self-contained primer on the mathematical techniques needed to work with the theory. Einstein's assumption was that his physicist readers were unacquainted with these mathematical techniques. Einstein himself had needed the assistance of his friend Marcel Grossmann in 1912 and 1913 to gain access to them. Grossmann drew heavily on the review article by Ricci and Levi-Civita (1901) of the absolute differential calculus, and he provided a primer on the novel mathematics needed for the new theory in his mathematical part of their joint work, Einstein and Grossmann (1913). The following year, Einstein felt that he had sufficiently mastered the mathematical techniques to write the primer himself and in such a way that it would "enable a complete understanding of the theory without the need to read other pure mathematical treatises" (Einstein 1914, p. 1040).

The compliance of Einstein with the older number manifold tradition is not immediately obvious from a reading of the above expositions. Unlike Klein, Minkowski, and others, neither Einstein nor Grossmann defined what their manifolds are. Rather, their unexplicated primitive is the space-time coordinate system, x_1, x_2, x_3, x_4, from which the expositions move immediately to the treatment of coordinate transformations and the transformation laws for vectors and tensors. The standard practice of modern readers is to interpret these coordinate systems as coordinate charts of another mathematical object, a general differentiable manifold, which in turn represents the actual or physically possible space-time. Thus, this modern reading requires the presence of an intermediate level of mathematical structure that is never explicitly addressed in the expositions. If we take Einstein at his word and accept that the expositions are intended to be self-contained, then we must assume that Einstein did not suppress a major level of mathematical structure, and we must read Einstein's coordinate systems as being just like those of Klein and Minkowski. They coordinate the actual or physically possible space-times with the mathematical structure used to represent them, R^4 or its open subsets.

In the appendix, I review in greater detail the mathematical traditions upon which Einstein and Grossmann drew in their work on general relativity. This review helps explain their failure to define the nature of their manifolds.

4. The Mathematical Structure of Einstein's Space-Time Theories

The use of a number manifold rather than a general differentiable manifold in the older tradition is just one manifestation of the fact that the older tradition used a much simpler set of mathematical machinery than we use now. To illustrate the point, I will consider the mathematical structures that Einstein used in his formulation of general relativity and to enable close comparison with modern methods, I will read Einstein's formulations in a way that mimics the modern extensional or model theoretic formulations of space-time theories.

4.1 Models

Suppressing the stress-energy tensor T_{ab}, in the modern formulation, general relativity has the models[7]

$$\langle M, g_{ab} \rangle. \tag{2}$$

To recover Einstein's models, we now know that we should replace the manifold M by an open set of R^4, for example, A, so that we have the models

$$\langle A, \text{metrical object} \rangle. \tag{3}$$

The second position in Eq. (3) requires a mathematical object representing a metrical structure. Einstein routinely introduced a non-Minkowskian metrical structure into his expositions by considering two infinitesimally close space-time points x_i and x_i+dx_i and writing the interval ds between them as

$$ds^2 = g_{ik}\, dx_i\, dx_k. \tag{4}$$

Following the standard definitions of Einstein and Grossmann (1913) and Einstein (1914 and 1916a), the "fundamental" or metric tensor corresponding to Eq. (4) is not the matrix g_{ik} but the equivalence class of all the matrices produced by the well-known tensor transformation law under all smooth transformations of the coordinate system. Since we seek an object peculiar to the coordinate system image set A for the second position in Eq. (3), we choose not the tensor but the matrix g_{ik}, so that the models are of the form

$$\langle A, g_{ik} \rangle. \tag{5}$$

Generalizing, Einstein's formulation of a space-time theory posits models of the form

$$\langle A, (O_1)_{ik}\cdots, (O_2)_{ik}\cdots, \cdots \rangle, \tag{6}$$

where A is an open set of R^4, the quantities $(O_n)_{ik\ldots}$ are matrices, and each matrix occupies the position corresponding to a geometric object of equivalent rank in the modern formulation.

4.2 Coordinate Transformations

We have seen that the set of models of a space-time theory in the modern formulation is divided into equivalence classes of diffeomorphic models. Two models $\langle M, g_{ab}, T_{ab}\rangle$ and $\langle M', g'_{ab}, T'_{ab}\rangle$ of general relativity, for example, belong to the same equivalence class just in case there is a diffeomorphism h for which $M' = hM$, $g'_{ab} = h^* g_{ab}$, and $T'_{ab} = h^* T_{ab}$.

Smooth maps from open sets of R^4 to open sets of R^4—that is, coordinate transformations—serve the same function for Einstein's models. In specifying a model Eq. (6), one must also specify the transformation law for each of the quantities $(O_n)_{ik\ldots}$ (covariant tensor, contravariant tensor, mixed tensor, etc.) A model of general relativity is the triple $\langle A, g_{ik}, T_{ik}\rangle$, where g_{ik} and T_{ik} transform as covariant tensors. Two models, $\langle A, g_{ik}, T_{ik}\rangle$ and $\langle A', g'_{ik}, T'_{ik}\rangle$, belong to the same equivalence class just in case there is a smooth map from A to A' under which g_{ik} and T_{ik} transform into g'_{ik} and T'_{ik}.

4.3 Special Relativity and General Relativity

The model theoretic version of Einstein's formulation of general relativity looks very much like the modern version. Its model set is the set of all triples $\langle A, g_{ik}, T_{ik}\rangle$ that was defined previously and satisfies the field equation

$$G_{ik} = kT_{ik}, \tag{1'}$$

corresponding to Eq. (1). The formulation of special relativity looks a little less like its modern counterpart. In his space-time formulations of special relativity, for example, Einstein (1922), introduced the metrical structure through the line element

$$ds^2 = -dx_1^2 - dx_2^2 - dx_3^2 + dx_4^2,$$

which corresponds only to a matrix of metrical components:

$$h_{ik} = \text{diag}\,(-1, -1, -1, +1), \tag{7}$$

so that, assuming the space-time is topologically R^4, special relativity has a single model:

$$\langle R^4, h_{ik}\rangle. \tag{8}$$

The more interesting case of multiple models arises if there are further matter fields defined on the space-time, such as a Maxwell field or dust cloud. To avoid unnecessary detail, I shall represent these matter fields by their stress-energy matrix T_{ik} so that the model set is the set of all triples of the form

$$\langle R^4, h_{ik}, T_{ik} \rangle, \qquad (8')$$

such that the T_{ik} satisfy the laws of the relevant matter theory and transform as a covariant tensor. A comparison of Einstein's formulation and the modern formulation of a space-time theory highlights the advantages of the modern formulation. In particular, if the space-time in question is not topologically R^n, then Einstein's formulation cannot represent it by a single model, but must combine several in a patchwise fashion, with all the attendant complications.

5. The Problem of Superfluous Manifold Structure

A general differentiable manifold is structurally a quite impoverished representation of a space-time. In modern formulations of space-time theories, if we wish to represent such familiar entities as frames of reference, we need to introduce them as further structures defined on the manifold. We represent a frame of reference by adding a congruence of curves. But, Einstein's use of number manifolds to represent space-time raised a quite different problem for him. Number manifolds have too much structure. Take some model $\langle R^4, g_{ik}, T_{ik} \rangle$ of a relativistic space-time. Since the model represents the space-time, the straightforward reading is that each of its mathematical properties represents a property of the space-time. Some of these properties are:

1. Inhomogeneity: each point of R^4 is distinct, so every space-time event is intrinsically different to every other;
2. Absolute simultaneity: x_4 is a time coordinate, so the hypersurfaces of constant x_4 represent hypersurfaces of simultaneity.
3. Absolute rest: the natural rest frame of the space-time is represented by the congruence of x_4 curves;
4. Set of inertial frames: each parallel congruence of straights with constant velocity V, where $V^2 = (dx_1/dx_4)^2 + (dx_2/dx_4)^2 + (dx_3/dx_4)^2$ represents a frame of reference of uniform velocity; and

5. Metrical significance of the coordinates: the x_1, x_2, x_3 coordinates are measures of length, and the x_4 coordinate is a measure of time.

The idea that number manifolds have such default physical interpretations played an important role in Einstein's developments of space-time theories. For example, it was not uncommon for him to introduce the notion of a space-time coordinate system by means of property (5), that is, by specifying the space and time measuring operations needed to define the coordinate values (see, for example, Einstein 1917, Chapters 2 and 3). But then, proceeding to general relativity and its wider class of coordinate systems, Einstein went to great pains to convince the reader that coordinate systems must forfeit their direct metrical significance (see, for example, Einstein 1916a, Section 4; 1917, Chapters 23–25).[8]

6. The Solution: Einstein's Covariance Principles Are Physically Significant Principles

What Einstein needed was some systematic method of denying physical significance to all of the superfluous structures imported into his space-time theories by his use of number manifolds. Einstein was not the only one to face this problem. Felix Klein had faced and solved it brilliantly some 40 years earlier. The central idea of his celebrated *Erlangen* program was to characterize geometric structures as the invariants of groups. Einstein applied this same idea to his space-time theories. Associated with each theory was a covariance group of coordinate transformations. The physically significant mathematical properties were those that remained invariant under the transformations of the group.

Einstein's solution admits a precise statement in a way that mirrors the requirements of general covariance/Leibniz equivalence, defined in Section 2 for the modern formulation. The main difference is that the modern view only needs to define covariance with respect to the group of smooth transformations, whereas for Einstein, the group varies in size with the theory.

> Covariance of a theory under a group G of transformations:
> If $\langle A, (O_1)_{ik} \cdots, (O_2)_{ik} \cdots, \cdots \rangle$ is a model of a space-time theory, then any tuple $\langle A', (O_1)'_{ik} \cdots, (O_2)'_{ik} \cdots, \cdots \rangle$ related to it by a transformation within G is also a model of the theory.
>
> Leibniz equivalence: If two models $\langle A, (O_1)_{ik} \cdots, (O_2)_{ik} \cdots, \cdots \rangle$ and $\langle A', (O_1)'_{ik} \cdots, (O_2)'_{ik} \cdots, \cdots \rangle$ of a space-time theory are related by a

> transformation within G, then the two models represent the same actual or physically possible space-time.

Following the pattern in Section 2, I have distinguished covariance from Leibniz equivalence, although the two are usually lumped together and simply labeled as a covariance principle. In particular, I shall read Einstein's assertions that a space-time theory is or should be covariant under some group as the assertion that it does or should satisfy both of these requirements. Einstein clearly understood Leibniz equivalence to be part of a covariance requirement, as he showed in his autobiographical notes. There, he imagined that one treats some generally-covariant field theory as if it were only Lorentz covariant. Recognizing its general covariance, he urged, led to the

> level of understanding corresponding to the general principle of relativity. For, from the standpoint of the Lorentz group, two solutions would incorrectly have to be viewed as physically different if they can be transformed into each other by a nonlinear transformation of coordinates, i.e., if from the point of view of the wider group they are merely different representations of the same field. (Einstein 1949, pp. 70–73)

The Lorentz and generally covariant approaches of his example have identical covariance in the strict sense of my definition above because they have identical sets of models ("solutions"), generated by the one set of equations. They differ just on the issue of the Leibniz equivalence requirement associated with each group, that is, on which models are to represent the same physical space-time.

Einstein also predicated the covariance property not directly to the model set but to the equations that define the model set. The two are clearly equivalent; if the equations defining a model set are covariant under a group G then the model set must also be covariant under that group and vice versa.

The covariance/Leibniz equivalence requirements provide a systematic method of depriving some of the mathematical properties of a theory's models of physical significance. If two intertransformable models represent the same space-time, then all mathematical properties upon which they differ can have no physical significance. That is, only those properties shared by the models—the invariants of the theory's covariance group—can be physically significant.

Finally and most important, notice that the satisfaction of a covariance/Leibniz equivalence requirement by a space-time theory is not merely a matter of mathematical definition. The requirement involves a judgment of whether certain mathematical properties of models are physically significant and such judgments are necessarily matters of physical contingency. In this regard, the status of the requirement is exactly the same as that of active covariance/Leibniz equivalence requirements for the modern formulation of space-time theories, so that the considerations of Section 2 apply equally here. The conclusion that Einstein's covariance/Leibniz equivalence principles have physical content after all is a direct result of the new reading that I urge here for Einstein's "coordinate systems." I stress that this conclusion is not possible under the standard modern reading of Einstein's coordinate systems as the coordinate charts of a differentiable manifold. For example, taking some model $\langle M, g_{ab}, T_{ab}\rangle$ of general relativity, and transforming between different coordinate charts of its manifold M simply transforms between what are *by mathematical definition* different component representations of the same model.[9]

7. Covariance Principles as Relativity Principles

The covariance/Leibniz equivalence properties of space-time theories formulated in the modern way usually attract little attention. The fact that they deprive certain mathematical properties of the space-time models of physical significance is rarely mentioned or even noticed since these mathematical properties are usually of minimal interest.[10]

For Einstein, however, the situation was very different. The covariance/Leibniz equivalence properties of his space-time theories contain the relativity principles of his theories. The mathematical properties of his models that are deprived of physical significance include those associated with preferred states of motion, properties 1–5 listed in Section 5. The extension of the principle of relativity in the transition from special to general relativity is embodied in a sequence of three covariance requirements: Lorentz covariance, its extension through the principle of equivalence, and general covariance.

7.1 Special Relativity

Special relativity is Lorentz covariant, that is, it is covariant under the transformation of the extended Lorentz group. By definition, the extended Lorentz group is the set of all transformations that map $\langle R^4, h_{ik}\rangle$ onto itself. To be a special relativistic matter theory, the laws that govern the T_{ik} of the models Eq. (8') must be Lorentz covariant; therefore, if $\langle R^4, h_{ik}, T_{ik}\rangle$ is a model of the theory, then so is any Lorentz transform of it, $\langle R^4, h_{ik}, T'_{ik}\rangle$ and under Leibniz equivalence, the two models must represent the same physically possible space-time.

Applying the discussion of Section 6, we can now read some of the properties of the models deprived of physical significance by Lorentz covariance. They are those that are not invariant under Lorentz transformation and include (1) inhomogeneity, (2) absolute simultaneity, and (3) absolute rest. The assertion that there is no physical significance accorded an absolute state of rest is the principle of relativity of special relativity.

But, those structures that are invariant under Lorentz transformation remain physically significant. The most notable of these is property 4, Section 5, the set of inertial frames. While each inertial frame of R^4 will in general be mapped onto a different frame, the set of inertial frames is mapped onto itself. This feature of the coordinate systems of special relativity makes Einstein's description of them as "inertial systems" a natural one.

7.2 General Relativity

An extension of the principle of relativity to accelerated motion requires a theory in which not just the state of rest but also the inertial frames of the number manifolds are deprived of physical significance. This result is obtained in general relativity through its general covariance. Its covariance group is the group of all smooth transformations. Under this group, none of properties 1–5 are invariant, including in particular the set of inertial frames, property 4. Thus, throughout his writings, Einstein characterized general relativity as the theory that achieved the elimination of the preferred inertial system (see, for example, Einstein 1913, p. 1260, footnote; Einstein 1953).

7.3 Principle of Equivalence

The basic strategy of the extension of the principle of relativity in the move from special to general relativity is the expansion of the covariance group of the theory. To motivate this expansion, Einstein routinely included the case of the principle of equivalence as an intermediate between special and general relativity.[11] The principle is nothing other than an expansion of the covariance group of special relativity. As formulated so far, the intertransformable models of special relativity are always related by a Lorentz transformation. The principle of equivalence requires that we add to the set of models all those models produced by transformations that map inertial frames into uniformly accelerated frames. Under these new transformations, a model

$$\langle R^4, h_{ik} \rangle \tag{8}$$

will be mapped to a model

$$\langle A, g_{ik} \rangle \tag{9}$$

where A is some open subset of R^4 and g_{ik} no longer has the simple constant values of $h_{ik} = \mathrm{diag}(-1, -1, -1, +1)$.

The covariance/Leibniz equivalence requirement asks that Eqs. (8) and (9) represent the same physically possible space-time. Now, Eq. (9) has an interesting physical interpretation. The trajectories of free test particles do not coincide with the curves of the inertial frames, as canonically defined for A via (property 4), Section 5. But, these accelerated trajectories of the free particles are governed solely by the matrix of coefficients g_{ik} and are independent of the particles' masses. So, in a much celebrated argument dependent on the equality of inertial and gravitational mass, Einstein urged that we interpret the acceleration of the particles as due to a homogeneous gravitational field represented by g_{ik}. The covariance/Leibniz equivalence requirement amounts to the assertion of the physical equivalence of Eqs. (8) and (9), one interpreted as gravitation free and the other as with a homogeneous gravitational field. If we use K to label the coordinate system of the inertial model, Eq. (8) and K' for the coordinate system of the accelerated model, Eq. (9), the covariance/Leibniz equivalence requirement becomes exactly Einstein's statement of the principle of equivalence:

> The assumption of the complete physical equivalence of the systems of coordinates *K* and *K'*, we call the "principle of equivalence"... (Einstein 1922, p. 56)

Finally, since the principle of equivalence is an intermediate between Lorentz and general covariance, we can understand why Einstein would claim:

> The requirement of general covariance of equations embraces the principle of equivalence as a quite special case." (Einstein 1916b, p. 641)

This is a claim that has hitherto resisted coherent interpretation by modern commentators.

8. Kretschmann's Objection

One of the earliest objections to Einstein's equating of general covariance with a general principle of relativity was raised in Kretschmann 1917.[12] Kretschmann's paper is remarkable not so much for this objection, for which the paper is usually remembered, or, as I urge in Section 8.2, *mis*remembered. Rather, as I explain in Section 8.3, it is remarkable for its prescient treatment of relativity principles and other ideas in space-time theories.

8.1 The Popular Version

Kretschmann's objection is now routinely represented (see, for example, Graves (1971, p. 137).) as the remark that one can take the equations expressing any given law and, with sufficient mathematical ingenuity, modify them until they take on generally-covariant forms. Thus, the achievement of general covariance is merely a challenge to our mathematical ingenuity and fails to express anything physical, let alone to express a relativity principle.

The objection, in this simplified form, has been widely accepted. But, it is not at all clear that *this* form of the objection is correct. Of course, one can always take a law-like equation and modify its form until it becomes generally-covariant. One can then *assert* that the modified equation has the same physical content as the original. But, this assertion might require physically contingent hypotheses. In modifying the original equations, one might, for example, introduce further mathematical structures to enable achievement of general covariance. To retain physical equivalence with the original equation, one must make the physically contingent assumption that the added structure is merely an auxiliary and has no independent physical significance.

A clear instance of this arises in the expansion of the covariance of special relativity by the principle of equivalence. The expansion was effected by adding models of the form Eq. (9). However, the assertion that the expanded theory has the same physical content as the original theory required an assumption. It was that a model of the form Eq. (8) represents the same space-time as a model of the form in Eq. (9) to which it could be transformed, even though the second model was considerably different in mathematical structure. The assumption needed, that these mathematical differences have no physical significance, is Leibniz equivalence, which, as I have repeatedly stressed, is a physically contingent hypothesis. Its physical character is shown very clearly in this case, because, in the guise of the principle of equivalence, it makes the assertion that a gravitation-free model, Eq. (8), represents the same space-time as the model in Eq. (9), which contains a homogeneous gravitational field.

Finally, if one accepts this form of the objection, I see no reasonable way that one can avoid analogous objections against the physical significance of just about any property of any given theory. To establish the physical vacuity of some given property of a theory, one need only represent it as a purely *formal* property and establish that the presence of that formal property can be guaranteed in formulations of a wide range of incompatible theories, a task so loosely defined that it is trivially achievable. Thus, the presence of $E = mc^2$ in the laws of special relativity can be shown to be devoid of physical content. If one modifies the formulation of special relativity so that the string $E = mc^2$ becomes a stipulation, the definition of a quantity m in terms of the energy E of some body, then the physical content formerly carried by $E = mc^2$ is transferred to the physically contingent $m' = m$, where m' is defined as the inertial mass of the body. With only a little mathematical ingenuity and the introduction of a few auxiliary terms, one can similarly guarantee the presence of $E = mc^2$ in a formulation of Newtonian mechanics[13] or just about any other theory.

8.2 Kretschmann's Version

What Kretschmann actually said was more cautious and more interesting than the simplified report of section 8.1. At the outset of his paper, after recalling the connection claimed by Einstein between covariance and relativity principles, Kretschmann summarized his first worry:[14]:

> ...imagine that all physical observations consist in the last analysis of the determination of pure topological relations (coincidences[a]) between space-time objects of perception and, hence immediately that no coordinate system is privileged[b] above any other by them, so that one is forced to the conclusion that each physical theory can be brought into accord with every arbitrary relativity postulate, including the most general, without alteration of the theory's freely chosen and observationally testable content, by a means associated at worst with mathematical difficulties: a purely mathematical transformation of the representing equations.[c] (Kretschmann 1917, pp. 575–576)

Kretschmann claimed that the laws of any space-time theory can be brought into generally-covariant form purely by mathematical manipulation and thus, by Einstein's lights, into accord with a generalized principle of relativity. But, the claim is dependent on an assumption: that "all physical observations consist in the last analysis of the determination of pure topological relations (coincidences[a])..." This assumption is immediately recognizable as the central assertion of Einstein's so-called point-coincidence argument, and of course, Kretschmann's footnote (a) to the word coincidences is to Einstein's best-known published version of the argument (Einstein 1916a, p. 776).

The point-coincidence argument is Einstein's version of argument (1) for Leibniz equivalence given in Section 2.[15] In brief and translated into the model theoretic terms of this chapter, he urged that all observables are, in the end, space-time coincidences, such as the world line intersections of particles in a simplified universe containing only particles or, more generally, that all measurement reduces to coincidences of material systems and measuring instruments. All such coincidences are preserved under coordinate transformation, so that intertransformable models must represent the same physically possible space-time, which is Leibniz equivalence (and, in turn, is included automatically in Einstein's requirement of general covariance).

This assumption, upon which the argument and Kretschmann's claim are based, amounts to a significant physical assumption. It amounts to requiring that

> (PC) *the physical content of a space-time theory is fully exhausted by the catalog of its space-time coincidences,*

so that we are enjoined to accord no physical significance to any property of a model of a space-time theory, if the property is not fully recoverable from this catalog of coincidences. That such a dramatic assumption is at the heart of Einstein's argument is seen more clearly in the versions of the argument written to his correspondents. To Besso in a letter of January 3, 1916, he explained the core of the argument:

> *Reality* is physically nothing other than the totality of space-time point coincidences. If, for example, all physical occurrences were constructed from the motion of material points alone, then the meetings of the points, i.e., the intersections of their world lines, would be the only reality, i.e., that which is in principle observable. (Speziali 1972, p. 64)

Following the pattern of his other presentations of the argument,[16] Einstein continued to argue that this assumption leads directly to general covariance:

> Naturally, these intersection points are preserved under all transformations (and nothing new is added), as long as certain uniqueness conditions remain true. Therefore it is most natural to require of the laws that they do not determine any *more* than the totality of time-space coincidences. According to what I have said above, this has already been achieved through generally covariant equations.

Thus, we have neither Einstein's nor Kretschmann's arguments warranting the claim that general covariance is physically vacuous. Rather, for both, it is a consequence of a profound physical assumption (PC) about the world, that the physical content of a space-time theory is exhausted by the catalog of its space-time coincidences.

I have been unable to develop a clear picture of Einstein's own attitude toward Kretschmann's arguments and, in particular, whether Einstein realized that Kretschmann had failed to demonstrate the physical vacuity of general covariance. Einstein's published response (Einstein 1918) to Kretschmann is clearest. There we find that Einstein conceded a little less than is usually thought. He listed the three principles upon which his general theory depends: (1) the principle of relativity, (2) the principle of equivalence and (3) Mach's principle. The principle of relativity *becomes* the assumption (PC) and the remark that general covariance follows from it:[17]

> *Principle of relativity*: The laws of nature are only assertions about time-space coincidences; therefore, they find their only natural expression in generally covariant equations.

On the following page (p. 242) he conceded to Kretschmann that general covariance "is only a requirement relating to the mathematical formulation [of laws]," but this concession is clearly dependent on the prior assumption of (PC). The concession is immediately explained by Einstein with a restatement of the assumption (PC) and the fact that it leads directly to general covariance.[18]

Einstein's later treatments of general relativity uniformly stress the fundamental role played by a requirement of general covariance and seem to remain essentially unaffected by whatever concession he might have made to the claimed physical vacuity of general covariance. Such a concession is not mentioned, for example, in his textbook-like exposition (Einstein 1922, see especially p. 60) or in a lengthy, unpublished exposition of special relativity and general relativity (Einstein 1920). It is mentioned in passing elsewhere (such as in Einstein 1924, pp. 90–91) and in greater detail in Einstein (1949, pp. 64–65). But, it is difficult to see that he took the objection seriously given, in the latter instance (Einstein 1949), the discussion that follows and especially his insistence on (Einstein 1949, p. 73) that "we have already given *physical* reasons for the fact that in physics invariance under the wider [general] group has to be required. . . " (Einstein's emphasis—not mine!).

8.3 Kretschmann's Formulation of Relativity Principles

That generally-covariant formulations should be available for all space-time theories (given PC) was only the first point of Kretschmann's critique. The bulk of his paper was devoted to the question of determining just what relativity principle was satisfied by any given space-time theory, given that this judgment could not simply be made from the covariance group of the theory's equations. The answer he gave and many of the results he derived turn out to be characteristic of much later work.

In modern terms, he recommended that we identify the relativity principle of a space-time theory with the symmetry group of the geometric structure of the theory. This viewpoint emerges most clearly when he gives a "geometric determination" of the theory's relativity principle. In essence, he sought that group of transformations that maps the conformal and affine structure of the theory back into itself, that is, their symmetries. His approach depended on the idea that these are the observable structures of the theory. For special relativity (Kretschmann 1917, pp. 581–583), he sought only the symmetries of the conformal structure, that is, of the "bundle of the light-like world lines" and arrived at a group of transformations that is independent of the equations describing the theory. He provided (Kretschmann 1917, pp. 607–611) a similar analysis for general relativity, first seeking the symmetries of the conformal structure and then the (time-like) affine structure to conclude that these structures jointly have no non-trivial symmetries, so that general relativity is "a completely absolute theory" and, moreover, that special relativity has the largest relativity group of space-time theories employing a Lorentz signature metric.

It is remarkable that as a part of his analysis of the extent to which observation can determine the mathematical structures of special relativity and general relativity, Kretschmann asked and answered the question of how much conformal and affine structures determine the metrical structure. He found (Kretschmann 1917, pp. 585–590) that the conformal structure determines the metric up to a conformal factor and that the affine structure forces this factor to take a constant value. These results and the general approach are usually attributed to a later tradition initiated by Weyl (Torretti 1983, pp. 192–193). Kretschmann's paper is also notable for its treatment of invariant methods of defining specialized coordinate systems in general relativity in terms of curvature invariants (Kretschmann 1917, pp. 591–606).

The precise response (Einstein 1918) to this latter and most significant part of Kretschmann's critique is less clear. Presumably, it is embodied in his distinction of Mach's Principle[19] from the principle of relativity, a distinction that he conceded in a footnote (Einstein 1918, p. 241) he had not previously made and that, he allowed, had caused confusion. Presumably, Einstein hoped that the satisfaction of Mach's principle by general relativity would distinguish it from other generally-covariant space-time theories and justify the idea that the theory embodies some kind of extension of the relativity of motion. It is now well known that these hopes were never fulfilled and that Einstein abandoned Mach's principle in later years. The story is told in many places; see, for example, Torretti 1983, pp. 194–202.

9. Conclusion

What has been established here is that the covariance principles of Einstein's formulations of space-time theories have physical content in a manner precisely analogous to the less important active covariance principles of the modern formulation of space-time theories. We have also seen that Einstein's covariance principles have the character of relativity principles since they deny physical significance to a series of mathematical structures interpreted as representing rest and inertial frames of reference. But, this relativity principle character is peculiar to the formulation of space-time theories chosen by Einstein. It depends on his use of number manifolds to represent space-time, which introduces considerably more mathematical structure into his models than is present in the models of a modern formulation of the same theory. The corresponding covariance principles of the modern formulations, then, have physical content, but they do not have the character of a relativity principle. Just as Kretschmann urged in 1917, the relativity principles of modern formulations are defined in terms of the symmetries of the geometrical structures defined on the manifold, and perhaps with the exception of special relativity, the identification of these principles has considerably less importance to the theory's formulation.

Perhaps the most important heuristic outcome of the comparison of Einstein's and the modern formulations of space-time theories is the recognition that the latter uses a considerably richer repertoire of mathematical structures. I argued that the failure to recognize this or to account for it properly had led to a systematic mistranslation of many of Einstein's ideas into the modern context. I have argued elsewhere, in a longer treatment of the issues raised here (Norton 1989), that another instance of this problem is the great difficulty modern commentators have faced in deciding whether to read Einstein's coordinate transformations actively or passively in the context of his hole and point-coincidence arguments, and I urge that the methods introduced here provide a complete solution to this problem.

Appendix: Two Traditions
Absolute Differential Calculus and Vector Analysis

Einstein and Grossmann drew their mathematical techniques for general relativity from two traditions so that these techniques are actually an amalgamation of the methods and standard procedures of these two traditions. A review of these traditions and how they were combined explains some of the apparent idiosyncrasies of Einstein and Grossmann's techniques and, to some extent, their neglect to define the nature of their manifolds.

First and most important was the tradition in differential geometry and the invariants of quadratic differential forms that began with Gauss's theory of surfaces and extended through the work of Riemann, Christoffel, Lie, Ricci, and Levi-Civita. For the advent of general relativity, the tradition's most important product was the Ricci and Levi-Civita (1901) review article of the absolute differential calculus upon which Einstein and Grossmann drew heavily. There were two distinct parts to this tradition. On the one hand was the investigation initiated by Gauss of surfaces characterized in the simplest two-dimensional case by the line element

$$ds^2 = E\,du^2 + 2F\,du\,dv + G\,dv^2 \tag{A1}$$

where u and v are the coordinates and E, F, and G are some functions of them. The study of such surfaces led directly to the investigation of the invariants of quadratic differential forms such as Eq. (A1). However, the theory of these invariants could in principle be developed as an autonomous theory; one need only think of u and v as variables subject to transformations and suppress the surface theoretic interpretation. The literature in the theory of invariants of quadratic differential forms, such as Wright (1908, p. 4), did go to some pains to point out that differential geometry provided just one interpretation of the theory.

Einstein and Grossmann's debt to this tradition has been frequently recalled (see, for example, Pais 1982, pp. 210–213) and is much celebrated as an instance of pure mathematics anticipating the needs of the physicist. This tradition is usually designated as the primary mathematical source for general relativity. Einstein (1915, p. 779) had introduced one of the final versions of the theory as "a veritable triumph of the method of general differential calculus founded by Gauss, Riemann, Christoffel, Ricci, and Levi-Civiter [sic]." More soberly in the introductory page of his 1916 review of the theory, he remarked:

> The mathematical aids necessary for the general theory of relativity lay ready made in the "absolute differential calculus," which rests on the

> research of Gauss, Riemann, and Christoffel on non-Euclidean manifolds and had been brought into a system by Ricci and Levi-Civita and already applied to the problems of theoretical physics. (Einstein 1916a, p. 769)

(Lamentably this page was omitted from the standard English translation, Einstein 1916b.)

It is easy to misunderstand Einstein's remarks here. What he pointed out are the mathematical methods required for general relativity that were not *already* a part of the standard repertoire of physicists and of relativists in particular. There was a second mathematical tradition upon which Einstein also drew heavily, the tradition of work in vector analysis. Einstein did not mention it since it was already incorporated into relativity theory by 1912 and 1913. (We shall see that Grossmann was clearer on the role of vector analysis in general relativity.)

This second tradition grew from the work of Hamilton and Grassmann, and Gibbs and Heaviside, and its methods and concepts were developed with a very careful eye on their applications in physics. (For a general historical survey, see Crowe 1967). Vector analysis entered the German physics literature by means of its prime application, the theory of electricity and magnetism, when August Föppl decided to use vector methods in his text (Föppl 1894) on Maxwell's theory. This text, and its later incarnations, soon became the most popular text in electricity and magnetism in the German physics community. Föppl's text offered a self-contained exposition of vector analysis, a practice that was repeated in the later incarnations, such as in Abraham and Föppl (1904) and the last version of Becker (1964).

Since there were close historical connections between special relativity and the theory of electricity and magnetism, it was natural for vector methods to become associated with special relativity, even though Einstein's 1905 article "On the Electrodynamics of Moving Bodies" (Einstein 1905) did not employ the latest vector methods. That association was made most firmly by Minkowski (1907), when he showed that the four-dimensional space-time formulation of special relativity enabled an especially simple formulation of the theory of electrodynamics. The vehicle of this new formulation of electrodynamics was the extension of the standard vector analysis of the period to a four dimensional vector analysis at whose heart lay (among others) the two new quantities "space-time vectors of the first kind" and "space-time vectors of the second kind," which soon came to be known as four- and six-vectors. Introductions to the mathematical techniques required for the space-time formulation of special relativity soon leaned heavily toward an exposition of four-dimensional vector analysis. Thus, Sommerfeld's (1910a, 1910b) introduction to these techniques is entirely devoted to an exposition of four-dimensional vector analysis, and the association between vectorial concepts and special relativity had become so strong that Sommerfeld formulated the principle of relativity simply as the assertion that "only *space-time vectors* are allowed to appear in physical equations" (1910a, p. 749; Sommerfeld's emphasis).

The two traditions described developed in essentially autonomous literatures, although it was clearly obvious to the researchers in both traditions that they shared common problems and that their methods could be applied in the other tradition. For example, the review article by Ricci and Levi-Civita (1901) concludes with a discussion of the application of the absolute differential calculus in physics to vectorial fields and the example of the theories of electrodynamics, heat, and elasticity in general coordinates. Conversely, Abraham in his 1901 *Teubner Encyklopädie* article on vector analysis, noted in passing that the absolute differential calculus could be used to represent vectors and tensors in curvilinear coordinates (Abraham 1901, p. 38). That a more substantial connection should be effected between the traditions became all the more likely after Minkowski (1907) stressed the importance to the Lorentz group of the invariance of the space-time interval

$$x^2 + y^2 + z^2 - t^2, \tag{A2}$$

an expression quadratic in the space-time coordinates x, y, z, t. This emphasis was close to the focus of the differential geometry of the time, the invariants of quadratic differential forms.

It was Einstein and Grossmann who effected the first systematic amalgamation of the two traditions. In developing general relativity, Einstein began with special relativity in the four-dimensional vector formulation given to it by Minkowski. In order to incorporate gravitation into the theory, Einstein replaced Minkowski's requirement of the invariance of Eq. (A2), by the invariance of the quadratic differential form:

$$dx^2 + dy^2 + dz^2 - dt^2,$$

or, more generally in the arbitrary curvilinear coordinate x^i (i = 1, 2, 3, 4),

$$g_{ik}\, dx^i dx^k. \tag{A3}$$

Through Grossmann, Einstein found that the mathematical techniques needed to understand the invariants of this quadratic differential form were already collected in the Ricci and Levi-Civita (1901) review article on the absolute differential calculus. Einstein and Grossmann took the techniques of this calculus and grafted them onto the vectorial methods of four-dimensional special relativity to yield a generalized vector analysis that became the standard for work in general relativity.

Grossmann explicitly acknowledged this procedure in the introduction to his mathematical part of Einstein and Grossmann (1913). He noted that the mathematical aids for developing a "vector analysis of the gravitational field" derive from the work of Christoffel and Levi-Civita and continued:

> However, the vector analysis of Euclidean space related to arbitrary curvilinear coordinates is formally identical to the vector analysis of an arbitrary manifold given through its line element. Therefore, there are no difficulties in extending the vector analytic conceptual system, as it had been developed in recent years by Minkowski, Sommerfeld, Laue, and others for relativity theory, to the general theory of Einstein given here.
>
> The *general vector analysis,* which one then recovers, proves with some practice to be just as simple to manage as the special vector analysis of three- or four-dimensional Euclidean spaces;. . . (Einstein and Grossmann 1913, p. 23; Grossmann's emphasis)

Elsewhere, Grossmann (1913) introduced the new mathematical system with an equation by equation comparison of the "usual vector analysis" of Minkowski and Sommerfeld, which was limited to orthogonal coordinate systems, with the "general vector analysis" of the new theory to which it had been generalized by introduction of the work of Christoffel, and of Ricci and Levi-Civita and now admitted arbitrary curvilinear coordinate systems.

As a result of this procedure of generalization, aspects of vector methods and concepts were accorded a prominence that they otherwise might not have enjoyed. The clearest illustration of this prominence lies in the nomenclature used. The Ricci and Levi-Civita (1901) review article is now commonly known by such titles as "Ricci and Levi-Civita's Tensor Analysis Paper" (Hermann, 1975). However, the remarkable fact is that the term tensor appears nowhere in Ricci and Levi-Civita's article. The quantities we would now call tensors (and vectors) are there called "contravariant and covariant systems." Moreover, the term tensor is not used in the modern way in standard texts in differential geometry and the invariants of quadratic differential forms at the time of Einstein's development of general relativity (e.g., Wright 1908, Bianchi 1910, Darboux 1914).

The term tensor comes directly from vector analysis. Tensors were routinely but certainly not prominently defined in the standard developments of vector analysis. However, in the older tradition, the scope of the definition was narrower than the modern definition. Thus, Abraham (1901, p. 28) defined as a tensor only those quantities whose components transformed like the components of what we would now call the outer product of two vectors. That is, Abraham's tensors were what we would now call second rank, symmetric tensors. The reason for this narrowness was simple. As the word tensor betrays, the quantities were defined with a specific application in mind, the stress tensors of the theory of elasticity, which were second rank and symmetric. The generality of the modern definition would simply have been superfluous. The term tensor found its way into the four-dimensional vector analysis of special relativity—eventually. Minkowski (1907) did not use the term, although he did present a matrix representation for the components of the four-dimensional analogs of tensors. Sommerfeld (1910a, p. 767) applied the term tensor in the four-dimensional vector analysis in as narrow a way as Abraham and for similar reasons. Sommerfeld's tensors were again what we would call second rank, symmetric tensors, and the word tensor was still entirely appropriate to their principal application as a stress energy tensor. Following Minkowski (1907, p. 82), Sommerfeld (1910a, p. 754) recognized that his six-vectors had components that formed a second rank, antisymmetric matrix, but he did *not* apply the term tensor or antisymmetric tensor to the quantity. Finally, Laue's (1911) exposition of special relativity offered as restricted a definition of his four-dimensional "world tensors" as Sommerfeld's definition of his "tensor," although deeper in his exposition, Laue admitted an "unsymmetric," three-dimensional stress tensor (p. 151).

It is easy to see why Einstein or Grossmann would categorize the coefficients g_{ik} in the fundamental form Eq. (A3) as representing a tensor within their generalized vector analysis, even though there is no longer any immediate connection between the quantity represented and the theory of elasticity. These coefficients transform exactly as required by the definition of the tensors of Sommerfeld and Laue. Yet, for a reason that is not clear to me, the term tensor was defined by Grossmann in Einstein and Grossmann (1913, p. 25) as applying not just to the quantity represented by the coefficients g_{ik}, but to all those represented by Ricci and Levi-Civita's contravariant and covariant systems. This broader usage justifies the labeling of the absolute differential calculus as a tensor calculus. It became the standard not just in the generalized vector analysis used in relativity theory, but in differential geometry and the theory of the invariants of quadratic differential forms as well. In 1925, Levi-Civita published a treatise called *Lezioni di Calcolo Differenziale Assoluto* (Levi-Cevita 1925). Its expanded English translation was published the following year [(Levi-Civita 1926) (see Levi-Civita 1925 in references)] with the appropriate title *The Absolute Differential Calculus.* But, to it, presumably in deference to the new nomenclature grafted onto it from vector analysis, was added parenthetically the new subtitle "Calculus of Tensors."

We can now turn to the question of Einstein and Grossmann's failure to define the nature of their manifolds. In brief, this failure simply follows the standard practice of many of the expositions in both of the mathematical traditions upon which they drew.

Expositions of vector analysis around 1900 saw their primary burdens to be the definition of vectorial quantities and the development of their properties. Thus, the first definitions given are typically of polar and axial vectors, which are characterized in terms of their behavior under coordinate transformation. There is no discussion of the nature of the physical space in which the vectors are defined; coordinate systems and right angled coordinate systems are simply introduced as primitive notions presumed intelligible to the reader. Sommerfeld (1910a, 1910b) took a similar approach in his exposition of the four-dimensional vector algebra and analysis. After a brief introductory page, he turned immediately to his first definitions, four- and six-vectors. The concepts of four-dimensional space-time and coordinate systems are presumed to be familiar to the reader.

In their exposition of the absolute differential calculus, Ricci and Levi-Civita noted:

> ... a manifold V_n is defined intrinsically in its metrical properties by n independent variables and by all of a class of quadratic forms of the differentials of these variables, such that any two of them are transformable from one to the other by a point transformation. (Ricci and Levi-Civita 1901, p. 482)

This definition of a manifold is different from the definition of a manifold as a number manifold, which we saw in Section 3 in the work of Klein, and Minkowski and, in 1925, of Levi-Civita.[20] Later in his 1925 exposition, Levi-Civita (1926, p. 119) defined what he called a "metric manifold," designated by V_n, as a number manifold in conjunction with a quadratic differential form, Eq. (A3). The manifolds V_n of the Ricci and Levi-Civita review article are actually equivalence classes of these 1925 V_n. Schouten (1924, p. 58) effectively gave the same definition as

Levi-Civita's 1925 definition for a V_n, a number manifold in conjunction with the specification of a fundamental tensor g_{ik}, and called it a "manifold with quadratic metric" or a "Riemannian manifold." Schouten (1924, p. 8) reserved the designation X_n for number manifolds.

For our purposes, the important point is that Ricci and Levi-Civita buried their definition of a manifold in the short preface to their paper within an account of the geometric ancestry of their absolute differential calculus. The formal exposition of their calculus begins in Chapter 1, with no mention of manifolds and in a way that seems to seek as much of a divorce from geometrical associations as possible. The first topic, for example, is the transformations of *variables*, a discussion that modern readers almost irresistibly read as "really" about the transformations of coordinate systems. They then proceeded to define their covariant and contravariant systems. Like the researchers in the theory of quadratic differential forms, Ricci and Levi-Civita clearly conceived their calculus as a very general instrument. Its obvious geometrical application in the theory of surfaces is just one of the applications alluded to in the paper's title, so that its discussion is isolated in Chapter IV. Chapters III, V, and VI deal with applications in analysis, mechanics, and physics.

Thus, by 1912, the precedent for Grossmann was clearly set. Whether he conceived his mathematical part of Einstein and Grossmann (1913) as the exposition of a generalized vector analysis, as the exposition of the absolute differential calculus of Ricci and Levi-Civita, or as the exposition of some combination of them, the concepts of manifold and coordinate system were to be taken as terms already known to the reader. At best, they were to be dismissed briefly in prefatory remarks. One is to proceed through the topic of transformations as rapidly as one can to the important definitions, the definitions of the vectorial or, correspondingly, covariant quantities. That is essentially what Grossmann did. After defining the differential form Eq. (A3), he proceeded through the topic of coordinate transformations to his first major definitions, the definitions of covariant, contravariant, and mixed tensors of arbitrary rank. Similarly the mathematical aids section of Einstein's 1914 article begins with the definitions of covariant and contravariant four-vectors and proceeds to define tensors, their algebra and the differential operations that can be applied to them (Einstein 1914, p. 1034). The section on mathematical aids in Einstein's 1916 article follows exactly the 1914 pattern with the insertion of a brief introduction to motivate the need for the excursion into tensor calculus (Einstein 1916a).

Einstein and Grossmann's precedents gave them no strong reason to define the nature of their manifolds in their expositions of the mathematical techniques needed for the new theory. But, if that definition is not given there, should it not be given elsewhere in the developments of the theory? A simple answer is that Einstein's developments of general relativity began with the presumption that the reader was already familiar with the four-dimensional formulation of special relativity and that whatever manifold concept was used there was to be carried over automatically to the new theory. While I do believe that this presumption was made, I think there is a simpler answer. That Einstein represented his physically possible space-times by number manifolds without any intervening mathematical structures is simply implicit in the way he used coordinate systems. I believe that he saw the point as such an elementary one that it bore no real discussion. The point is only not immediately obvious to modern readers because they approach Einstein's writings with a far more complicated concept of the manifold already in hand. But, how else, for example, are we to read Einstein's discussion in the early pages of his introductory relativity text, *Meaning of Relativity*, when, after presuming certain elementary properties for space, he wrote:

> ... it is easy to say what we mean by the three dimensionality of space; to each point three numbers, x_1, x_2, and x_3 (coordinates), may be associated in such a way that this association is uniquely reciprocal and that x_1, x_2, and x_3 vary continuously when the point describes a continuous series of points (a line). (Einstein 1922, pp. 3–4)

In introducing the three-dimensional coordinate system, Einstein simply gave us a recipe for expressing more precisely the topological properties of a physical space: one represents the space with the mathematical structure of the number manifold R^3.[21]

Addendum: I am grateful to John Stachel for pointing out to me an affirmation in the mathematical literature of 1914 of the novelty of the Einstein–Grossmann use of the term "tensor." E. Budde, in *Tensoren und Dyaden im Dreidimensionalen Raum: Ein Lehrbuch* (Budde 1914), concluded with a brief discussion of the extension of the subject of his book to higher dimension spaces and higher ranks (pp. 245–46). The latter extension is due to W. Voigt and invokes quantities of the "*n*th rank," which "transform as a combination of the *n*th dimension of the coordinates." He described Voigt's nomenclature:

> The quantities of the first rank then are vectors, the symmetric, second rank quantities are the tensors, those of the third rank Voigt calls trivectors [*Triivektoren*]; those of the fourth rank bitensors [*Bitensoren*].

A sentence later, the section concludes by contrasting this usage with that of Grossmann:

> Recently, Mr. Grossmann (see bibliography) has proposed a still further reaching generalization. He denotes quantities of arbitrary rank as "tensors," so that vectors, trivectors, and bitensors are also subsumed by the term "tensor"; the generalization consists in extending his definitions to structures of *n*th rank in *m*-dimensional space.

The only work by Grossmann in Budde's bibliography is "*A. Einstein und M. Grossmann: Entwurf einer verallgemeinerten Relativitätstheorie.* Leipzig und Berlin 1913." Thus, the generalized use of the term tensor in the *Entwurf* paper was novel at least to Budde, who was sufficiently acquainted with the mathematical literature available in 1914 to write a textbook on three-dimensional tensors.

NOTES

[1] See, also, for example, Einstein 1916, p. 776 and Einstein 1917, pp. 97–98.

[2] Of course, it has proven possible to find reinterpretations of Einstein's ideas that do make sense to modern readers. The most successful of these attempts is based on the notions of absolute and dynamic objects, best known from the work of Anderson (1967), and explored most recently by Stachel (1986).

[3] For elaboration on the difference between these active versions of the requirements and the corresponding passive versions, see Norton 1987 or Norton 1989.

[4] Imagine, for example, a space-time theory that requires two manifolds with disjoint point sets to represent two different space-times, even though the models that host the manifolds may be diffeomorphic. For a concrete example, take a Euclidean three space and foliate it into a family of flat, two-dimensional hypersurfaces S_i, with i a real valued index. We can model each hypersurface by the unique manifold-metric pair $\langle N_i, h_{ab}^{(i)}\rangle$, where by stipulation, two diffeomorphic models $\langle N_i, h_{ab}^{(i)}\rangle$ and $\langle N_j, h_{ab}^{(j)}\rangle$ differ only by having disjoint point sets.

[5] A major deficiency of the tradition was that spaces that were not topologically R^n could only be represented in a patchwise fashion.

[6] I translate Minkowski's "Mannigfaltigkeit" as "manifold" where the standard translation (Minkowski 1908a, p. 76) has "multiplicity."

[7] I adopt the following convention. Indices a, b, c, ... are to be read according to the abstract index notation (Wald 1984, Section 2.4), so that g_{ab} is a second rank covariant tensor, i.e., a bilinear map from the tangent space of the manifold to the reals. Indices i, k, l, ... take real values 1, 2, 3, 4 so that g_{ik} represents a 4×4 matrix of reals, which could be, for example, the components of g_{ab} in some coordinate chart.

[8] This procedure recapitulates Einstein's historical pathway to general relativity. This fascinating story can be found in Stachel 1980.

[9] For further discussion of this problem of the modern reading, see Norton 1989.

[10] The most important exception arises in the context of the Cauchy problem in which these mathematical properties engender a gauge freedom that threatens the determinism of the space-time theory.

[11] For an extensive survey of Einstein's statements of the principle and its role in relativity theory, see Norton 1985.

[12] Kretschmann's mathematical methods also lie within the number manifold tradition of Klein and Minkowski. This fact is expressed neatly in Kretschmann's use of the term "coordinate manifold" (see, for example, Kretschmann 1917, pp. 581–82, p. 583). In 1915, he was even more explicit. He announced that he would "conceive of the space-time reference system of physics as a four-dimensional manifold of pure numbers" (Kretschmann 1915, p. 917) and he devoted considerable analysis to the "representation postulates" of physical theories, which "relate empirical space and empirical time with the spaces and time coordinates of theoretical physics" (Kretschmann 1915, p. 979).

[13] For example, guarantee the presence of the string $E = mc^2$ by defining $m = E/c^2$, where E is the kinetic energy of a body with inertial mass m_i and velocity v; m then enters the physically contingent law $m = m_i v^2/2c^2$. I thank Cory Juhl for discussion.

[14] At the point marked by my superscript letters *a, b,* and *c,* Kretschmann's text has footnote references: (a) Einstein 1916a, p. 766; (b) for details, Kretschmann (1915, p. 914–924); (c) compare Ricci and Levi-Civita (1901, p. 125). In the place cited in (b) and pages 924–26 following, Kretschmann argued at length for what is essentially just the point-coincidence argument: observation provides only "topological" results, such as the coincidence of parts of measuring instruments and subjects, so that the choice between coordinate systems is made by convention and arbitrary stipulation. Kretschmann's paper was submitted on October 21, 1915, two months before the point-coincidence argument even appeared in Einstein's correspondence. This fact leaves room for speculation on the priority and circumstances surrounding its discovery.

[15] For a more detailed treatment of Einstein's formulations of the point-coincidence argument and its role in his work on general relativity, see Norton 1987.

[16] For example, Einstein 1916a, p. 776 or Einstein to P. Ehrenfest, December 26, 1915, in Norton 1987, pp. 168–169.

[17] This is not the place to analyze the ambiguity lurking in the notion of "space-time coincidence." My view is that, in the last analysis, the assumption (PC) can only be made precise by replacing it by nothing less than the requirement of general covariance/Leibniz equivalence.

[18] He only then turned to the remark best known from his reply. *Even if,* he said, all empirical laws can take on generally-covariant forms, his relativity principle still has heuristic power because for two empirically equivalent theories, we should prefer the one whose generally-covariant formulation (absolute

differential calculus) is simpler and more transparent.

[19] This name appears here for the first time and the formulation given begins with the now familiar words "The G-field is determined *without residue* by the masses of bodies...."

[20] While Grossmann gave no definition of a manifold in Einstein and Grossmann 1913, this Ricci–Levi-Civita definition is clearly the one he had in mind when he wrote of "the vector analysis of an arbitrary manifold given through its line element" (p. 23) and again in a similar remark on p. 31.

[21] I believe that Minkowski also found the point too elementary to bear sustained discussion. His technical exposition (Minkowski 1907) does not contain the definition of "the world" quoted in Section 3. He limited himself to the remark that "a single system of values x, y, z, t or x_1, x_2, x_3, x_4, is to be called a *space-time point*" (p. 57, Minkowski's emphasis), with the more elementary discussion of "the world" delegated to his popular lecture (Minkowski 1908).

References

Abraham, Max (1901). "Mechanik der deformierbaren Körper. Geometrische Grundbegriffe." In *Encyklopädie der mathematischen Wissenschaften mit Einschluss ihrer Anwendungen.* Vol. 4, *Mechanik,* Part 3. Felix Klein and Conrad Müller, eds. Lepzig: B.G. Teubner, pp. 3–47.

Abraham, Max and Föppl, August (1904). *Einführung in die Maxwellsche Theorie der Elektrizität.* 2nd rev ed. Leipzig: B.G. Teubner.

Anderson, James L. (1967). *Principles of Relativity Physics.* New York: Academic Press.

Becker, Richard (1964). *Electromagnetic Fields and Interactions.* Fritz Sauter, ed. Glasgow and London: Blackie; reprint New York: Dover, 1982.

Bianchi, Luigi (1910). *Vorlesungen über Differentialgeometrie.* 2nd ed. Max Lukat, trans. Leipzig: B.G. Teubner.

Budde, E. (1914). *Tensoren und Dyaden im dreidimensionalen Raum: Ein Lehrbuch.* Braunschweig: Friedrich Vieweg und Sohn.

Crowe, Michael J. (1967). *A History of Vector Analysis: The Evolution of the Idea of a Vectorial System.* Notre Dame, Indiana: University of Notre Dame Press.

Darboux, Gaston (1914). *Leçons sur la théorie générale des surfaces et les applications géométriques du calcul infinitésimal.* Vol. 1, 2 2nd ed.; Vol. 3, 4 1st ed. New York: Chelsea, first published 1914, 1915, 1894, 1896.

Earman, John S., and Norton, John D. (1987). "What Price Space-Time Substantivalism? The Hole Argument." *British Journal for the Philosophy of Science,* 38: 515–525.

Einstein, Albert (1905). "Zur Elektrodynamik bewegter Körper." *Annalen der Physik* 17: 891–921.

—— (1913). "Zum gegenwärtigen Stande des Gravitationsproblems." *Physikalische Zeitschrift* 14: 1249–66.

——— (1914). "Die formale Grundlage der allgemeinen Relativitätstheorie." *Königlich Preussische Akademie der Wissenschaften* (Berlin) *Sitzungsberichte*: 1030–1085.

——— (1915). "Zur allgemeinen Relativitätstheorie." *Königlich Preussische Akademie der Wissenschaften* (Berlin) *Sitzungsberichte*: 778–86.

——— (1916a). "Die Grundlage der allgemeinen Relativitätstheorie." *Annalen der Physik* 49: 769–822. Translated without p. 769 as the "The Foundation of the General Theory of Relativity." In Lorentz, Hendrik Antoon, et al. *The Principle of Relativity.* Arnold Sommerfeld, ed. W. Perrett, and G.B. Jefferey, trans. London: Methuen, 1923; reprint New York: Dover, 1952, pp. 109-164.

——— (1916b). "Über Friedrich Kottlers Abhandlung 'Über Einsteins Äequivalenzhypothese und die Gravitation.' " *Annalen der Physik* 51: 639–642.

——— (1917). *Über die spezielle und die allgemeine Relativitatätstheorie (gemeinverstandlich).* Braunschweig: Friedrich Vieweg und Sohn. Trans. as *Relativity: The Special and the General Theory.* Robert W. Lawson, trans. London: Methuen, 1977.

——— (1918). "Prinzipielles zur allgemeinen Relativitätstheorie." *Annalen der Physik* 55: 241–44.

——— (1920). "Grundgedanken und Methoden der Relativitätstheorie in ihrer Entwicklung dargestellt." Unpublished manuscript, EA 2-070.

——— (1922). *The Meaning of Relativity: Four Lectures Delivered at Princeton University.* London: Chapman & Hall, 1976.

——— (1924). "Über den Aether." *Schweizerische Naturforschende Gesellschaft. Verhandlungen* 105: 85–93.

——— (1949). *Autobiographical Notes.* Open Court, 1979.

——— (1953)."Foreword." In Max Jammer. *Concepts of Space.* Cambridge, Massachusetts: Harvard University Press, 1969.

Einstein, Albert, and Grossmann, Marcel (1913). *Entwurf einer verallgemeinerten Relativitätstheorie und einer Theorie der Gravitation. I. Physikalischer Teil von Albert Einstein. II. Mathematischer Teil von Marcel Grossmann.* Leipzig and Berlin: B.G. Teubner. Reprinted with added "Bemerkungen" by Einstein, *Zeitschrift für Mathematik und Physik* 62 (1913): 225–261.

Föppl, August (1894). *Einführung in die Maxwell'sche Theorie der Elektricität.* Leipzig: B.G. Teubner.

Graves, John C. (1971). *The Conceptual Foundations of Contemporary Relativity Theory.* Cambridge, Massachusetts: MIT Press.

Grossmann, Marcel (1913). "Mathematische Begriffsbildung zur Gravitationstheorie." *Naturforschende Gesellschaft Zürich. Vierteljahrsschrift* 58: 291–97.

Hausdorff, Felix (1914). *Grundzüge der Mengenlehre.* Leipzig: B.G. Teubner.

Hawking, Stephen W., and Ellis, George F.R. (1973). *The Large-Scale Structure of Space-Time.* London and New York: Cambridge University Press.

Hermann, Robert (1975). *Ricci and Levi-Civita's Tensor Analysis Paper: Transla-*

tion, Comments, and Additional Material. Brookline, Massachusetts: Math Sci Press.

Klein, Felix (1873). "Über die sogenannte nicht-Euklidische Geometrie (2te Aufsatz)." In *Gesammelte Mathematische Abhandlungen.* Vol. 1. Berlin: J. Springer, 1921–1923, pp. 311–43.

—— (1927). *Vorlesungen über nicht-Euklidische Geometrie.* W. Roseman, ed. Berlin: J. Springer.

Kretschmann, Erich (1915). "Über die prinzipielle Bestimmbarkeit der berechtigten Bezugssysteme beliebiger Relativitätstheorien." *Annalen der Physik* 48: 907–982.

—— (1917). "Über den physikalischen Sinn der Relativitätspostulate, A. Einsteins neue und seine ursprüngliche Relativitätstheorie." *Annalen der Physik* 53: 575–614.

Laue, Max von (1911). *Die Relativitätstheorie.* Braunschweig: Friedrich Vieweg und Sohn.

Levi-Civita, Tullio (1925). *Lezioni di calcolo differenziale assoluto.* Roma: Alberto Stock.

—— (1926). *The Absolute Differential Calculus.* M. Long, trans. Glasgow and London: Blackie, 1926; reprint New York: Dover, 1977.

Minkowski, Hermann (1907). "Die Grundgleichungen für die elektromagnetischen Vorgänge in bewegten Körpern." *Königliche Gesellschaft der Wissenschaften zu Göttingen. Mathematisch-physikalische Klasse. Nachrichten* 1908: 53–111 (presented December 21, 1907).

—— (1908). "Raum und Zeit." In Lorentz, Hendrik A., et al., *Das Relativitätsprinzip.* Leipzig: B.G. Teubner, pp. 56–68.

—— (1908a). "Space and Time." In Lorentz, Hendrik A., et al., *The Principle of Relativity: A Collection of Original Memoirs on the Special and General Theory of Relativity.* Arnold Sommerfield, ed. W. Perrett and G. B. Jeffery, trans. London: Methuen, 1923; reprint New York: Dover, 1952, pp. 72–91.

Norton, John D. (1985). "What was Einstein's Principle of Equivalence?" *Studies in History and Philosophy of Science* 16: 203-246.

—— (1987). "Einstein, the Hole Argument and the Reality of Space." In *Measurement, Realism and Objectivity.* John Forge ed., Dordrecht: Reidel, pp. 153–188.

—— (1989). "Coordinates and Covariance: Einstein's View of Space-Time and the Modern View." *Foundations of Physics* 19: 1215–1263.

Pais, Abraham (1982). *"Subtle is the Lord ...": The Science and the Life of Albert Einstein.* London and New York: Oxford University Press.

Ricci, Gregorio and Levi-Civita, Tullio (1901). "Méthodes de calcul differentiel absolu et leurs application." *Mathematische Annalen* 54: 125–201. Reprinted in Tullio Levi-Civita. *Opere Matematiche,* Vol. 1, Bologna: Nichola Zanichelli, 1954, pp. 479–559.

Riemann, Bernard (1854). "Über die Hypothesen, welche der Geometrie zu Grunde liegen." *Königliche Gesellschaft der Wissenschaften zu Göttingen. Ab-*

handlungen 13 (1867): 133–152. Trans. as "On the Hypotheses Which Lie at the Foundations of Geometry," Henry S. White, trans., in *A Source Book in Mathematics*, Vol. 2. David E. Smith, ed. New York: McGraw-Hill, 1929; reprint New York: Dover, 1959, pp. 411–425.

Sachs, Rainer K., and Wu, Hung-Hsi (1977). *General Relativity for Mathematicians*. New York: Springer.

Schouten, Jan A. (1924). *Der Ricci-Kalkül.* Berlin: J. Springer.

Sommerfeld, Arnold (1910a). "Zur Relativitätstheorie. I. Vierdimensionale Vektoralgebra." *Annalen der Physik* 32: 749–776

——— (1910b). "Zur Relativitätstheorie. II. Vierdimensionale Vektoranalysis." *Annalen der Physik* 33: 649–689.

Speziali, Pierre, ed. (1972). *Albert Einstein–Michele Besso. Correspondance 1903–1955*. Paris: Hermann.

Stachel, John (1980). "Einstein and the Rigidly Rotating Disk." In *General Relativity and Gravitation: A Hundred Years after the Birth of Einstein.* Vol. 1. Alan Held, ed., New York: Plenum, pp. 1–15.

——— (1986). "What a Physicist Can Learn from the History of Einstein's Discovery of General Relativity." In *Proceedings of the Fourth Marcel Grossmann Meeting on General Relativity.* Remo Ruffini ed. Amsterdam: Elsevier, pp. 1857–1862.

Torretti, Roberto (1983). *Relativity and Geometry.* Oxford: Pergamon.

Wald, Robert (1984). *General Relativity.* Chicago, Illinois: University of Chicago Press.

Weyl, Hermann (1913). *Die Idee der Riemannschen Fläche.* Leipzig: B.G. Teubner.

Wright, Joseph E. (1908). *Invariants of Quadratic Differential Forms.* Cambridge: Cambridge University Press.

PART III

Technical Aspects of General Relativity

Albert Einstein's 1907 *Jahrbuch* Paper: The First Step from SRT to GRT

Arthur I. Miller

In its historic and scientific contexts, Albert Einstein's 1907 review paper, "On the Relativity Principle and the Consequences That Follow from It" (Einstein 1907c) is remarkable. When Einstein embarked on writing it, apparently unassailable empirical evidence weighed against what everyone assumed to be a key result from the "Lorentz–Einstein theory of the electron," the electron's transverse mass, and so, too, against the "Lorentz–Einstein principle of relativity."[1] Undeterred, Einstein set about to widen what he interpreted to be a principle of relativity to include measurements taken in accelerating reference systems. His master stroke was to realize the equivalence between the phenomena of gravitation and acceleration.

How this realization came about and how Einstein first applied it in the 1907 *Jahrbuch* paper are the topics addressed here. I take the approach of seeking connections between concepts that were important to Einstein in inventing the 1905 relativity theory (e.g., the relative nature of electric and magnetic fields) with his efforts to generalize the mass-energy equivalence and, then, with a 1907 *Gedanken* experiment that led to his discovery of the equivalence between the effects of a uniform gravitational field and an oppositely directed constant acceleration. All the while, I emphasize continuities between Einstein's approach to discovering relativity theory in 1905 with the one he took in 1907 to achieve the theory's generalization to include acceleration and gravity.

1. Einstein's Principle of Relativity in 1907

In 1907, physicists judged the results in Einstein's 1905 paper, "On the Electrodynamics of Moving Bodies" (Einstein 1905a), to offer improvements and generalizations for H.A. Lorentz's theory of the electron, which was in conflict with empirical data. The so-called Lorentz–Einstein principle of relativity was the basis of a version of the sought-after electromagnetic world-picture in which the laws of mechanics are reduced to those of Lorentz's electromagnetic theory.

But, when Einstein referred to a principle of relativity, it was of a different sort: it was axiomatic, that is, beyond experimental proof; it covered electromagnetism and mechanics equally, without trying to reduce one to the other; and it included no assumptions on the constitution of electrons. In publications, Einstein referred to his own view on the electrodynamics of moving bodies with the term "relativity theory," coined by Max Planck (1906), albeit with the intent of separating the Lorentz–Einstein theory of the electron from other electron theories.

2. Derivations of the Mass-Energy Equivalence

Sometime after having submitted the 1905 relativity paper to the *Annalen der Physik*, Einstein realized that it implicitly contained the mass–energy equivalence in the equation for the kinetic energy of a point electron in an external electrostatic field acting in the electron's direction of motion:

$$K_0 = m_0c^2(\gamma - 1), \tag{1}$$

where $\gamma = 1/\sqrt{(1 - v^2/c^2)}$ and m_0 is the electron's mass in its rest system.[2] Suffice it to say that Einstein's derivations of the mass-energy equivalence in 1905 and 1906[3] were inconclusive for what he referred to in the title of a May 1907 paper as "The Consequences of the Relativity Principle for the Inertia of Energy" (Einstein 1907b). "An assumption of such unusual generality," wrote Einstein of the mass-energy equivalence, "inevitably invites challenge for proving its necessity or correctness in the most general manner....Are there not special cases in which this assumption leads to incompatible consequences?" One special case that Einstein had in mind was the "dynamics of parallel translation" of rigid bodies, which he treated "from the point of view of the present *Relativitätselektrodynamik*," that is, with no assumptions on the constitution of matter.

He considered a rigid body of arbitrary shape with a continuous volume charge distribution. Although acted on by an external electromagnetic force, the body remains in equilibrium in its rest system k. Relative to the laboratory system K, the external force increases the body's energy by the amount

$$\Delta E = \int dt \int \rho\vec{v} \cdot \vec{E}\, dV, \tag{2}$$

where the vector quantity $\vec{E}$ is the external electric field, the scalar quantity E is the body's energy measured in K, ρ is the charge distribution in K, and dV is a volume element of the body measured in K. A Lorentz transformation into the body's rest system k yields

$$\Delta E = \gamma v \int d\tau(\Sigma K_\xi), \tag{3}$$

where τ is the time coordinate in k to which other primed quantities also refer, and ΣK_ξ is the total effect over the charged body of the ξ component of the electric field along k's x-axis. Although the quantity ΣK_ξ vanishes in k where the body is in equilibrium, it need not vanish in K because, owing to the relativity of simultaneity, the integration limits depend on the coordinate ξ as follows:

$$\tau_0 = \frac{t_0}{\gamma} - \frac{v}{c^2}\xi; \quad \tau_1 = \frac{t_1}{\gamma} - \frac{v}{c^2}\xi. \tag{4}$$

So, relative to K the body's energy is increased by an amount

$$\Delta E = -\left(\frac{v}{c}\right)^2 \gamma[\Sigma\xi K_\xi], \tag{5}$$

where, in order to evaluate the integral in K, Einstein took the force to be either slowly varying or constant during the time interval considered and to vanish at $t = t_0$. The quantity $\Delta E > 0$ because K and ξ are directly opposite. Eq. (5) is an "unexpected result," wrote Einstein, because

> if we subject a rigid body, on which originally no forces acted, to the influence of forces that impart no acceleration to the body, these forces perform—from the viewpoint of a coordinate system moving relatively to the body—a work ΔE upon the body, which depends on the force's distribution over the body and the velocity v. According to the energy principle, the kinetic energy of a rigid body acted on by an external force is ΔE greater than the same kinetic energy of a fast-moving body that is acted on by no force. (Einstein 1907b, pp. 376–377)

In order to demonstrate another unexpected property of Eq. (5), Einstein referred to the example in which the charged body persists in inertial motion under the influence of its self-electromagnetic forces only. One would expect that the body's total kinetic energy K_T would be

$$K_T = K_0 + (E^e - E_0^e), \tag{6}$$

where K_0 is as in Eq. (1), E^e is the charged body's self-electromagnetic energy, and E_0^e is the static or rest energy. But, Eq. (6) depends on the rigid body's orientation, which is incorrect. The error is trying to calculate the kinetic energy of a rigid body as if the body "were acted on by no force." In this case, Eq. (5) is the additional energy from the body's self-electromagnetic forces. Combining Eq. (5) with Eq. (6) yields

$$K_T = (m_0 + E_0^e c^2)(\gamma - 1), \tag{7}$$

which is the version of Eq. (1) for charged bodies with extension:[4] "The electrostatically charged body possesses an inertial mass that exceeds that of the uncharged body by the electrostatic energy divided by the square of the velocity of light." Einstein had not yet realized why he should also extend this result to the gravitational mass. He went on to remind the reader of the several simplifying assumptions that permitted the calculation of the result in Eq. (7) (e.g., the uniform volume charge distribution for a *rigid* body).

What is the nature of the quantity ΔE that is a consequence of the relativity of simultaneity and, even more basically, of the principle of relativity? On this point, Einstein could go no further because, he wrote, we are still far from possessing a "dynamics of parallel translation of rigid bodies [since we] do not yet possess a world picture [*Weltbild*] according to the principle of relativity," that is, the means to analyze the effects caused by the constitution of matter. He went on to relate the quantity ΔE to an unknown *Qualität* that propagates with a "finite velocity" in bodies that are observed from another inertial reference system.[5] For example, this *Qualität* is necessary because according to the principle of relativity if a body is in equilibrium in one inertial system, such as the body in the above example, then it must be in equilibrium in every inertial system. Equilibrium cannot be Lorentz-transformed away.

3. Einstein's 1907 *Jahrbuch* Paper

This brings us to September 1907, when Einstein received the invitation from Johannes Stark to write a review paper on the *Relativitätstheorie* for the *Jahrbuch*.[6] Stark received the completed *Jahrbuch* paper on December 4, 1907.

The outstanding difference between Einstein's presentations on relativity theory in 1905 and in the 1907 *Jahrbuch* paper is that in the *Jahrbuch* he emphasized repeatedly that the reference systems in which clocks are synchronized and the principle of relativity is valid are non-accelerated. Perhaps Einstein wanted to prepare the reader for the final section in which he widened the principle of relativity to include accelerated reference systems. Einstein's review of relativity theory concluded with a critical analysis of Walter Kaufmann's 1906 data that were at variance with the Lorentz–Einstein theory of the electron. This issue has been explored in some detail elsewhere.[7]

3.1 The Mass-Energy Equivalence as an Axiom

New results for the theory of relativity appeared in the section "On the Dependence of Mass on Energy," in which Einstein improved his earlier 1907 derivation of the mass-energy equivalence. He replaced the artifice of a rigid electrified body with a system of charges surrounded by a massless envelope impenetrable to radiation. This device enabled him to make contact with certain new results from Planck on the pressure, temperature, and entropy of moving systems. Planck's results may have given Einstein some insight into the "unknown *Qualität*," although Einstein did not pursue this point in the *Jahrbuch*[8] because in the interim he realized how to avoid it and yet go right to the heart of the mass-energy equivalence.

Einstein assumed that the external forces do not act outside the time interval in K and that they vanish at the temporal boundaries t_0 and t_1. Then, the contribution from the force ΣK_ξ also vanishes in K.[9] From Eq. (3), Einstein obtained

$$dE = \gamma\, dE', \tag{8}$$

which means that the energy of a uniformly moving system can be a function only of two variables, the relative velocity v and the energy content E_0, relative to the rest system k. From Eq. (8), we have

$$E = (m_0 + E_0/c^2)\gamma c^2. \tag{9}$$

Therefore, continued Einstein, for all intensive purposes the system acts as if it were a "material point" of mass:

$$M = m_0 + E_0/c^2. \tag{10}$$

He proceeded immediately to develop and generalize the physical content of Eq. (10), which is model independent: "This result is of extraordinary theoretical importance because a physical system's inertial mass and energy content appear to behave like the same thing," that is, mass is confined energy, which is the general meaning of the mass-energy equivalence that he had sought since 1905.

There ensued an analysis of the feasibility of testing the mass-energy equivalence with radioactive substances which led Einstein to conclude that five-figure accuracy of atomic masses is required. This was unavailable in 1907. He did not belabor this shortcoming because for his purposes it was beside the point. The essence of Eq. (10) lay in how mass variation is measured, even in principle, and this was the means used to generalize Eq. (10) from a "physical system's inertial mass" to its gravitational mass.

> In the preceding, we have assumed implicitly that such a measurement of mass variation can be carried out with the usually employed instrument, the scale; therefore, the relation [Eq. (10)] holds not only for the inertial mass, but also for the gravitational mass or, in other words, inertia and gravity [*Schwere*] are under all circumstances exactly proportional.... The proportionality between inertial and gravitational masses is valid universally for all bodies with an accuracy as precisely as can currently be determined; so that unless proven otherwise, we must accept it as universally valid. (Einstein 1907c)

In this dazzling passage, Einstein raised the equivalence of gravitational and inertial mass from a statement valid to "an accuracy as precisely as can currently be determined" to a principle "universally valid." He had applied a similar approach to a key concept in the 1905 relativity paper in which he raised the "conjecture" that to order v/c there is absolute motion neither in mechanics nor in electromagnetism to the status of a principle beyond experimental proof, the principle of relativity.

He continued in the 1907 *Jahrbuch* paper, "We shall find in the final section of this communication a new argument supporting this assumption." How Einstein developed this new argument in the section "Principle of Relativity and Gravitation" requires the discussion of a *Gedanken* experiment.

3.2 A *Gedanken* Experiment

There is an effect that almost everyone has the capacity to experience when writing a review paper. The integrated material can take on a new meaning; it can be better understood, perhaps even at a deeper level. Previously unconnected parts can become connected in ways unexpected, ways that can lead to new results. This is what happened to Einstein sometime between September and December 1907.

Gedanken experiments occur to the prepared mind. A key realization for Einstein toward inventing the 1905 relativity theory had been the relative nature of the electric and magnetic fields, for which analysis of the phenomenon of electromagnetic induction had been important. In an unpublished manuscript of 1919 [brought to light by Gerald Holton (1971)], Einstein recalled this work: [The] "difference between [observers on the magnet and on the conductor] could not be a real difference but rather, in my conviction, only a difference in the choice of the reference point.... The existence of an electric field was therefore a relative one, depending on the state of motion of the coordinate system being used." He continued in 1919 to recollect his work on the *Jahrbuch* paper: "When, in 1907, I was working on a summary essay concerning the special theory of relativity for the *Jahrbuch für Radioaktivität und Elektronik,* I had to try to modify Newton's theory of gravitation in such a way that it would fit into the theory [of relativity]. Attempts in this direction showed the possibility of carrying out this enterprise, but did not satisfy me because they had to be supported by hypotheses without physical basis." While mulling over the failure of this apparently most direct route to the theory of gravity, there coalesced for Einstein the relativity of the electric and magnetic fields (from 1905) and that the energy of these fields possess inertia (from 1907).

> At this point, there occurred to me the happiest thought of my life. Just as is the case with the electric field produced by electromagnetic induction, the gravitational field has similarly only a relative existence. *For if one considers an observer in free fall, e.g., from the roof of a house, there exists for him during this fall no gravitational field—at least not in his immediate vacinity* [italics in original]. Indeed, if the observer drops some bodies, then these remain relative to him in a state of rest or in uniform motion, independent of their particular chemical or physical nature (in this consideration the air resistance is, of course, neglected). The observer therefore has the right to interpret his state as 'at rest'.... The experimentally known matter independence of the acceleration of fall is therefore a powerful argument for the fact that the relativity postulate has to be extended to coordinate systems that, relative to each other, are in nonuniform motion. (Einstein 1919, in Holton 1971)

Like the observers who rode the magnet and conductor in 1905, the falling observer of 1907 "has the right to interpret his state as 'at rest'."

We may speculate that Einstein's mathematical support for the "happiest thought of my life" went along the following line. But, no matter what path one conjectures for Einstein, what is astonishing is that even in its historic context, the physics of the 1907 *Gedanken* experiment is so incredibly straightforward. Everything turns on an assumption that all physicists make, particularly astronomers: the gravitational and inertial masses are exactly equal. Einstein took this assumption and widened it to include acceleration and gravity. Assume that the observer falling freely from the roof carries a reference system K'. An observer at rest on the ground has a reference system K. The y axes of K' and K are aligned. Let all velocities involved be small enough so that Newtonian mechanics can be used. Then, relative to K, the equation of motion for the freely falling observer is

$$m_I \, d^2y/dt^2 = -m_G g + F, \tag{11}$$

where $m_I(m_G)$ is the inertial (gravitational) mass of the falling body, g is the acceleration of free fall due to the external gravitational field, and F represents any other forces exerted on the falling body. Set

$$y' = y - at^2/2, \tag{12}$$

where a is the body's constant acceleration relative to K and $t' = t$. Then,

$$m_I \, d^2y'/dt'^2 + m_I a = -m_G g + F'. \tag{13}$$

Whatever Einstein's mathematical analysis of the 1907 *Gedanken* experiment was, he had to arrive at an equation similar to Eq. (13). Assume now that $m_I = m_G = m$. At this juncture, one final assumption must be made for which is central the theory-laden visual imagery of a freely falling observer in whose immediate surroundings there is "no gravitational field [and so the observer] has the right to interpret his state as 'at rest' "—namely, that $a = -g$, with the result

$$m \, d^2y'/dt'^2 = F'. \tag{14}$$

The freely falling observer who uses y' and the observer at rest who uses y will both agree on the Newtonian laws of mechanics, even though the freely falling observer will experience no gravitational field. Consequently, just as the electric field in electromagnetic induction has a relative existence, so is the gravitational field a relative quantity.

It has been well established that in 1907 Einstein was unaware of Eötvos's experiments. So, we may wonder what clues were available to him with regard to explicit comments to the effect that one assumes equality of the gravitational and inertial masses? One clue is in a book that we know Einstein had read before 1905, Henri Poincaré's *Science and Hypothesis* (1902). A theme in this book is to ferret out tacit hypotheses in scientific research. For example, Poincaré drew attention to the way in which we measure the masses of planets with satellites using Newton's gravitational law: "What we measure in this manner is not the mass in the ratio of force to acceleration, it is the attracting mass. It is not the inertia of a body, it is the attracting power." Never one to spell out anything directly, implicit in Poincaré's analysis is that when we "weigh" a planet that has a satellite we cancel the mass of the satellite from the right- and left-hand sides of the equation $F = m_I a$, where we use Newton's gravitational law for F. The tacit assumption is that the gravitational and inertial masses of the satellite are exactly equal. Einstein, as we have seen, extended this tacit assumption to the equivalence between a uniform gravitational field and an oppositely directed constant acceleration.

3.3 Acceleration, Gravitation, and Coordinate Transformations

This being the case, then just as in special relativity, there ought to be coordinate and time transformations from a moving system (i.e., accelerating) to a resting system (i.e., one with a uniform gravitational field). In this way, a dynamics problem can be transformed into a statics problem, just like the procedure in electrodynamics known since 1892. In fact, for this purpose, Lorentz, in *Treatise on a Theory of Electrical and Optical Phenomena in Moving Bodies (1895)*, used three different reference systems. Einstein learned well from Lorentz because he did likewise in the 1905 electrodynamics paper and again in the 1907 *Jahrbuch* paper, as we will see.

In Part V of the *Jahrbuch* paper, Einstein presented the problematic "Is it conceivable that the principle of relativity is also valid for systems that are accelerated relative to one another?" His reply began with a version of the *Gedanken* experiment. A reference system Σ_1 moves in the direction of its x-axis, with a constant acceleration a relative to a reference system Σ_2 that remains at rest. In Σ_2, there is a constant gravitational field in which all bodies fall with an acceleration $-a$. Assuming the equivalence between inertial and gravitational masses, Newton's laws predicted the same motions in Σ_1 and Σ_2; hence, Einstein concluded "we have no cause from our experience to distinguish between Σ_1 and Σ_2; thus, in what follows we shall accept the full physical equivalence of the gravitational field and the corresponding acceleration of the coordinate system." (In 1912, Einstein referred to this statement as the "equivalence principle.") Acceleration lost its absoluteness and, with the gravitational field, became a relative quantity, at least for constant acceleration. Einstein next used the 1905 relativity theory as a guide to seek a new definition of simultaneity in a uniformly accelerating reference system Σ.

We recall that since Maxwell the method to deal with problems in the electrodynamics of moving bodies was to first write the equations of electromagnetism relative to an ether-fixed reference system and then apply the Galilean transformations to move the problem situation into an inertial reference system. Effects appeared that were, however, not measured in ether-drift experiments. So, Lorentz (1892, 1895, 1904) sought further transformations on the Galilean coordinates that could somehow also incorporate the various hypotheses for the causes of countereffects that were supposed to explain the failure of ether-drift experiments, such countereffects as the contraction of moving bodies and the appearance of proper charge distributions on the exterior of conducting magnets and charged bodies at rest on the moving earth. During 1895–1904, Lorentz used three different coordinate systems: an ether-based system $S(x, y, z, t)$, an inertial reference system $S_r[x_r, y_r, z_r, t_r(= t)]$, and a fictitious or mathematical reference system $S'(x', y', z', t_L)$, in which the equations of electromagnetism have the same form as they do in S, where $t_L = t - vx/c^2$ is the mathematical "local time coordinate."

In the 1905 relativity paper, Einstein likewise used three coordinate systems to derive the relativistic transformation equations for spatial and time coordinates: $K(x, y, z, t)$ and $k(\xi, \eta, \zeta, \tau)$ are the inertial reference systems whose space and time coordinates agree with the principle of relativity and the light axiom, and S' is the coordinate system whose axes are aligned with those of k and whose spatial coordinates are Galilean $(x', y' = y, z' = z)$, but whose time coordinate is relativistic $t' = \tau$. In 1907, Einstein used a similar tactic, but with the variation that S' is a reference system in which the principles of the special theory of relativity are valid. The reference system k became the system Σ, which is uniformly accelerated relative to a nonaccelerated system S. Einstein set out to find the definition of simultaneity in Σ.

Assume that at times $t = \tau = 0$, Σ and S are at relative rest and their axes coincide. Then, Σ begins to accelerate at a constant rate a and $v = at$. At a particular instant of time, the inertial system S' coincides with Σ. Einstein assumed that the acceleration a is small enough so that the shape of bodies and rate of clocks in Σ are unaffected. Consequently, at the instant t', when S' and Σ coincide, clocks in Σ can be adjusted so that their time is t'. Einstein asked, "What can we say about the rate of the clocks in the next time element τ ?" Since effects of acceleration are neglected, in the next instance of time τ, the light axiom applies, provided that only small light paths are considered. Call the "local time" σ the time on the clocks of Σ that was set at $t = 0$ when S and Σ coincided and were at relative rest. But σ is not the "time" of Σ because, although the clocks in Σ remain synchronous with those in S, according to the special theory of relativity, they are not synchronous with those in S' and so they are not synchronous with each other. Rather, Einstein defined the time τ in Σ to be the readings of a single clock at Σ's origin, for which $\tau = t$. What is the relationship between times σ and τ of an event? For this purpose, the local Lorentz frame S' is used. Two events are simultaneous in S', and so they are also simultaneous in Σ when

$$t_1 - vx_1/c^2 = t_2 - vx_2/c^2. \tag{15}$$

Small intervals of time are considered, so that

$$x_2 - x_1 = x_2' - x_1' = \xi_2 - \xi_1, \tag{16}$$

$$t_1 = \sigma_1; \quad t_2 = \sigma_2; \quad v = at = a\tau. \tag{17}$$

So, from Eq. (15),

$$\sigma_2 - \sigma_1 = a\tau(\xi_2 - \xi_1)/c^2. \tag{18}$$

setting $\sigma_1 = \tau$ and $\xi_1 = 0$, Einstein obtained

$$\sigma = \tau\left(1 + \frac{a\xi}{c^2}\right). \tag{19}$$

From the equivalence principle, this equation becomes

$$\sigma = \tau\left(1 + \frac{\Phi}{c^2}\right), \tag{20}$$

where Φ is the gravitational potential between the origin and ξ.

One prediction follows immediately. Consider clocks that are the "generators" of spectral lines, i.e., atoms. Then, light coming from the sun's surface "possesses a wavelength that is greater than the light generated by the same material on the earth's surface." In a footnote, Einstein wrote that he assumed this result also held for a nonuniform gravitational field.

Einstein went on to attain covariance for the equations of electromagnetism in Σ, provided that c is replaced with

$$c\left(1 + \frac{a\xi}{c^2}\right) = c\left(1 + \frac{\Phi}{c^2}\right), \tag{21}$$

which means that light rays not propagating in the ξ direction are bent by the gravitational field by the amount

$$\frac{a}{c^2} \sin \phi, \tag{22}$$

where ϕ is the angle between c and the direction of the gravitational field.

Einstein continued by pointing out that the predicted effects of the earth's gravitational field on optical and electrical phenomena are too small to be detected. In 1911, he found a means to test his ideas on gravitation by using a variant of Eq. (21) and Huygens' principle to predict a bending of light rays in the field of a gravitational point source (Einstein 1911).

Then, using the standard method for investigating energy conservation or Poynting's theorem in electromagnetism, Einstein provided the promised "new argument supporting" the assumption that the mass-energy equivalence should be extended to include the gravitational energy of a position as follows. After extending the volume integration limits to infinity, he obtained

$$\int \left(1 + \frac{\Phi}{c^2}\right) \rho \vec{v} \cdot \vec{E}\, dV + \frac{\partial}{\partial \tau} \int \left(1 + \frac{\Phi}{c^2}\right) \left(\frac{E^2 + B^2}{8\pi}\right) dV = 0, \tag{23}$$

which is the usual Poynting's theorem plus energy flow $\mathcal{E}$ multiplied by Φ/c^2, that is, $\mathcal{E}\Phi/c^2$: "This equation [Eq. (23)] contains a very remarkable result.... To every energy $\mathcal{E}$ belongs thus in the gravitational field an energy of position which is as large as the energy of position of a ponderable mass of magnitude $\mathcal{E}/c^2$." The caveat must be added here, he wrote, that a uniform gravitational field can be replaced with an oppositely directed constant acceleration, as he had assumed earlier.

Although Einstein did not publish again on gravitation until 1911, the problem still concerned him. For example, on December 24, 1907, he wrote to Conrad Habicht: "At this time, I am [again] busy with considerations on relativity theory in connection with the law of gravitation.... I hope to clear up the so-far unexplained secular changes of the perihelion length of Mercury... [but] so far it does not seem to work." "It" would not work until 1915.

To tie some loose ends together, Einstein's "unknown *Qualität*" was explicated in 1911 by Max von Laue with a statement based on the axiomatic status of Einstein's principle of relativity. In order to ensure the equilibrium of a body that is in inertial motion, one should write

$$\sum_{\nu=1}^{4} \frac{\partial(\Sigma T_{\mu\nu})}{\partial x_\nu} = 0, \tag{24}$$

where $\Sigma T_{\mu\nu}$ contains as many stress-energy tensors as is necessary to "close" the system.[10]

What about the problem of the motion of rigid bodies? From correspondence in the Einstein Archives, John Stachel (1980) has shown in his paper "Einstein and the Rotating Disk" that, beginning in 1908, Einstein became concerned with the problems and paradoxes of rotating rigid bodies, which led him to question the validity of Euclidean geometry in rotating reference systems, a key point toward the general relativity theory.

In conclusion, how the path to the general theory of relativity led Einstein away from the one that he had taken in 1907 is beyond the scope of this chapter, which explores how Einstein obtained key insights that coalesced in the 1907 *Gedanken* experiment. To paraphrase Einstein's recollection of the 1895 *Gedanken* experiment that was of key importance to his invention of special relativity, the *Gedanken* experiment of 1907 "contained the germ of the general theory of relativity," the principle of equivalence. Despite preoccupation with the light quantum during 1908–1911 and setbacks and blind alleys in his subsequent research during 1911–1915 toward the general theory of relativity,[11] Einstein's persistence and tenacity kept him moving toward the theory's final achievement. As Einstein wrote to Arnold Sommerfeld on October 29, 1912, "Compared to this problem [general relativity], the original relativity is child's play."[12]

NOTES

[1] This important phase in the history of physics during the first decade of the 20th century has been analyzed in some depth. Here, I take the liberty to cite only my 1981 book, in which the interested reader can find an extensive bibliography to the secondary literature.

[2] As Einstein wrote to Conrad Habicht sometime after submitting the 1905 paper to the *Annalen*, "However, a consequence of the work in electrodynamics has suddenly occurred to me ... the mass of a body is a direct measure of its energy content" (see Miller 1981, p. 353).

[3] In 1905b, Einstein considered the case of a moving body that emits radiation pulses back to back with respect to the laboratory reference system. Using the transformation equations for the energy of a light pulse, Einstein showed that to order $(v/c)^2$ if the body emitted an amount of radiation energy E its mass diminishes by E^2. In 1906, Einstein demonstrated that to order (v/c) the validity of the principle of conservation of the center of mass motion depends on the mass-energy equivalence.

[4] See Miller 1981 for analyses of Einstein's derivation of Eq. (7) and on the possibility that Einstein attempted a variant of this analysis toward clarification within the relativity theory of the stress hypothesized by Henri Poincaré for the

purpose of rendering stable Lorentz's deformable electron. The interesting point is that if Einstein had actually tried to do this (and he very probably did) then he would have obtained the wrong answer because it turned out that Einstein's formalism for how bodies are stressed in their rest system is not always covariant, as von Laue showed in 1911.

5 In order to pursue this point further, he went on to explore a second, less model-dependent example that cut right to the core of the previous calculation, serving also to critique the concept of the rigid body. Consider a rigid rod of length l (as measured in its rest system k) moving along the x axis of a reference system K. Along the rod's axis, let equal and opposite impulses be exerted simultaneously relative to k. Relative to K, however, the impulses are not simultaneous; rather, the impulse striking A at time t_A acts first, and then at a later time t_B, the impulse at B acts, where

$$t_B = t_A + \gamma(vl/c^2).$$

Since the impulse applied at A is not compensated instantaneously by the impulse applied at B, the impulse at A does work on the rod, thereby increasing the rod's energy during the interval $\Delta t = t_B - t_A$. But, the observer in K registers no change in the rod's velocity, which apparently violates the principle of conservation of energy. A proposal that Einstein offered for removing this difficulty is that there arises in the rod an "unknown *Qualität* " that propagates with a "finite velocity" along the rod, causing the acceleration required for compensating the force at A, thereby maintaining the rod's equilibrium relative to$K$.

6 Besides the fact that Einstein was indeed an expert on the *Relativitätstheorie*, according to Hermann (1966) what may also have led Stark to solicit an article from him is that earlier Einstein had written flatteringly of Stark's research on canal rays in a letter to Stark dated April 13, 1907, and again in an *Annalen* paper, in which Einstein proposed an empirical test of the principle of relativity based on the Doppler shifted spectra of canal rays (Einstein 1907a). During September and October 1907, some correspondence followed between them that is of interest mainly for noting what scientific literature Einstein claimed was available to him while he worked in the patent office in Berne.

7 See Miller 1981 for details and references to the secondary literature. Suffice it to say that after finding nothing glaringly wrong with Kaufmann's data, except perhaps for some unknown and thus far unnoticed systematic errors, Einstein echoed Planck's published reservations and joined him in calling for more data. Einstein then offered his own critical opinion that these data support theories of the electron (Alfred Bucherer's and Max Abraham's) that have a rather "small probability because ... these theoretical systems [do not] embrace a greater complex of phenomena."

8 From correspondence between Stark and Einstein during September and October 1907, we know that in September 1908 Einstein was unaware of Planck's paper "On the Dynamics of Moving Systems" (Planck 1907), in which he essentially put the finishing touches on a relativistic thermodynamics. For example,

from the principle of least action in conjunction with the principles of energy conservation and relativity, Planck deduced a new transformation equation for the force that was independent of any model, such as a rigid body:

$$F_x = F_\xi + v/c^2\left(V'\,dp_0/dt' + T'\,dS/dt'\right),$$

where F_ξ is the component of the net force along k's ξ axis; primed quantities refer to the body's instantaneous rest system k, so T' is the temperature and V' is the volume; S is the Lorentz invariant entropy; and p_0 is the Lorentz invariant pressure exerted on the body. Planck did not explore the physical content of the force transformation equation that, in fact, contains Einstein's "remarkable result": even if no net forces act in a body's rest system, a change in the body's internal state leads to a measurable force in K. There are no rigid bodies.

9 In the *Jahrbuch* paper, Einstein went on to treat the more general case in which forces continually operate in order to ensure the system's equilibrium in K. Then, $\Sigma\xi K_\xi = -p_0 V'$, where p_0 is a hydrostatic-like pressure that acts on the impermeable membrane in order to maintain the system's equilibrium. It is the same sort of pressure as Poincaré's unknown pressure that ensures the stability of Lorentz's electron. It was along these lines that Einstein made contact with some of Planck's results in relativistic thermodynamics.

10 Von Laue's work depended on Abraham's 1902 results (the possible non-colinearity of velocity and momentum), Planck's 1908 results (extension of the mass-energy equivalence to the flow of any sort of energy momentum), and Hermann Minkowski's 1908 four-vector formalism. Minkowski showed that the Lorentz force and the energy conservation law can be expressed as

$$f_\mu = \sum_{\nu=1}^{4} \frac{\partial(\Sigma T_{\mu\nu})}{\partial x_\nu}, \tag{A}$$

where f_μ is a four-vector with components $(\vec{f}, i\vec{f}\cdot\vec{v}/c)$, and $T_{\mu\nu}$ is a traceless antisymmetric tensor that can be represented as a 4×4 matrix whose spatial components are the Maxwell stress tensor and the mixed space-time components, the momentum density and Poynting vector. From the Lorentz covariance of Eq. (A), if $f_\mu = 0$ in one inertial reference system, then it is zero in every inertial reference system that is, equilibrium cannot be Lorentz-transformed away. Von Laue generalized this result in the sharpest possible manner, that is, one based on the axiomatic stature of Einstein's principle of relativity, namely, Eq. (24). See Miller 1981, Chapter 12.

11 See Pais 1982 for a discussion of Einstein's research during 1911–1915 toward the general theory of relativity.

12 From Pais 1982, p. 216.

References

Einstein, Albert (1905a). "Zur Elektrodynamik bewegter Körper." *Annalen der Physik* 17: 891–921. Translated in Miller 1981, pp. 392–415.

——— (1905b). "Ist die Trägheit eines Körpers von seinem Energieinhalt abhängig?" *Annalen der Physik* 18: 639–641. English translation in Lorentz, et al. 1923, pp. 69–71. Volume number is stated incorrectly and title misspelled in reprint volume.

——— (1906). "Das Prinzip von der Erhaltung der Schwerpunktsbewegung und die Trägheit der Energie." *Annalen der Physik* 20: 627–633.

——— (1907a). Über die Möglichkeit einer neuen Prüfung des Relativitätsprinzips." *Annalen der Physik* 23: 197–198.

——— (1907b). "Über die vom Relativitätsprinzip geforderte Trägheit der Energie." *Annalen der Physik* 23: 371–384.

——— (1907c). "Über das Relativitätsprinzip und die aus demselben gezogenen Folgerungen." *Jahrbuch der Radioaktivität und Elektronik* 4: 411–462.

——— (1911). "Über den Einfluss der Schwerkraft auf die Ausbreitung des Lichtes." *Annalen der Physik* 35: 898– 908. English translation in Lorentz, et al. 1923, pp. 99–108.

——— (1912). "Lichtgeschwindigkeit und Statik des Gravitationsfeldes." *Annalen der Physik* 38: 355–369.

Hermann, Armin (1966). "Albert Einstein und Johannes Stark: Briefwechsel und Verhältnis der beiden Nobelpreisträger." *Sudhoffs Archiv für Geschichte der Medizin und der Naturwissenschaften, der Pharmazie und der Mathematik* 50: 267–285.

Holton, Gerald (1971). "On Trying to Understand Scientific Genius." *The American Scholar* 41: 95–110.

Kaufmann, Walter (1906). "Über die Konstitution des Elektrons." *Annalen der Physik* 19: 487–553.

Laue, Max von (1911). *Die Relativitätstheorie.* Braunschweig: Friedrich Vieweg und Sohn.

Lorentz, Hendrik Antoon (1892). "La théorie électromagnétique de Maxwell et son application aux corps mouvants." *Archives Néerlandaises des Sciences Physiques et Naturelles* 25: 363.

——— (1895). *Versuch einer Theorie der electrischen und optischen Erscheinungen in bewegten Körpern.* Leiden: E.J. Brill.

——— (1904). "Electromagnetic Phenomena in a System Moving with any Velocity Less than that of Light." *Koninklijke Akademie van Wetenschappen te Amsterdam. Section of Sciences. Proceedings* 6: 809–831. Reprinted in part in Lorentz, et al. 1923, pp. 11–34.

Lorentz, Hendrik Antoon, Einstein, Albert, Minkowski, Hermann, and Weyl, Hermann (1923). *The Principle of Relativity: A Collection of Original Memoirs on the Special and General Theory of Relativity.* Arnold Sommerfeld, ed. W. Perrett and G. B. Jeffery, trans. London: Methuen. Reprint New York: Dover, 1952.

Miller, Arthur I. (1981). *Albert Einstein's Special Theory of Relativity: Emergence (1905) and Early Interpretation (1905–1911).* Reading, Massachusetts: Addison-Wesley.

Pais, Abraham (1982). *"Subtle is the Lord ..." : The Science and Life of Albert Einstein.* London and New York: Oxford University Press.

Planck, Max (1906). "Die Kaufmannschen Messungen der Ablenkbarkeit der β-Strahlen in ihrer Bedeutung für die Dynamik der Elektronen." *Physikalische Zeitschrift* 7: 753–761.

—— (1907). "Zur Dynamik bewegter Systeme." *Königlich Preussische Akademie der Wissenschaften* (Berlin). *Sitzungsberichte*: 542–570.

Poincaré, Henri (1902). *La science et l'hypothèse.* Paris: Flammarion.

Stachel, John (1980). "Einstein and the Rigidly Rotating Disk." In *General Relativity and Gravitation One Hundred Years After the Birth of Albert Einstein.* Vol. 1. Alan Held, ed. New York: Plenum, pp. 1–15.

On the Origin of the Concept of an Einstein Space

Gunnar Berg

1. Introduction

The creation of the general theory of relativity by Einstein in the years prior to 1916 was, as is well known, heavily dependent on previous developments in the field of differential geometry. It is also true, and this is perhaps less well known, that the general theory of relativity in its turn strongly influenced the subsequent development of geometry. In order to illuminate this influence, I will discuss the emergence of a concept of great importance in present-day mathematics: the concept of an Einstein space.

Starting by giving some general remarks on the relation between mathematics and physics in general and between general relativity and Riemannian geometry in particular, I then move to a short description of the necessary mathematical terms. In the next section, I explain the historical background to the terminology, and finally, I discuss an ironic twist as to the aptness of the naming of the concept as given.

2. Mathematics and General Relativity

As a characteristic feature, mathematical thought is often seen in its total freedom to create its concepts independently of sense impressions or external experiences. This was expressed very clearly by T.H. Huxley:

> Mathematics is that study that knows nothing of observation, nothing of experiment, nothing of induction, nothing of causation.[1]

Now, compare Einstein's remark:

> How is it possible that mathematics, which after all is a product of human thought, independent of all experience, is so admirably adapted to real objects? (Einstein 1921, p. 124)[2]

From this point of view, it makes sense to talk with Wigner about "the unreasonable effectiveness of mathematics,"[3] and it becomes a matter of profound surprise that this world of concepts and connections between concepts, purportedly created with a view only to inner beauty and consistency, can serve as an efficient toolbox for physicists and other scientists.

On the other hand, if one wishes to argue for a more direct and intimate relation between mathematics and physics, general relativity seems to me to provide perhaps the most illuminating example. Indeed, rather than saying that Einstein found his tools in Riemannian geometry and tensor analysis fortuitously worked out, as it were, and ready to be used, one might argue that he and Riemann were to a certain extent concerned with the same problem, and that Einstein succeeded in vindicating Riemann's vision of a space(–time) singled out by physical principles of a larger class, in its turn selected by epistemological principles from an even larger class of logically possible spaces.

In the following, I would like to concentrate on the effect of general relativity on the development of Riemannian geometry, an effect that was considerable:

> Through Albert Einstein's theory of gravitation has the long neglected geometry of Riemann... again come to the fore.[4]

How did it come about that Einstein's creation had this fertilizing effect on Riemannian geometry? In general, it is, of course, true that the way in which physical theorizing influences mathematics can be intricate and tortuous; compare, for example, the route from heat conduction via Fourier series to Cantor's theory of transfinite numbers or from the physics of strong interactions via Yang–Mills theory to exotic $\mathbf{R}^4$. Since the relation between differential geometry and general relativity is particularly intimate—witness for example the work by Weyl (Weyl 1918), in which he, on the basis of general relativity, simultaneously creates an (abortive) unified theory of gravitation and electromagnetism, and a "truly infinitesimal" generalization of Riemannian geometry—it might, however, be possible to disentangle the strands in this case. As for the mechanism of the influence, I would like to suggest some possibilities.

The progressing 19th century had witnessed a proliferation of different geometrical theories and structures of which Riemann's was just one example—and not always one particularly highly regarded. The reason why mathematicians largely stayed away might have been that Riemannian manifolds cannot be included in Klein's group-theoretic Erlangen program, whereas the shyness of physicists could have been caused by a conviction with Helmholtz that manifolds, wherein distance measurements can be performed using rigid rods, by necessity have constant curvature, making the full theory of Riemannian manifolds with varying curvature a mere formal game. The use of this geometry in a revolutionary and exciting physical theory obviously showed that it was more than a formal exercise, and even if we are in principle free to choose our axioms any way we like, in order for us to feel sufficient interest in working out the consequences, perhaps they have to correspond to some inner intuitive truth, a common source for all our conceptual thinking—mathematical as well as physical. Examine, for instance, Polanyi's statement:

> All these difficulties are but consequences of our refusal to see that mathematics cannot be defined without acknowledging its most obvious feature, namely, that it is interesting. (Polanyi 1958, p. 188)

Related to this is the fact that an application can enable physical intuition to interfere and make the theory more visual and hence excite the imagination. For instance, Poincaré stated

> experience leaves us a free choice, but it guides us by helping us to perceive the most convenient route. (Poincaré 1902, p. 3)[5]

He also stated

> physics does not only give us the opportunity to solve problems; it helps us to find the means and in two ways. It leads us to anticipate the solutions; it suggests arguments. (Poincaré 1898, p. 89)[6]

An illuminating example of this visualization is Klein's use of electromagnetic fields to illustrate the properties of Riemann surfaces (Klein 1882).

3. A Mathematical Interlude

On a Riemannian manifold, there are essentially three notions of curvature:

1. sectional curvature given analytically by the Riemann–Christoffel (RC) tensor,
2. Ricci curvature (given analytically by the contracted RC or Ricci tensor), and

3. scalar curvature (given analytically by the contracted Ricci tensor).

Of these objects, the one having the most geometrical interpretation is the first. It is directly related to the Gaussian curvature of certain two-dimensional submanifolds. The other two can be seen as certain averages of sectional curvatures. If Ric(X, Y) is the (index-free) description of the Ricci tensor, where X and Y are tangent vectors, the *Ricci curvature in the direction* X is given by Ric(X, X)$/g(X, X)$. The central mathematical object of this paper can now be defined: an *Einstein space* is a Riemannian manifold such that the Ricci curvature at a given point is independent of the direction.

Now, note the following:

1. One can easily show that if the Ricci curvature is constant at every point it is in fact constant on the whole manifold, i.e., independent of the chosen point. This was first demonstrated by Herglotz (Herglotz 1916), a paper devoted to a geometrical interpretation of the unmodified Einstein field equations. There is a corresponding result by Schur concerning manifolds with constant sectional curvature, and both these results can be seen as statements to the effect that "isotropy implies homogeneity."
2. The condition above implies Ric(X, Y) $= kg(X, Y)$ for arbitrary tangent vectors X and Y, so in component form the condition for a manifold to be an Einstein space is $R_{ij} = kg_{ij}$.
3. One can show that constant sectional curvature implies constant Ricci curvature, which in its turn implies constant scalar curvature.

The present intensive study of Einstein spaces seems to be largely motivated by a question put by René Thom in 1958, namely, given a differentiable manifold, are there any particularly attractive Riemannian metrics on it? The general consensus among geometers today seems to be that *Einstein metrics*, i.e., metrics with constant Ricci curvature, answers Thom's question. The reason is among other things that there are too few manifolds having metrics of constant sectional curvature (essentially only three in each dimension), and that, on the other hand, there are too many metrics of constant scalar curvature on any given manifold (an infinite dimensional amount on every compact manifold of dimension greater than two, for example). A nice introduction to Einstein manifolds can be found in the paper by Berger (Berger 1980), whereas the full story, as it is known today, is told in Besse's monumental work (Besse 1987).

4. Why Einstein Space?

The first time Einstein spaces appear in the literature under this name seems to be in the work by H.W. Brinkmann (Brinkmann 1924). The problem that Brinkmann attacked was to decide when a Riemannian manifold could be mapped conformally onto an Einstein space and is of a purely mathematical nature. The reason given for the name Einstein space is the following. Referring to Einstein's work from 1919, in which Einstein made his first attempt at a unified field theory by trying to account for the structure of matter by purely gravitational means, and introducing a modified set of field equations (Einstein 1919), Brinkmann noticed that in dimension n, the vacuum equations take the form

$$R_{ij} = (R/n)g_{ij},$$

so it is precisely the condition for a space to have constant Ricci curvature. This is in contrast to the original covariant field equations introduced by Einstein in 1916, where vacuum solutions are *Ricci flat*, i.e., they have vanishing Ricci curvature.

The influence from general relativity on Brinkmann's work was not restricted to giving him an interesting space with which to work, however. In fact, the problem that he treats concerning conformal maps has roots both in purely mathematical considerations and physics. In 1921, E. Kasner started to investigate the following problem: to what extent does the behavior of light rays determine the paths of massive point particles and vice versa (Kasner 1921a)? He showed first that if the behavior of light rays is the same as in Euclidean space, then we must have a Euclidean space to start with, or in other words, a four-dimensional space that satisfies Einstein's (original) vacuum equations, i.e., is Ricci flat and can be mapped conformally onto Euclidean space, is Euclidean. In a second paper, Kasner demonstrated the corresponding result for the "solar gravitational field," i.e., for the Schwarzschild solution (Kasner 1922). It might be remarked that Kasner suggested the term "Einstein type" for a vacuum solution of the original equations, a terminology that did not catch on. In a third paper (Kasner 1921b), he generalized his results to what later came to be called Einstein spaces and showed, for example, that a four-dimensional Einstein space, which can be mapped conformally onto Euclidean space, has a constant sectional curvature. The work of

Brinkmann followed naturally upon that of Kasner, who is also referred to by the former author. As for the purely mathematical background to Brinkmann's work, see the introduction to Schouten (Schouten 1921).

Kasner's work can be seen as an occupation with the foundations of general theory of relativity as opposed to the working out of its consequences, and it might be compared to the result obtained by Weyl at about the same time (Weyl 1921), which showed that the behavior of massive point particles and light rays, or more precisely, the projective and conformal structure on a Riemannian manifold determine its metric uniquely. This kind of work related to the axiomatic formulation of the theory seems to have lain largely dormant until Ehlers, Pirani, and Schild, following in Weyl's footsteps, took it up again in the early 1970s (Ehlers, Pirani, and Schild 1972).

The canonization of the term Einstein space in Brinkmann's sense seems to have resulted from its use in Eisenhart's important monograph (Eisenhart 1926). It might be remarked that Einstein's theory has continued to play a role in the development of the mathematical theory of Einstein spaces up to recent times, as shown by the important example of Page (see Besse 1987 for reference).

5. A Final Irony

Prior to the work of Einstein, which led Brinkmann to name and study the spaces, the former had already modified his original field equations by adding a term containing the cosmological constant (Einstein 1917). Whereas the original reason behind this term was to obtain a static cosmological solution, it later came to be associated with the attempt to implement Mach's principle in the sense that matter should completely determine the gravitational field. This would mean that there should be *no* vacuum solutions to the field equations with a nonvanishing cosmological term. For instance,

> Apparently there is no singularity-free space-time continuum with an everywhere vanishing energy tensor for matter.[7] (Einstein 1918)

But a vacuum solution in this case is exactly an Einstein space (with nonvanishing Ricci curvature), so clearly Einstein would have been happier without the spaces named after him, at least in the Lorentzian, four-dimensional case.

Notes

[1] Quoted by James J. Sylvester in an address to the British Association, 1869 (see Newman 1956, p. 1759).

[2] "Wie ist es möglich, dass die Mathematik, die doch ein von aller Erfahrung unabhängiges Produkt des menschlichen Denkens ist, auf die Gegenstände der Wirklichkeit so vortrefflich passt?"

[3] Part of the title of a paper by Wigner 1960.

[4] "Durch Albert Einsteins Theorie der Schwere ist die lang vernachlässigte Geometrie Riemanns... wieder in den Vordergrund gerückt worden." (From the review by Blaschke of a paper by T. Levi-Cevita in *Jahrbuch über die Fortschritte der Mathematik* 46 (1916–1918): 1125. This is the important paper in which Levi-Cevita introduced parallel transport as a metric-independent procedure. It might be noted that this work, in the introduction of which Levi-Cevita explicitly mentioned general relativity as motivating his work, was the first paper on Riemannian geometry by him since 1900.)

[5] "L'expérience nous laisse notre libre choix, mais elle le guide en nous aidant à discerner le chemin le plus commode."

[6] "La physique ne nous donne pas seulement l'occasion de résoudre des problèmes; elle nous aide à en trouver les moyens, et de deux manières. Elle nous fait pressentir la solution; elle nous suggère des raisonnements."

[7] "Ein singularitätenfreies Raum-Zeit-Kontinuum mit überall verschwindendem Energietensor der Materie scheint es... nicht zu geben." This quotation seems to be explicit enough to go counter to the claim by Pais concerning this problem (Pais 1982, p. 287).

References

Berger, Marcel (1980). "Rapport sur les variétés d'Einstein." *Astérisque* 80: 5–19.

Besse, Arthur L. (1987). *Einstein Spaces*. Berlin and New York: Springer-Verlag.

Brinkmann, H. W. (1924). "Riemannian Spaces Conformal to Einstein Spaces." *Mathematische Annalen* 91: 269–278.

Ehlers, Jürgen, Pirani, Felix and Schild, Adolf (1972). "The Geometry of Free Fall and Light Propagation." In *General Relativity*. Lochlainn O'Raifeartaigh, ed. London and New York: Oxford University Press (Clarendon), pp. 63–84.

Einstein, Albert (1917). "Kosmologische Betrachtungen zur allgemeinen Relativitätstheorie." *Königlich Preussische Akademie der Wissenschaften* (Berlin). *Sitzungsberichte*: 142–152.

—— (1918). "Prinzipielles zur allgemeinen Relativitätstheorie." *Annalen der Physik* 55: 241–244.

—— (1919). "Spielen Gravitationsfelder im Aufbau der materiellen Elementarteilchen eine wesentliche Rolle?" *Königlich Preussische Akademie der Wissenschaften* (Berlin). *Sitzungsberichte*: 349–356.

——— (1921). "Geometrie und Erfahrung." *Königlich Preussische Akademie der Wissenschaften* (Berlin). *Sitzungsberichte*: 123–130.

Eisenhart, Luther P. (1926): *Riemannian Geometry.* Princeton, New Jersey: Princeton University Press.

Herglotz, Gustav (1916). "Zur Einsteinschen Gravitationstheorie." *Königlich Sächsischen Gesellschaft der Wissenschaften zu Leipzig. Mathematisch-physische Klasse. Berichte über die Verhandlungen* 68: 199–203.

Kasner, Edward (1921a). "Einstein's Theory of Gravitation: Determination of the Field by Light Signals." *American Journal of Mathematics* 43: 20–28.

——— (1921b). "Geometrical Theorems on Einstein's Cosmological Equations." *American Journal of Mathematics* 43: 217–221.

——— (1922). "The Solar Gravitational Field Completely Determined by Its Light Rays." *Mathematische Annalen* 85: 227–236.

Klein, Felix (1882). *Über Riemann's Theorie der algebraischen Functionen und ihrer Integrale.* Leipzig: B. G. Teubner.

Newman, James R. (1956). *The World of Mathematics.* New York: Simon and Schuster.

Pais, Abraham (1982). *"Subtle is the Lord ...": The Science and the Life of Albert Einstein.* London and New York: Oxford University Press.

Poincaré, Henri (1898). "Sur les rapports de l'analyse pure et de la physique mathématique." In *Verhandlungen des ersten internationalen Mathematikerkongresses.* Ferdinand Rudio, ed. Leipzig: B. G. Teubner, pp. 81–90.

——— (1902). *La science et l'hypothèse.* Paris: Flammarion.

Polanyi, Michael (1958). *Personal Knowledge.* Chicago, Illinois: Univ. of Chicago Press.

Schouten, Jan A. (1921). "Über die konforme Abbildung n-Dimensionaler Mannigfaltigkeiten mit quadratischer Massbestimmung auf eine Mannigfaltigkeit mit euklidischer Massbestimmung." *Mathematische Zeitschrift* 11: 58–88.

Weyl, Hermann (1918). "Reine Infinitesimalgeometrie." *Mathematische Zeitschrift* 2: 384–411.

——— (1921). "Zur Infinitesimalgeometrie: Einordnung der projektiven und konformen Auffassung." *Königliche Gesellschaft der Wissenschaften zu Göttingen. Nachrichten:* 99–112.

Wigner, Eugene P. (1960). "The Unreasonable Effectiveness of Mathematics in the Natural Sciences." *Communications on Pure and Applied Mathematics* 13: 1–14.

H. A. Lorentz's Attempt to Give a Coordinate-Free Formulation of the General Theory of Relativity

Michel Janssen

1. Introduction[1]

In the course of 1916, the Dutch physicist H.A. Lorentz submitted a series of four papers on Einstein's general theory of relativity to the Amsterdam Academy of Sciences (Lorentz 1916a, 1916b).[2] To my knowledge, this four-part article comprises the first published attempt to arrive at a coordinate-free geometrical formulation of general relativity. Lorentz's basic strategy was to construct the Lagrangians for both the metric field and matter (material points and electromagnetic fields) in a geometrical way.

The main purpose of this paper is to elucidate Lorentz's rather cryptic discussion of the geometrical interpretation of the curvature scalar, i.e., the Lagrangian for the gravitational field. In order to understand Lorentz's reasoning, it is crucial to realize that his article was written prior to Levi-Civita's introduction of the notion of parallel displacement. Levi-Civita first introduced this concept into differential geometry in 1917 (Levi-Civita 1917). So, when Lorentz wrote his article, the modern interpretation of Riemannian curvature in terms of parallel displacement was not at his disposal. Lorentz drew on older ideas, essentially due to Gauss and Riemann, the founding fathers of differential geometry. This explains why Lorentz's reasoning looks so unfamiliar to a modern reader. At the same time, it may help to explain why Lorentz devoted only three short sections to the issue and why he did not give references to the literature on differential geometry he must have consulted. Lorentz presumably took it for granted that his readers were familiar with the basic concepts he used.

Although it looks as though Lorentz was able to give a geometrical interpretation of Riemannian curvature without invoking the notion of parallel displacement at any point, he ran into difficulties when he went on to construct the Lagrangian for the electromagnetic field in a geometrical way. There he used a naive coordinate dependent definition of distant parallelism. From a modern point of view, this, of course, completely destroys the allegedly coordinate-free character of his approach. At the time, however, Lorentz could hardly have realized that this was a fatal flaw. It was Lorentz's misfortune that he set himself the difficult task of casting general relativity in a coordinate-free geometrical form just one year before Levi-Civita provided what seems to be an indispensable tool for such an enterprise.

2. The Ether and General Covariance

Before I turn to the main topic of this paper, Lorentz's geometrical interpretation of the curvature scalar, I want to make some comments on the apparent discrepancy between Lorentz's attitude toward general relativity on the one hand and his attitude toward special relativity on the other. For a more extensive discussion of these matters, the reader is referred to Kox (1988). It is well known that to the end of his life, Lorentz persisted in his belief in the existence of an ether. In Lorentz's ether-based theory of electrons, which in its final form is empirically fully equivalent to special relativity (cf. Lorentz 1909, sections 189–194), all kinematical effects of special relativity are explained either as dynamical effects or as artifacts of measurements having to do with motion through the ether. One might even doubt whether Lorentz ever came to understand that in special relativity, phenomena such as length contraction and time dilation can simply be understood in terms of changes of perspective in a Minkowski space-time (Dorling 1968). Given this attitude toward special relativity, it is somewhat puzzling that Lorentz, of all people, should worry about a geometrical formulation of general relativity.

In order to understand Lorentz's concern, we will have to take a closer look at his motivation for trying to cast general relativity in a coordinate-free geometrical form [cf. sections 1–3 of Lorentz's 1916 paper (Lorentz 1937, pp. 246–248)]. Today, the purpose of eliminating all reference to coordinates from a theory is to bring out more clearly the geometrical content of the theory under consideration. Lorentz, however, was not—at least not primarily—interested in this intrinsic value of a coordinate-free formulation. For him, the geometrical approach was just a way to guarantee general covariance right from the start. In section 3 of his article, Lorentz wrote:

> In such a geometric treatment the introduction of coordinates will be of secondary importance;. . . In the discussion of the general principles coordinates play no part; and it is thus seen that the formulation of these principles can take place in the same way whatever be our choice of coordinates. So we are sure beforehand of the general covariancy of the equations that was postulated by Einstein.[3] (Lorentz 1937, p. 248)

Why did Lorentz feel so strongly about general covariance? From Lorentz's point of view, there seems to be nothing wrong with having privileged frames of reference. As a matter of fact, in their extensive correspondence on general relativity, Lorentz more than once pointed out to Einstein that in almost every physical situation there are privileged frames of reference (cf. Kox 1988, pp. 69–70). Lorentz, however, clearly realized that reference frames that are physically different can still be empirically equivalent. At least for Lorentz, it is this empirical equivalence of frames of reference that provides the rationale for requiring general covariance.

What seems to have convinced Lorentz of the need for general covariance is a remark made by Einstein in the correspondence Einstein, Lorentz, and Ehrenfest had on general relativity in late 1915 and early 1916 (cf. Kox 1988, pp. 72–73). In section 2 of his article, after he had introduced the concept of a "coincidence," simply the intersection of two world lines, Lorentz wrote:

> Now Einstein has made the striking remark[3)] that the only thing we can learn from our observations and with which our theories are essentially concerned, is the existence of these coincidences.[4] (Lorentz 1937, p. 247)

The footnote belonging to this passage reads: "in a correspondence I had with him" (Lorentz 1937, p. 247).[5]

Einstein's remark had to do with the resolution of the so-called "hole argument" (*Lochbetrachtung*); (cf. Norton 1984, pp. 286–291). This argument shows that generally-covariant field equations cannot uniquely determine the metric field. The argument seems to have temporarily reconciled Einstein to the restricted covariance of the Einstein–Grossmann theory. In the course of 1915, however, Einstein had come to realize that the hole argument does not rule out the possibility that the gravitational field equations are generally-covariant. As Einstein explained in a letter to Ehrenfest early in 1916, which Ehrenfest enclosed in one of his letters to Lorentz,[6] only the set of point coincidences, not the metric field, needs to be uniquely determined by the field equations. Since the set of point coincidences does not depend on the choice of coordinates, it should be possible to formulate a theory in a generally-covariant way. This argument quickly convinced Lorentz, who had recently thought of an objection against general covariance virtually identical to the hole argument. It therefore need not surprise us that the argument for general covariance in Lorentz's article, as the passage quoted above in fact illustrates, should center on the fundamental importance of point coincidences.

In summary, Lorentz's involvement with general relativity did not force him to give up the cherished notion of an ether or, more generally, the notion of privileged frames of reference. Einstein, however, convinced him that physical theories, since they only deal with point coincidences, should be generally-covariant. In order to guarantee general covariance right from the start, Lorentz set out to give a coordinate-free geometrical formulation of Einstein's theory.

3. A Brief Survey of Lorentz's 1916 Paper on the General Theory of Relativity

The paper "On Einstein's Theory of Gravitation" (Lorentz 1916b),[7] in which Lorentz set forth his geometrical formalism for the general theory of relativity, is rather lengthy. Only a small number of the 64 sections of the paper will be discussed in detail here. In order to place things in proper perspective, however, I will first give a brief survey of the article as a whole.

I have, in fact, already covered the most important points of the introductory sections 1–3. In section 4, Lorentz set out to develop his geometrical formalism for general relativity. First, he introduced the concept of *indicatrices,* which are to take over the role of the metric tensor in his formalism. The idea is very simple and straightforward. To every point P of space-time, two hyperspheres are assigned, $ds^2 = g_{\mu\nu}\, dx^\mu dx^\nu = \varepsilon^2$ and $ds^2 = g_{\mu\nu}\, dx^\mu dx^\nu = -\varepsilon^2$, where ε is an infinitesimally small positive constant. These hyperspheres are called the *indicatrix* and the *conjugate indicatrix*, respectively. As Lorentz explained in section 5 of his article, the indicatrices and conjugate indicatrices provide the "natural units" for evaluating the length of lines, the area of hypersurfaces, and the volume of three- or four-dimensional regions in space-time. Consider an infinitesimal time-like line element PQ through an arbitrary point P of space-time that intersects the local indicatrix at P'. The length of the line element is defined as the number of times the segment PP' fits into PQ, multiplied by ε, the length ascribed to the radius of the indicatrix. The length of an infinitesimal space-like line element through P is defined in a completely analogous manner with the help of the conjugate indicatrix, the only difference being that the radius of the conjugate indicatrix has an imaginary length $i\varepsilon$ ascribed to it. These definitions of length are easily seen to be equivalent to the usual notions of proper time and proper length.

A large portion of Lorentz's article (sections 6–31) is then devoted to the geometrical construction of various Lagrangians—or "principal functions" as Lorentz called them—with the help of lengths, areas, and volumes expressed in natural units. First, in section 6, the Lagrangian for a system of point masses is defined. As usual, the Lagrangian is obtained simply by adding the proper lengths of the world lines of all particles (insofar as these world lines lie within a region enclosed by some chosen boundary surface) weighted by their rest masses, with an overall minus sign. In sections 7 through 9, Lorentz turned to the Lagrangian for the metric field itself. It is these three sections with which I will be mostly concerned in the rest of this paper.

After some preliminary remarks in sections 10 through 12, Lorentz set out to construct the Lagrangian for the electromagnetic field. Whereas it took Lorentz only 4 short sections to define both the Lagrangian for a system of point masses and the Lagrangian for the metric field, the geometrical construction of the Lagrangian for the electromagnetic field takes up no less than 19 sections, the last two sections of the first part of the article (sections 13 and 14) and the entire second part (sections 15–31). The kind of geometry used by Lorentz looks rather laborious and artificial. It seems though that this was just how this type of mathematics was being handled at the time. The mathematical apparatus used by Lorentz is not all that different from the apparatus used, e.g., by Weyl (Weyl 1918) or Pauli (Pauli 1958). An exposition of Lorentz's approach that is still reasonably accessible can be found in Fokker's textbook on relativity (Fokker 1929).

Having defined the various Lagrangians, Lorentz, in the third part of his article (sections 32–46), applied a variational principle to derive the field equations for a gravitational field generated either by a system of point masses or by some electromagnetic field configuration. In the remainder of the article (sections 47–64), Lorentz tried to find a quantity to represent the energy-momentum of the gravitational field. The fact that the energy-momentum of the gravitational field cannot be represented by a generally-covariant tensor, as we know today, must have been very confusing at the time. Both Lorentz and Einstein struggled with this problem, but their considerations remained inconclusive (cf., e.g., Norton 1984, p. 285).

It is somewhat disappointing to find that in these later parts of Lorentz's article the discussion is entirely in terms of coordinate representations of the relevant quantities. This is especially true for the way in which it handles variational calculus. In the first part of his article, Lorentz wrote that he would derive the gravitational field equations for the two cases mentioned above by "the consideration of infinitely small variations of the indicatrices, by which the numerical values of all quantities that are measured by these surfaces will be changed" (Lorentz 1937, p. 260).[8] This suggests that Lorentz had in mind some geometrical approach to variational calculus. However, in the third part of his article, when Lorentz got down to the actual calculations, it turns out that he took the usual approach of looking at variations of the components of the metric tensor in some arbitrary system of coordinates. As a matter of fact, in the third and fourth part of Lorentz's paper, even the word indicatrix no longer occurs. We should keep in mind here that several months elapsed between writing the first and the last part of the paper. What seems to have happened is that somewhere along the line Lorentz decided to relax his self-imposed constraint of introducing generally-covariant concepts in a coordinate-free geometrical way.[9]

4. Lorentz's Geometrical Interpretation of the Curvature Scalar

After this brief overview of the contents of Lorentz's article, I now focus on the geometrical construction of the Lagrangian for the gravitational field in sections 7 through 9 (Lorentz 1937, pp. 251–255). At the beginning of section 7, Lorentz wrote that he knew from his correspondence with Einstein and from a paper by Hilbert that one should take the curvature scalar as the Lagrangian for the gravitational field. While Einstein just gave the analytical expression for this quantity in terms of the components of the metric tensor and their derivatives, Hilbert also pointed out that it could actually be regarded as a measure of the curvature of the manifold to which this metric belonged. Lorentz, following up on Hilbert's suggestion, wanted to give a detailed geometrical interpretation of this quantity.

It is interesting to notice that, with the obvious exception of the metric tensor, Einstein was not concerned with the geometrical meaning of the quantities he borrowed from differential geometry in his work leading up to the final version of the general theory of relativity. He only looked at their transformation properties and in some cases checked whether they fit with a natural generalization of Newton's gravitational theory. This attitude is nicely illustrated by the way in which Einstein in 1913 briefly considered, and then rejected the Ricci tensor as a serious contender for the key position in the gravitational field equations.

Einstein and Grossmann's *Entwurf* paper cites a result obtained by Christoffel, who completed the work begun by Riemann, showing that all covariant quantities containing only components of the metric tensor and their derivatives can be constructed out of the curvature tensor. Since a generalization of the Poisson equation of the Newtonian theory obviously had to involve derivatives of the metric tensor, it was only natural for Einstein and Grossmann to turn to the curvature tensor (or rather to its second rank contraction now known as the Ricci tensor) when they began constructing field equations for the metric field (cf. Einstein and Grossmann 1913, pp. 35–36). Geometrical considerations did not play any role here. Likewise, the story of the rejection of the Ricci tensor and the story of its ultimate return in November 1915 can be told without any reference to the geometrical meaning of curvature (cf. Norton 1984, pp. 255–257).

I would like to venture that this attitude at least partially may have been due to the fact that Einstein and Grossmann seem to have taken their approach to differential geometry from papers by Christoffel and by Ricci and Levi-Civita (cf. Einstein and Grossmann 1913, p. 23). These papers laid the foundations for modern tensor calculus. Traditional geometrical methods largely gave way to more analytical ones. In order to show that some quantity has geometrical substance and is not just an artifact of some arbitrary choice of coordinates, it sufficed to invoke its covariance properties. One no longer needed to show that the quantity under consideration represents a visualizable geometrical entity.

As we will see, Lorentz is most likely to have learned differential geometry from a book by Bianchi. The original Italian edition of this book is dated 1886. A slightly updated and extended German translation appeared in 1899 (Bianchi 1899). Its nature is much more geometrical (in the traditional sense alluded to above) than the work by Christoffel, Ricci, and Levi-Civita. This may help to explain why Lorentz's work is much more geometrically oriented than Einstein's.

The geometrical ideas used by Lorentz in his interpretation of the curvature scalar were about to be eclipsed—at least as far as their application to physics is concerned—by the notion of parallel displacement introduced by Levi-Civita in 1917. This makes it difficult for a modern reader to understand Lorentz's reasoning. On the other hand, since Einstein only dealt with the analytical expressions for the curvature scalar, the Ricci tensor, and the curvature tensor, his writings remain as accessible as they were before the introduction of parallel displacement.

Levi-Civita's introduction of parallel displacement apparently had a great and immediate impact. For example, already in the first edition of Hermann Weyl's influential *Raum, Zeit, Materie* of 1918, Weyl introduced Riemannian curvature in the context of parallel displacement (Weyl 1918, pp. 110–111). In the preface, Weyl acknowledged the influence of Levi-Civita's 1917 paper on his presentation of tensor analysis. In 1919, in the preface to the third edition of his book, Weyl stated that Levi-Civita's findings had led him to reexamine the foundations of Riemannian geometry.[10]

Not surprisingly then, Lorentz's pre-1917 reasoning looks rather obsolete to a modern reader. The situation is made worse by the fact that Lorentz neither gave any detailed calculations nor referred to the mathematical literature he used for his ideas. Presumably, he expected his audience to be thoroughly familiar with the concepts he used. He may, in fact, have been well justified in this belief since most of these concepts derive directly from the work of Gauss and Riemann, the founding fathers of differential geometry.

Fortunately, in Pauli's famous treatise on relativity, in which Lorentz is credited to have been the first to study the geometrical meaning of curvature in the context of general relativity, there is a footnote informing us that shortly after, and seemingly independently of Lorentz's work, the German physicist and mathematician Gustav Herglotz worked out a geometrical interpretation of Riemannian curvature that proceeds along very much the same lines as Lorentz's interpretation (Pauli 1958, p. 47). To begin with, this supports my suggestion that the concepts Lorentz used were well known among his contemporaries. More important, Herglotz, unlike Lorentz, gave some detailed calculations. Moreover, his article contains a key reference to the literature on differential geometry (Herglotz 1916, p. 199). For one of the crucial relations needed for both his and Lorentz's interpretation of the curvature scalar, Herglotz referred to the textbook by Bianchi mentioned earlier (Bianchi 1899). From Lorentz's notebooks, deposited at the Algemeen Rijksarchief at The Hague, we know that Lorentz studied Bianchi's book. Like Herglotz, Lorentz presumably took this central relation from Bianchi's book.

Both Lorentz and Herglotz wanted to reduce the Riemannian curvature scalar of the four-dimensional space-time manifold to a sum of Gaussian curvatures of six two-dimensional submanifolds. From that point, however, their paths diverged. Herglotz went on to show how the various components of the Einstein tensor can be written as the sum of such Gaussian curvatures. In Lorentz's variational approach, the field equations inherited their geometrical meaning from the Lagrangian and Lorentz did not bother to spell out the details of the legacy. Instead, he wanted to reduce the curvature scalar to still more primitive notions by trying to generalize a theorem by Gauss relating Gaussian curvature to the so-called angular excess in geodetic triangles. Lorentz apparently failed to deliver on his promise to do the necessary calculations in a future article (Lorentz 1937, p. 254). I have not been able to fully reconstruct this second step in Lorentz's interpretation of the curvature scalar. However, with the help of Herglotz (1916), Bianchi (1899), and two more recent texts, Haantjes (1954) and Lanczos (1970), I was able to fill in the details of the first step.

In order to reduce the curvature scalar R at some arbitrary point P of the space-time manifold M to a sum of Gaussian curvatures, we start by choosing a time-like vector ξ_0 and three space-like vectors ξ_1, ξ_2 and ξ_3, together forming an orthonormal basis of $T_P(M)$, the tangent space of M at P. The orthonormality is with respect to the standard Minkowski inner product. We take an arbitrary pair $\{\xi_a, \xi_b\}$ from this set and consider the two-dimensional submanifold made up of geodesics in the directions spanned by ξ_a and ξ_b. We will denote this hypersurface by (ab). There are six different hypersurfaces of this kind: (01), (02), (03), (12), (13), and (23). We compute the Gaussian curvature of these geodetic hypersurfaces (ab), which we will denote by $K(ab)|_P$. Now, the curvature scalar of M at P turns out to be equal to the double sum of these six Gaussian curvatures $K(ab)|_P$. This is the basic idea of Lorentz and Herglotz's geometrical interpretation of the curvature scalar, an idea that goes back to Riemann (cf., e.g., Bianchi 1899, p. 571; Pauli 1958, p. 46; Torretti 1978, p. 101).

We will now prove this relation between Riemannian and Gaussian curvature. As is well known, the curvature scalar can be obtained by contracting the first and the third as well as the second and the fourth indices of the curvature tensor:

$$R = R_{\rho\sigma\mu\nu} g^{\rho\mu} g^{\sigma\nu}, \tag{1}$$

where the $R_{\rho\sigma\mu\nu}$ are the components of the fully covariant curvature tensor and the $g^{\mu\nu}$ are the contravariant components of the metric tensor on M, in some arbitrary coordinate system. Now, suppose the vectors

$\xi_a \in T_P(M)$ have components ξ_a^μ in this particular coordinate system. The fact that they form an orthonormal basis can then be expressed as

$$g_{\mu\nu}|_P \xi_a^\mu \xi_b^\nu = \eta_{ab}, \tag{2}$$

where η_{ab} is the standard diagonal Minkowski metric $(1, -1, -1, -1)$. This is equivalent to

$$g^{\mu\nu}|_P = \xi_a^\mu \xi_b^\nu \eta^{ab}. \tag{3}$$

Inserting Eq. (3) into Eq. (1), we have

$$R|_P = R_{\rho\sigma\mu\nu}|_P \left(\xi_a^\rho \xi_{a'}^\mu \eta^{aa'}\right) \left(\xi_b^\sigma \, \xi_{b'}^\nu \, \eta^{bb'}\right). \tag{4}$$

We now introduce the quantities δ_a defined as

$$\delta_a = \begin{cases} 1 & \text{for } a = 0 \\ -1 & \text{for } a = 1, 2, 3 \end{cases} \tag{5}$$

to rewrite Eq. (4) in the following way:

$$R|_P = \sum_{a,b} R_{\rho\sigma\mu\nu}|_P \frac{\xi_a^\rho \, \xi_a^\mu}{\delta_a} \frac{\xi_b^\sigma \, \xi_b^\nu}{\delta_b}. \tag{6}$$

From the antisymmetry of $R_{\rho\sigma\mu\nu}$ in both its first and its second index pair, it follows that all terms in this sum are symmetrical under permutation of a and b and that terms with $a = b$ vanish. That leaves us with only six different terms. What we will have to prove now is that those six terms correspond to the Gaussian curvatures $K(ab)|_P$ of the corresponding (ab) geodetic hypersurfaces.[11]

It is convenient to talk about the (12) hypersurface whatever our actual choice of (ab) may be. We denote the coordinates on this (12) hypersurface by u^1 and u^2. The components of quantities defined on the (12) hypersurface in terms of u^1 and u^2 are written with Latin indices, assuming only the values 1 and 2, to distinguish them from the components of the corresponding quantities defined on the four-dimensional space-time manifold, which are written with Greek indices. The actual proof is best split into three parts.

1. When we embed a surface in three-dimensional Euclidean space and introduce the Gaussian curvature κ of the surface at some arbitrary point P as the product of the so-called principal curvatures of the surface at that point, it can be proved that κ is equal to the ratio of the determinants of the so-called second and first fundamental forms of the surface

at that point (cf. Haantjes 1954, pp. 67–69, pp. 77–78; Lanczos 1970, pp. 176–179):

$$\kappa = \frac{\det(h_{ij})}{\det(g_{ij})}. \tag{7}$$

Here, h_{ij} and g_{ij} are the coefficients of the second fundamental form $d\sigma^2 = h_{ij}\, du^i du^j$ (more frequently written as $d\sigma^2 = D\, du^2 + 2D'\, du\, dv + D'' dv^2$) and of the first fundamental form $ds^2 = g_{ij}\, du^i du^j$ (the metric), respectively.[12]

2. It can be proved that $\det(h_{ij})$ is the (only independent component of the) curvature tensor of the surface (cf. Lanczos 1970, pp. 180–181):

$$\det(h_{ij}) = R_{1212}. \tag{8}$$

The argument is as follows. We add a third coordinate to our arbitrary coordinates on the surface to obtain a coordinatization of the embedding Euclidean space in a thin layer on both sides of the surface. Using these coordinates, we derive an expression for the (1212) component of the curvature tensor in this part of the three-dimensional embedding space. We find that it is given by $R_{1212} - \det(h_{ij})$, where R_{1212} now refers to the curvature tensor *of the surface.* Eq. (8) then follows from the fact that since the embedding space is flat its curvature tensor will vanish.

Combining Eqs. (7) and (8), we see that the Gaussian curvature is an intrinsic quantity, depending only on the surface's metric. This is remarkable since it was obtained by multiplying together the two principal curvatures, which are extrinsic quantities (in defining them, we necessarily make reference to the embedding space). This is, in fact, the purport of Gauss's famous *Theorema Egregium* (cf., e.g., Torretti 1978, p. 80).

3. The third link in our proof is the relation between the curvature tensor of the four-dimensional space-time manifold[13] and the curvature tensor of the two-dimensional geodetic hypersurfaces we have been considering[14] (cf. Bianchi 1899, 571–574):

$$R_{1212}|_P \; = \; R_{\rho\sigma\mu\nu}|_P\, \xi_1^\rho\, \xi_1^\mu\, \xi_2^\sigma\, \xi_2^\nu. \tag{9}$$

Whereas Eqs. (7) and (8) hold for arbitrary surfaces with arbitrary coordinates, Eq. (9) only holds for geodetic hypersurfaces and only for a special choice of coordinates on these hypersurfaces. We must, in fact, use subsets of the so-called Riemann normal coordinates (cf., e.g., Eisenhart 1925, pp. 53–55). We choose the origin of this coordinate system to be the point P at which we had the orthonormal set of vectors $\{\xi_0, \xi_1, \xi_2, \xi_3\}$.

Consider an arbitrary point Q on the (12) hypersurface on the geodesic in the direction $\alpha\xi_1 + \beta\xi_2$. Let t be the arc length of the segment PQ of this geodesic. Now, the coordinates ascribed to Q as part of the geodetic hypersurface will be ($u^1 = \alpha t$, $u^2 = \beta t$), which is a subset of the Riemann normal coordinates $(0, \alpha t, \beta t, 0)$ that would be ascribed to Q as part of the four-dimensional manifold. In Eq. (9), the coordinates on the four-dimensional manifold are arbitrary. We will only assume that P is the origin of this coordinate system. The relation between the arbitrary x^μ coordinates of Q and its u^i coordinates is obtained by parameterizing the geodesic through P and Q in x^μ-coordinates by the arc length s and then Taylor expanding it around P:

$$x^\mu(s) = x^\mu(0) + \frac{dx^\mu}{ds}\Big|_{s=0}\, s + \tfrac{1}{2}\,\frac{d^2x^\mu}{ds^2}\Big|_{s=0}\, s^2 + \ldots . \tag{10}$$

Using $x^\mu(0) = 0$, substituting $\alpha\xi_1^\mu + \beta\xi_2^\mu$ for $(dx^\mu/ds)|_{s=0}$ and using the geodesic equation

$$\frac{d^2x^\mu}{ds^2} + \Gamma^\mu_{\rho\sigma}\frac{dx^\rho}{ds}\frac{dx^\sigma}{ds} = 0 \tag{11}$$

to evaluate $(d^2x^\mu/ds^2)|_{s=0}$, we find that, up to second order in u^i, the relation between x^μ and u^i is given by

$$x^\mu = \xi_i^\mu\, u^i - \tfrac{1}{2}\,\Gamma^\mu_{\rho\sigma}|_P\, \xi_i^\rho \xi_j^\sigma\, u^i\, u^j + \ldots . \tag{12}$$

This is the basic relation that we need in order to derive Eq. (9). Notice that we can use both the u^i coordinates and the x^μ coordinates to evaluate the interval ds between two infinitesimally close points on the geodetic hypersurface:

$$ds^2 = g_{ij}\, du^i du^j = g_{\mu\nu}\, dx^\mu dx^\nu . \tag{13}$$

From Eq. (13), it follows that

$$g_{ij} = \frac{\partial x^\mu}{\partial u^i}\frac{\partial x^\nu}{\partial u^j}\, g_{\mu\nu}. \tag{14}$$

Using Eqs. (12) and (14), we can express any quantity containing only g_{ij} and its derivatives, taken at P, in terms of $g_{\mu\nu}$ and its derivatives, taken at P. When we do this for the curvature tensor of the (12) hypersurface, a tedious but completely straightforward derivation gets us to Eq. (9).[15]

We now want to combine the results of items Eqs. (7), (8), and (9) to show that the curvature scalar at some arbitrary point P of space-time can indeed be obtained as a sum of Gaussian curvatures at P. In order to do this, we need to close one more little gap. We need to show that

$$\det(g_{ij})|_P = \delta_1 \delta_2. \tag{15}$$

Starting from Eq. (14) and using Eqs. (12), (2), and (5), we can easily verify that this relation, in fact, obtains:

$$g_{ij}|_P \overset{(14)}{=} \left(\frac{\partial x^\mu}{\partial u^i} \frac{\partial x^\nu}{\partial u^j} g_{\mu\nu} \right) |_P \overset{(12)}{=} \xi_i^\mu \xi_j^\nu \, g_{\mu\nu}|_P \overset{(2)}{=} \eta_{ij}. \tag{16}$$

Now, $\det(\eta_{ij})$ is equal to +1 if ξ_1 and ξ_2 are both space-like, and it is equal to -1 if one of them happens to be the time-like vector ξ_0 [remember that it was for the sake of convenience that we labeled these vectors 1 and 2; they can be any pair (ξ_a, ξ_b) out of the set of four]. Looking back at the definition of δ_a in Eq. (5), we see that instead of $\det(\eta_{ij})$ we might just as well use the product $\delta_1 \delta_2$. This proves Eq. (15).

We now have everything we need in order to show that the Gaussian curvature $K(12)|_P$ of the hypersurface made up of geodesics spanned by ξ_1 and ξ_2 is equal to the (12) term in the sum on the right-hand side of Eq. (6) for the curvature scalar. First, we apply Eqs. (7) and (8), valid for arbitrary surfaces, to our particular case of geodetic hypersurfaces coordinatized by subsets of Riemann normal coordinates. We then apply Eqs. (15) and (19), which were derived explicitly for this particular case. In this way, we end up with the following expression for $K(12)|_P$:

$$K(12)|_P \overset{(7)}{=} \left(\frac{\det(h_{ij})}{\det(g_{ij})} \right) |_P \overset{(8)(15)}{=} \frac{R_{1212}|_P}{\delta_1 \, \delta_2} \overset{(9)}{=} R_{\rho\sigma\mu\nu}|_P \, \frac{\xi_1^\rho \xi_1^\mu}{\delta_1} \frac{\xi_2^\sigma \xi_2^\nu}{\delta_2}. \tag{17}$$

Now, what is true for $K(12)|_P$ is true for any other $K(ab)|_P$ as well; so, we have

$$K(ab)|_P = R_{\rho\sigma\mu\nu}|_P \, \frac{\xi_a^\rho \xi_a^\mu}{\delta_a} \frac{\xi_b^\sigma \xi_b^\nu}{\delta_b}. \tag{18}$$

Comparing Eq. (18) with Eq. (6), we conclude that

$$R|_P = \sum_{a \neq b} K(ab)|_P. \tag{19}$$

Since the point P was chosen completely arbitrarily, this relation holds at every point of the space-time manifold. Hence, we have vindicated the first part of Lorentz and Herglotz's geometrical interpretation of the curvature scalar: the curvature scalar can be reduced to a sum of Gaussian curvatures.

Now, the reduction of Riemannian to Gaussian curvature only makes sense if the geometrical meaning of the latter is somehow more readily grasped than the geometrical meaning of the former. The way in which Gaussian curvature was introduced earlier, namely, as the product of the two so-called principal curvatures, is not very satisfactory in this respect. What Lorentz had in mind, however, as I have mentioned, was to generalize a theorem by Gauss that would allow him to reduce the still highly abstract concept of Gaussian curvature to angles and areas of geodetic triangles.

Consider an arbitrary point P on a surface S embedded in three-dimensional Euclidean space. Draw three geodesics on S, forming an infinitesimally small triangle enclosing P. Determine the area Δ of this triangle as well as the sum of its angles. For nonflat surfaces, this sum will deviate from its Euclidean value π. The difference is called the angular excess e. Now, Gauss proved that the Gaussian curvature κ of S at P is equal to the ratio of e and Δ:

$$\kappa = \frac{e}{\Delta}. \tag{20}$$

For a simple modern proof of this theorem, see Lanczos (1970, pp. 203–210, pp. 186–188).

When we want to generalize this theorem to hypersurfaces in space-time, we first need to define what is meant by the angle between space-like and time-like lines, or rather between their tangent vectors. Lorentz did this in section 7 of his paper. Insisting on the familiar rule for inner products,

$$g_{\mu\nu}\,\xi_1^\mu\,\xi_2^\nu = |\xi_1|\;|\xi_2|\;\cos\alpha_{12}, \tag{21}$$

Lorentz obtained an implicit definition of the angle α_{12} between the vectors ξ_1 and ξ_2 (Lorentz 1937, p. 251). It can easily be shown that, with this definition, the angles of a geodetic triangle in flat Minkowski space-time add up to π, just as in ordinary Euclidean space (cf. Fokker 1929, p. 89). In a curved space-time, Lorentz stated in section 9 of his paper, there will be an angular excess e, satisfying the relation $e = \kappa\Delta$ derived by Gauss. For geodetic triangles formed by space-like geodesics alone, this follows from Gauss's theorem; for geodetic triangles involving both space-like and time-like geodesics, Lorentz wrote: "the proof can be given by direct calculation. The considerations necessary for this ...

will be communicated in a later paper" (Lorentz 1937, p. 254).[16] As I mentioned, Lorentz never wrote this paper.

In Fokker's textbook on relativity, however, this very problem is set as an exercise for the reader (Fokker 1929, p. 224). The derivation suggested by Fokker is roughly as follows. Once again, consider the construction we used in our derivation of Eq. (19), i.e., the geodetic hypersurface coordinatized by the Riemann normal coordinates u^1 and u^2. First, we have to find the equations $u^i(s)$ for three geodesics on the geodetic hypersurface intersecting near P to form an infinitesimally small geodetic triangle enclosing P. We then evaluate the derivatives $[du^i(s)/ds]$ at the vertices of the triangle to obtain the components of the tangent vectors to the three geodesics at those points. Taking the inner products of the two tangent vectors we have at each vertex and using Eq. (21), we can find the angles of the triangle and, finally, its angular excess.

For the inner products, we need the components g_{ij} of the surface's metric tensor in u^i coordinates. In order to find an expression for g_{ij}, we start with Eq. (14). It is convenient to choose the x^μ coordinates to be locally geodetic at P, in which case we have $\Gamma^\mu_{\rho\sigma}|_P = 0$. This choice greatly simplifies the relation between the x^μ coordinates and the u^i coordinates. Since we now need this relation up to third order in u^i, this is an important gain. Proceeding as we did to obtain Eq. (12), we find

$$x^\mu = \xi^\mu_i\, u^i - \tfrac{1}{6}\Gamma^\mu_{\rho\sigma,\tau}|_P\, \xi^\rho_k\, \xi^\sigma_l\, \xi^\tau_m\, u^k u^l u^m + \ldots . \tag{22}$$

Inserting Eq. (22) into Eq. (14), we find

$$g_{ij} = \eta_{ij} + \tfrac{1}{3}R_{\rho\sigma\mu\nu}|_P\, \xi^\rho_i\, \xi^\nu_j\, \xi^\sigma_k\, \xi^\mu_l\, u^k u^l = \eta_{ij} + \tfrac{1}{3}R_{iklj}|_P\, u^k u^l, \tag{23}$$

where in the last step we used Eq. (9). Eq. (23) clearly shows how curvature causes the sum of the angles in a geodetic triangle to deviate from its Minkowskian value π.

Unfortunately, I have not been able to solve Fokker's exercise in its entirety. For our purposes, however, the most important thing is that the derivation outlined above clearly does not involve the concept of parallel displacement at any point. Even though we just have to take it on faith that the derivation really leads to the correct result, i.e., $e = \kappa\Delta$, we can safely conclude that Lorentz's geometrical interpretation of the curvature scalar can be made to work without using the notion of parallel displacement.

Modern proofs of Gauss's theorem, such as the proof in Lanczos (1970) mentioned previously, do make use of the notion of parallel displacement. To find the angular excess, we transport a vector parallel to itself along the sides of the triangle. When we get back to the point from which we started, we look at the angle the parallel transported vector makes with the original vector. This angle is the angular excess of the triangle. So, Lorentz's interpretation of Riemannian curvature in terms of the angular excess in geodetic triangles can eventually be subsumed under the modern interpretation in terms of parallel displacement. The modern interpretation has some obvious advantages over Lorentz's. First, it works regardless of the exact nature of the circuits used. There is absolutely no need to restrict ourselves to geodetic triangles. Second, using parallel displacement, we have no trouble dealing with the full four-dimensional space-time manifold when we want to specify its curvature at any point, whereas in Lorentz's approach, we always have to slice the four-dimensional manifold into six two-dimensional submanifolds first. This clearly shows that one is seriously handicapped when trying to give geometrical meaning to Riemannian curvature without knowing about parallel displacement.

Even though Lorentz could do without parallel displacement in his geometrical interpretation of the curvature scalar, the lack of a coordinate-independent definition of distant parallelism completely ruined his coordinate-free approach to general relativity later in his paper. In the first passage that I quoted from Lorentz's paper, I left out a seemingly innocent clause: "the introduction of coordinates will be of secondary importance; with a single exception (section 13), it only serves for short calculations which we have to intercalate " (Lorentz 1937, p. 248).[17] The exception is not nearly as innocent as this passage suggests. In section 13 of his paper, at the beginning of his considerations concerning the Lagrangian for the electromagnetic field, Lorentz observed the following:

> It must be expressly remarked that if an equation... in which we are concerned with the composition of vectors at *different* [Lorentz's italics (MJ)] points of the field-figure, shall have a definite meaning we must know which components are to be considered as having the same direction, so that they can be added. This has been determined by the introduction of coordinates.[18] (Lorentz 1937, p. 259)

This passage shows that Lorentz was clearly aware of the need for a definition of distant parallelism. At the same time, it shows that he did not fully realize the importance of this concept. After all, he obviously had no qualms about introducing it into his avowedly coordinate-free formalism in a coordinate-dependent way. Lorentz wrote these lines the year before Levi-Civita introduced the concept of parallel displacement into differential geometry. Without this concept, it seems, Lorentz's project to cast general relativity in a coordinate-free geometrical form was doomed from the outset.

NOTES

[1] This paper is a condensed version of part of the thesis I submitted to obtain a master's degree in theoretical physics from the University of Amsterdam. I would like to thank A.J. Kox, Gerard Bäuerle, Jon Dorling, Xaveer Leijtens, Hadass Eviatar, Giel Halberstadt, and John Culler for their helpful comments on the manuscript of my master's thesis and on earlier drafts of the present paper.

[2] These papers appeared in the 1915–1916 and 1916–1917 volumes of the Dutch version of the Academy's proceedings (Lorentz 1916a) and in the 1916–1917 and 1917–1918 volumes of the English version (Lorentz 1916b). The page references to the Dutch version of the paper are to the four communications as they appeared in the proceedings. The communication being referred to is denoted by Roman numerals. For the English version, I refer to the reprint of the paper in Vol. 5 of Lorentz's collected papers (Lorentz 1937). Presumably, this version will be more widely available than the original.

[3] "Bij zulk een meetkundige beschouwingswijze zal de invoering van coordinaten van ondergeschikt belang zijn;. . . In de uiteenzetting der algemeene grondbeginselen spelen zij geen rol; daarin ligt onmiddellijk de waarborg ervoor dat de formuleering dezer beginselen op een wijze kan geschieden, die bij elke willekeurige keus der coördinaten dezelfde is, d.w.z. de waarborg voor de door Einstein gewenschte en tenslotte ook bereikte algemeene covariantie der vergelijkingen." (Lorentz 1916a, I, p. 1392). A literal translation of "gewenscht en tenslotte ook bereikt" would be "desired and in the end achieved," which is not really matched by "postulated."

[4] "Einstein heeft nu[3)] de treffende opmerking gemaakt, dat het eenige waarover onze waarnemingen ons uitsluitsel geven en waarmede in den grond der zaak onze theorieën zich hebben bezig te houden, het bestaan dezer coïncidenties is" (Lorentz 1916a, I, p. 1390).

[5] "In een briefwisseling" (Lorentz 1916a, I, p. 1390).

[6] We may safely assume that the remark Lorentz is referring to in the passage quoted above comes from this letter of Einstein to Ehrenfest. It is tempting to cite the absence of the explicit "I had with him" in the Dutch version of the footnote belonging to this passage as further evidence for this assumption, although I think this is really just a coincidence.

[7] "Over Einstein's theorie der zwaartekracht" (Lorentz 1916a). Cf. note 2.

[8] ". . . door de beschouwing van oneindig kleine variaties van de indicatrices, waardoor de getalwaarden van alle met behulp van die oppervlakken gemeten grootheden zullen veranderen" (Lorentz 1916a, I, p. 1402).

[9] As Lorentz stated in a footnote at the beginning of the third part of his article, he rewrote the original version of the third part, a process that took him several months. Going by just the submission dates, one would expect a discontinuity, if any, between the third and the fourth part. The submission dates

on the first three communications are February 26, March 25, and April 28, respectively. The submission date on the fourth communication is October 28 (Lorentz 1916a). To the submission date on the third communication, however, the following footnote was added: "Published September 1916, a revision having been found desirable" (Lorentz 1916b, III, p. 2). [A similar footnote occurs in the Dutch version of the paper: "Na dien tijd kwam mij eene omwerking wenschelijk voor, waarmede ik eerst thans, September 1916, gereed ben gekomen" (Lorentz 1916a, III, p. 468).] This footnote was omitted in the reprint of the article in Lorentz's collected papers (Lorentz 1937). Given this footnote, it need not surprise us that we find a discontinuity between the second and third communications.

10 Weyl made some important contributions to the development of the concept of parallel displacement (cf. Pauli 1958, p. 37).

11 Herglotz took it on the authority of Bianchi that the Gaussian curvature of the (ab) geodetic hypersurfaces is given by $(R_{\rho\sigma\mu\nu}\,\xi^{\rho}_{a}\,\xi^{\sigma}_{b}\,\xi^{\mu}_{a}\,\xi^{\nu}_{b})/(\delta_a\,\delta_b)$, and he then showed that when we sum over a and b we obtain the curvature scalar. So, Herglotz basically went through Eqs. (1) through (6) in reversed order (Herglotz 1916, pp. 199–201).

12 Since the derivation of Eq. (7), like the derivation of Eq. (8), involves the embedding of the surface in three-dimensional Euclidean space, the question arises whether we can apply the result to hypersurfaces in space-time that are made up of both time-like and space-like geodesics and that therefore do not have a definite metric. Lorentz had no trouble doing this since he introduced imaginary space-like coordinates, in which case definiteness of the metric is at least formally retained.

13 Bianchi derived this relation for an n-dimensional Riemannian manifold rather than for a four-dimensional *semi*-Riemannian manifold, but nothing in his derivation depends on definiteness of the metric.

14 As Bianchi explained in the preface to Bianchi (1899), the original 1886 Italian edition of his book only dealt with the theory of surfaces. To the German edition, he added two chapters on n-dimensional manifolds. It is in one of these later chapters that Bianchi derived the relation that we will be looking at here between the curvature tensor of an n-dimensional manifold and the curvature tensor of two-dimensional submanifolds made up of geodesics. The geometrical meaning of the curvature tensor R_{1212} for a two-dimensional manifold is explained earlier in Bianchi's book (Bianchi 1899, p. 52, p. 105). In these earlier parts, it is shown that R_{1212} is essentially equal to the Gaussian curvature [cf. our equations (7) and (8)].

15 Eqs. (9) through (14) would still be valid if instead of taking the subset (u^1, u^2) of the Riemann normal coordinates we were to take the full set (u^0, u^1, u^2, u^3), i.e., if the Latin indices were to run from 0 to 3 just like the Greek ones. In that case, Eq. (9) would follow from the fact that, according to Eq. (12), $(\partial x^{\mu}/\partial u^{i})|_P = \xi^{\mu}_{i}$. At first sight, this seems to turn Bianchi's derivation of Eq. (9) into an unnecessarily cumbersome detour. Notice, however, that with the Latin indices

running from 0 to 3, R_{1212} in Eq.(9) refers to the curvature tensor of the four-dimensional manifold rather than to the curvature tensor of the two-dimensional submanifold. What Bianchi's derivation in conjunction with the more trivial result I just mentioned, in fact, shows is that when we use Riemann normal coordinates the curvature tensor of the two-dimensional (ab) geodetic hypersurface is equal to the $(abab)$ component of the curvature tensor of the four-dimensional space-time manifold.

[16] "kan men het bewijs door rechtstreeksche berekening geven. Deze berekening... hoop ik elders mede te deelen" (Lorentz 1916a, I, p. 1397).

[17] "met een enkele uitzondering (§13) doen zij slechts dienst voor kleine tusschenberekeningen..." (Lorentz 1916a, I, p. 1392).

[18] "Hierbij moet worden opgemerkt dat een vergelijking... waarin van de samenstelling van vectoren in *verschillende* punten sprake is, eerst dan zin heeft als men weet welke componenten als 'gelijk gericht' beschouwd zullen worden, en dus bij elkaar zullen worden opgeteld. Dit hebben wij door de invoering van coördinaten vastgesteld" (Lorentz 1916a, I, p. 1402).

References

Bianchi, Luigi (1899). *Vorlesungen über Differentialgeometrie.* Leipzig: B.G. Teubner.

Dorling, Jon (1968). "Length Contraction and Clock Synchronization: the Empirical Equivalence of the Einsteinian and Lorentzian Theories." *British Journal for the Philosophy of Science* 19: 67–69.

Einstein, Albert, and Grossmann, Marcel (1913). *Entwurf einer verallgemeinerten Relativitätstheorie und einer Theorie der Gravitation.* Leipzig: Teubner. (The paper, with some additional remarks, subsequently appeared in *Zeitschrift für Mathematik und Physik* 62: 225–261.)

Eisenhart, Luther P. (1925). *Riemannian Geometry.* Princeton, New Jersey: Princeton University Press.

Fokker, Adriaan D. (1929). *Relativiteitstheorie.* Groningen: Noordhoff.

Haantjes, J. (1954). *Inleiding tot de differentiaal-meetkunde.* Groningen and Djakarta: Noordhoff.

Herglotz, Gustav (1916). "Zur Einsteinschen Gravitationstheorie." *Sitzungsberichte der naturforschenden Gesellschaft Leipzig. Mathematisch-physikalische Klasse* 69: 199–203. (Reprinted in G. Herglotz, *Gesammelte Schriften,* pp. 356–360. Göttingen: Vandenhoeck & Ruprecht, 1979.)

Kox, A. J. (1988). "Hendrik Antoon Lorentz, the Ether, and the General Theory of Relativity." *Archive for the History of Exact Sciences* 38: 67–78.

Lanczos, Cornelius (1970). *Space through the Ages: The Evolution of Geometrical Ideas from Pythagoras to Hilbert and Einstein.* San Diego, California: Academic Press.

Levi-Civita, Tullio (1917). "Nozione di parallelismo in una varietà qualunque." *Rendiconti del Circolo Matematico di Palermo* 42: 173–205.

Lorentz, Hendrik A. (1909). *The Theory of Electrons.* Leipzig: B.G. Teubner.
——— (1916a). "Over Einstein's theorie der zwaartekracht, I–IV." *Koninklijke Akademie van Wetenschappen te Amsterdam. Verslagen van de Gewone Vergaderingen der Wis- en Natuurkundige Afdeeling* 24: 1389–1402, 1759–1774; 25: 468–486, 1380–1396. English translation: Lorentz (1916b).
——— (1916b). "On Einstein's Theory of Gravitation, I–IV." *Koninklijke Akademie van Wetenschappen te Amsterdam. Proceedings of the Section of Sciences* 19: 1341–1354, 1354–1369; 20: 2–19, 20–34. Translation of Lorentz (1916a). Reprinted as Lorentz (1937).
——— (1937). "On Einstein's Theory of Gravitation." In *Collected Papers,* Vol. 5. The Hague: Nijhoff, pp. 246–313. Reprint of Lorentz 1916b.
Norton, John (1984). "How Einstein found his Field Equations: 1912–1915." *Historical Studies in the Physical Sciences* 14: 253–316.
Pais, Abraham (1982). *"Subtle is the Lord ...": the Science and the Life of Albert Einstein.* London and New York: Oxford University Press.
Pauli, Wolfgang (1958). *Theory of Relativity.* Oxford: Pergamon. (Original German edition: "Relativitätstheorie." In: *Encyklopädie der mathematischen Wissenschaften, mit Einschluss ihrer Anwendungen,* Vol. 5, *Physik,* Part 2. Arnold Sommerfeld, ed. Leipzig: Teubner, 1904–1922, pp. 539–775.)
Torretti, Roberto (1978). *Philosophy of Geometry from Riemann to Poincaré.* Dordrecht and London: Reidel.
Weyl, Hermann (1918). *Raum, Zeit, Materie: Vorlesungen über allgemeine Relativitätstheorie.* Berlin: J. Springer (1st ed.).

Quantization of the Gravitational Field, 1930–1988

Peter G. Bergmann

Today's theoretical physics is largely built on two giant conceptual structures: quantum theory and general relativity. As the former governs primarily the atomic and subatomic worlds, whereas the latter's principal applications so far have been in astronomy and cosmology, our failure to harmonize quanta and gravitation has not yet stifled progress on either front. Nevertheless, the possibility that there might be some deep dissonance has caused physicists an esthetic unease, and it has caused a number of people to explore avenues that might lead to a quantum theory of gravitation, no matter how many decades away the observation of "gravitons" might lie in the future.

To my knowledge, the earliest attempt to construct a quantum theory of the gravitational field consists of two papers by Léon Rosenfeld (Rosenfeld 1930, 1932), the latter of which precedes by just one year his famous paper with Niels Bohr on quantum electrodynamics (Bohr and Rosenfeld 1933). An essential difference between the Bohr–Rosenfeld analysis and any corresponding treatment of quantum gravity is based on the principle of equivalence. Bohr and Rosenfeld could introduce tools and mechanical restraints without affecting the electromagnetic fields relevant to the behavior of their test particles, but this is not possible with gravitational fields.

In 1939, Wolfgang Pauli and Markus Fierz (Pauli and Fierz 1939) established that the graviton—if it exists at all—has a spin 2 and two modes of polarization. After World War II, Paul A.M. Dirac (Dirac 1950, 1951; Pirani and Schild 1950) and those of us at Syracuse (Bergmann et al. 1950) initiated the canonical formulation of general relativity as an initial step toward quantum gravity. These efforts culminated in Dirac's hands in complete success (Dirac 1958, 1959). The Lie algebra initiated by Dirac was closed more than 10 years later (Bergmann and Komar 1972).

The canonical formulation of general relativity was achieved on a classical basis. It did not mean the quantization was complete, or even within easy reach. There remains, for one, the close relationship between the "gauge" group of general relativity and its dynamics, which leads to the fusion of the Hamiltonian of the theory with the constraints that generate coordinate transformations. The problem of factor ordering of the constraints, so that their commutators would represent a closed Lie algebra, which was considered insoluble in the 1950s and 1960s, appears to yield to the introduction of tetrads, or triads, or corresponding spinor structures (Ashtekar 1987, Ashtekar, Mazur, and Torre 1987; Goldberg 1988). Finally, the construction of a ground state of the Hamiltonian and a Fock space of excited states, or some similar steps leading to a Hilbert space of acceptable state vectors, remains an unsolved problem.

References

Ashtekar, Abhay (1987). "New Hamiltonian Formulation of General Relativity." *Physical Review D* 36: 1587–1602.

Ashtekar, Abhay, Mazur, Pawel, and Torre, Charles G. (1987). "B. R. S. T. Structure of General Relativity in Terms of New Variables." *Physical Review D* 36: 2955–2962.

Bergmann, Peter G., Penfield, Robert, Schiller, Ralph, and Zatzkis, Henry (1950). "The Hamiltonian of the General Theory of Relativity with Electromagnetic Field." *Physical Review* 80: 81–88.

Bergmann, Peter G. and Komar, Arthur (1972). "The Coordinate Group Symmetries of General Relativity." *International Journal of Theoretical Physics* 5: 15–28.

Bohr, Niels and Rosenfeld, Léon (1933). "Zur Frage der Messbarkeit der elektromagnetischen Feldgrössen." *Det Kongelige Danske Videnskabernes Selskab. Matematisk-Fysiske Meddelelser* 12, No. 8, pp. 65.

Dirac, Paul A. M. (1950). "Generalized Hamiltonian Dynamics." *Canadian Journal of Mathematics* 2: 129–148.

——— (1951). "The Hamiltonian Form of Field Dynamics." *Canadian Journal of Mathematics* 3: 1–23.

——— (1958). "The Theory of Gravitation in Hamiltonian Form." *Royal Society of London. Proceedings A* 246: 333–343.

——— (1959). "Fixation of Coordinates in the Hamiltonian Theory of Gravitation." *Physical Review* 114: 924–930.

Goldberg, Joshua N. (1988). "Triad Approach to the Hamiltonian of General Relativity." *Physical Review D* 37: 2116–2120.

Pauli, Wolfgang, and Fierz, Markus E. (1939). "Über relativistische Feldgleichungen von Teilchen mit beliebigen Spin im elektromagnetischen Feld." *Helvetica Physica Acta* 12: 297–304.

Pirani, Felix A. E., and Schild, Adolf (1950). "On the Quantization of Einstein's Gravitational Field Equations." *Physical Review D* 79: 986–991.

Rosenfeld, Léon (1930). "Zur Quantelung der Wellenfelder." *Annalen der Physik* 5: 113–152.

—— (1932). "La théorie quantique des champs." *Institut Henri Poincaré. Annales* 2: 25–51.

The First Steps of Quantum Gravity and the Planck Values

Gennady E. Gorelik

Every contemporary discussion of quantum gravity inevitably includes the so-called Planck values, which are combinations of three fundamental constants: c (the velocity of light), G (the gravitational constant), and h (Planck's constant):

$$X_{Pl} = c^x G^y h^z,$$

where the exponents x, y, z are rational numbers. Depending upon the choice of x, y, z the Planck values can have any dimensionality—length, time, density, and so forth.

Today, the Planck values are connected with the quantum limits of general relativity (GR). Such an interpretation of these values is reached in different ways—from arguments within the framework of tentative variants of quantum-gravitational theory to simple dimensional considerations. Because the latter approach does not require any complex theoretical constructions, one can surmise that the quantum-gravitational role of the Planck values may have been known to Planck himself.

In reality, however, Planck introduced his cGh values in 1899, without any connection to quantum gravity. Quantum limits to the applicability of general relativity (and, implicitly, their Planck scale) were first discovered in 1935 by the Soviet theorist Matvey P. Bronsteyn (1906–1938). It was not until the 1950s that the explicitly quantum-gravitational significance of the Planck values was pointed out almost simultaneously by several physicists. In the following, the history of the Planck values will be reviewed. The topic is an important one, because the Planck scale is almost the only element certain to be a part of a future synthesis of general relativity and quantum theory (Misner, Thorne, Wheeler 1973).

1. The Birth of the Planck Values

In the fifth installment of his continuing study of irreversible radiation processes (Planck 1899), Max Planck introduced two new universal physical constants, a and b, and calculated their values from experimental data. The following year, he redesignated the constant b by the famous letter h (and in place of a, he introduced $k = b/a$, the Boltzmann constant).

In 1899, the constant b (that is, h) did not yet have any quantum theoretical significance, having been introduced merely in order to derive Wien's formula for the energy distribution in the black-body spectrum. However, Planck had previously described this constant as universal. During the six years of his efforts to solve the problem of the equilibrium between matter and radiation, he clearly understood the fundamental, universal character of the sought-for spectral distribution.

It was perhaps this universal character of the new constant that stimulated Planck, in that same paper of 1899, to consider a question that was not directly connected with the paper's main theme. The last section of the paper is entitled "Natural Units of Measure" ["Natürliche Maasseinheiten"]. Planck noted that in all ordinary systems of units, the choice of the basic units is made not from a general point of view "necessary for all places and times," but is determined solely by "the special needs of our terrestrial culture" (Planck 1899, p. 479). Then, basing himself upon the new constant h and also upon c and G, Planck suggested the establishment of

> units of length, mass, time, and temperature that would, independently of special bodies and substances, necessarily retain their significance for all times and all cultures, even extraterrestrial and extrahuman ones, and which may therefore be designated as natural units of measure. (Planck 1899, pp. 479–480)

These new units are chosen in such a way that in the new system of units every mentioned constant (c, G, h) is equal to one. Thus, Planck introduced the following units:

$$
\begin{aligned}
l_{Pl} &= (hG/c^3)^{1/2} = 10^{-33}\,\text{cm} \\
m_{Pl} &= (hc/G)^{1/2} = 10^{-5}\,\text{g} \\
t_{Pl} &= (hG/c^5)^{1/2} = 10^{-43}\,\text{s}
\end{aligned}
$$

(where modern notation and values are used). In cosmology, a "natural unit" of density is also now very popular:

$$\rho_{Pl} = c^5/\hbar G^2 = 10^{94}\,\text{g/cm}^3.$$

While this last section of the paper bears only a slight connection to the rest, it is strongly connected with Planck's broader philosophy of physics, in particular, with his attention to the absolute elements of the scientific world picture and his opposition to anthropomorphism in physics (Goldberg 1976). Eddington (1918) seems to be the first who attempted to uncover a theoretical significance in the Planck length, a significance that may be the clue to some essential structure of gravitation. But, both Planck's and Eddington's suppositions were ignored or even ridiculed (see, for example, Bridgman 1922).

2. Gravitation and Quanta

The quantum-gravitational meaning of the Planck values could be revealed only after a relativistic theory of gravitation had been developed. As soon as that was done, Einstein pointed out the necessity of unifying the new theory of gravitation with quantum theory. In 1916, having obtained the formula for the intensity of gravitational waves, he remarked:

> Because of the intra-atomic movement of electrons, the atom must radiate not only electromagnetic but also gravitational energy, if only in minute amounts. Since, in reality, this cannot be the case in nature, then it appears that the quantum theory must modify not only Maxwell's electrodynamics but also the new theory of gravitation. (Einstein 1916, p. 696).

(For a similar comment, see Einstein 1918.) Einstein obviously had in mind the problem of the instability of the atom. But his conclusion was hardly based on quantitative estimates. Atomic radiation, calculated according to classical electrodynamics, results in the collapse of the atom in a characteristic time interval on the order of 10^{-10} s (in stark contradiction with reality), whereas atomic gravitational radiation, calculated according to Einstein's formula, is characterized by an enormous time collapse on the order of 10^{37} s. Thus, there was no reason to worry that the relativistic theory of gravitation would contradict empirical data, but the analogy with electrodynamics determined the direction of Einstein's thinking.

The "cosmological" value of the time of atomic gravitational outradiation reminds one that during these years Einstein was also thinking about cosmological problems. Even though the effect of gravitational radiation is small, Einstein had to reject its possibility, apparently because of his prerequisites for a cosmology. In a static, eternal universe, any instability of the atom is inadmissible, wholly independently of its magnitude. It is interesting to compare this attitude, quite natural for that time, with the current, more tolerant attitude toward the instability of the proton (Salam 1979). The change is explained by the acceptance of an evolutionary picture of the universe.

For two decades after Einstein pointed out the necessity of a quantum-gravitational theory in 1916, only a few remarks about this subject appeared. There were too many other more pressing theoretical problems (including quantum mechanics, quantum electrodynamics, and nuclear theory). And, the remarks that were made were too superficial, which is to say that they assumed too strong an analogy between gravity and electromagnetism. For example, after discussing a general scheme for field quantization in their famous 1929 paper, Heisenberg and Pauli wrote:

> One should mention that a quantization of the gravitational field, which appears to be necessary for physical reasons, may be carried out without any new difficulties by means of a formalism wholly analogous to that applied here. (Heisenberg and Pauli 1929, p. 3)

They grounded the necessity of a quantum theory of gravitation on Einstein's mentioned remark of 1916 and on Oskar Klein's remarks in an article of 1927 in which he pointed out the necessity of a unified description of gravitational and electromagnetic waves, one taking into account Planck's constant h.

Heisenberg and Pauli obviously intended that quantization techniques be applied to the linearized equations of the (weak) gravitational field (obtained by Einstein in 1916). Being clearly approximative, this approach allows one to hope for an analogy with electromagnetism, but it also allows one to disregard some of the distinguishing properties of gravitation—its geometrical essence and its nonlinearity. Just such an approach was employed by Léon Rosenfeld, who considered a system of quantized electromagnetic and weak gravitational fields (Rosenfeld 1930), studying the mutual transformations of light and "gravitational quanta" (a term that he was the first to use).

The first really profound investigation of the quantization of the gravitational field was undertaken by Matvey P. Bronsteyn. The essential results of his 1935 dissertation, entitled "The Quantization of Gravitational Waves," were contained in two papers published in 1936. The dissertation was mainly devoted to the case of the weak gravitational field, where it is possible to ignore the geometrical character of gravitation, that is, the curvature of space-time. However, Bronsteyn's work also contained an important analysis revealing the essential difference between quantum electrodynamics and a quantum theory of gravity not thus restricted to weak fields and "nongeometricness." This analysis demonstrated that the ordinary scheme of quantum field theory and the ordinary concepts of Riemannian geometry are not sufficient for the formulation of a consistent theory of quantum gravity. At the same time, Bronsteyn's analysis led to the limits of quantum-gravitational physics (and to Planck's cGh-values).

3. Bronsteyn and His Way to Quantum Gravity

In the 1930s, the name of the Leningrad theorist Matvey Petrovich Bronsteyn (1906–1938) was well known in Soviet physics, and this in spite of Bronsteyn's youth. He was a friend and colleague of the bright young physicists George Gamow, Lev Landau, and Dmitriy Ivanenko, as well as the astronomers Victor Ambarzumian and Nikolai Kosyrev. Bronsteyn's papers were published in the main physical journals. In 1932, Dirac's *Principles of Quantum Mechanics* was published in his translation and with many of his footnotes. At Leningrad University and the Leningrad Physical-Technical Institute, the students and faculty appreciated Bronsteyn's participation in the journal *Physikalische Dummheiten*, which circulated in a single copy. Many more readers knew Bronsteyn's popular scientific articles and books and his three wonderful books for children. The latter are being reprinted today, fifty years after the death of their author. In 1937, Bronsteyn was arrested and then killed by Stalinism. For two decades, his name was banned as the name of an "enemy of the people."

Bronsteyn had broad scientific interests—his papers dealt with astrophysics, semiconductors, cosmology, and nuclear physics. But, the investigation of quantized gravity became his most important achievement (see Gorelik 1983; Gorelik and Frenkel 1985, 1990).

One of his first papers of 1926 contained evidence of his acquaintance with general relativity, but the full measure of his knowledge of the topic was in an extensive and profound survey article on relativistic cosmology that he published in 1931. The appearance of this review, which contains sections on the foundations of general relativity, on the formulation of the cosmological problem, and on all of the then-known cosmological models, was stimulated by Hubble's discovery of 1929.

In this article, one finds no doubts about the possibility of constructing a cosmological theory based solely on general relativity. In Bronsteyn's subsequent papers concerning cosmology, such doubts are characteristic, and they are based on his general conception of physics. Bronsteyn looked at physics through the "magic cGh cube." He drew a chart describing the structure of theoretical physics based on the role played by the three fundamental constants c, G, and h, and on the corresponding limits of applicability of different theories (ignoring some of these constants). His next paper on the topic (Bronsteyn 1933b) contains a chapter entitled "Relations of Physical Theories to Each Other and to Cosmological Theory," the basic analysis of which was illustrated by the following scheme:

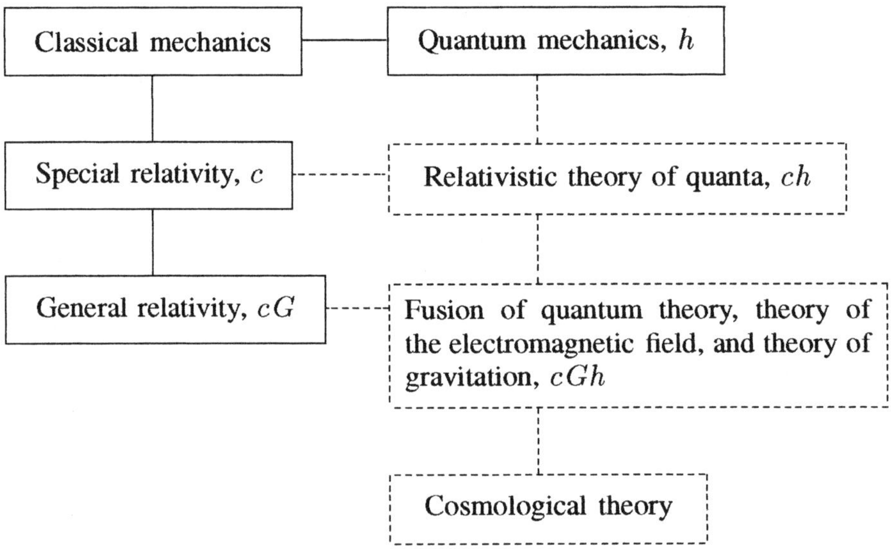

Here, as Bronsteyn wrote, "solid lines correspond to existing theories, dotted ones to problems not yet solved" (Bronsteyn 1933b, p. 15). Relativistic quantum theory, that is, ch theory, was then the center of attention. But, as a rule, gravity was ignored.

Bronsteyn also wrote about the quantum limits of general relativity in other papers, in the context of astrophysics and cosmology (see, for example, Bronsteyn 1933a). One phrase from Bronsteyn's popular article of 1929 (which concerned Einstein's attempt to unify gravitation and electromagnetism) reveals his general attitude to the problem of the quantum generalization of general relativity: "The construction of a space-time geometry that could result not only in laws of gravitation and electromagnetism but also in quantum laws is the greatest task ever to confront physics" (Bronsteyn 1929, p. 25). Thus, it is no surprise that Bronsteyn chose the quantization of gravitation as the topic of his dissertation (when the system of scientific degrees was introduced in the USSR). However, this choice was rather surprising in the contemporary scientific environment, because in the 1930s, fundamental theorists were concentrating on quantum electrodynamics and nuclear physics.

Bronsteyn defended his thesis in November 1935. His examiners, the prominent Soviet theorists Vladimir Fock and Igor Tamm, praised the thesis highly as "the first work on the quantization of gravitation resulting in a physical outcome" (Gorelik and Frenkel 1985, p. 317). Bronsteyn's dissertation and two corresponding publications (Bronsteyn 1936a, 1936b) were mainly devoted to the quantization of weak gravitational fields, but one of the most important results is an analysis of the compatibility of quantum concepts and classical general relativity in the general case (this latter topic will be examined seperately).

Bronsteyn considered gravitation in the weak-field approximation (where it is described by a tensor field in Minkowski space) in accordance with Heisenberg and Pauli's general scheme of field quantization (Heisenberg and Pauli 1929). From quantized weak gravitation, he deduced two consequences: (1) a quantum formula for the intensity of gravitational radiation coinciding in the classical limit with Einstein's formula, and (2) the Newtonian law of gravitation as a consequence of quantum-gravitational interactions.

One might think that what we see here is merely an expression of the simple requirement of the correspondence principle. In fact, Bronsteyn's results had real significance, because the peculiar character of the gravitational field (namely, its identification with the space-time metric) gave rise to doubts about the possibility of synthesizing quantum concepts and general relativity. For example, Yakov I. Frenkel (the head of the theoretical department at the Leningrad Physical-Technical Institute, where Bronsteyn worked) was very sceptical about the possibility of quantizing gravitation, because he, like some other physicists, considered gravitation as a macroscopic, effective property of matter. It should be mentioned that even in the 1960s, Léon Rosenfeld supposed that the quantization of gravitation might be meaningless since the gravitational field probably has only a classical, macroscopic nature (Rosenfeld 1963)—and Rosenfeld was the first to consider a formula for the expression of quantized gravitation. On the other hand, Einstein's attitude was also well known. He believed that the correct full theory was separated from general relativity by a much smaller distance, so to speak, than from quantum theory.

Bronsteyn's attitude was near the golden mean, since he was sure that one should look for the synthesis of both great theories—general relativity and the quantum theory. His investigation brought to light deep connections between the classical and quantum ways of describing gravitation and so gave testimony to the effect that the quantum generalization of general relativity is both possible and necessary.

4. "...an essential difference between quantum electrodynamics and the quantum theory of the gravitational field." Quantum Limits of General Relativity

Before obtaining the two mentioned results and just after deducing commutation relations, Bronsteyn analyzed the measurability of the gravitational field, taking into account quantum restrictions.

The question of measurability had occupied an important place in fundamental physics since Heisenberg had discovered the restrictions on measurability resulting from the indeterminacy relations (restrictions on the measurability of conjugate parameters). As physicists looked toward the development of ch-theory (relativistic quantum theory), the question of measurability attracted great attention, especially after Landau and Peierls rejected in their 1931 paper the concept of an "electromagnetic field at a point," based upon its immeasurability within the ch-framework. This paper led to the detailed and careful analysis of the situation that was undertaken in 1933 by Bohr and Rosenfeld. They saved a local field description in quantum electrodynamics, but at the high price of assuming arbitrarily high densities of mass and electrical charge.

Bronsteyn had an eye for this subject, and just after Bohr and Rosenfeld's paper, he summed up the situation in a short but clear note (Bronsteyn 1934). So it was quite natural that in considering quantum gravity Bronsteyn decided to analyze the problem of the measurability of the gravitational field within the cGh-framework.

He proceeded from the weak-field approximation, where the metric $g_{ik} = \delta_{ik} + h_{ik}$, and δ_{ik} is the Minkowski metric, with $h_{ik} \ll 1$. In this case, the Einstein field equations have the form

$$\Box h_{ik} = \kappa(T_{ik} - \tfrac{1}{2}\delta_{ik}T), \tag{1}$$

where T_{ik} is the energy-momentum tensor, and $\kappa = 16\pi G/c^2$. In order to analyze the measurability of the gravitational field, Bronsteyn considered the following thought experiment for measuring the value of the Christoffel symbol Γ^i_{jk} (gravitational field strength) by observing the movement of a test body.

In the weak-field approximation, the equation of a geodesic

$$\frac{d^2x^1}{ds^2} + \Gamma^1_{ik}\frac{dx^i}{ds}\frac{dx^k}{ds} = 0$$

can be written as ($x^1 = x$)

$$\frac{d^2x}{c^2\,dt^2} = \Gamma_{1,00} = \frac{\partial h_{01}}{c\,\partial t} - \frac{1}{2}\frac{\partial h_{00}}{\partial x}. \tag{2}$$

To measure $\Gamma_{1,00}$ averaged over the volume V and the time interval T, one should measure the component of momentum p_x of a test body with volume V at the beginning and at the end of the interval T:

$$\Gamma_{1,00} = \frac{d^2x}{c^2\,dt^2} = \frac{p_x(t+T) - p_x(t)}{c^2\rho VT},$$

where ρ is the density of the test body. If the measurement of the momentum has indeterminacy $\sim \Delta p_x$, then

$$\Delta\Gamma_{1,00} \sim \Delta P_x/c^2\rho VT. \tag{3}$$

The indeterminacy Δp_x is the sum of two parts: the ordinary quantum mechanical term $(\Delta p_x)_1 = h/\Delta x$, where Δx is the indeterminacy in the coordinate x, and "the term connected with the gravitational field that is created by the measurer due to the recoil in the measurement of momentum." The latter term can be estimated with help of Eq. (1):

$$\Box h_{01} = \kappa T_{01} = \kappa\rho v_x/c.$$

If a separate measurement of momentum requires a time interval $\Delta t(\ll T)$, then the indeterminacy in h_{01}, connected with the indeterminacy in the recoil velocity $v_x \sim \Delta x/\Delta t$, is of the order

$$\Delta h_{01} \sim \kappa\rho\,\frac{\Delta x}{c\,\Delta t}(c\,\Delta t)^2,$$

and, in accordance with Eq. (2), $\Delta\Gamma_{1,00} \sim \kappa\rho\,\Delta x$. So the indeterminacy in the momentum that is connected with the gravitational field of the test body would be

$$(\Delta p_x)_2 \sim c^2\rho V\Delta\Gamma_{1,00}\cdot\Delta t = c^2\kappa\rho^2 V\Delta x\Delta t.$$

The total indeterminacy

$$(\Delta p_x) = (\Delta p_x)_1 + (\Delta P_x)_2 \sim \hbar/\Delta x + c^2\kappa\rho^2 V\Delta x\Delta t \tag{4}$$

is minimized by

$$\Delta x = (\hbar/\kappa c^2\rho^2 V\Delta t)^{1/2}. \tag{5}$$

Then,

$$(\Delta p_x)_{\min} \sim (\hbar\kappa c^2\rho^2 V\Delta t)^{1/2},$$

$$(\Delta\Gamma_{1,00})_{\min} \sim \frac{1}{c^2T}\left(\frac{\hbar\kappa c^2\Delta t}{V}\right)^{1/2}. \tag{6}$$

Two conditions determine the lower bound on the duration Δt of the measurement of momentum. First, it is necessary that $\Delta t > \Delta x/c$ in order for the recoil velocity to be less than c. Hence, from Eq. (5), we get

$$\Delta t > \tau_1 = (\hbar/\kappa c^4\rho^2 V)^{1/3}. \tag{7}$$

Second, from the very meaning of the measurement of the field in volume V, it follows that the indeterminacy $\Delta x < V^{1/3}$. Hence,

$$\Delta t > \tau_2 = \hbar/c^2\kappa\rho^2 V^{5/3}. \tag{8}$$

Having obtained these two lower limits, Bronsteyn remarked that the ratio

$$\frac{\tau_1}{\tau_2} = \left(\frac{c\kappa\rho^2 V^2}{\hbar}\right)^{2/3} \equiv \left(\frac{c\kappa m^2}{\hbar}\right)^{2/3} \tag{9}$$

"depends on the mass of the test body, being quite insignificant in the case of the electron and becoming of order 1 in the case of a dust particle weighing one hundredth of a milligram" (Bronsteyn 1936b, p. 216). There are then two limits for $\Delta\Gamma$, respectively

$$(\Delta\Gamma)_{\min 1} \gtrsim \frac{1}{c^2 T}\left(\frac{\hbar^2\kappa c}{\rho V^2}\right)^{1/3}, \tag{10}$$

$$(\Delta\Gamma)_{\min 2} \gtrsim \frac{\hbar}{c^2 T\rho V^{4/3}}. \tag{11}$$

As we have seen, to measure Γ in a certain volume V as precisely as possible, one should use a test body with as large a mass (density) as possible. Hence, only the first limit is essential.

Bronsteyn wrote that the preceding considerations are analogous to those in quantum electrodynamics. But there arises here the essential difference between quantum electrodynamics and quantum gravity, because

> in formal quantum electrodynamics, which does not take into consideration the structure of the elementary charge, there is no consideration limiting the increase of density ρ. With sufficiently high charge density in the test body, the measurement of the electrical field may be arbitrarily precise. In nature, there are probably limits to the density of the electrical charge... but formal quantum electrodynamics does not take these limits into account.... The quantum theory of gravitation represents a quite different case: it has to take into account the fact that the gravitational radius of the test body ($\kappa\rho V$) must be less than

its linear dimensions

$$\kappa \rho V < V^{1/3}. \tag{12}$$

(Bronsteyn 1936b, p. 217)

It follows that Eq. (10) gives "the absolute minimum for the indeterminacy":

$$\Delta\Gamma > \frac{1}{c^2 T}\left(\frac{\hbar^2 \kappa^2 c}{V^{4/3}}\right)^{1/3}. \tag{13}$$

Bronsteyn understood that "the absolute limit is calculated roughly" (in the weak-field framework), but he believed that "an analogous result will be valid also in a more exact theory." He formulated the fundamental conclusion as follows:

> The elimination of the logical inconsistencies connected with this requires a radical reconstruction of the theory, and in particular, the rejection of a Riemannian geometry dealing, as we see here, with values unobservable in principle, and perhaps also the rejection of our ordinary concepts of space and time, modifying them by some much deeper and nonevident concepts. *Wer's nicht glaubt, bezahlt einen Taler.* (Bronsteyn 1936b, p. 218)

(The same German phrase concludes one of the Grimm brothers' very improbable fairy tales).

In such a way, the quantum limits of general relativity were revealed for the first time. The only thing lacking for the modern theorist is the Planck scale of these limits. Of course the Planck values were implicitly present, because all three constants c, G, and h, were involved in Bronsteyn's analysis. One such value, the Planck mass $(hc/G)^{1/2}$, appeared in Bronsteyn's text, "a dust-particle weighing one hundredth of a milligram."

It is not difficult to supplement Bronsteyn's reasoning in order to demonstrate the Planck scale of quantum gravity, if only one tries to measure the gravitational field as precisely as possible in a volume as small as possible (Gorelik 1983). Then, one can obtain, for example, the indeterminacy of the metric

$$\Delta g = l_{Pl}/cT.$$

The true value of Bronsteyn's fundamental result concerning the quantum limits of general relativity was not recognized by his contemporaries. For example, Vladimir Fock had, on the whole, a very high opinion of Bronsteyn's dissertation, but in a summary written for an abstract journal, Fock gave only the following vague characterization of this important result: "a few observations on the measurability of the field quantities are appended" (Fock 1936, p. 88).

In the 1930s, it was much easier to unite the constants c and h with e, m_e, and m_p than with G (Heisenberg 1930). Only a deep understanding of the structure of theoretical physics allowed one to single out just the three constants c, G, and h. It was clear to Bronsteyn that a synthesis of general relativity and quantum theory would be inevitable for astrophysics, cosmology, and fundamental physics in general (Gorelik and Frenkel 1985, 1990). Today, when such a synthesis is eagerly sought, Bronsteyn's analysis of measurability does not lose its significance.

5. The Planck Values and Quantum Gravity

For two decades after Bronsteyn's work, there was calm in the field of quantum gravity. Only in the mid-1950s did the length $l_0 = (Gh/c^3)^{1/2}$ appear almost simultaneously in a few different forms in a few papers. For example, in 1954, Landau pointed out that the length $l = G^{1/2}h/ce(= \alpha^{-(1/2)}l_0$, very near to the Planck length) is *the limit of the region outside of which quantum electrodynamics cannot be considered as a self-consistent theory because of the necessity of taking into account gravitational interactions* ($Gm^2/r \sim e^2/r$, when $m \sim p/c \sim h/lc$) (Abrikosov, Landau, Khalatnikov 1954).

In a report at the Bern relativity jubilee conference in 1955, Oskar Klein remarked that l_0 corresponds to *the gravitational limit of the region of applicability of special relativistic quantum theory.* He reasoned in the following way. A relativistic particle represented by a wave packet in volume λ^3 has energy hc/λ and mass $h/c\lambda$. A difference between the gravitational potentials at the center and the edge of the wave packet $\Delta\varphi \sim Gh/c\lambda^2$ would not change the metric significantly if $\Delta\varphi \ll c^2$ or $\lambda \gg l_0$ (Klein 1954, 1956, p. 61). In his discussion of Klein's report, Wolfgang Pauli also referred to Landau's remark and conjectured that there is a connection between the problem of the quantization of gravity and the problem of divergences in quantum field theory (Pauli 1956, p. 69). Such a possibility continues to attract the attention of theorists.

The most famous quantum-gravitational interpretation of the length l_0 was developed in John Archibald Wheeler's 1955 paper entitled "Geons." The main subjects of the paper are hypothetical gravitational-electromagnetic pure field configurations (geons). In addition, Wheeler evaluated the manner in which quantum fluctuations of the metric Δg_{ik} depend on a characteristic space scale $\mathcal{L}$. Using Feynman quantization, he showed that $\Delta g_{ik} \sim l_0/\mathcal{L}$, that is, the fluctuations become significant when $\mathcal{L} \sim l_0$. Thus, the length $l_0 = (G\hbar/c^3)^{1/2}$ corresponds to the *quantum limits of general relativity*. This is the approach that became the most popular.

The term "Planck values," which is now generally accepted, was introduced later (Misner and Wheeler 1957). According to Wheeler, he did not know in 1955 about Planck's "natural units" (private communication).

6. Conclusion

A history of the Planck values provides interesting material for reflections on timely and premature discoveries in the history of science. Today, the Planck values are more a part of physics itself than of its history. They are mentioned in connection with the cosmology of the early universe as well as in connection with particle physics. In considering certain problems associated with a unified theory (including the question of the stability of the proton), theorists discovered a characteristic mass $\sim 10^{16} m_p (m_p$ is the proton mass). To ground such a great value, one first refers to the still greater mass $10^{19} m_p$. In the words of Steven Weinberg:

> This is known as the Planck mass, after Max Planck, who noted in 1900 that some such mass would appear naturally in any attempt to combine his quantum theory with the theory of gravitation. The Planck mass is roughly the energy at which the gravitational force between particles becomes stronger than the electroweak or the strong forces. In order to avoid an inconsistency between quantum mechanics and general relativity, some new features must enter physics at some energy at or below 10^{19} proton masses. (Weinberg 1981, p. 71).

The fact that Weinberg takes such liberties with history in this quotation is evidence of the need to describe the real historical circumstances in which the Planck mass arose. As we saw, when Planck introduced the mass $(ch/G)^{1/2} (\sim 10^{19} m_p)$ in 1899, he did not intend to combine the theory of gravitation with quantum theory; he did not even suppose that his new constant would result in a new physical theory. The first "attempt to combine the quantum theory with the theory of gravitation," which demonstrated that "in order to avoid an inconsistency between quantum mechanics and general relativity, some new features must enter physics," was made by Bronsteyn in 1935. That the Planck mass may be regarded as a quantum-gravitational scale was pointed out explicitly by Klein and Wheeler twenty years later. At the same time, Landau also noted that the Planck energy (mass) corresponds to an equality of gravitational and electromagnetic interactions.

Theoretical physicists are now confident that the role of the Planck values in quantum gravity, cosmology, and elementary particle theory will emerge from a unified theory of all fundamental interactions and that the Planck scales characterize the region in which the intensities of all fundamental interactions become comparable. If these expectations come true, the present report might become useful as the historical introduction for the book that it is currently impossible to write, *The Small-Scale Structure of Space-Time.*

Acknowledgment: The author is thankful to Jean Eisenstaedt and Don Howard for their help in preparing this report.

REFERENCES

Abrikosov, Alexei A., Landau, Lev D., and Khalatnikov, Isaac, M. (1954). "Asimptoticheskoe vyrazhenie dlya fotonnoy funktsii Grina v kvantovoy elektrodinamike." ["Asymptotic Expression for the Green's Function of a Photon in Quantum Electrodynamics."] *Doklady Akademii Nauk SSSR* 95: 1177–1180.

Bohr, Niels, and Rosenfeld, Léon (1933). "Zur Frage der Messbarkeit der elektromagnetischen Feldgrössen." *Det Kongelige Danske Videnskabernes Selskab. Mathematisk-Fysiske Meddelelser* 12 (No. 8): 3–65.

Bridgman, Percy W. (1922). *Dimensional Analysis.* New Haven, Connecticut: Yale Univ. Press.

Bronsteyn, Matvey Petrovich (1929). "Vsemirnoe tyagotenie i elektrichestvo (novaya teoriya Eynshteyna)." ["Universal Gravity and Electricity (New Einstein Theory)."] *Chelovek i priroda* 8: 20–25.

——— (1931). "Sovremennoe sostoyanie relyativistskoy kosmologii." ["The Modern State of Relativistic Cosmology."] *Uspekhi Fizicheskikh Nauk* 11: 124–184.

——— (1933a). "Vnutrennee stroenie zvezd i istochniki zvezdnoy energii." ["Inner Structure of Stars and Sources of Stellar Energy."] *Uspekhi Astronomicheskikh Nauk. Sbornik* 2: 84–103.

——— (1933b). "K voprosy o vozmozhnoy teorii mira kak tselogo." ["On the Question of a Possible Theory of the World as a Whole."] *Uspekhi Astronomicheskikh Nauk. Sbornik* 3: 3–30.

——— (1934). "K voprosy o relyativistskom obobshchenii printsipa neopredelennosti." ["On the Question of a Relativistic Generalization of the Indeterminacy Principle."] *Doklady Akademii Nauk SSSR* 1: 388–390.

——— (1936a). "Quantentheorie schwacher Gravitationsfelder." *Physikalische Zeitschrift der Sowjetunion* 9: 140–157.

——— (1936b). "Kvantovanie gravitatsionnykh voln." ["Quantization of Gravitational Waves."] *Zhurnal Eksperimentalnoy i Teoreticheskoy Fiziki* 6: 195–236.

Eddington, Arthur (1918). *Report on the Relativity Theory of Gravitation.* London: Fleetway Press.

Einstein, Albert (1916). "Näherungsweise Integration der Feldgleichungen der Gravitation." *Königlich Preussische Akademie der Wissenschaften* (Berlin). *Sitzungsberichte*: 688–696.

—— (1918). "Über Gravitationswellen." *Königlich Preussiche Akademie der Wissenschaften* (Berlin). *Sitzungsberichte*: 154–167.

Fock, Vladimir A. (1936). Abstracts of Bronsteyn 1936a, 1936b. *Zentralblatt für Mathematik und ihre Grenzgebiete* 14: 87–88.

Goldberg, Stanley (1976). "Max Planck's Philosophy of Nature and His Elaboration of the Special Theory of Relativity." *Historical Studies in the Physical Sciences* 7: 125–160

Gorelik, Gennady, E. (1983). *Razmernost'' prostranstva: istoriko-metodologicheskiy analiz. [Dimensionality of Space: Historical and Methodological Analysis.]* Ch. 5, "Kvantovye granitsy klassicheskoy geomtrii OTO." ["Quantum Limits of Classical Geometry of General Relativity."] Moscow: University Press.

Gorelik, Gennady E., and Frenkel, Victor Ja. (1985). "M. P. Bronshteyn i kvantovaya teoriya gravitatsii." ["M. P. Bronshteyn and the Quantum Theory of Gravitation."] In *Eynshteynovskiy Sbornik 1980–1981.* Moscow: Nauka, pp. 291–327.

—— (1990). *Matvey Petrovich Bronshteyn (1906–1938).* Moscow: Nauka.

Heisenberg, Werner (1930). *Die physikalischen Prinzipien der Quantentheorie.* Stuttgart: S. Hirzel.

Heisenberg, Werner, and Pauli, Wolfgang (1929). "Zur Quantendynamik der Wellenfelder." *Zeitschrift für Physik* 56: 1–61.

Klein, Oskar (1927). "Zur fünfdimensionalen Darstellung der Relativitätstheorie." *Zeitschrift für Physik* 46: 188–208.

—— (1954). "Aktuella Problem Kring Fysikens Små och Stora Tal." *Kosmos (Sweden)* 32: 33.

—— (1956). "Generalization of Einstein's Theory of Gravitation Considered from the Point of View of Quantum Field Theory." In Mercier and Kervaire, eds. 1956, pp. 58–68.

Landau, Lev (1955). "On the Quantum Theory of Fields." In *Niels Bohr and the Development of Physics.* Wolfgang Pauli, Léon Rosenfeld, and Victor Weisskopf, eds., Oxford: Pergamon, pp. 52–69.

Landau, Lev, and Peierls, Rudolf (1931). "Erweiterung des Unbestimmtheitsprinzips für die relativistische Quantentheorie." *Zeitschrift für Physik* 69: 56–69.

Mercier, André, and Kervaire, Michel, eds. (1956). *Fünfzig Jahre Relativitätstheorie. Bern, 11.–16. Juli 1955.* (*Helvetica Physica Acta. Supplementum* 4.) Basel: Birkhaüser.

Misner, Charles, W., and Wheeler, John Archibald (1957). "Classical Physics as Geometry: Gravitation, Electromagnetism, Unquantized Charge, and Mass as Properties of Curved Empty Space." *Annals of Physics* 2: 525–603.

Misner, Charles, W., Thorne, Kip S., and Wheeler, John Archibald (1973). *Gravitation.* San Francisco: Freeman.

Pauli, Wolfgang (1956). Discussion following Klein 1956. In: Mercier and Kervaire, eds., 1956, pp. 68–69.

Planck, Max (1899). "Über irreversible Strahlungsvorgänge. Fünfte Mitteilung." *Königlich Preussische Akademie der Wissenschaften* (Berlin). *Sitzungsberichte*: 440–480.

Rosenfeld, Léon (1930). "Über die Gravitationswirkungen des Lichtes." *Zeitschrift für Physik* 65: 589–599.

——— (1963). "On Quantization of Fields." *Nuclear Physics* 40: 353–356.

Salam, Abdus (1979). "Gauge Unification of Fundamental Forces. The 1979 Nobel Prize Address." *Reviews of Modern Physics* 52 (1980): 525–538.

Weinberg, Steven (1981). "The Decay of the Proton." *Scientific American* 244 (June): 64–75.

Wheeler, John Archbald (1955). "Geons." *Physical Review* 97: 511–536.

Dirac Equations in Curved Space-Time

S. Kichenassamy

1. Introduction

More than 60 years after Dirac (1928a, 1928b) formulated the special relativity equations (SRD) for the spinning electron (Uhlenbeck and Goudsmit 1925, Pauli 1927), no consensus exists as to what should be the coupling of electron and gravitation. Essentially, there exist two descriptions, one in the space-time of general relativity (GR) and the other in an Einstein–Cartan space-time with torsion, such that when the gravitational field is switched off, the first goes over to SRD and the latter does not. Indeed, measurement of the precession of the spinning particle or a Stern–Gerlach-type experiment may, in principle, decide between these two descriptions, but such tests remain beyond feasibility. We note that general relativity has the advantage of presently being the only relativistic theory of gravitation accounting for all known observational tests, while Einstein–Cartan theory is attractive because it is in the trend of gauge theories.

Now, Dirac's equations are four linear partial differential equations for what was in 1927 a new object, the 4-spinor, although the word "spinor" was coined only in 1929, following Ehrenfest's suggestion, by B.L. Van der Waerden, who also developed spinor calculus (Van der Waerden 1929). Though already implicit in the work of Lipschitz on sums of square roots and more explicit in that of Felix Klein or Sophus Lie (see Van der Waerden 1980), spinors were discovered by Elie Cartan (1913) in his classification of simple Lie algebras; their nature was investigated by many authors (for example, Bargmann 1932, Whittaker 1937, and Givens and Veblen 1935–1936) and clearly elucidated from a global point of view by Brauer and Weyl (1935). Spinors are the geometrical objects defining the irreducible representation space of the Clifford group Spin(n), the universal covering group of SO(n). Moreover, in the case of $n = 4$ (or more generally even), the 4-spinor may be constructed out of the reduced spinors or half spinors (Cartan 1938a and 1938b, Chevalley 1954).

Thus, as early as 1928, the extension of Dirac theory to GR was considered, even before its standard interpretation was established, and one had to face two chief problems: the definition of spinor fields on a manifold and the derivation of the equations they obey. The evolution of these problems owes much to H. Weyl, and we now sketch it here.

2. Spinor Fields

Two different approaches are used to define on a manifold spinor fields and their covariant derivatives: the orthonormal tetrad approach (ONTA), respecting the group properties of spinors, and the general covariant approach (GCA), using classical Riemannian techniques.

In the ONTA, first used by V. Fock (Fock 1929a, 1929b, Fock and Iwanenko 1929), the transformation law of a spinor under rotation derives from that of the vector corresponding to this spinor and its conjugate, thus introducing an indetermination leading in the differentiation of spinors to four new quantities, which Fock hoped to relate to the electromagnetic potential. However, Weyl (1928, 1929a, 1929b) connected this phase transformation of the spinor ψ with the gauge transformation of the electromagnetic field and showed the necessity of replacing in the Dirac Lagrangian of an electron in an electromagnetic field the derivatives $\partial_k \psi$ by $(\partial_k + iA_k)\psi$, where A_k plays the same role as the connection in GR, and thus he appears to be the first promoter of gauge theories.

The GCA was initiated by H. Tetrode (1928) and was developed by E. Schrödinger (1932) so as to reconcile it with Fock's attempt. However, L. Infeld and B. Van der Waerden (1933) took a more radical view; they considered spinors and tensors as autonomous objects with respect to transformation laws, but connected via some mixed quantities $\sigma^{\mu A}_{B}$ on the basis of a 2-spinor space, thus allowing a rich structure on the manifold. This view was nevertheless amended by V. Bargmann (1932, 1947) and W. L. Bade and H. Jehle (1953, see also the bibliography in Parke and Jehle 1965), who introduced some ad hoc conditions that made these spinors equivalent to those obtained by the ONTA.

Cartan (1938a, 1938b), like Weyl (Weyl 1931, Brauer and Weyl 1935), insisted on the determination of spinors with respect to the orthogonal group and not the linear group (as it follows from the GCA); indeed, the last theorem of Cartan's book asserted the impossibility for those spinors to define the linear representation space of the general linear group. This impossibility, added Cartan, may explain why Infeld and Van der Waerden's point of view was "geometrically and physically so disagreeable." He again stressed these difficulties in posthumous notes:

> This stems from the fact that the spinors provide a linear representation of the orthogonal group but not a linear representation of the general linear group; moreover, one can justify this impossibility with purely topological reasons. (Cartan 1974, p. 112)[1]

Presumably, these topological reasons are those given by A. Haefliger (1956) and J. Milnor (1963), namely, the vanishing of the first two Stiefel–Whitney characteristics (see, for example, Chern 1967). A systematic introduction of spinors on a Lorentzian manifold, according to Cartan, was given by André Lichnerowicz (1964a).

However, M. Riesz (1953) found Cartan too severe toward the physicist, who stuck to general covariance as a guiding principle, and probably for analogous reasons, D. Brill and J.A. Wheeler (1957) and R. Penrose (1960) adopted the amended GCA. R. Penrose and W. Rindler (1984) found it even more useful to consider the two spinors as primitive elements; they also gave an axiomatic approach to the covariant differentiation of spinors and showed that its uniqueness implies the *conservation of spinor metric* and the *vanishing of torsion.*

On the other hand, Y. Ne'eman (1978) and Dj. Sijacki with Ne'eman (1979) claimed that Cartan's objection was not valid for an infinite dimensional representation of GL(n, R) and, using a procedure for obtaining irreducible representations of semi-simple Lie groups due to Harish-Chandra (1953, 1954a, 1954b), constructed bivalued linear representations of the group of general coordinate transformations, the bandors or band spinors. This attempt should therefore be compared to that of Infeld and Van der Waerden.

3. Field Equations

If the first attempts obtained general relativistic Dirac equations (GRD) formally by replacing ordinary derivatives ∂_k by covariant derivatives ∇_k in SRD, that of Weyl (1929a, 1929b) gave GRD and energy-momentum contributions of Dirac fields through a variational principle of *Hilbert type*, that is using a Lagrangian formalism with tetrad fields $\{e_a\}$ and matter fields ψ as independent variables, the connection γ^a_{bc} having been expressed in terms of $\{e_a\}$ and their derivatives. Otherwise, if we exclude studies of Dirac equations in pentadimensional or projective theories (for example, Schouten 1931a, 1931b; Pauli 1933a, 1933b) or in

Cartan's book (1938b), there was a pause until Weyl (1948) returned to the problem and displayed a different set of field equations invalidating the Riemannian structure of the space-time by using the *mixed variation*, that is a Lagrangian formalism in which tetrad fields $\{e_a\}$, matter variables ψ and connection γ^a_{bc} are varied independently. Then, GRD can only be recovered by adding some new terms to the Lagrangian. Note that since many misprints in his paper (Weyl 1948) obscured its content, Weyl published it again in *Physical Review* (Weyl 1950).

Unnoticed by P.G. Bergmann and R. Thomson (1953) in their study of spin in GR or by Brill and Wheeler (1957) in their paper on the neutrino, Weyl's contribution was combined with Tetrode's suggestion of the spin giving rise to a nonsymmetric energy tensor by D.W. Sciama (1962) to rederive Weyl's equations relating spin and torsion. This theory was independently obtained by T.W.B. Kibble (1961) by extending the Yang–Mills procedure to the Lorentz group. The new equations presuppose an Einstein–Cartan space-time with a metric preserving nonsymmetric connection, of which the antisymmetric part, the torsion, is nevertheless *arbitrary*. Note also that only the totally antisymmetric part of the torsion couples to the Dirac field (see, for example Smalley and Ray 1986). One may remark that (1) when no matter is present, GRD holds well and, (2) the ECSK (Einstein–Cartan–Sciama–Kibble) Lagrangian may be reduced to the GR Lagrangian plus quadratic terms in torsion, so that subtracting them one gets GRD.

These ideas were more or less formulated with almost no real progress until the early 1970s (Kibble 1963; Lemmer 1965; Kannenberg and Arnowitt 1967; Hehl and Datta 1971; Trautman 1972), as if the scientific community was as perplexed as Weyl in 1950. Moreover, GRD continues to be an object of study (see, for example, Lichnerowicz 1964b).

The first interesting step toward the understanding of the theory was made by W. Kopczyński (1975), who studied through the use of Lagrange multipliers the influence of different constraints (euclidicity of the connection, vanishing of the torsion, etc.) in an Einstein–Cartan space-time, although euclidicity was already considered by M. Lenoir (1971). The technique of introducing constraints was exploited in a number of papers by Hehl and his co-workers (Hehl et al. 1976; Hehl et al. 1981) to establish the rationale of the spin-torsion relation, while J. Nester (1977) displayed an "effective equivalence" between ECSK and GRD, so that Lorentzian and Einstein–Cartan space-times could equally serve for the description of spin particles.

In fact, I have shown elsewhere (Kichenassamy 1986) that the systematic use of Lagrange multipliers in a space-time with a metric-preserving connection leads to

1. General relativity when the matter part of the Lagrangian does not depend on the connection γ^a_{bc},
2. ECSK when this part depends on γ^a_{bc} as a consequence of minimal coupling (replace ∂_k by D_k in special relativistic expressions), and
3. GRD for the Dirac Lagrangian depending on γ^a_{bc} when the torsion is vanishing, this last constraint leading to the Belinfante–Rosenfeld symmetrization of the energy tensor.

When the gravitational field is switched off ($g_{ij} \to \eta_{ij}$), GRD goes over to SRD, whereas ECSK gives a teleparallelism theory (curvature = 0, torsion $\neq 0$) that has in contrast to SRD *no experimental basis*; torsion introduces the unwanted Heisenberg–Pauli-type terms (Peres 1962) in the Dirac equations. However, some kind of local equivalence principle may be preserved by requiring that the ECSK Lagrangian reduces to the SR Lagrangian when the connection vanishes (von der Heyde 1975).

On the other hand, in order to settle the issue between these descriptions, the motion of spin particles has been studied through different methods: conservation laws (Hehl 1971, Adamowicz and Trautman 1975), the Mathisson–Papapetrou approach (Yasskin and Stoeger 1980, Liebscher 1980), post-Newtonian methods (Nitsch 1980), the classical limit of quantum equations of motion through Corben's algebraic method (Rumpf 1980), and WKB (Audretsch 1981). It appears that a Stern–Gerlach-type experiment or a measurement of the spin precession of a Dirac particle may give the right answer, although the relation of the torsion responsible for these effects to the source is far from clear.

Therefore, GRD has the whole weight of GR in its favor, while ECSK seems attractive as one hopes that Einstein–Cartan space-time may be the basis for Poincaré gauge theories. However, as was shown by M. Schweizer (1980), the existing so-called Poincaré gauge theory (Hehl 1980) combines local Lorentz invariance with general covariance to get local Poincaré invariance.

Moreover, the arbitrariness of the torsion in the $(g \cdot \Gamma)$ relation and the nonuniqueness of covariant differentiation of spinors when torsion is not vanishing reduce the choice between GRD and ECSK to a philosophical one: should spin be directly related to the gravitational potentials or to some new quantity, such as torsion? This question will not be further discussed here. I would rather conclude by observing that it is indeed amazing to me that Weyl's contributions to this subject are now being ignored by the present protagonists of ECSK theory.

Notes

[1] "Cela tient à ce que les spineurs fournissent une représentation linéaire du groupe des rotations euclidiennes, mais ne peuvent fournir aucune représentation linéaire du groupe des rotations affines; on peut du reste justifier cette impossibilité par des raisons purement topologiques."

References

Adamowicz, W. and Trautman, A. (1975). "Equivalence between the Einstein–Cartan and General Relativity Theories in Linear Approximation for a Classical Model of Spin." *Académie Polonaise des Sciences. Série des Sciences Mathématiques, Astronomiques et Physiques. Bulletin* 23: 1203–1205.

Audretsch, Jürgen (1981). "Dirac Electron in Space-Times with Torsion: Spinor Propagation, Spin Precession, and Non-Geodesic Orbits." *Physical Review* D 24: 1470–1477.

Bade, William L., and Jehle, Herbert (1953). "An Introduction to Spinors." *Reviews of Modern Physics* 25: 714–728.

Bargmann, Valentin (1932). "Bemerkungen zur allgemein-relativistischen Fassung der Quantentheorie." *Preussische Akademie der Wissenschaften* (Berlin). *Sitzungsberichte*: 346–354.

—— (1947). "Irreducible Unitary Representations of the Lorentz Group." *Annals of Mathematics* 48: 568–640.

Bergmann, Peter G., and de Sabbata, Venzo, eds. (1980). *Cosmology and Gravitation. Spin, Torsion, Rotation, and Supergravity.* New York and London: Plenum.

Bergmann, Peter G., and Thomson, Robb (1953). "Spin and Angular Momentum in General Relativity." *Physical Review* 89: 400–407.

Brauer, Richard, and Weyl, Hermann (1935). "Spinors in *n*-Dimensions." *American Journal of Mathematics* 57: 425–449.

Brill, Dieter R., and Wheeler, John A. (1957). "Interaction of Neutrinos and Gravitational Fields." *Reviews of Modern Physics* 29: 465–479.

Cartan, Elie (1913). "Les groupes projectifs qui ne laissent invariante aucune multiplicité plane." *Société Mathématique de France. Bulletin* 41: 53–96.

—— (1938a). *Leçons sur la théorie des spineurs. I. Les spineurs de l'espace à trois dimensions.* Actualités Scientifiques et Industrielles, No. 643. Paris: Hermann.

—— (1938b). *Leçons sur la théorie des spineurs. II. Les spineurs de l'espace à $n \geq 3$-dimensions. Les spineurs en géométrie riemannienne.* Actualités Scientifiques et Industrielles, No. 701. Paris: Hermann.

—— (1974). *Notice sur les travaux scientifiques.* Collection "Discours de la Méthode." Paris: Gauthier-Villars.

Chern, Shiing-Shen (1967). *Complex Manifolds without Potential Theory.* Princeton, New Jersey: Van Nostrand-Reinhold.

Chevalley, Claude C. (1954). *The Algebraic Theory of Spinors*. New York: Columbia University Press.

Dirac, Paul A. M. (1928a). "The Quantum Theory of the Electron." *Royal Society of London. Proceedings A* 117: 610–624.

—— (1928b). "The Quantum Theory of the Electron. Part II." *Royal Society of London. Proceedings A* 118: 351–361.

Fock, Vladimir A. (1929a). "Geometrisierung der Diracschen Theorie des Elektrons." *Zeitschrift für Physik* 57: 261–277.

—— (1929b). "Über den Begriff der Geschwindigkeit in der Diracschen Theorie des Elektrons." *Zeitschrift für Physik* 55: 127–140.

Fock, Vladimir A., and Iwanenko, D. (1929). "Géométrie quantique linéaire et déplacement parallèle." *Académie des Sciences* (Paris). *Comptes Rendus* 188: 1470–1472.

Givens, Wallace J., and Veblen, Oswald. Seminar 1935–1936. *Geometry of Complex Domains*. Princeton, New Jersey: The Institute for Advanced Study. Reissued with corrections, 1955.

Haefliger, André (1956). "Sur l'extension du groupe structural d'un espace fibré." *Académie des Sciences* (Paris). *Comptes Rendus* 243: 558–560.

Harish-Chandra (1953). "Representations of a Semisimple Lie Group on a Banach Space. I. " *American Mathematical Society. Transactions* 75: 185–243.

—— (1954a) "Representations of Semisimple Lie Groups. II." *American Mathematical Society. Transactions* 76: 26–65.

—— (1954b) "Representations of Semisimple Lie Groups. III." *American Mathematical Society. Transactions* 76: 234–253.

Hehl, Friedrich W. (1971). "How Does One Measure Torsion of Space-Time ?" *Physics Letters A* 36: 225–226.

—— (1980). "Four Lectures on Poincaré Gauge Field Theory." In Bergmann and de Sabbata, eds., 1980, pp. 5–61.

Hehl, Friedrich W., and Datta, B. K. (1971). "Nonlinear Spinor Equation and Asymmetric Connection in General Relativity." *Journal of Mathematical Physics* 12: 1334–1339.

Hehl, Friedrich W., Lord, Eric A., and Smalley, Larry L. (1981). "Metric-Affine Variational Principles in General Relativity. II. Relaxation of the Riemannian Constraint." *General Relativity and Gravitation* 13: 1037–1056.

Hehl, Friedrich W., von der Heyde, Paul, and Kerlick, David G. (1976). "General Relativity with Spin and Torsion: Foundations and Prospects." *Reviews of Modern Physics* 48: 393–416.

Infeld, Leopold, and Van der Waerden, Bartel L. (1933). "Die Wellengleichung des Elektrons in der allgemeinen Relativistätstheorie." *Preussische Akademie der Wissenschaften* (Berlin). *Sitzungsberichte*: 380–401.

Kannenberg, L., and Arnowitt, Richard L. (1967). "Stress-Tensor Operator for Fermion Field Interacting with Classical Gravitational Field." *Annals of Physics* (New York) 45: 416–444.

Kibble, T. W. B. (1961). "Lorentz Invariance and the Gravitational Field." *Journal of Mathematical Physics* 2: 212–221.

——— (1963). "Canonical Variables for the Interacting Gravitational and Dirac Fields." *Journal of Mathematical Physics* 4: 1433–1437.

Kichenassamy, S. (1986). "Lagrange Multipliers in Theories of Gravitation." *Annals of Physics* (New York) 168: 404–424.

Kopczyński, W. (1975). "The Palatini Principle with Constraints." *Académie Polonaise des Sciences. Série des Sciences Mathématiques, Astronomiques et Physiques. Bulletin* 23: 467–473.

Lemmer, G. (1965). "The Canonical Form of Interacting Gravitational and Spin Fields." Il *Nuovo Cimento* 37: 1647–1658.

Lenoir, Marcel (1971). "Sur les équations des particules de spin 1/2." *Académie des Sciences* (Paris). *Comptes Rendus, A* 273: 943–946.

Lichnerowicz, André (1964a). "Champs spinoriels et propagateurs en relativité générale." *Société Mathématique de France. Bulletin* 92: 11–100.

——— (1964b). "Champ de Dirac, champ du neutrino et transformations *C, P, T* sur un espace-temps courbe." *Institut Henri Poincaré. Annales A* 1: 233–290.

Liebscher, Dierck, E. (1980). "The Motion of Test Particles in Non-Riemannian Space-times." In Bergmann and de Sabbata, eds. 1980, pp. 125–137.

Milnor, John (1963). "Spin Structures on Manifolds." *Enseignement Mathématique* 9: 198–203.

Ne'eman, Yuval (1978). "Spinor-Type Fields with Linear, Affine and General Coordinate Transformations." *Institut Henri Poincaré. Annales* A. 28: 369–378.

Ne'eman, Yuval, and Sijacki, Dj. (1979). "Unified Affine Gauge Theory of Gravity and Strong Interactions with Finite and Infinite *GL*(4,*R*) Spinor Fields." *Annals of Physics* 120: 292–315.

Nester, James M. (1977). "Effective Equivalence of the Einstein-Cartan and Einstein Theories of Gravity." *Physical Review D 16*: 2395–2401.

Nitsch, Jürgen (1980). "The Macroscopic Limit of the Poincaré Gauge Field Theory of Gravitation." In Bergmann and de Sabbata, eds. 1980, pp. 63–91.

Parke, William C., and Jehle, Herbert (1965). "Covariant Spinor Formulation of Relativistic Wave Equations under the Homogeneous Lorentz Group." In *Lorentz Group*. Lectures in Theoretical Physics, No. 7A. Wesley E. Brittin, and Asim O. Barut, eds. Boulder: Univ. of Colorado Press, pp. 297–386.

Pauli, Wolfgang (1927). "Zur Quantenmechanik des magnetischen Elektrons." *Zeitschrift für Physik* 43: 601–623.

——— (1933a). "Über die Formulierung der Naturgesetze mit fünf homogenen Koordinaten. Teil I: Klassische Theorie. *Annalen der Physik* 18: 305–336.

——— (1933b). "Über die Formulierung der Naturgesetze mit fünf homogenen Koordinaten. Teil II: Die Diracschen Gleichungen für die Materiewellen." *Annalen der Physik* 18: 337–372.

Penrose, Roger (1960). "A Spinor Approach to General Relativity." *Annals of Physics* 10: 171–201.

Penrose, Roger, and Rindler, Wolfgang (1984). *Spinors and Space-Time.* Vol. 1. *Two-Spinor Calculus and Relativistic Fields.* London and New York: Cambridge University Press.

Peres, Asher (1962). "Spinor Fields in Generally Covariant Theories." *Il Nuovo Cimento,* 24 (Suppl.): 389–452.

Riesz, Marcel (1953). "L'équation de Dirac en relativité générale." *Tolfte Skandinaviska Matematikerkongressen. Comptes Rendus du Douzième Congrès des Mathématiciens Scandinaves. Tenu à Lund 10–15 Août 1953.* Lund: Håkan Ohlssons Boktryckeri, pp. 241–259.

Rumpf, Helmut (1980). "Quasi-classical Limit of the Dirac Equation and the Equivalence Principle in the Riemann–Cartan Geometry." In Bergmann and de Sabbata, eds. 1980, pp. 93–104.

Schouten, J.A. (1931a). "Dirac Equations in General Relativity. 1. Four-Dimensional Theory." *Journal of Mathematics and Physics* 10: 239–271

——— (1931b) "Dirac Equations in General Relativity. 2. Five-Dimensional Theory." *Journal of Mathematical Physics* 10: 272–283.

Schrödinger, Erwin (1932). "L'électron de Dirac dans la théorie de la relativité générale." In *Actes du Congrès International d'Electricité Paris, 1932.* Reprinted in Schrödinger, Erwin, *Collected Papers.* Vol. 3, *Contributions to Quantum Theory.* Vienna: Österreichischen Akademie der Wissenschaften: Braunschweig/Wiesbaden: Friedrich Vieweg und Sohn, 1984, pp. 470–480.

Schweizer, Martin A. (1980). "The Gauge Symmetries of Gravitation." In Bergmann and de Sabbata, eds. 1980, pp. 117–124.

Sciama, Dennis W. (1962). "On the Analogy between Charge and Spin in General Relativity." In *Recent Developments in General Relativity.* New York: Pergamon; Warsaw: PWN-Polish Scientific Publishers, pp. 415–439

Smalley, Larry L., and Ray, John R. (1986). "Geometrization of Spin and the Weyssenhoff Fluid Conjecture." *General Relativity and Gravitation* 18: 549–553.

Tetrode, H. (1928). "Allgemein-relativistische Quantentheorie des Elektrons." *Zeitschrift für Physik* 50: 336–346.

Trautman, Andrzej (1972). "On the Einstein-Cartan Equations. I." *Académie Polonaise des Sciences. Série des Sciences Mathématiques, Astronomiques et Physiques. Bulletin* 20: 185–190.

Uhlenbeck, George E., and Goudsmit, Samuel (1925). "Ersetzung der Hypothese vom unmechanischen Zwang durch eine Forderung bezüglich des inneren Verhaltens jedes einzelnen Elektrons." *Die Naturwissenschaften* 13: 953–954.

Van der Waerden, Bartel L. (1929). "Spinoranalysis." *Gesellschaft der Wissenschaften zu Göttingen. Nachrichten*: 100–109.

——— (1980). *A History of Algebra.* Berlin and New York: Springer-Verlag.

von der Heyde, Paul (1975). "The Equivalence Principle in the U_4 Theory of Gravitation." *Lettere al Nuovo Cimento* 14: 250–252.

Weyl, Hermann (1928). *Gruppentheorie und Quantummechanik.* Stuttgart: Hirzel.

——— (1929a). "Gravitation and the Electron." *National Academy of Sciences. Proceedings* 15: 323–334.

——— (1929b). "Elektron und Gravitation. I." *Zeitschrift für Physik* 56: 330–352.

——— (1931). *Gruppentheorie und Quantummechanik,* 2nd. rev. ed. Leipzig: S. Hirzel.

——— (1948). "A Remark on the Coupling of Gravitation and Electron." *Academia de Ciencias Exactas Fisicas y Naturales de Lima. Actas* 11: 1–17.

——— (1950). "A Remark on the Coupling of Gravitation and Electron." *Physical Review* 77: 699–701.

Whittaker, Sir Edmund T. (1937). "On the Relations of the Tensor-Calculus to the Spinor-Calculus." *Royal Society of London. Proceedings A* 158: 38–46.

Yasskin, Philip B., and Stoeger, William R. (1980). "Propagation Equations for Test Bodies with Spin and Rotation in Theories of Gravity with Torsion." *Physical Review D* 21: 2081–2094.

Levi-Civita and the General-Relativistic Problem of Motion

Thibault Damour and Gerhard Schäfer

In his late years, Tullio Levi-Civita (March 29, 1873–December 29, 1941) became interested in the problem of the motion of several bodies in general relativity. He was well aware of the earlier investigations by Droste and De Sitter (Droste 1916, De Sitter 1916),[1] but (rightly) felt the incompleteness of these works, notably with respect to the treatment of the self-field contributions of the bodies. Around the year 1936, he started an ambitious research program on the general-relativistic problem of motion aimed at: (1) clarifying the mathematical and physical assumptions underlying the (now-called) post-Newtonian approximation method, (2) studying the self-field effects, and (3) deriving the observable relativistic effects (periastron advance and acceleration of the center of mass) in binary systems comprising objects of comparable masses.[2] However, in spite of his very careful presentation of the issues at hand, the final results of Levi-Civita's calculations obviously show a diversity of self-contradictory features.

Regarding the self-field contributions of the bodies, Levi-Civita at first (Levi-Civita 1937a) claimed that they could be absorbed into the mass parameters of the bodies only in the two-body case (with Brillouin 1922, Levi-Civita called this phenomenon of renormalizability the "effacing principle"; see Damour 1987 for a recent discussion of this issue). In his later book (Levi-Civita 1950)[3], he claimed to show that the absorption of the self-field contributions could be extended to the n-body case.

Regarding the relativistic periastron advance, Levi-Civita had obtained three different results. At the Harvard Tercentenary Conference of Arts and Sciences, September 1936, he gave (Robertson 1938) a preliminary report on his work and announced an incorrect result of $(1+5p/3)$ times the correct one; $p = m_1 m_2/(m_1+m_2)^2$, where m_1 and m_2 are the mass parameters of the bodies. In his second paper (Levi-Civita 1937b), one can read the correct result. In his book,[4] he had $(1 + p/3)$ times the correct result. Strangely enough, in his book, he did not comment on these differences nor on the differences with papers (Robertson 1938, Eddington and Clark 1938) that were well known to him and where two independent derivations of the correct result were given. On the contrary, he felt secure enough about his last-found (incorrect) result to comment on the deviation with respect to the familiar (since Einstein's Mercury perihelion calculation) restricted two-body result (with the replacement, "mass of the central body" changes into "sum of the masses of the two bodies"). One can interpret, to the credit of Levi-Civita, this confidence in his results, notwithstanding their discrepancy with other results, from his probable feeling that the other approaches contained some basic methodological obscurities. Indeed, on the one hand, the calculation of Robertson was based on a paper (Einstein et al. 1938) that introduced a complicated and ill-justified method,[5] and led to calculations so complex that they could not be published in detail. On the other hand, the calculation of Eddington and Clark, far from following from a consistent method, consisted of grafting onto the previous works of Droste and De Sitter some purely heuristic arguments concerning the relativistic concept of mass and the "effacement" of self-field effects (i.e., precisely two of the issues to which Levi-Civita wanted to provide well-justified answers by means of a logically consistent method).

Regarding the secular acceleration of the center of mass motion of a binary system, Levi-Civita published two different results. In his second article (Levi-Civita 1937b), one finds a nonzero secular acceleration and in his later book he claimed to prove the absence of any acceleration, in accordance with Roberston, and Eddington and Clark. In his book, he stated that only "une erreur matérielle de calcul" was responsible for his (incorrect) first result.

A superficial look at the various results of Levi-Civita that we have quoted above may lead one to believe that Levi-Civita was the first one to derive (in 1937) the correct result for the relativistic periastron advance and that he confirmed (in his book), after Robertson and Eddington and Clark, that there is no secular acceleration of the center of mass of a binary system. Furthermore, he seems to have proved, in his book, the effacing principle for Einstein's theory at the first post-Newtonian order of approximation. However, in view of the diversity of Levi-Civita's results, a thorough investigation of his contributions is obviously called for to clarify what has been correctly derived by him and what are the kinds of errors involved, calculational or methodological. The two articles (Levi-Civita 1937a, 1937b) do not give enough details for enabling one to trace back the origin of his different results, therefore one must make use of his book.

Levi-Civita first thought that it was possible to absorb the self-field contributions into the mass parameters of the bodies only for the two-body problem. This seems to have been the consequence of a calculational error that has not been made in the book [compare the correct equation, Eq. (V. 47), of the book with Eq. (16) of the first article]. In his book, he formally showed that one can efface self-field effects for a system of pressureless bodies. However, as had previously been remarked by Einstein (see Levi-Civita 1937a), a dust model is unable to yield dynamically stable bodies.[6] The first physically satisfactory treatment of self-field effects is to be found in the works by Fock and co-workers (Fock 1939, Petrova 1940)[7] who worked throughout with an ideal fluid model for the matter.

In Chapter VI, Section 3.2, of his book, Levi-Civita concluded in favor of the absence of secular acceleration. A careful examination shows that this conclusion is inconsistent with the equations of motion that he wrote in the book, which still predict a secular acceleration. There are in fact several misprints, and forgotten coefficients, in his crucial sequence of Eq. (VI. 80), but as they stand and as they should read, according with his previous result in Eq. (VI. 39), they lead to a nonzero secular acceleration. Obviously, these are calculational errors. The main error is located between Eq. (VI. 37) and Eq. (VI. 38), where Levi-Civita computed $B_{h|i}D_h^{(2)}$. There, he omitted the term proportional to the spatial gradient of $\beta_r^2 = (\vec{n} \cdot \vec{v})^2$, possibly because of the notation, which makes it resemble a pure velocity-dependent term.[8] In our opinion, this was a methodological error, linked to the difficulties of implementing his way[9] of finding a "constant-dependent" Lagrangian. Indeed, he looked for a function $L(\vec{r}, \vec{v}, e, a)$ such that $(\partial_{\vec{r}} - d/dt\, \partial_{\vec{v}})L$ would give the correct equations of motion when, after differentiation, e and $1/a$ are equated to the energy and the angular momentum (modulo some coefficients). In such an approach, it is easy to make mistakes by replacing $\vec{v}^2$ or $(\vec{n} \cdot \vec{v})^2$ in terms of e, $1/a$ and $1/r$ too soon. Moreover, the term proportional to $r^{-1}\partial_i \beta_r^2$ that he omitted is the most difficult to compute correctly with his approach. No wonder he never got it right. It is certainly possible, by treating this term differently (and still incorrectly) and making some slight calculational errors, to get his earlier Lagrangian (Levi-Civita 1937b). On the other hand, by treating this term correctly, we found (Damour and

Schäfer 1988) that the first post-Newtonian correction to the Newtonian relative-motion Lagrangian should read (in his notation):

$$D^{\text{correct}} = \Pi - (\tfrac{1}{2}\beta^2 - \gamma - e)p\gamma,$$

where Π is given in Eq. (5) of the 1937b paper. This also differs from Eq. (VI. 42) in his book. A further effect related to this correction is the absence of any secular acceleration of the center of mass motion. The surprising result that Levi-Civita coincidentally obtained the correct periastron advance in his 1937b paper (and a correct potential function for the relative spatial trajectory, as already verified by Eddington and Clark) lies in the fact that the correction to Π in our equation above happens to drop out for this effect (however, it does contribute to the computation of the relative motion as a function of time).

In conclusion, we may say that Levi-Civita was the first to stress the need for setting up a mathematically consistent framework for the general-relativistic problem of the motion of n weakly self-gravitating bodies, including a clear treatment of self-field effects. Moreover, he was also the first to emphasize the physical importance of deriving several observable relativistic effects in binary systems comprising objects of comparable masses, namely, the relativistic periastron advance and the a priori possible acceleration of the center of mass motion. Concerning the general framework, in his book, he gave a careful presentation of the many subtle (mathematical and physical) issues involved in the relativistic problem of motion. This beautiful model of what a relativistic theory of the n-body problem should look like has, alas, rarely been matched by subsequent works in this field. Particularly remarkable is Levi-Civita's clear recognition of the importance of the "effacement" properties of self-field effects in general relativity.

However, as far as explicit new results are concerned, one must admit that Levi-Civita fell short of achieving the ambitious aims of his research program. He went beyond Droste and De Sitter in correctly expressing an n-body metric in terms of a coordinate-conserved mass density (p. 55 of his book; the density μ is defined so that $\int \mu \, d^3x$ is constant). If he had been willing to ignore heuristically self-field effects, he could have probably preceded Eddington and Clark in deriving explicitly the correct metric for n-point masses and their equations of motion. However, his (self-contradictory) use of a pressureless model for the matter rendered his proof of the effacement of self-field effects physically unsatisfactory. The merit for a satisfactory demonstration must be credited to Fock and his school (Fock 1939, Petrova 1940), although there exists a long forgotten forerunner (Lorentz and Droste 1917) that had explicitly achieved this (for a particular incompressible fluid model). On the other hand, the priority for the correct derivation of the relativistic periastron advance, and the absence of secular acceleration of the Newtonian center of mass, of a binary system belongs to Robertson (Eddington and Clark coming about five months after Robertson). Fock proved, within his approach, the absence of acceleration of a suitably defined relativistic center of mass in 1941 (Fock 1941). The explicit, fully effaced form of the n-body Lagrangian was first published in 1950, the year of the posthumous publication of Levi-Civita's book (Fichtenholz 1950). It must, however, be remarked that this Lagrangian is, after a small substitution for the "renormalized" masses, identical to the 1917 result of Lorentz and Droste.[10]

As a final cautionary remark, and as a tribute to the vision of Levi-Civita, it must be said that no complete and logically consistent treatment of the relativistic theory of n extended bodies at the first post-Newtonian approximation level has been published to date.[11]

NOTES

[1] In the 1930s, these works were considered the standard approach to the relativistic problem of motion and were, e.g., discussed in detail in Chazy (Chazy 1930). The fact that (because of an error that originated in Droste 1916) they lead to incorrect equations of motion was apparently first noted by Eddington and Clark (Eddington and Clark 1938). The early correct treatment of Lorentz and Droste in 1917 was forgotten in the literature until much later.

[2] In all previous applications, one had considered only the "restricted" relativistic two-body (or n-body) problem, i.e., the case where one of the masses is much bigger than the other one(s).

[3] Apart from some misprints, Levi-Civita 1964 is, as we have checked, a mere translation of Levi-Civita 1950.

[4] The completion of this book took place between 1939 (last quoted reference) and 1941 (Levi-Civita's death).

[5] In effect, the method used by Einstein, Infeld, and Hoffmann was logically flawed. In subsequent publications, Einstein and Infeld admitted this and more or less corrected the method, but introduced, so doing, a technical mistake. In fact, it is only through the recent research on the motion of strongly self-gravitating bodies that the justification and realm of applicability of the Einstein–Infeld–Hoffmann approach have been clarified (see Damour 1987 and references therein).

[6] Both in his first article and in Chapter V, Section 4 (in particular, Hypothesis II), of his book, Levi-Civita assumed an internal structure for his bodies that he could never dynamically realize within his dust model.

[7] The work by Petrova was published only much later (Petrova 1949).

[8] On the other hand, his treatment of the $\dot{\beta}_j$ acceleration term (of the other body) as constant is correct because, at this stage, he dealt with the geodesic Lagrangian of a test particle in the field of the other body.

[9] In his book, Levi-Civita said (Chapter VI, Section 2.2) that unfortunately he did not otherwise "succeed."

[10] Lorentz and Droste did not try to make contact with the previous (incompatible) results of Droste and De Sitter, nor did they discuss the difference in one coefficient of g_{44} between Droste 1916a and Droste 1916b.

[11] Written in July 1990.

REFERENCES

Brillouin, Marcel (1922). "Gravitation einsteinienne. Statique. Points singuliers. Le point matériel. Remarques diverses." *Académie des Sciences* (Paris). *Comptes Rendus* 175: 1009–1012.

Chazy, Jean (1930). *La théorie de la relativité et la mécanique céleste.* Vol. 2. Paris: Gauthier-Villars.

Damour, Thibault (1987). "The Problem of Motion in Newtonian and Einsteinian Gravity." In *Three Hundred Years of Gravitation.* Stephen W. Hawking and Werner Israel, eds. Cambridge and New York: Cambridge University Press, pp. 128–198.

Damour, Thibault and Schäfer, Gerhard (1988). "Higher-Order Relativistic Periastron Advances and Binary Pulsars." *Il Nuovo Cimento B* 101: 127–176.

De Sitter, Willem (1916). "On Einstein's Theory of Gravitation and Its Astronomical Consequences. Second Paper." *Royal Astronomical Society. Monthly Notices* 77: 155–184.

Droste, Johannes (1916). "Het veld van *n* bewegende centra in Einstein's theorie der zwaartekracht." *Koninklijke Akademie van Wetenschappen te Amsterdam. Wis- en Natuurkundige Afdeeling. Verslagen van de Gewone Vergaderingen* 25: 460–467. English translation "The Field of *n* Moving Centers in Einstein's Theory of Gravitation." *Koninklijke Akademie van Wetenschappen te Amsterdam. Section of Sciences. Proceedings* 19 (1916–1917): 447–455.

Eddington, Arthur S. and Clark, Gordon L. (1938). "The Problem of *n* Bodies in General Relativity Theory." *Royal Society of London. Proceedings A* 166: 465–475.

Einstein, Albert, Infeld, Leopold, and Hoffmann, Banesh (1938). "The Gravitational Equations and the Problem of Motion." *Annals of Mathematics* 39: 65–100.

Fichtenholz, I.G. (1950). "Lagranzheva forma uravneniy dvizheniya vo vtorom priblizhenii teorii tyagoteniya Eynshteyna" ["Lagrange Form of the Equations of Motion in the Second Approximation of Einstein's Theory of Gravitation"]. *Zhurnal Eksperimentalnoi i Teoreticheskoi Fiziki* 20: 233–242.

Fock, Vladimir A. (1939). "Sur le mouvement des masses finies d'après la théorie de gravitation einsteinienne." *Journal of Physics* (Moscow) 1: 81–116.

——— (1941). "Sur les intégrales du centre de gravité dans le problème relativiste de deux masses finies." *Académie des Sciences* (Moscou). *Comptes Rendus (Doklady)* 32: 25–27.

Levi-Civita, Tullio (1937a). "The Relativistic Problem of Several Bodies." *American Journal of Mathematics* 59: 9–22.

——— (1937b). "Astronomical Consequences of the Relativistic Two-Body Problem." *American Journal of Mathematics* 59: 225–234.

——— (1950). *Le problème des n corps en relativité générale.* Mémorial des Sciences Mathématiques, No. 116. Paris: Gauthier-Villars. English translation *The n-Body Problem in General Relativity.* Dordrecht: D. Reidel, 1964.

Lorentz, Hendrik A., and Droste, Johannes (1917). "De beweging van een stelsel lichamen onder den invloed van hunne onderlinge aantrekking, behandeld volgens de theorie van Einstein I, II." *Koninklijke Akademie van Wetenschappen te Amsterdam. Wis- en Natuurkundige Afdeeling. Verslagen van de Gewone Vergaderingen* 26 (1917–1918): 392–403; 649–660. English translation in Lorentz 1934–1939, Vol. 5, pp. 330–355.

Lorentz, Hendrik A. (1934–1939). *Collected Papers.* 9 vols. The Hague: Nijhoff.

Petrova, N. M. (1940). "On the Equations of Motion and the Mass Tensor for Systems of Finite Masses in the General Theory of Relativity." PhD dissertation, Leningrad.

——— (1949). "Ob uravnenii dvizheniya i tenzore materii dlya sistemy konechnylkh mass v obshchey teorii otnositel'nosti." ["On the Equations of Motion and the Mass Tensor for Systems of Finite Masses in the General Theory of Relativity."] *Zhurnal Eksperimentalnoi i Teoreticheskoi Fiziki* 19: 989–999.

Robertson, Howard P. (1938). "Note on the Preceding Paper: The Two-Body Problem in General Relativity." *Annals of Mathematics* 39: 101–104.

Myron Mathisson's and Jan Weyssenhoff's Work on the Problem of Motion in General Relativity

Bronisław Średniawa

1. Mathisson's Early Work

Myron Mathisson (1897–1940) was born in Warsaw on December 14, 1897 (Średniawa 1980, 1982, 1985). He began his studies at Warsaw Technical University. Having obtained his first diploma after three years of study, he transferred to the Philosophical Faculty of Warsaw University. He earned a living by doing calculations for the construction of buildings. In 1930, Mathisson obtained his doctorate; the degree was conferred by Professor Czesław Białobrzeski. Then, while he continued his engineering calculations, Mathisson started giving lectures on theoretical physics at Warsaw University. He lectured on quantum mechanics (following Weyl's *Gruppentheorie und Quantenmechanik*) and the theory of relativity. In June 1932, he took his qualifying exams in Warsaw, and in 1935–1936, he spent a year at the University of Kazan in the Soviet Union.

Mathisson's whole scientific activity was devoted to the problem of motion in the theory of relativity, in both the general and special theories. Since his work has already been discussed in various publications (Średniawa 1980, 1982, 1985; Havas 1986), we shall confine ourselves to short remarks on his papers.

In his first paper (Mathisson 1931a), Mathisson derived the equations of geodesics from the equations of the gravitational field, assuming, as did his predecessors, that the particle as the source of a gravitational field is centrally symmetric (in a suitably chosen frame of reference). But, in his second paper, (Mathisson 1931b) he expressed the opinion that the equations of motion could depend on the structure of the particles, and he made the first attempt to abandon the too stringent assumption of central symmetry of the field of the particle.

We shall present Mathisson's results as presented in his most important paper (Mathisson 1937a), entitled "On the New Mechanics of Material Systems," for the special case of a weak gravitational field, with a Minkowski metric as the background metric. Mathisson considered the particle as a singularity of the field and surrounded its world line with a world tube. Then, he replaced the singularity by a regular gravitational field inside the tube, which was the extension to its interior of the field outside the tube. The structure of the particle was taken into account by abandoning the spherical symmetry of its gravitational field. Then, the gravitational field of the particle was split into the sum of a background field and a small deviation from it. This weak field satisfied in first approximation an appropriate system of differential equations in which the energy-momentum tensor occurred. To these equations, Mathisson attached (similarly to Einstein and Grommer 1927) the "normalization condition" expressing that the four-divergence of the energy-momentum tensor vanishes. We present this for the Minkowski background $\overset{0}{g}_{\alpha\beta}$ with

$$g_{\alpha\beta} = \overset{0}{g}_{\alpha\beta} + \gamma_{\alpha\beta'} \quad \alpha, \beta = 0,\ 1,\ 2,\ 3.$$

The energy-momentum tensor $T^{\alpha\beta}$ satisfies the conservation-law

$$\partial_\beta T^{\alpha\beta} = 0, \quad T^{\alpha\beta} = T^{\beta\alpha}.$$

Then, Mathisson developed two methods to derive the equations of motion. The first one consisted of the formulation of a variational principle that is equivalent to the field equations. He chose a four-dimensional region $D^{(4)}$ surrounding the tube representing the motion of the particle and an arbitrary vector field $\xi^\alpha(x^\beta)$ that vanishes on the boundary of $D^{(4)}$ Then, the variational principle has the form

$$\int_{D^{(4)}} T^{\alpha\beta} \partial_\beta \xi_\alpha \, d^4x = 0,$$

or, by defining

$$\partial_\beta \xi_\alpha = p_{\alpha\beta},$$

the form

$$\int_{D^{(4)}} T^{\alpha\beta} p_{\alpha\beta} \, d^4x = 0. \tag{1}$$

Let us choose for the region $D^{(4)}$ a narrow tube surrounding the primary singularity of the field and let us consider the free particle, assuming

$T^{\alpha\beta} = 0$ outside this tube. Let us take in this tube a time-like line L, which we regard as the world line of the particle:

$$x^\alpha = x^\alpha(\tau) \tag{2}$$

(τ is the proper time). Let us construct in each point of the line L the section $\Sigma(\tau)$ perpendicular to it and expand $p_{\alpha\beta}$ in a Taylor series. This allows us to transform the four-dimensional integral on the left-hand side of Eq. (1) into a curvilinear one and to obtain in this way "Mathisson's variational principle":

$$\int_{\tau_1}^{\tau_2} (m^{\alpha\beta} p_{\alpha\beta} + d^{\lambda\alpha\beta}\, \partial_\lambda p_{\alpha\beta} + q^{\lambda\mu\alpha\beta}\, \partial_\lambda \partial_\mu p_{\alpha\beta})\, d\tau = 0. \tag{3}$$

Here, in the coordinate system, where $u^i = 0$ $(i, j, \ldots = 1, 2, 3)$, we have

$$m^{\alpha\beta} = \int_{\Sigma(\tau)} T^{\alpha\beta}\, d^3x,$$
$$d^{ix\beta} = \int_{\Sigma(\tau)} x^i T^{\alpha\beta}\, d^3x, \quad d^{0\alpha\beta} = 0, \ldots$$

In lowest approximation, we get from the variational principle Eq. (3),

$$m^{\alpha\beta} = m_0 u^\alpha u^\beta, \quad m_0 = \text{const.},$$

and the equations of geodesics

$$\dot{u}^\alpha \equiv \frac{du^\alpha}{d\tau} = 0,$$

as the equation of motion for a free particle in Minkowski's world. In the next approximation, it can be shown that

$$d^{\lambda\alpha\beta} = \tfrac{1}{2}(s^{\alpha\lambda} u^\beta + s^{\lambda\beta} u^\alpha) + n^\lambda u^\alpha u^\beta, \tag{4}$$

where

$$s^{\alpha\beta} = -s^{\beta\alpha}, \quad s^{\alpha\beta} u_\beta = 0, \quad n^\lambda u_\lambda = 0$$

and

$$n^i = \int_{\Sigma(\tau)} x^i T^{00}\, d^3x,$$

for the coordinate system where $u^i = 0$. The particle is then characterized by its mass m_0, spin bivector $s^{\alpha\beta}$, and gravitational dipole moment n^λ.

By putting $n^\lambda = 0$, Mathisson limited himself to positive masses and obtained the equations of motion of the "spin particle":

$$m_0 \dot{u}^\alpha + s^{\alpha\beta} \ddot{u}_\beta = 0, \tag{5}$$

with the condition

$$\dot{s}^{\alpha\beta} = s^{\alpha\lambda} \dot{u}_\lambda u_\beta - s^{\beta\lambda} \dot{u}_\lambda u^\alpha. \tag{6}$$

Mathisson solved these equations (Mathisson 1937b) for the nonrelativistic case, i.e., for a slowly moving particle and obtained in an appropriate coordinate system a uniform circular motion of arbitrary radius.

Mathisson's second method of obtaining equations of motion from the field equations consisted of expanding retarded gravitational potentials into multipole terms and applying the "normalization condition" (Mathisson 1931b).

2. The Collaboration between Mathisson and Weyssenhoff

Mathisson's work aroused the interest of Jan Weyssenhoff (1889–1972), who in 1935 became professor of theoretical physics at Jagellonian University in Krakow. When Mathisson returned from Kazan, Weyssenhoff invited him to Krakow, having obtained a research grant for him that was provided by a group of rich Polish businessmen. This grant would enable Mathisson to devote himself exclusively to scientific work. In 1937, Mathisson moved to Krakow and joined Weyssenhoff and his collaborators, the physicist Józef Lubański, who had came with Weyssenhoff to Krakow from Vilna (where Weyssenhoff was professor until 1935), and the mathematician Adam Bielecki. Their collaboration lasted until the spring of 1939, when Mathisson went to France and then to England and Lubański went to Holland, and it resulted in three publications. The first was a paper by Lubański, which was written under Mathisson's direction (Lubański 1937). In it, Lubański derived Mathisson's equations of motion, Eqs. (5) and (6), for a spin particle, using the method of expanding the gravitational potentials of the linearized gravitational field into potentials of gravitational multipoles. The second publication was a short letter to *Nature* by Weyssenhoff about the nonradiating motion of a spinning electron with an electric charge and a magnetic moment (Weyssenhoff 1938). In the third paper (Bielecki, Mathisson and Weyssenhoff 1939), the authors formulated and proved in a precise way the principal theorem of Mathisson's earlier paper (Mathisson 1937a) concerning the transformation of the fourfold integral on the left-hand side of Eq. (1) into the single integral that appears in Eq. (3). In Krakow, Mathisson began to work with Lubański on the motion of spin particles in gravitational and electromagnetic fields. Mathisson continued this work on his own after he left Poland in 1939. His results were published in England in 1940 (Mathisson 1940, 1942). Mathisson died in Cambridge on September 13, 1940.

3. The Work of Weyssenhoff and Raabe on Spin Particles

In the fall of 1939, World War II began. Because of the war events, in September 1939, Weyssenhoff went to Lvov, where he stayed until 1941 and lectured on experimental physics at Lvov Technical University. There, he met the young Warsaw physicist Antoni Raabe. In 1941, Weyssenhoff returned with Raabe to Krakow. There, they began to continue and develop Mathisson's ideas. Of course, their work took place in secrecy, since during the German occupation, secondary schools and universities in Poland were closed. Teaching or doing scientific work was punishable by a prison sentence and deportation to a concentration camp. Weyssenhoff and Raabe's work could not be published during the German occupation. It appeared only after the war (Weyssenhoff and Raabe 1947a, 1947b; Weyssenhoff 1947).

Weyssenhoff and Raabe defined the momentum vector of the spin particle as

$$G^{\alpha} = m_0 u^{\alpha} + s^{\alpha\beta}\dot{u}_{\beta}. \tag{7}$$

With the help of this expression, Mathisson's equations (5) and (6) for the spin particle could be written in the simple forms

$$\dot{G}^{\alpha} = 0 \tag{8}$$

and

$$\dot{s}^{\alpha\beta} = G^{\alpha}u^{\beta} - G^{\beta}u^{\alpha}, \tag{9}$$

which should be called the "Weyssenhoff–Raabe equations." Making the very plausible assumption that the vector G^{α} is time-like, they could solve these equations generally. In a coordinate system where $G^{i} = 0$ they obtained a uniform circular motion with arbitrary velocity $v < c$; the radius of the circle, which decreased with increasing velocity v, could be arbitrary as well.

Weyssenhoff wanted to consider the whole "circle" as a particle of microscopic size. He postulated that the velocity of the circular motion should be equal to that of light. Weyssenhoff and Raabe deduced this limiting case from Eqs. (8) and (9) and solved the equations found in this way (Weyssenhoff and Raabe 1947b). They obtained a circle with a radius that depended on the mass and the spin of the particle.

In June 1942, Raabe fell victim to a German manhunt in the streets of Krakow and was sent to Auschwitz, where he died four months later. Weyssenhoff continued the work alone. He compared the properties of a spin particle moving with the speed of light with those of the Dirac electron. He found the following analogies.

1. In both theories, the momentum G^i is not parallel to the velocity v^i.
2. The equality of the kinematical velocity v to the speed of light c corresponds to the fact that the eigenvalues of the velocity operator in Dirac's theory are $\pm c$.
3. In both theories, angular momentum is not a constant of the motion.
4. The circular motion is a "picture" of the *Zitterbewegung*.
5. The size of the relativistic circle is of the order of the Compton wavelength.
6. The condition of stability of the relativistic circle in an electromagnetic field is the same as the condition of the nonformation of electron pairs in this field.

These analogies suggest that a spin particle that moves with the speed of light could in a certain sense be regarded as a relativistic model of Dirac's electron.

Further problems concerning relativistic spin particles were considered after the war by Weyssenhoff and his collaborators in Krakow.

References

Bielecki, Adam, Mathisson, Myron, and Weyssenhoff, Jan (1939). "Sur un théorème concernant une transformation d'intégrales quadruples en intégrales curvilignes dans l'espace de Riemann." *Académie Polonaise des Sciences. Classe des Sciences Mathématiques et Naturelles. Bulletin A:* 21–28.

Einstein, Albert, and Grommer, Jakob. (1927). "Allgemeine Relativitätstheorie und Bewegungsgesetz." *Preussische Akademie der Wissenschaften* (Berlin). *Physikalisch-Mathematische Klasse. Sitzungsberichte*: 2–13.

Havas, Peter (1986). "The Early History of the 'Problem of Motion' in General Relativity." In *Einstein and the History of General Relativity.* Don Howard and John Stachel, eds. Boston: Birkhäuser, pp. 234–276.

Lubański, Jósef (1937). "Neue Bewegungsgleichungen materieller Systeme in der Minkowskischer Welt." *Acta Physica Polonica* 6: 356–370.

Mathisson, Myron (1931a). "Die Beharrungsgesetze in der allgemeinen Relativitätstheorie." *Zeitschrift fur Physik* 67: 270–277.

——— (1931b). "Die Mechanik des Materieteilchens in der allgemeinen Relativitätstheorie." *Zeitschrift fur Physik* 67: 826–844.

——— (1937a). "Neue Mechanik materieller Systeme." *Acta Physica Polonica* 6: 163–200.

——— (1937b). "Das Zitternde Elektron und Seine Dynamik." *Acta Physica Polonica* 6: 218–227.

——— (1940). "The Variational Equation of Relativistic Dynamics." *Cambridge Philosophical Society. Proceedings* 36: 331–350.

——— (1942). "Relativistic Dynamics of a Spinning Magnetic Particle." *Cambridge Philosophical Society. Proceedings* 38: 40–60.

Średniawa, Bronisłav (1980). "Relativistic Equations of Motion of 'Spin Particles'." In *Cosmology and Gravitation. Spin, Torsion, Rotation, and Supergravity.* Peter G. Bergmann and Venzo de Sabbata, eds. New York and London: Plenum Press, pp. 423–434.

——— (1982). "Myron Mathisson (1897–1940)." *Postpy Fizyki* 33: 373–383

——— (1985). "History of Theoretical Physics at Jagellonian University in Krakow in the XIXth Century and in the First Half of XXth Century." *Zeszyty Naukowe Uniwersytetu Jagiellońskiego Prace Fizyczne* 24. Warsaw and Krakow: Polish Scientific Publishers, pp. 165–166, 198–201.

Weyssenhoff, Jan W. (1938). "A Non-Radiating Motion of a Spinning Electron." *Nature* 141: 328.

——— (1947). "On Two Relativistic Models of Dirac's Electron." *Acta Physica Polonica* 9: 46–53.

Weyssenhoff, Jan W., and Raabe, Antoni (1947a). "Relativistic Dynamics of Spin Fluids and Spin Particles." *Acta Physica Polonica* 9: 7–18.

——— (1947b). "Relativistic Dynamics of Spin Particles Moving with the Velocity of Light." *Acta Physica Polonica* 9: 19–25.

The Cauchy Problem in General Relativity—The Early Years

John Stachel

1. Introduction

A recent review article on the Cauchy or initial-value problem in general relativity remarks on the "resurgence of interest" in "the initial value problem, the canonical formalism, linearization stability and the space of gravitational degrees of freedom," and notes "the intimate relationship between these topics of current interest (Fischer and Marsden 1979, p. 138). Neither this article, nor other standard surveys of the same topic over the last quarter-century (see, for example, Bruhat 1962, Choquet-Bruhat and York 1980, York 1979), give references to any literature on the subject earlier than Lichnerowicz 1939, which seems to have become the standard cutoff for references in this field (the earliest reference in York 1979 is actually to Lichnerowicz 1944).

Yet, leaving aside global problems, almost all the significant conceptual advances in this field—as opposed to technical-mathematical ones—had been made well before 1939. It is the purpose of this paper to outline the "prehistory," so to speak, of this important topic, including not only work on the usual Cauchy problem on a space-like hypersurface, but also on the null initial-value problem for characteristic hypersurfaces. As is so often the case in general relativity (not to speak of other fields of physics), we shall see that a number of investigators approached these problems from various angles, without the later ones indicating any awareness of earlier work on the topic. (Indeed, one may find a number of examples in this area of a metatheorem that I once proposed: Anything worth discovering once in general relativity has been discovered independently at least twice!)

2. Mathematical Preliminaries

Before turning to the subject of this paper, let me briefly review some mathematical preliminaries, starting with the classic results on the Cauchy problem for systems of partial differential equations, based on the work of Augustin Cauchy (1789–1857; see Cauchy 1842) and Sofia Kovalevskaia [Sonia Kovalevsky] (1850–1891; see Kovalevskaia 1875). Consider a system of p partial differential equations for p functions, which depend on $q + 1$ independent variables. Such a system is said to be in *normal form* when the highest-order partial derivatives of the p dependent functions with respect to *one* of the independent variables (say x_1) are all equated to p given functions F_1, F_2, ..., F_p of (at most) the independent variables, the dependent variables, and their lower-order derivatives with respect to the independent variables. If the system is not initially in normal form, it is possible that it can be put into this form by a suitable transformation of the independent variables. We shall suppose the system under consideration to be in normal form. For some value of the first independent variable, say x_1^0, let the values of all the dependent functions and their derivatives with respect to this variable, up to but not including the highest-order derivatives, be prescribed; these values are collectively known as the initial values. Then, in the analytic case, in which the functions F_1, F_2, ..., F_p and the initial values are analytic functions of all their arguments, the Cauchy–Kovalevskaia existence theorem asserts that a unique solution, itself analytic, always exists in some finite neighborhood of x_1^0 that assumes the given initial values for $x = x_1^0$.

Of course, even if a system of partial differential equations is in Cauchy normal form, this does not mean that the Cauchy problem is the relevant problem to pose for this system. To decide whether it is, we need the concept of a well-posed problem, introduced by Jacques Hadamard (1865–1963). Hadamard's work is part of a long tradition of work on partial differential equations in France, which includes work on the Cauchy problem and other initial-value, boundary-value and mixed problems by many other prominent mathematicians, such as Elie Cartan (1869–1951), Edouard Goursat (1858–1936), Maurice Janet (1888–1983), Emile Picard (1856–1941), and Charles Riquier (1853–1929). It is sufficient here to recall Hadamard's studies of the Cauchy problem, culminating in his book on the subject (Hadamard 1923; see also Hadamard 1903, Hadamard 1964, Hadamard 1968), which first clearly formulated the idea of a well-posed mathematical problem. A problem is said to be well posed if a solution exists (possibly subject to certain other conditions), is unique, and depends continuously on the data used to specify such a solution. A unique solution to the Cauchy problem for a system of partial differential equations may exist, but the solution will generally not depend continuously on the initial data.

For linear hyperbolic-normal partial differential equations, such as the wave equation, however, the Cauchy problem on a space-like hypersurface is well posed, as Hadamard proved by actually constructing the solution as a functional of the initial data. He also showed that, for such hyperbolic-normal partial differential equations (linear or nonlinear), there exist real hypersurfaces in the space of independent variables, called characteristic hypersurfaces, on which the Cauchy problem cannot be well-posed: Uniqueness of the solution fails no matter how high the order of derivatives of the dependent functions specified on such hypersurfaces. Hadamard also demonstrated that the value of a solution to the field equations at some point exterior to the space-like hypersurface on which initial data are specified depends exclusively on the values of that initial data within and on the boundary of the intersection of that hypersurface with the characteristic conoid, i.e., the null hypersurface that has the point in question as its vertex.

When considered as a system of second-order partial differential equations, the field equations of general relativity raise several interesting problems. First of all, as is the case for any system of equations that is invariant under a gauge group depending on one or more arbitrary functions (here the group of diffeomorphisms depending on four such functions), the equations are not in Cauchy normal form. This gives rise to the mathematical problem of finding the *degree of indetermination* of the system (see, e.g., Janet 1927): i.e., ascertaining what supplementary data are needed to completely determine a unique equivalence class of solutions to the system, when two solutions related by a gauge transformation are regarded as equivalent. This mathematical problem is obviously very closely related to the physical problem of finding the *degrees of freedom* of the gravitational field, or *true observables*, as they have come to be called (see, e.g., Bergmann 1957). Then, there is the question of the characteristic hypersurfaces of the Einstein field equations. Here the signature of the metric is crucial. For the Lorentzian signature, the characteristics are real, and coincide with the null hypersurfaces. It is also necessary to demonstrate that all physical effects, as opposed to gauge-dependent effects, propagate within the characteristic or null conoids. The early work on solving these problems was done by four men: Hilbert, Darmois, Lanczos, and Stellmacher.

3. David Hilbert

David Hilbert (1862–1943) was certainly the first person to discuss the Cauchy problem for the Einstein equations. He did so in his second memoir on "The Foundations of Physics," dated November 1916 (Hilbert 1917). He emphasized that the general covariance of the field equations precludes their being of Cauchy normal form, and noted the serious problem for physics that this mathematical circumstance poses. His discussion of this question is almost certainly a response to Einstein's earlier rejection of generally-covariant field equations on the grounds that they were incompatible with the "law of causality" ("Kausalgesetz," see Einstein 1914, p. 1066). Einstein had developed a general argument leading to this conclusion in 1913, and only found the way around this argument late in 1915 when he had returned to generally-covariant gravitational field equations (for a discussion of Einstein's so-called "hole argument," see Stachel 1989). However, in these papers Einstein was always a bit vague about just what he meant by causality.

Hilbert's paper gave a quite precise meaning to the concept. For a physical system obeying laws involving no higher than second-order time derivatives, he noted that causality in physics had hitherto taken the form of the following principle: knowledge of certain physical quantities, and their first-order time derivatives if the law involves second-order time derivatives, at some initial time uniquely determines the entire future evolution of the system. The validity of this principle was assured by the circumstance that the physical laws in question could be expressed mathematically in Cauchy normal form (see Hilbert 1917, p. 59).

Hilbert showed that this form of the causality principle no longer holds in general relativity. He was looking for a general-relativistic generalization of Mie's special-relativistic, nonlinear electrodynamics (an early attempt at a unified field theory) in the hope of explaining the properties of matter (the existence of the electron, in particular) from a purely field-theoretical standpoint. So he used not the empty-space Einstein equations, but the Einstein equations coupled to electromagnetic field equations. The exact form of the Lagrangian leading to combined gravito-electromagnetic field equations is left unspecified by Hilbert throughout most of his first memoir (Hilbert 1915), but the combined field is represented by fourteen quantities: the electromagnetic potential four-vector, and the ten components of the metric tensor, which act as the gravitational potentials. Like Mie's original theory, Hilbert's generalization is not (necessarily) gauge-invariant. However, it is generally covariant, resulting in four identities linking the fourteen field equations (see Hilbert

1915, Theorem I, p. 397). For the empty-space Einstein equations, these identities are actually the contracted Bianchi identities, as was only later realized (see, e.g., Levi-Civita 1917). As a consequence, the field equations are not in Cauchy normal form (and cannot be put into that form without the introduction of coordinate conditions appropriately restricting their general covariance). Indeed—and Hilbert constructs an example to illustrate this point—by means of a coordinate transformation that leaves the field quantities and their first-time derivatives unchanged on some initial space-like hypersurface, the form on that hypersurface of the higher time derivatives of a solution to the field equations may be changed completely (pp. 59–60).

Hilbert concludes that the formulation of the principle of causality must be modified:

> Let the physical quantities and their time derivatives be known at the present [i.e., initial] time: then an assertion will only have physical meaning if it is invariant with respect to all those [coordinate] transformations, for which precisely the coordinates used for the present time remain unchanged; I maintain that assertions of this type are uniquely determined for the future, i.e., that the principle of causality holds in this formulation: *From knowledge of the 14 physical* [i.e., metric and electromagnetic] *potentials... at the present time, all assertions about them for the future follow necessarily and uniquely, insofar as they have physical meaning.* (Hilbert 1917, p. 61)

Hilbert here has dropped reference to the time derivatives of the 14 potentials, but this is clearly a slip. It is also not quite clear what he means by "the coordinates used for the present time" ["die für die Gegenwart benutzten Koordinaten"], but presumably he means the three-dimensional coordinate system on the initial space-like hypersurface. He goes on to prove his assertion by adopting "Gaussian coordinates," as he named them, also known today as geodesic normal coordinates (pp. 58–59). In these coordinates $g_{44} = 1$, and $g_{4i} = 0$ ($i = 1, 2, 3$), and 10 of the 14 field equations allow the determination of the remaining six components g_{ij} of the metric tensor and the four electromagnetic potentials. Hilbert does not explicitly say so, but he evidently realized that the six Einstein equations involving G_{ij} can be put in Cauchy normal form for the g_{ij} when x_4 is a time-like variable. He concludes: "Since the Gaussian coordinate system is itself uniquely determined [as he has previously shown], all assertions about these potentials [i.e., the six g_{ij} and the four electromagnetic potentials] in this coordinate system are of invariant character."

As far as it goes, Hilbert's discussion is fine. He realized the need to modify the statement of the principle of causality for generally-covariant theories since all solutions related by a coordinate transformation must be regarded as physically equivalent. He also realized that the use of Gaussian, or geodesic normal, coordinates gives one a normal system of equations for the remaining gravitational potentials. However, he neglected a complex of related problems. He does not discuss the remaining four equations in the set of fourteen.[1] Consequently, he did not consider the constraints they impose on the specification of the components of the metric and their time derivatives. Nor does he consider what happens if geodesic normal coordinates are not used. He also neglects the question of the significance of coordinate transformations on the initial hypersurface, curiously restricting himself in the definition of physically significant quantities to coordinate transformations that do not affect these hypersurface coordinates. He does not give a geometrical interpretation of his results, nor discuss the propagation of gravitational effects. As we shall see, later workers did consider these questions individually before 1939, although no one seems to have considered them all together.

The first reaction to the discussion of the Cauchy problem in Hilbert's paper appears to be Vessiot 1918. It is difficult to be certain, since Vessiot does not mention Hilbert. He mentions Einstein, but does not cite his papers, or any other papers on general relativity. Perhaps this may be attributed to the fact, remarkable in itself, that Vessiot's paper was presented to the French Académie des Sciences on February 25, 1918, while World War I was still raging and German savants were not exactly popular in France. At any rate, his discussion of ten of the fourteen equations for the gravitational and electromagnetic potentials in Gaussian normal coordinates parallels that of Hilbert (see p. 351 of Vessiot's paper). The novel result that Vessiot derived is that discontinuities in the derivatives of these potentials of higher order than first on space-like initial hypersurfaces are excluded by the field equations; such discontinuities can only occur on what he calls "wave surfaces" ("surfaces d'onde"), now called null hypersurfaces, along which both gravitational and electromagnetic waves can propagate.

4. Georges Darmois

George Darmois (1888–1960) seems to have been the next person to consider the Cauchy problem in general relativity. He first discussed it in Darmois 1923a, 1923b, which treat the exterior and interior problems (i.e., in the absence and presence of matter), respectively. Darmois 1924 gave a more detailed mathematical treatment of a number of points, while Darmois 1927 summarizes his approach and extends it in a few respects—most notably, it contains the first treatment of the initial value problem for null hypersurfaces, the characteristics of the Einstein field equations.

Darmois corrected Hilbert's oversight, noting that the four identities among the ten field equations for empty space (which he called the exterior problem) do not allow elimination of four of the equations; they do guarantee that, if four equations $G_{4\beta} = 0$ are satisfied on the initial (space-like) hypersurface $x_4 = 0$, then they will be satisfied everywhere. He rewrote these four equations in terms of the initial data for the remaining six equations, and gave the geometrical interpretation of this initial data: it consists of the first and second fundamental forms of the hypersurface (see Darmois 1923a). Thus, he first wrote down and correctly interpreted what are now called the constraint equations (see Darmois 1923a, and 1924, sec. 56, 57, pp. 67–69). He also realized that, in addition to the initial data, the four metric components $g_{4\beta}$ (β = 1, 2, 3, 4) may be specified arbitrarily on and off the initial hypersurface. He noted that geodesic normal coordinates (g_{44} = 1, g_{4i} = 0), although they especially simplify the problem, represent only one possible choice of these four components (see Darmois 1923a, p. 647).

Darmois also considered the interior problem for the cases of source stress energy tensors of a dust and of a perfect fluid, and the problem of correctly matching interior and exterior solutions on a time-like world tube (see Darmois 1923b, and Darmois 1924, secs. 58–61, pp. 70–77); we shall not discuss this work further. Darmois 1927 summarizes his earlier work on the Cauchy problem and fills in certain details and equations missing from the earlier papers (see pp. 7–10). But its most important novel contribution to the theory of initial value problems in general relativity is a discussion of the singular case, i.e., the case of a null initial hypersurface null (see pp. 10–14). Without going into detail here, it may be noted that many of the results of the most important recent paper on the null initial value problem, Sachs 1962, are clearly presented 35 years earlier in this monograph.

Darmois clearly formulated the constraints on the Cauchy data and their geometrical interpretation as conditions on the first and second fundamental forms on a space-like hypersurface. However, he did not discuss the question of how much these constraints actually limit the geometry of the hypersurface.

5. Kornel Lanczos

Kornel (or Cornelius) Lanczos (1893–1974) seems to have been the first person to do this (see Lanczos 1932). He also appears to have undertaken his investigation in total ignorance of the previous work of Hilbert and Darmois, duplicating many of their results without mentioning their papers.

Lanczos set out to investigate the question of whether the gravitational field equations, without further conditions, uniquely determine the geometry of space-time: in particular whether there are any solutions to the equations $R_{ik} = 0$ $(i, k = 1, 2, 3, 4)$ other than flat space-time. He adapted Gaussian normal coordinates that start from an initial time-like hypersurface $x_4 = 0$. He recognized that, in these coordinates, the $g_{ik,0}$ form a tensor on the initial hypersurface, a tensor that he denotes by τ_{ik}, but he did not seem to have realized that this tensor is essentially the second fundamental form of the hypersurface. While he did state that τ_{ik}, together with g_{ik}, determines the metric of the first neighboring hypersurface $x_4 = \epsilon$, he did not note that this hypersurface is geodesically parallel to the initial one, $x_4 = 0$.

He then showed that the six field equations $R_{ik} = 0$ $(i,k = 1, 2, 3)$ determine the second time derivatives of the six g_{ik} on the original hypersurface, given g_{ik} and $g_{ik,0} = \tau_{ik}$ on that hypersurface. Lanczos now returned to the remaining four equations, $R_{i4} = 0$ $(i = 1, 2, 3)$. He showed that, as a consequence of the divergence identity for the Einstein tensor, these four equations will automatically be satisfied off the initial hypersurface if they are satisfied on it and the other six equations are satisfied everywhere. He then wrote down these four constraint equations, as we now call them, in terms of g_{ik} and τ_{ik}, noting that they consist of one three-dimensional vector and one scalar relation between the g_{ik} and the τ_{ik}. Although he did not appear to know it, so far he was basically duplicating previous work.

Next he made his original contribution to the problem, by applying the concept of "degrees of freedom" of a system to the gravitational field. He observed that "We have three degrees of freedom in prescribing the geometry" of the initial hypersurface, but that, as a consequence of the constraint equations "only one of them is lost, so that there still remain *two* freedoms at every point [of the hypersurface] (p. 628)."

The argument he gives for this conclusion is not correct. He maintained that the vectorial condition $(\tau_{ik} - \tau g_{ik})_{;k} = 0$ does not constitute a restriction on the intrinsic geometries of the two neighboring surfaces $x_4 = 0$ and $x_4 = \epsilon$, which are described by g_{ik} and $g_{ik} + \epsilon\tau_{ik}$, respectively; but rather conditions on the "fitting together" of the two hypersurfaces, which can always be satisfied by an appropriate choice of coordinates on the second surface. He has forgotten that his choice of Gaussian normal coordinates deprives him of the freedom of "fitting together" the coordinates on the two neighboring surfaces. The vectorial constraint does constitute a restriction on the geometry of the second, geodesically-parallel surface. Nevertheless, his conclusion is correct. The gravitational field can be shown to have two degrees of freedom at each point of a space-like hypersurface (see, for example, York 1979, pp. 102–103).

6. Karl Stellmacher

By 1932 all the formal aspects of the Cauchy problem had been investigated, the geometrical interpretation of the initial data had been given, and it had been established that the gravitational field has two degrees of freedom per space-time point. The work was formal in the sense that formal power series expansions were invoked to construct solutions, without any serious effort to investigate existence theorems for such solutions. At best, such methods could claim mathematical validity only for the analytic case. But for hyperbolic normal systems of partial differential equations, it is just the nonanalytic solutions that are of special interest, since they are the ones that carry new information. Indeed, one way of defining the characteristic hypersurfaces of such a system of equations is as those hypersurfaces on which nontrivial discontinuities in the highest-order derivatives of the field components can occur. (In the case of generally-covariant equations, trivial discontinuities are those that can be produced or removed by coordinate transformations.)

The first nonformal study of the Cauchy problem for the gravitational field equations, in this sense of formal, was published in 1938 by Karl Stellmacher (b. 1909). Stellmacher 1938 is based on recent progress in methods for proving existence theorems for hyperbolic equations. Stellmacher recalled the work of Vessiot (Vessiot 1918), who showed that null hypersurfaces S, i.e., solutions of the equations $g^{ik}S_{,i}\,S_{,k} = 0$, are the characteristic hypersurfaces of the Einstein equations. (Stellmacher also cited Hilbert's work on the Cauchy problem cited above.) He based his work on the use of the so-called de Donder coordinate conditions,[2] which transform the Einstein equations into a system in Cauchy-normal form. He then applied the methods of Friedrichs and Lewy 1928 to demonstrate what he called "the validity of the causal postulate" for nonanalytic initial data (p. 138). By causal postulate, he means the assertion that the field at a point P anterior to the initial hypersurface depends only on that part of the initial data that lies on and within the intersection of the initial hypersurface with the posterior characteristic conoid with vertex at P.

With Stellmacher's work, we have reached the end of what one may call the "prehistoric period" of work on the Cauchy problem in general relativity. André Lichnerowicz made it clear in his thesis (Lichnerowicz 1939) that he was well aware of the work of Darmois (which is not surprising, since he was Darmois's student), and of that of Stellmacher; but he does not mention Hilbert or Lanczos. Lichnerowicz and his students went on to make a number of important mathematical advances in the treatment of the Cauchy problem, but consideration of this later part of the story will have to await another occasion.

Acknowledgment: I thank Professor James W. York, Jr. for pointing out an error in my original discussion of Lanczos's work.

Notes

[1] This neglect seems to be connected with a misunderstanding of the role of the contracted Bianchi identities, which he believed allowed the complete elimination of four field equations.

[2] They are also called isothermal coordinates and harmonic coordinates. For a discussion of their use, with historical references, see Arzeliès 1961, pp. 274–279.

References

Arzeliès, Henri (1961). *Relativité généralisée. Gravitation.* Part 1, *Principes généraux: équations d'Einstein. Dynamique et optique. Réperages non einsteiniens.* Rabat: Travaux de l'Institut Scientifique Chérifien, Série Sciences Physiques, No. 7. Paris: Gauthier-Villars.

Bergmann, Peter G. (1957). "Topics in the Theory of General Relativity." In *Brandeis University Summer Institute of Theoretical Physics*, n.p., n.d.[1957], pp. 1–44.

Bruhat, Yvonne (1962). "The Cauchy Problem." In *Gravitation: An Introduction to Current Research.* Louis Witten, ed. New York: Wiley, pp. 130–168.

Cauchy, Augustin (1842). "Mémoire sur l'emploi du calcul des limites dans l'intégration des équations aux dérivées partielles." *Académie des Sciences* (Paris). *Comptes Rendus* 15: 44–59.

Choquet-Bruhat, Yvonne, and York, James W. (1980). "The Cauchy Problem." In *General Relativity and Gravitation: One Hundred Years After the Birth of Albert Einstein.* Vol. 1. Alan Held, ed. New York: Plenum, pp. 99–172.

Darmois, Georges (1923a). "Sur l'intégration locale des équations d'Einstein." *Académie des Sciences* (Paris). *Comptes Rendus* 176: 646–648.

—— (1923b)."Intégration locale des équations d'Einstein (problème intérieur)." *Académie des Sciences* (Paris). *Comptes Rendus* 176: 731–733.

—— (1924). "Eléments de géométrie des espaces. Introduction aux théories de la relativité générale." *Annales de Physique* 1: 5–87.

—— (1927). *Les équations de la gravitation einsteinienne*. Mémorial des Sciences Mathématiques No. 25. Paris: Gauthier-Villars.

Einstein, Albert (1914). "Die formale Grundlage der allgemeinen Relativitätstheorie." *Königlich Preussische Akademie der Wissenschaften* (Berlin). *Sitzungsberichte*: 1030–1085.

Fisher, Arthur E. , and Marsden, Jerrold E. (1979). "The Initial Value Problem and the Dynamical Formulation of General Relativity." In *General Relativity. An Einstein Centenary Survey.* Stephen W. Hawking and Werner Israel, eds. Cambridge: Cambridge University Press, pp. 138–211.

Friedrichs, Kurt, and Lewy, Hans (1928). "Über die Eindeutigkeit und das Abhängigkeitsgebiet der Lösungen beim Anfangswertproblem linearer hyperbolischer Differentialgleichungen." *Mathematische Annalen* 98: 192–204.

Hadamard, Jacques (1903). *Leçons sur la propagation des ondes et les équations de l'hydrodynamique.* Paris, Hermann.

—— (1923). *Lectures on Cauchy's Problem in Linear Partial Differential Equations.* New Haven: Yale University Press. Second ed.: *Le problème de Cauchy et les équations aux dérivées partielles linéaires hyperboliques.* Paris: Hermann, 1932.

—— (1964). *La théorie des équations aux dérivées partielles linéaires hyperboliques.* Peking: Foreign Language Press.

—— (1968). *Oeuvres.* Paris: Editions du Centre National de la Recherche Scientifique, vol. 3.

Hilbert, David (1915). "Die Grundlagen der Physik. (Erste Mitteilung)." *Königliche Gesellschaft der Wissenschaften zu Göttingen. Mathematisch-physikalische Klasse. Nachrichten*: 395–407.

—— (1917). "Die Grundlagen der Physik. (Zweite Mitteilung)." *Königliche Gesellschaft der Wissenschaften zu Göttingen. Mathematisch-physikalische Klasse. Nachrichten* : 55–76.

Janet, Maurice (1927). *Les systèmes d'équations aux dérivées partielles.* Mémorial des Sciences Mathématiques No. 21. Paris: Gauthier-Villars.

Kovalevskaia, Sofia [Kowalevsky, Sophie von] (1875). "Zur Theorie der partiellen Differentialgleichungen." *Journal für die reine und angewandte Mathematik* 80: 1–32.

Lanczos, Cornelius (1932). "Zur Frage der regularen Lösungen der Einsteinschen Gravitationsgleichungen." *Annalen der Physik* 13: 621–635.

Levi-Civita, Tullio (1917). "Sulla espressione analitica spettante al tensore gravitazionale nella teoria di Einstein." *Accademia dei Lincei. Rendiconti* 26: 381–391.

Lichnerowicz, André (1939). *Sur certains problèmes globaux relatifs au système des équations d'Einstein.* Paris: Hermann. Also issued as *Problèmes globaux en mécanique relativiste.* Actualités Scientifiques et Industrielles No. 833. Paris: Hermann, 1939.

—— (1944). "L'intégration des équations de la gravitation relativiste et le problème des *n* corps." *Journal des Mathématiques Pures et Appliquées* 23: 37–63.

Sachs, Rainer K. (1962). "On the Characteristic Initial Value Problem in Gravitational Theory." *Journal of Mathematical Physics* 3: 908–914.

Stachel, John (1989). "Einstein's Search for General Covariance, 1912–1915." In *Einstein and the History of General Relativity.* Don Howard and John Stachel, eds. Boston: Birkhaüser, pp. 63–100.

Stellmacher, Karl (1938). "Zum Anfangswertproblem der Gravitationsgleichungen." *Mathematische Annalen* 115: 136–152.

Vessiot, Ernest (1918). "Sur la propagation par ondes et sur la théorie de la relativité générale." *Académie des Sciences* (Paris). *Comptes Rendus* 166: 349–351.

York, James W. , Jr. (1979). "Kinematics and Dynamics of General Relativity." In *Sources of Gravitational Radiation. Proceedings of the Battelle Seattle Workshop, July 24–August 4, 1978.* Larry L. Smarr, ed. Cambridge and New York: Cambridge University Press, pp. 83–126.

Spherical Coordinates in General Relativity from 1915 to 1960: A Physical Interpretation

Christiane Vilain

1. Introduction

This chapter gives a brief survey of the spherically symmetric coordinate systems that were used during the first 45 years of general relativity, from Eddington to Kruskal. Although all the papers mentioned here will deal with Schwarzschild's solutions, we are not directly concerned with horizons or singularities, but rather with various choices for coordinate systems and their implicit or explicit motivations. These motivations will be shown to move from what we may call an "operational standpoint" to more geometrical considerations, leading later to coordinate-free studies for global solutions.

Here, we emphasize the reason why relativists did not begin to study the global properties of Einstein solutions for spherical spaces until 1960, although coordinate-free methods were available long before then. The constraint on these solutions was created by an early requirement for a strong link between coordinates and observers, and thus for the necessity of a physical interpretation, which was later dropped. When one deals primarily with the relationships between an observer and external phenomena, it does not seem worth studying the whole manifold, that is, its maximal extension.

Coordinates were first denied all physical meaning because of the theory's requirements for general covariance. Since coordinates are merely charts or maps of the real number sets used to specify the location of the manifold, they can be modified through "allowed" transformations (sufficiently smooth) without any physical consequences. Allowed transformations will then leave unchanged the curvature invariants related to the material content of the universe. But, they will not obviously lead to the same topology of the solution because the links between that topology and the distribution of matter are far form being determined by the theory.

Schwarzschild's horizon is known to be meaningless with regard to the curvature invariants, which remain regular, but it is nevertheless a true horizon for any static observer. Schwarzschild's coordinates may thus be considered physically meaningful for such an observer.

Although coordinate charts are today given a quite different mathematical status from that of the frames of reference belonging to tangent spaces, they were previously considered suitable coordinates for a set of observers rather than for an extended view of the whole manifold. In this chapter, definitions for "suitable" coordinates and for transformations from Schwarzschild's system, as given or suggested in various books and publications, are reviewed and roughly classified according to operational viewpoints. Let us first recall some statements about the relations between coordinates and covariance given during the early years of general relativity.

2. Coordinates and Covariance

At the end of 1916, following the creation of general relativity, Schwarzschild's coordinates were given as a solution for empty spherical static spaces (Schwarzschild 1916). Our first observation is concerned with A. Eddington and his deep conceptual and didactical preoccupations.

As everyone knows, space is not defined prior to its physical content, and all experiments aimed at the intrinsic geometry are known to test the behavior of physical objects as related to the whole content of the universe. The consequence of such a criticism led H. Poincaré to continue with Euclidean geometry, but led Eddingon to follow Einstein and choose a more appropriate one. With curved spaces, the new problem of suitable coordinates emerged.

In his first book, *Space, Time and Gravitation* (1920), Eddington distinguished between physical phenomena, measure, and representation. His first attempt to illustrate covariance with regard to coordinates referred to the idea, previously expressed by Einstein (1916), that the physical facts are only mere coincidences of world lines inside spacetime. Space is thus a sort of "jelly," inside which regular meshes are deformed in any way:

> If we deform the jelly in any way, the intersections will still occur in the same order along each world line, and no additional intersections will

> be created.... Suppose now we introduce space and time partitions, which we might do by drawing rectangular meshes in the jelly. We have now two ways of locating the world lines and events in space and time, both on the same absolute footing. But, clearly it makes no difference in the result of thc location whether we first deform the jelly and then introduce regular meshes, or whether we introduce irregular meshes in the underformed jelly. So, all the mesh systems are on the same footing. (Eddington 1920, p. 88)

This illustration of covariance is very attractive, except that with such deformations one could retain the same topology but not the same curvature. The argument is therefore unsuitable to general relativity, where the curvature invariants are endowed with a physical meaning. That is why Eddington (1923) had to drop his "jelly" as far as the mathematical framework of the theory was concerned. His new statement about coordinate transformations asserted the physical meaning of distances alone, without reference to location or orientation:

> Extension is now fundamental; and the location of an object is a computational result summarizing the physical fact that it is at certain intervals from other objects in the world. Any idea contained in the concept location that is not expressible by reference to distances from other objects must be dismissed from our minds. To put the conclusion rather crudely, space is not a lot of points close together, it is a lot of distances interlocked. (Eddington 1923, p. 9)

According to Eddington as soon as we describe a figure with coordinates we add the irrelevant fact of its orientation. Only distances have to be considered.

This modification about what is relevant for the description of the physical world comes from the preeminence of metrical properties assigned to space through Einstein's equations. Physical facts are no longer restricted to coincidences once general relativity becomes consistent with the special theory for any local inertial observer. The curvature thus measures the relative acceleration of nearby freely falling observers. This requirement gives a new meaning to coordinates, as was pointed out by Einstein. Some coordinates might acquire a direct physical meaning through the special theory.

The harsh criticism of measurement methods is softened by reference to previous physical laws, and consequently, although all systems are allowed, they are not all equivalent. Thus appears an operational viewpoint governing the choice of coordinates, which we will now state explicitly.

3. Operational Coordinates

Operational coordinates can be split into two groups, "proper" coordinates and radio-geodesic or "null" coordinates.

3.1. Proper Coordinates

Local triangulation with solid rods will provide, according to Eddington, "isotropic" coordinates, although they are usually described as "natural" ones:

> This system of coordinates (isotropic) is naturally arrived at when we partition space by rigid scales or by light triangulations in a small region, e.g., in terrestrial measurements. Since the ultimate measurements involved in any observation are carried out in a terrestrial laboratory, we ought, strictly speaking, always to employ the isotropic system that conforms to assumptions made in those measurements. (Eddington 1923, p. 93)

Such coordinates lead to a Minkowskian metric at a given point, so that proper distances can be measured by coordinate variation in the neighborhood of that point.

Proper time, used as the time coordinate, gives us "geodesic coordinates," but it is not generally valid for more than one geodesic since it does not provide a unique function at every point. A tetrad is carried by parallel transport along that particular geodesic, whose vectors are tangents to the coordinate mesh.

We will see later that coordinates used by Lemaître (1933) and Robertson (1939) are related to proper times, a choice allowed by spherical symmetry and inspired by cosmological models. Coordinates exhibited by Synge (1950) can be said to be related to proper spatial distance, as we will show later, although in a quite different way from what he calls "Gaussian coordinates" in his subsequent book (1960). Both Gaussian and geodesic coordinates are valid along one single geodesic, except when particular symmetry conditions hold, such as spherical symmetry.

The need to obtain a direct description in terms of proper distances or proper time is greatest in cosmological studies in which the observer is a moving particle of the same kind as the others, and Lemaître and Robertson were obviously mainly interested in cosmological models. This is no longer the case for the field of a single particle when the observer is far enough away to be considered as static. Such coordinates may thus be related to operational devices in a rather fictitious way, for example, by considering a set of freely falling observers. This assumption may be the

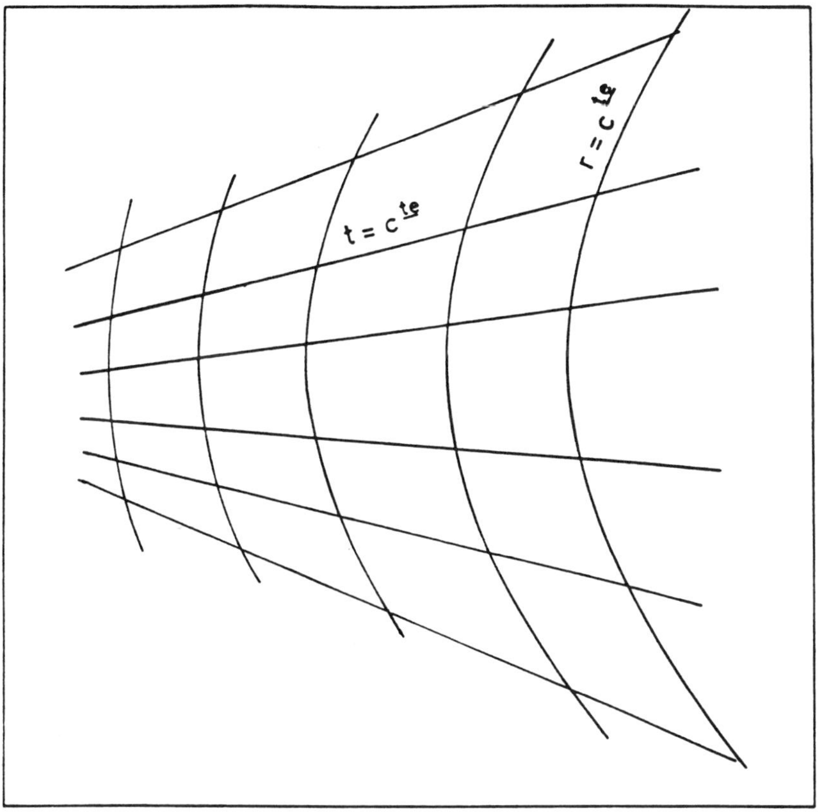

FIGURE 1

reason why comoving coordinates prove more useful as a step toward a global view of the manifold.

Although proper coordinates were not used by Eddington except at a given point, a similar interest in a proper distances throughout the manifold is attested to by Figure 1, notable for its qualitative similarity to Synge's and Kruskal's diagrams. Clearly, Eddington's motivation was to get a true "picture" of the space, instead of a "map" given by coordinates.

3.2. RADIO-GEODESIC OR NULL COORDINATES

Clocks and light rays can be used for measuring time-like or space-like distances provided we use three clocks. The time taken on those clocks could be given as coordinate numbers for any event in the manifold, as was proposed by J.L. Synge (1921). Then, the only metric quantity is the time of oscillation of the atoms.

Three atoms A, B, and C emit signals at times t'_A, t'_B, and t'_C, which are reflected by the particle or the event. That event will then be assigned the coordinates:

$$
\begin{aligned}
x_0 &= \tfrac{1}{2}(t'_A + t''_A),\\
x_1 &= \tfrac{1}{2}(t''_A - t'_A),\\
x_2 &= \tfrac{1}{2}(t''_B - t'_B),\\
x_3 &= \tfrac{1}{2}(t''_C - t'_C).
\end{aligned}
$$

This can be considered as the closest approach to an actualization of the "point coincidence" argument, although it is based on the possibility of using three synchronized distinct clocks adequately fixed in space, which is again a sort of fiction.

This procedure was again given by Synge in his 1950 publication, but it was used afterwards for his new line element or in his book (Synge 1960), where it was applied to spherical symmetry. This exposition is similar to that given by Robertson in the book written by Noonan from his notes of Robertson's lectures (Robertson and Noonan 1968). Because of spherical symmetry, no more than one clock is needed, and the "fiction" becomes reality.

Let us give the transformation between these null coordinates and Schwarzschild's system (see Figure 2; Robertson and Noonan 1968, p. 234):

$$
\begin{aligned}
x_0 &= \tfrac{1}{2}(t''_A + t'_A),\\
x_1 &= \tfrac{1}{2}(t'' - t'_A),\\
x_2 &= \theta,\\
x_3 &= \phi.
\end{aligned}
$$

The labels r and t are, throughout this work, Schwarzschild coordinates defined through the well-known radial line element:

$$ds^2 = (1 - a/r)\,dt^2 - dr^2/(1 - a/r)$$

with

$$a \equiv \frac{2GM}{c^2}.$$

Radial null geodesics are easily obtained by writing

$$ds^2 = 0 \rightarrow dt = \pm\frac{dr}{1 - a/r}.$$

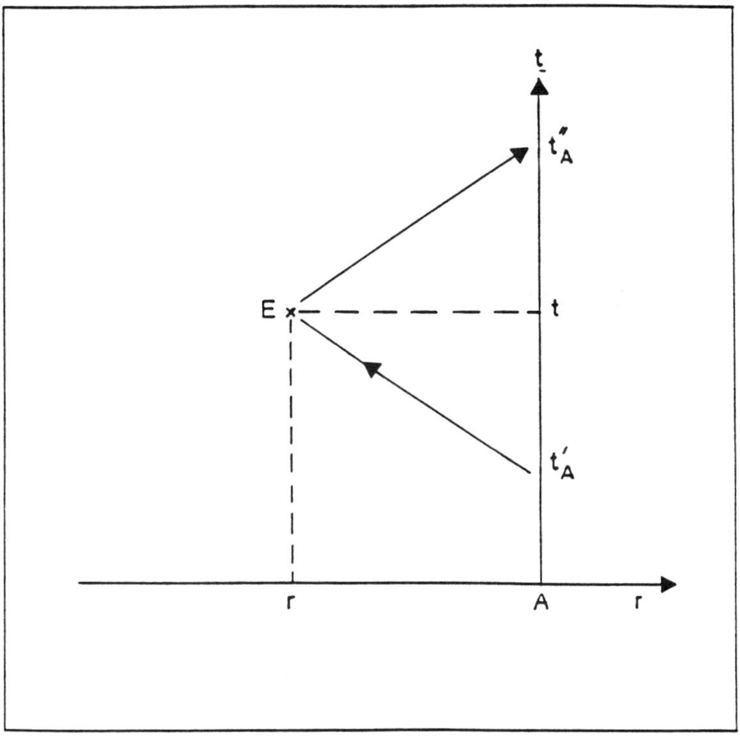

FIGURE 2

Hence, one gets explicitly

$$t''_A - t = f(r_A) - f(r),$$
$$t - t'_A = f(r_A) - f(r)$$

with

$$f(r) = a\left[(r/a + \ln(r/a - 1)\right].$$

Then,

$$x_0 = t,$$
$$x_1 = f(r) - f(r_A) = g(r).$$

Let us now look at the new line elements:

$$ds^2 = (1 - a/r)\, dt' dt'',$$
$$ds^2 = (1 - a/r)\, (dx_0^2 - dx_1^2).$$

It is worth mentioning that the second form was never written either by Synge or Robertson, although that form of the radial part of the line element is quite interesting because it is a Minkowskian form for the null cones similar to Kruskal's and it removes the "singularity." It does not allow any extension of Schwarzschild's external space ($r \geq a$), however, since x goes to infinity as r goes to a. Another static clock is then needed for the description of the internal space ($r \leq a$).

4. New Line Elements for Spherical Symmetry

New coordinate systems were found successively by A. Eddington (1924), G. Lemaître (1933), A. Einstein and N. Rosen (1935), H.P. Robertson (1939), V.V. Narlikar and K.R. Kamarkar (1946), J.L. Synge (1950), A. Raychaudhuri (1953), D. Finkelstein (1958), M.D. Kruskal (1960), and G. Szekeres (1960). The aim of the search for new line elements was to remove or study the Schwarzschild horizon, except in the first proposal by Eddington, whose purpose was to make a comparison between Whitehead's theory and Eddington's coordinate time:

$$t_w = t - a \ln(r - a).$$

That time can be compared with our previous time t'', given by null coordinates. This transformation was considered again by Finkelstein (1958), who discussed the time asymmetry of the solution involved, an asymmetry that had been previously noticed by Eddington. Moreover, he ascertained the analyticity of the manifold thus obtained except at $r = 0$, thus extending Schwarzschild's space.

Finkelstein was unaware of the incompleteness of his solution until it was pointed out by Schild, as he mentioned in a footnote. Since the solution is time asymmetric, there is an equally effective but different solution obtained by time reversal, which shows the same analytic properties, except that light rays would be able to enter the horizon but not exit; this is the opposite of what happened in the first solution. That reversed solution could obviously be related to a choice of t' instead of t'', according to our previous analysis.

An almost identical configuration was found in the solutions given by Lemaître (1933) and Robertson (1939). In the first case, proper time along an outgoing geodesic was taken as the time coordinate, while in the second case, proper time along a freely falling radial geodesic was used. These two solutions, each of which is asymmetric with respect to time, are symmetric with respect to each other under time reversal.

In both cases, there is an extension through the horizon in one direction, and there remains an incomplete geodesic in the other direction, just as occurs in Finkelstein's work. The transformations given independently in papers by Narlikar and Kamarkar (1946) and Raychaudhuri (1953) belong to this type of transformation, being very close to Robertson's work. Raychaudhuri's work was nevertheless original in its search for a global extension of internal coordinates comoving with some perfect fluid.

We give Robertson's transformation, his new line element, as an illustration of what has already been said:

$$\tau = t + f(r), \qquad \rho = r + \tfrac{3}{2}\left[(f(r) + t)\right],$$

where

$$f(r) = 2r + \ln\left[(r^{1/2} - 1)/(r^{1/2} + 1)\right].$$

Then,

$$ds^2 = d\tau^2 - (\rho/r)\, d\rho^2.$$

An analysis of this form of the solution was given by J. Eisenstaedt (1987).

Einstein and Rosen (1935) obtained a quite different transformation through a new coordinate u:

$$u^2 = \pm(r - a).$$

By keeping both negative and positive solutions, they obtained a space that could be mapped twice on Schwarzschild's external space, passing without a singularity from one exterior to another exterior (or the same) by a "bridge."

This first attempt toward a new space can no longer be related to operational coordinates, but it is formally similar to a transformation used for an embedding of Schwarzschild's space by Flamm (1916).

The remaining works, those by Synge (1960), Kruskal (1960), and Szekeres (1960), show the same sort of duplication of the initial space, even more than twice for J.L. Synge's space.

This latter paper, written in 1950 (Synge 1950), was almost unknown at the time and has remained so. The reason why it was never mentioned as an important search for new coordinates may be its technical sophistication. Synge used a polynomial series that led to a parametrical representation for new coordinates. There was no justification for the whole process except through this announcement:

> I have removed the Schwarzschild singularity in a different way... [from Lemaître (1933) and Robertson (1939), according to his lecture in Toronto]. (Synge 1950)

That difference was mainly expressed as follows:

> The singularity $r = 0$ in Lemaître's work in a single space-like curve, each part of which is of infinite length, and at each point on it there is but one null direction. In my work, the singularity $r = 0$ consists of an infinite set of space-like curves, each having the properties mentioned above for the Lemaître singularity.... As was pointed out by Robertson, a time-like geodesic reaches $r = a$ in a finite proper time and without becoming tangent to a null line. Thus, the geodesic may be continued. That is in fact what has been done by Lemaître and me, but we have made different continuations of space-time. (Synge 1950)

Synge began his search by introducing two rectangular Cartesian coordinates u and v and, moreover, several functions of the variable $u^2 - v^2$. Such a function $Z(u^2 - v^2)$ is, for instance, fixed implicitly by the formulas

$$u^2 - v^2 = a\left[\xi + \sinh(\xi)\cosh(\xi)\right],$$

where

$$\xi^2 = z \qquad \text{for} \quad Z > 0 \quad \text{or} \quad u^2 - v^2 > 0,$$
$$u^2 - v^2 = a\left[\eta + \sin(\eta)\cos(\eta)\right] \qquad \text{for} \quad Z < 0 \quad \text{or} \quad u^2 - v^2 < 0.$$

Then, the function r is given:

$$r = a\cosh^2(\xi), \quad u^2 - v^2 \geq 0,$$
$$r = a\cos^2(\xi), \quad u^2 - v^2 \leq 0.$$

This leads to a new function $r(u^2 - v^2)$ through the preceding one $Z(u^2 - v^2)$.

Two more functions $F(u^2 - v^2)$ and $h(u^2 - v^2)$ are given by similar procedures, and a new line element can then be fixed.

$$\Phi = du^2 - dv^2 + (v\,du - u\,dv)^2 F + r^2(d\theta^2 + \sin^2\theta\, d\phi^2)$$

An analysis of this new line shows that its signature is the one that is required in general relativity and that Schwarzschild's line element can be retrieved through the following transformations.

For $u - v > 0$,

$$u = R\cosh(t/2a),$$
$$v = R\sinh(t/2a),$$

and for $u - v < 0$,

$$u = R\sinh(t/2a),$$
$$v = R\cosh(t/2a).$$

It is thus ascertained that the new line element is a solution of the field equations for spherical symmetry.

The next step gives the equations for time-like and null geodesics from the new line element. It becomes obvious that geodesics can go through the first barrier, $r = a$, just as in Robertson's space, but they also go through the singularity $r = 0$:

> The test particle acquires the velocity of light at $r = 0$. The question may now be raised: is it conceivable that a test particle should travel right through the singularity $r = 0$? ... We may say that in our representation of the field of a gravity particle, the passage of a test particle through the singularity $r = 0$ may be uniquely defined. (Synge 1950)

This uniqueness is obtained by some energetic condition.

Thus, these are roughly the lines of Synge's paper that we should remember, along with Kruskal's better known work. Their main formulas, displayed here, will prove useful for a comparison and some conclusions. From Synge's 1950 article, we have, for $r \leq a$,

$$\begin{aligned} dR &= -\frac{dr^2}{(1 - a/r)}, \\ R(r) &= \pm a\left[\arccos(r/a) + (r/a)(1 - r/a)\right], \\ u &= R(r)\ \sinh(t/2a), \\ v &= R(r)\ \cosh(t/2a). \end{aligned}$$

For $r > a$,

$$\begin{aligned} dR &= +\frac{dr^2}{(1 - a/r)}, \\ R(r) &= \pm a\left\{\ln\left[(r/a) + (r/a - 1)\right] + (r/a)(r/a - 1)\right\}, \\ u &= R(r)\ \cosh(t/2a), \\ v &= R(r)\ \sinh(t/2a). \end{aligned}$$

Then,

$$ds^2 = du^2 - dv^2 + (v\,du - u\,dv)F(R^2).$$

From Kruskal's 1959 article, we have, for $r < a$,

$$\begin{aligned} \frac{dR}{R} &= \frac{dr}{(1 - a/r)}, \\ R(r) &\propto (1 - r/a)e^{r/2a}, \\ u &= \pm R(r)\ \sinh(t/2a), \\ v &= \pm R(r)\ \cosh(t/2a). \end{aligned}$$

For $r > a$,

$$\frac{dR}{R} = \frac{dr}{(1 - a/r)},$$

$$R(r) \propto (r/a - 1)e^{r/2a},$$

$$u = \pm R(r) \cosh(t/2a),$$

$$v = \pm R(r) \sinh(t/2a).$$

Then,

$$ds^2 = (du^2 - dv^2)h(r).$$

The intermediate function $R(r)$ is obviously the proper spatial distance in Synge's transformation, and the time $t(r)$ along a null geodesic is that in Kruskal's. These properties are not explicitly given by the authors. They both use quite heuristic methods for justifying the extension of Schwarzschild's solution, and, although meaningful, they are not relevant here. These equations are given here to clarify Figures 3, 4, 5, and 6.

The multivaluedness of Synge's function $R(r)$ allows an infinite duplication of the internal space, a strange procedure that was not noticed at that time, nor later. A particle is thus allowed to go through the "true" singularity $r = 0$, although traveling at the speed of light through the region of infinite curvature.

Hence, Synge wrote:

> A test particle may enter the sphere $r = a$ and oscillate infinitely often inside it, or it may be emitted from the sphere after an infinite number of oscillations inside. (Synge 1950)

In Kruskal's work, the particle is not allowed to go through $r = 0$ because of the impossibility of extending the manifold through infinite curvature.

The new coordinates u and v seem arbitrarily obtained from the intermediate R function and hyperbolic functions of time t. Such transformations cannot be related to any of our previous operational considerations. But, they do remind us of the coordinates obtained by embedding Schwarzschild's space in a Euclidean space of higher dimensions. A similar form of new coordinates was given by Kasner as early as 1921 (Kasner 1921), using trigonometric functions instead of hyperbolic ones, but this was not mentioned by Synge, who seems to be the first one to choose hyperbolic functions.

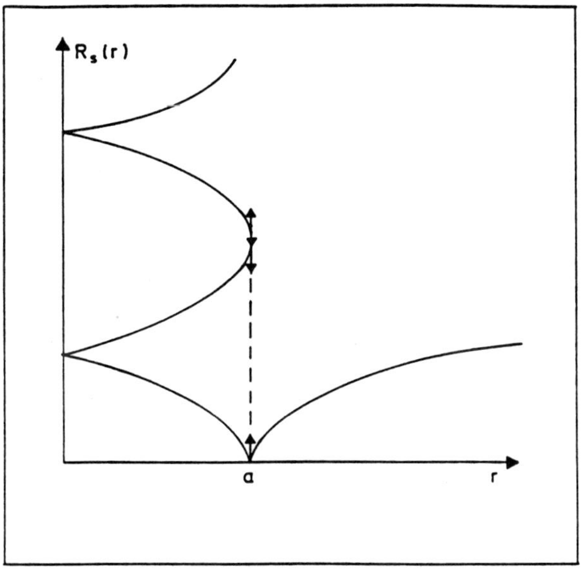

FIGURE 3 Synge's proper spatial distance.

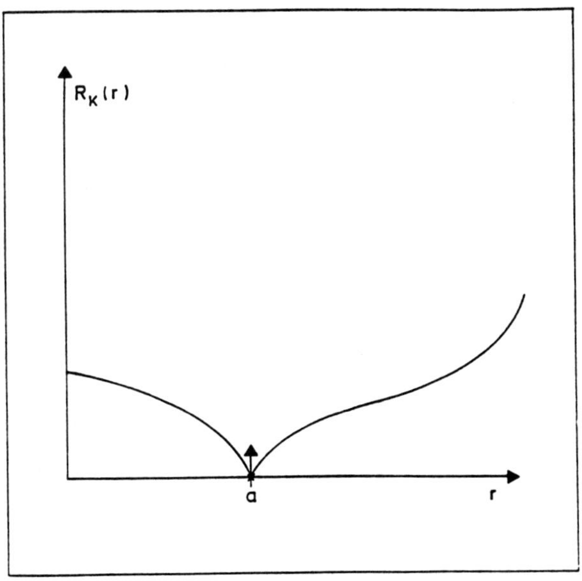

FIGURE 4 Kruskal's time along a null geodesic.

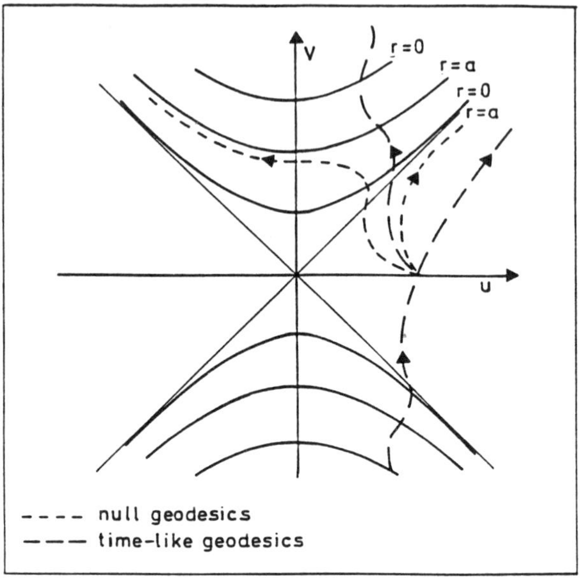

FIGURE 5

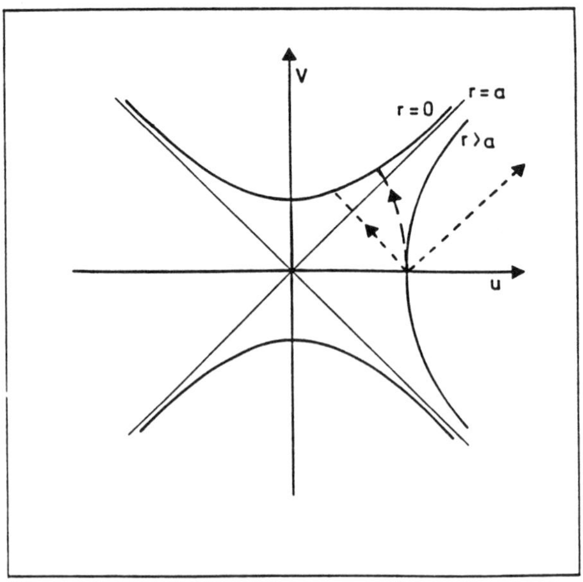

FIGURE 6

It is obvious that embeddings, such as those of Fronsdal (1959), although leading to formulas similar to Kruskal's, aim at a sort of visual picture rather than at a better mapping for the whole manifold. This last assertion is supported by Fronsdal's endeavor to retrieve time and space coordinates at the end of his work, considering u and v as mere tools for representation.

We hope to have brought some enlightenment to the evolution of the relationship between operative and representative forces. The uncertain status of the observers in these models is clearly exhibited, unless quite implicit. The observers can be considered as duplicated or altogether neglected; they may be encountered only outside the whole picture, or perhaps they are everywhere.

REFERENCES

Eddington, Arthur S. (1920). *Space, Time and Gravitation.* Cambridge: Cambridge University Press.

——— (1923). *The Mathematical Theory of Relativity.* Cambridge: Cambridge University Press.

——— (1924). "A Comparison of Whitehead's and Einstein's Formulas." *Nature* 113: 192.

Einstein, Albert (1916). "Die grundlage der allgemeinen Relativitätstheorie." *Annalen der Physik* 49: 769–822. Translated as "The Foundation of the General Theory of Relativity." In Lorentz, Hendrik Antoon et al. *The Principle of Relativity.* Arnold Sommerfeld, ed. W. Perrett and G. B. Jeffery, trans. London: Methuen, 1923. Reprint New York: Dover, 1952, pp. 109–164.

Einstein, Albert, and Rosen, Nathan (1935). "The Particle Problem in the General Theory of Relativity." *Physical Review* 48: 73–77.

Eisenstaedt, Jean (1987). "Trajectoires et impasses de la solution de Schwarzschild." *Archive for History of Exact Sciences* 37: 275–357.

Finkelstein, David (1958). "Past–Future Asymmetry of the Gravitational Field of a Point Particle." *Physical Review* 110: 965–967.

Flamm, Ludwig (1916). "Beitrage zur Einsteinschen Gravitationstheorie." *Physikalische Zeitschrift* 17: 448–454.

Fronsdal, C. (1959). "Completion and Embedding of the Schwarzschild Solution." *Physical Review* 116: 778–781.

Kasner, Edward (1921). "Finite Representation of the Solar Gravitational Field in Flat Space of 6 Dimensions." *American Journal of Mathematics* 43: 130–133.

Kruskal, M. D. (1960). "Maximal Extension of Schwarzschild Metric." *Physical Review* 119: 1743–1745.

Lemaître, Georges (1933). "L'univers en expansion." *Société Scientifique de Bruxelles. Annales A* 53: 51–85.

Narlikar, V. V., and Karmarkar, K. R. (1946). "Geodesic Form of Schwarzschild's External Solution." *Nature* 108: 515–516.

Raychaudhuri, Amalkumar (1953). "Arbitrary Concentration of Matter and the Schwarzschild Singularity." *Physical Review* 89: 417–421.

Robertson, Howard P., and Noonan, Thomas W. (1968). *Relativity and Cosmology.* Philadelphia, Pennsylvania: Saunders.

Schwarzschild, Karl (1916). "Über das Gravitationsfeld eines Massenpunktes nach der Einsteinschen Theorie." *Königlich Preussische Akademie der Wissenschaften* (Berlin). *Sitzungsberichte*: 189–196.

Synge, John L. (1921). "A System of Coordinate Space-Time." *Nature* 108: 275.

—— (1950). "The Gravitational Field of a Particle." *Royal Irish Academy* (Dublin). *Proceedings* 53: 83–114.

—— (1960). *Relativity: The General Theory.* Amsterdam: North-Holland.

Szekeres, Gy (1960). "On the Singularities of a Riemannian Manifold." *Publicationes Mathematicae* (Debrecen) 7: 285–301.

PART IV

Relativistic Cosmology

Contributions of Lemaître to General Relativity (1922–1934)

Odon Godart

1. Some Biographical Preliminaries

Georges Lemaître was born at Charleroi in Belgium on July 17, 1894. He was admitted to the College of Engineering at the University of Louvain, where he obtained his BA in July 1913. At the same time, he also enrolled for the philosophy courses offered by the Institut Supérieur de Philosophie, founded by Désiré Mercier (the future Cardinal of Malines, who became famous in World War I). In 1914, the war interrupted Lemaître's studies in mechanical engineering. He joined the Belgian army and fought in the 3rd artillery regiment. After 53 months of war ordeals and military camps, he lost interest in a professional career and decided to become a priest.

Before entering the major seminary at Malines at the end of 1920 for religious studies, he shifted his enrollment at the university from engineering to mathematical sciences, and in 1920, he obtained his doctorate with the greatest distinction with a dissertation entitled "L'approximation de fonctions réelles à plusieurs variables."

Lemaître then became interested in the theory of relativity. The program of religious studies left him sufficient time to acquaint himself with the literature on special and general relativity and to prepare three essays that he submitted in 1923 to the commission in charge of allocating scholarships for stays abroad.

In September 1923, Lemaître was ordained a priest; with a scholarship from the Belgian government, he was accepted at the Observatory of Cambridge University by its director Arthur Eddington, where as his research subject he pursued the study of the concept of simultaneity in general relativity. Pleased with the results, Eddington added a foreword to a manuscript submitted to *Philosophical Magazine* entitled "The Motion of a Rigid Body According to the Relativity Principle," which was published in July 1924 (Lemaître 1924).

The following academic year was spent at the Harvard Observatory in Cambridge, Massachusetts, under the tutorship of Harlow Shapley. Lemaître's main interest continued to be relativistic cosmology. In particular, at a meeting of the American Physical Society in April 1925, he showed how, by reformulating De Sitter's solution of Einstein's equations, he could remove its spurious inhomogeneity, and how the introduction of new coordinates separating space and time led to a linear relation between velocity and distance.

2. The Genesis of the Hypothesis of the Expanding Universe

The features of the De Sitter universe model (De Sitter 1917) were quite intriguing, and their considerations influenced Lemaître in his conception of the expanding universe. We give here the essentials of Lemaître's paper (Lemaître 1925), together with some manuscript notes in notebooks of the same epoch (1924–1926).[1] Einstein started the development of relativistic cosmology by pointing out that the field equations in their general form with the cosmological constant allowed the possibility of a homogeneous finite universe that is closed and has no boundary. Moreover, Einstein took it for granted, as did the majority of the scientists at that time, that the universe as a whole must be unchanging with time (Einstein 1917).

The work of De Sitter was a critique of Einstein's cosmological hypothesis. To distinguish both solutions let us write the metric for Einstein's solution Eq. (1) and De Sitter's solution Eq. (2):

$$ds^2 = c^2\,dt^2 - R_E^2[d\chi^2 + \sin^2\chi\,d\theta^2 + \sin^2\chi\sin^2\theta d\phi^2], \qquad (1)$$

$$ds^2 = c^2\cos^2\chi\,dt^2 - R_S^2[d\chi^2 + \sin^2\chi\,d\theta^2 + \sin^2\chi\sin^2\theta\,d\phi^2], \qquad (2)$$

where R_E and R_S are two constants, t is the time coordinate, c is the speed of light in vacuum, θ and ϕ are the colatitude and longitude in spherical coordinates, and r is the radial coordinate. In these expressions, the coefficients of dt, $d\chi$, $d\theta$, $d\phi$ are independent of t. The difference between both metrics is the presence in Eq. (2) of the term $\cos^2\chi$ as the coefficient of dt^2. This completely changes the meaning of t, which for Einstein was really the cosmic time, understood as the time measured by a clock permanently associated with a frame in which matter was on the average at rest.

Moreover, the field equations show the universe of De Sitter to be empty, so that the equivalence principle of inertia is inapplicable; there is a metric structure without matter, which was an essential argument used by Einstein for its rejection. But, it might in fact be presumed that De Sitter considered his solution as asymptotic to the real universe and as a better approximation than Einstein's universe, which was associated with too large a density.

The strangeness of Eq. (2) becomes clear when we put $\chi = \pi/2$. For an observer located at the origin, this corresponds to the set of points situated at a distance of a quarter of the universe's circumference. One gets the hypersurface

$$ds^2 = -R_S^2[d\chi^2 + d\theta^2 + \sin^2\theta\, d\phi^2]. \tag{3}$$

For the trajectories of light rays (given by $ds = 0$), one then has $d\chi = d\theta = d\phi = 0$. On this hypersurface, light and thus any physical quantity will be unable to move. It is an inseparable limit that stops all signals and all motion; it is a kind of horizon the origin of which could in principle be chosen arbitrarily.

This horizon has not only a geometrical but also a dynamical meaning. The end of things on the hypersurface $\chi = \pi/2$ is equivalent to a mass accumulation on this sphere. The universe, presented as empty, is empty everywhere except there. It is not homogeneous and there is in fact a false symmetry between Eq. (1) and Eq. (2).

As Lemaître showed in his paper of 1925, homogeneity could be reestablished in the De Sitter metric by the introduction of a new time that depends on the radial coordinate,

$$t = \tau - \frac{R_S}{c}\log\cos\chi, \tag{4}$$

and by the use of a new radial coordinate

$$r = R_S \sin\chi e^{-(c\tau/R_S)}, \tag{5}$$

giving the classical metric of the De Sitter universe :

$$ds^2 = c^2 d\tau^2 - e^{2c\tau/R_S}\,[(dr^2 + r^2 d\theta^2 + r^2\sin^2\theta\, d\phi^2)]. \tag{6}$$

In this form, the geometry of the space is Euclidean.

Moreover, the properties of this manifold may to a certain extent be regarded as intrinsically independent of the new variables, since the transformation $r' = re^{-(c\tau/R_S)}$, $\tau' = \tau - \tau_0$, which may be considered as a change in spatial scale combined with a change to a new zero point for the time-like variable, leaves the form of the line element unaltered. This line element could then be called stationary.

The field is no longer static, and Lemaître had envisaged (Lemaître 1925) a model asymptotic to the Einstein universe in the past and to the De Sitter universe in the future.

3. The Instability of the Einstein Universe

Taking advantage of the Gordon–Mackay agreements, Lemaître, while at Harvard Observatory, also registered as a graduate student in the Department of Physics at the Massachusetts Institute of Technology. Finally, at the suggestion of Eddington, he presented as a doctoral dissertation a memoir entitled "The Gravitational Field in a Fluid Sphere of Uniform Invariant Density According to the Theory of General Relativity," which in fact dealt with paradoxes encountered by the Schwarzschild metric in a De Sitter universe. This was the very first step toward a theory of the expanding universe. The dissertation was never published (it was accepted at the beginning of 1926).

On his return from the United States, Lemaître was appointed associate professor in the Department of Mathematics at the University of Louvain (1926). Initially, his very light teaching load gave him ample opportunity for research, which mainly focused on the De Sitter cosmology. Unaware of the paper published by Aleksander Friedmann (1922), he was the first to show how to introduce a radius varying with time in the De Sitter metric (Lemaître 1925); then, he deduced from the modified metric a linear relation between radial velocities and distances for galactic nebulas (Lemaître 1926).

In 1927, he published three papers on relativistic cosmology in the *Annales de la Société Scientifique de Bruxelles*, improving some of his previous works: one on relativistic time (Lemaître 1927a), the other concerning the motion of a rigid body in the theory of relativity (Lemaître 1927b). The third and most important paper is entitled "Un univers homogène de masse constante et de rayon croissant rendant compte de la vitesse radiale des nébuleuses extra galactiques" (Lemaître 1927c) and relates a solution of general relativity to observational astronomy.

In fact, Lemaître was aiming at a theory of the real universe. He was guided by physical conditions and had the astronomical observations continually in mind. Lemaître likened the galaxies to the molecules of a rarefied gas. One thus has a homogeneous picture of the universe, which allows one to speak of an energy density ρ, the main part of which is due to the mass at rest. There will also be a uniform pressure p arising mainly from electromagnetic radiation and for an almost negligible part from the motion of matter. Let R be the radius of the universe varying with time, so that the line element is $ds^2 = -R^2 d\sigma^2 + c^2 dt^2$, where $R d\sigma$ is the elementary distance in three-dimensional space. Natural units will be used where $c = 1$ and $G = 1$. From Einstein's field equations, Lemaître deduced

$$3\frac{\dot{R}^2}{R^2} + \frac{3}{R^2} = \Lambda + 8\pi\rho, \tag{7}$$

$$2\frac{\ddot{R}}{R} + \frac{\dot{R}^2}{R^2} + \frac{1}{R^2} = \Lambda - 8\pi p. \tag{8}$$

Here, Λ is the unknown cosmological constant and the overdot refers to the time derivative. These equations are the same as the Friedmann equations with elliptical geometry ($k = 1$), except for the pressure term p, which Friedmann did not take into account. They lead to the equation of the conservation of energy

$$\frac{d\rho}{dt} + 3\frac{\dot{R}}{R}(\rho + p) = 0. \tag{9}$$

To solve the problem completely, we must have at our disposal an equation of the form

$$p = p(\rho). \tag{10}$$

Taking with Lemaître the case of a universe filled with dust ($p = 0$), we have

$$\rho R^3 = M/\pi^2 = \text{const.}, \tag{11}$$

where M is the total mass. The relationship between R and t can now be worked out:

$$t = \int \frac{dR}{\sqrt{(\Lambda R^2/3) - 1 + (4M/3\pi R)}}. \tag{12}$$

For the static Einstein universe,

$$R = R_E, \quad \Lambda = R_E^{-2}, \quad 4M/\pi = R_E. \tag{13}$$

For the De Sitter stationary universe,

$$\Lambda = 3R_S^{-2}, \quad M = 0. \tag{14}$$

We can write Eq. (12) in the following way:

$$t = R_E\sqrt{3}\int \frac{dR}{R - R_E}\sqrt{\frac{R}{R + 2R_E}}, \tag{15}$$

which can be integrated to give

$$t = R_E\sqrt{3}\log\left(\frac{1+x}{1-x}\right) + R_E\log\left(\frac{\sqrt{3}x - 1}{\sqrt{3}x + 1}\right) + \text{const.}, \tag{16}$$

where

$$x^2 = \frac{R}{R + 2R_E}.$$

Its initial value for $t = -\infty$ is R_E; the solution makes R increase steadily as t increases.

There is thus a steady expansion from the start. The predicted expansion leads to a red shift for distant matter, which one can easily determine:

$$\frac{d\lambda}{\lambda} = \frac{\dot{R}}{R}r = \frac{r}{R_E\sqrt{3}}\sqrt{1 - 3\left(\frac{R_E}{R}\right)^2 + 2\left(\frac{R_E}{R}\right)^3}. \tag{17}$$

When Lemaître learned of the existence of important red shifts of distant galaxies—although Hubble's law had not been clearly stated yet—he decided to publish his results (Lemaître 1927c), and from the available red shifts, he deduced

$$R/R_E = 21.5.$$

This model shows the instability of the static solution of Einstein. Going back in time, the rate of expansion tends to zero logarithmically; it proceeds at an accelerated rate, reaching a maximum of $1/(R_E\sqrt{3})$ in an infinite future, as in the De Sitter universe. He tried to draw the attention of Einstein to his results when he was attending the 5th Solvay Conference on physics in Brussels (October 1927). Einstein was most abrupt: "Your calculations are correct, but your physical insight is abominable." Likewise, Lemaître failed with De Sitter at the 3rd General Assembly of the International Astronomical Union in Leiden (July 1928). But, later at a meeting of the Royal Astronomical Society (January 1930), De Sitter endorsed the careful observations made by Edwin P. Hubble at Mount Wilson Observatory and concurred with the majority in holding the linear relation between radial velocity and distance as a law of nature. He confessed, though, that he did not know how to assimilate it in his cosmology. This statement was published in the February issue of *Observatory*. Lemaître read it and wrote at once to Eddington to remind him that he had already solved the problem and also to ask him to send a reprint of his 1927 note to De Sitter. This time Eddington paid attention to Lemaître's contribution, dispatched a copy of it to De Sitter and Shapley, and reworked his communication to the Royal Society to make it a critical review of Lemaître's theory of the expanding universe. A slightly amended English translation of Lemaître's paper was published in the *Monthly Notices* (Lemaître 1931c).

In a lecture at the British Mathematical Association, published in *Nature* under the title "The End of the World from the Standpoint of Mathematical Physics," Eddington conceived the expansion as starting from a state of equilibrium given by the Einstein model (Eddington 1931). He argued that the second law of thermodynamics can be extrapolated in the future toward a state of complete disorganization expressed by an infinite entropy. In the past, however, the extrapolation would probably have to start from zero entropy, and this was a situation that he could not accept.

4. Universe Expanding from a Singularity

In the meantime, in a reconsideration of the physics related to the geometric model of 1927, Lemaître considered the logarithmic solution too artificial to describe the evolution of the material universe. In fact, it meant a slowing down of all physical processes in the past. Radioactivity and degradation of energy seem to indicate just an opposite evolution. Looking back at the two parameters of Eq. (12), Λ and M, there was no physical reason to give them the values chosen by Einstein. For $M > 0$, solutions will exist where the scaling factor R equals zero, meaning the reduction of the universe to a point-like singularity at a certain time considered as the origin. For a mass such that $M < (\pi/4\sqrt{\Lambda})$ (type A), the scaling factor will increase indefinitely in the future. For $M > (\pi/4\sqrt{\Lambda})$, (type B), the scaling factor after reaching a maximum recollapses to zero. In the Einstein case $M = (\pi/4\sqrt{\Lambda})$ (type C), we get another branch of the logarithmic solution with the static solution as a maximum. Then, Lemaître started to evaluate the integral of the second member of Eq. (17) by means of Weierstrass elliptic functions. He had the opportunity to show them to Einstein during his stay on the Belgian coast (at the end of 1930). Einstein informed him that calculations of that type had already been made by Friedmann (1922). At the time, Lemaître did not know of that work. Lemaître did not publish his calculations, but left drafts of these solutions which were published by Heller (1979). They were used, however, in his proposal of a model of the universe with an initial singularity (type A), as well as in his scheme for the formation of galaxies (type B) and clusters of galaxies (type C). For that purpose, he considered one particular condensation of spherical symmetry. He arrived at an equation similar to that for the expansion of a homogeneous universe containing clusters of galaxies enclosed by a surface (type C) and reaching a stable configuration. Inside, islands of denser matter would collapse (type B), forming galaxies.

When Lemaître read about the difficulties experienced by Eddington in connection with the variation of entropy in the past, his imagination was aroused. Convinced that quantum theory must play a significant role in the early universe, on March 21, 1931, he sent a letter to *Nature* entitled "The Beginning of the World from the Point of View of Quantum Theory" (Lemaître 1931a). He introduced the revolutionary idea of a supreme initial concentration of all matter in a primeval atom, identifying matter and space-time, with the atom splitting irreversibly into smaller and smaller pieces, thus giving a first version of what was later called the big bang theory.

Lemaître was invited to participate in a meeting of the British Association on the question of the relation of the physical universe to life and mind. In addition to Lemaître, the following persons took part in the discussion: J. Jeans, W. De Sitter, A.S. Eddington, R.A. Millikan, General G.E. Smits, Bishop E.W. Barnes, and D. Lodge. The communications were published in October 1931 in the supplement to *Nature* (Lemaître 1931b). During the discussion of his hypothesis, Lemaître suggested that big stars were remnants of the successive splittings of the primeval atom and that, with their fireworks of radiation, they were the source of cosmic rays of high energy.

5. Lemaître's Models

As it was received with interest, but also with some doubts, Lemaître's hypothesis needed to be worked out in more scientific detail. On the general scenario of Lemaître's proposal, the universe expands rapidly from the singularity, then decelerates to a period of stagnation favorable for local condensations and the formation of galaxies, which is then followed by another expansion that is slightly but indefinitely accelerated. In contrast to a linear expansion, this scenario will retard the moment of singularity to allow enough time for the formation of galaxies, stars, and planets. Looking for a proof of the initial firework, Lemaître thought he could find it in cosmic rays. His studies extended over 20 years and were carried out mainly in collaboration with Professor Sandoval Vallarta of the Massachusetts Institute of Technology (MIT). As this chapter is restricted to topics related to general relativity, we do not mention this work any further.

At the end of 1932 and the beginning of 1933, Lemaître was in California, where he met numerous astronomers, in particular, Hubble and Russell; he also had a lively discussion with Einstein in Pasadena. This meeting was more cordial than the preceding ones, although Einstein was very skeptical about a singular beginning of the universe, in spite of his acceptance of the expansion of the universe. He thought that the singularity was due to the isotropy of the model studied and a reflection of the religious dogma of the creation in the thinking of a priest. But, the discussion concentrated mainly on the cosmological constant (lambda term) that Lemaître kept in his cosmology against the strong opposition of Einstein and, after him, of most other cosmologists.

The conception of the cosmological constant as expressed by Lemaître (Lemaître 1934) was that the vacuum would exercise a repulsive force giving rise to a negative pressure term. In Belgium, he felt that he had to present a structured version of his theoretical researches in relativistic cosmology. In his 1932 article (Lemaître 1932), the proof of the nonsingularity of the so-called "Schwarzschild singularity" was fully expressed, although it was already implicit in the criticism of the De Sitter universe (Lemaître 1925) and in one of the manuscripts of his thesis at MIT. In the same paper, he showed, in answer to Einstein, that an anisotropy will not preclude a singular beginning. The proof is as follows. Lemaître started with the quite general quadratic form

$$ds^2 = a_1^2\, dx_1^2 + a_2^2\, dx_2^2 + a_3^2\, dx_3^2 + a_4^2\, dx_4^2. \tag{18}$$

In the particular case of a homogeneous universe where the a_i are functions of $x_4 = t$ only, we rewrite Eq. (18)

$$ds^2 = -b_1^2\, dx_1^2 - b_2^2\, dx_2^2 - b_3^2\, dx_3^2 + dx_4^2. \tag{19}$$

The quantity $b_1 b_2 b_3 = R^3$ measures the space volume. Introducing the variation of the anisotropy by the parameter

$$I^2 = \left(\frac{\dot{b}_1}{b_1} - \frac{\dot{b}_2}{b_2}\right)^2 + \left(\frac{\dot{b}_2}{b_2} - \frac{\dot{b}_3}{b_3}\right)^2 + \left(\frac{\dot{b}_3}{b_3} - \frac{\dot{b}_1}{b_1}\right)^2, \tag{20}$$

and applying the Einstein equations, Lemaître obtained

$$3\frac{\ddot{R}}{R} = \frac{\kappa}{2}(T_1^1 + T_2^2 + T_3^3 - T_4^4) - \tfrac{1}{3}I^2. \tag{21}$$

In physical applications, the pressures T_1^1, T_2^2, T_3^3 will in general be negative and in any case smaller in absolute value than $T_4^4 = \rho$; $\ddot{R}$ will then be essentially negative, and R will reach the value zero in the past and the volume will vanish. This result precedes by more than 30 years (and for a quite general case as well) the proof of the existence of singularities in the solutions of Einstein's equations given by Hawking (1966).

6. The Singularity of the Schwarzschild Exterior Field

As we have mentioned in the biographical introduction, the spurious character of the singularity of the Schwarzschild field can be traced back to the note on De Sitter's model of the universe (Lemaître 1925), but the strict proof was given in "L'univers en expansion" (Lemaître 1932), and we will analyze it in that context.

Lemaître starts by considering the a_i ($i = 1, 2, 3, 4$) of Eq. (18) as functions of the four coordinates x_i. He explicitly gave the equations of the gravitational field:

$$\kappa T_\mu^\nu + \lambda g_\mu^\nu = -R_\mu^\nu + \tfrac{1}{2}\lambda g_\mu^\nu R, \tag{22}$$

where κ is the gravitational constant, λ is the cosmological constant, and g_μ^ν, R_μ^ν, and T_μ^ν are the tensor metric, the tensor of curvature, and the energy-momentum tensor, respectively.

In a following paragraph, he introduced spherical symmetry in his result and expressed ds^2 in a new form, introducing the colatitude θ and the longitude ϕ:

$$ds^2 = -a^2\,d\chi^2 - r^2[d\theta^2 + \sin^2\theta\,d\phi^2] + c^2\,dt^2, \tag{23}$$

where $a^2 = -a_1^2$, $-r^2 = a_2^2$, $c^2 = a_4^2$ are functions of χ and t. He also wrote $T_4^4 = \rho$ (energy density), $T_1^1 = -p$ (pressure), and $T_2^2 = T_3^3 = -\tau$. One may choose x_1 constant along any universe line; the lines $x_4 = t =$ constant are trajectories perpendicular to x_1. Then, all the rectangular components T_{ij} ($j = i$) vanish.

From the conservation law, we get

$$\frac{\partial}{\partial x_1}\left(T_1^1 a_2^2 \frac{\partial a_2}{\partial x_4}\right) = \frac{\partial}{\partial x_4}\left(T_4^4 a_2^2 \frac{\partial a_2}{\partial x_1}\right). \tag{24}$$

From Eq. (24), a scalar potential m can be introduced, and we may write

$$4\pi\rho r^2 \frac{\partial r}{\partial \chi} = \frac{\partial m}{\partial \chi}, \qquad 4\pi p r^2 \frac{\partial r}{\partial t} = -\frac{\partial m}{\partial t}. \tag{25}$$

Writing Einstein's constant in the form $\kappa = (8\pi G/c_0^2)$ (c_0 is the speed of light), Lemaître deduced the following from the fourth component of the Einstein equations:

$$\frac{c_0^2}{c^2}\left(\frac{\partial r}{\partial t}\right)^2 = -c_0^2\left[1-\left(\frac{\partial r}{a\partial \chi}\right)^2\right] + \frac{2Gm}{r} + \frac{\lambda c_0^2}{3}r^2. \tag{26}$$

Now, $T_{14} = 0$ can be written as

$$\frac{\partial}{\partial t}\left(\frac{1}{a}\frac{\partial r}{\partial \chi}\right) = \frac{1}{ac}\frac{\partial r}{\partial t}\frac{\partial c}{\partial \chi}. \tag{27}$$

The conservation theorem $T^{\nu}_{1,\nu} = 0$ gives

$$\frac{\partial p}{\partial \chi} + \frac{2}{r}\frac{\partial r}{\partial \chi}(p-\tau) + \frac{1}{c}\frac{\partial c}{\partial \chi} = 0. \tag{28}$$

In the case of equilibrium, p and τ can be neglected; c is a function of t only, and it can be reduced to c_0 by a change of variable. From Eq. (27), we get

$$\frac{1}{a}\frac{\partial r}{\partial \chi} = f(\chi), \tag{29}$$

and the metric becomes

$$ds^2 = -\left(\frac{\partial r}{\partial \chi}\right)^2 \frac{d\chi^2}{f^2(\chi)} - r^2(d\theta^2 + \sin^2\theta\, d\phi^2) + c_0^2\, dt^2. \tag{30}$$

Here r is a function of χ and t satisfying Eq. (26):

$$\left(\frac{\partial r}{\partial t}\right)^2 = -c^2[1-f^2(\chi)] + \frac{2Gm}{r} + \frac{\lambda c_0^2 r^2}{3}. \tag{31}$$

In the reasoning followed by Lemaître for the Schwarzschild exterior field (Lemaître 1933, p. 80), λ is considered different from zero. It remains valid when we pass to the limit $\lambda = 0$.

To simplify the presentation of his method, we take $\lambda = 0$ from the start and use a system of units where $G = c_0 = 1$. In addition, we assume the space to be Euclidean, $f(\chi) = 1$, and m a constant in vacuum. Then, Eq. (31) can be written as

$$r\left(\frac{\partial r}{\partial t}\right)^2 = 2m.$$

By integration,

$$\tfrac{2}{3}r^{3/2} = \sqrt{2m}\,(t - \chi). \tag{32}$$

With this linear choice for the function of χ, we have $\partial r/\partial t = -(\partial r/\partial \chi)$, and thus,

$$ds^2 = -2m\,\frac{d\chi^2}{r} = -r^2[d\theta^2 + \sin^2\theta\, d\phi^2] + dt^2, \tag{33}$$

which is a field solution in a vacuum.

The space remains Euclidean, and there is no singularity except for $r = 0$. If we take r as a coordinate, it is possible to define a coordinate in such a way as to express the field in the Schwarzschild form. We have

$$dr^2 = \frac{2m}{r}(dt - d\chi)^2. \tag{34}$$

Then,

$$\frac{2m}{r}(d\chi)^2 = dr^2 - \frac{2m}{r}(dt^2 - 2\,d\chi dt),$$

and

$$ds^2 = -dr^2 - r^2(d\theta^2 + \sin^2\, d\phi^2) + \left(1 + \frac{2m}{r}\right)\, dt^2 - \frac{4m}{r}\, d\chi\, dt.$$

Eliminating

$$d\chi = dt - \sqrt{\frac{r}{2m}}\, dr,$$

we get

$$ds^2 = -dr^2 - r^2(d\theta^2 + \sin^2 d\phi^2) + \left(1 - \frac{2m}{r}\right) dt^2 - 2\left(\frac{2m}{r}\right)^{1/2} d\chi\, dt.$$

Putting

$$dt = d\tau - \frac{\sqrt{2m/r}}{1 - (2m/r)}\, dr, \tag{35}$$

one gets

$$ds^2 = -\frac{dr^2}{1 - (2m/r)} - r^2(dt^2 + \sin^2\theta\, d\phi^2) + \big(1 - (2m/r)\big)\, d\tau^2, \tag{36}$$

which is the Schwarzschild form for a point-like mass. Now, integrating Eq. (35), we obtain

$$t = \tau - \sqrt{2mr} + 2m \log\left(\frac{\sqrt{r} - \sqrt{2m}}{\sqrt{r} + \sqrt{2m}}\right), \tag{37}$$

which is not valid for $r < 2m$. The new form of the field without the spurious singularity $r = 2m$ is expressed by Eq. (33), where the new coordinate r is given by Eq. (32). In the conclusion to his paper, Lemaître added that "the singularity of the Schwarzschild field is thus a fictitious singularity similar to the one of the horizon of the center in the original form of the universe of De Sitter."

7. The Relationship Between Science and Religion in Lemaître's Cosmological Hypothesis

As mentioned in Section 5, when Lemaître during his 1933 meeting with Einstein talked of the primeval atom, he was interrupted: "No, not that, that suggests too much the creation." In fact, the introduction by a Catholic priest of a singularity vaguely related to a creative action at a first instant was viewed by some people as the use of apparently scientific reasonings to defend a dogma of the church.

One might say that a pure scientist does not exist. Scientists are persons of their time, educated according to the ideas that prevailed in the environment during youth. In their research, they usually first accept certain philosophical options, in the broad sense of the word, until scientific critics convince them that they must change their views. This was also the case with Einstein with regard to his proposal of a static universe: after the discovery of the red shift of the galaxies spectra, he accepted models of an expanding universe.

When Lemaître started his investigations in relativity, he had at his disposal only two scientific models (the static models of Einstein and De Sitter), and he tried to find an intermediary model that started logarithmically from the Einstein model in the past and reached the corrected De Sitter model in the future. A critical study of the physics of such a universe led him to a solution of the Einstein equations with a past singularity.

For modern scientific cosmologists, although they may feel uneasy about this primordial singularity, the objectivity of the thinking of its initiator is beyond doubt. That was not always the case in the past, in particular, during the postwar period when the competing cosmological theory of "continuous creation" was favored, although such a theory raises even more philosophical difficulties.

Following the concept of Lemaître, our present theory of the universe breaks down at the singularity, and since our only access to the universe is through theories we construct ourselves, it looks as if the history of the universe stops at the singularity. Strictly speaking, however, it is only a gap in our knowledge. A gap, to be sure, that does not arise from accidental flaws in our reasoning, but appears in the very foundations of our science. We do not know at present whether we shall be able to overcome this gap or not, but in any case, it would be a dangerous methodological mistake to fill the gap in our knowledge with the "hypothesis of God." Even if the universe were indeed created by God in the singularity, we cannot prove this by means of the scientific method, i.e., by means of experimentally verifiable mathematical theories. This is a purely philosophical or theological question.

In other words, science is neutral with respect to any philosophical or religious doctrine. Of course, some of the scientific theories (for instance, those of the origin of the universe) may be, and in fact often are, interpreted either in a materialistic or theological manner. However, none of those interpretations belong to the scope of science.

For a long time, in fact until his Solvay talk of 1958, Lemaître kept silent about the postwar suspicions of subjectivity. Speaking in his Solvay talk about the philosophical implications of the primeval atom hypothesis, he emphasized:

> As far as I can see, such a theory remains entirely outside any metaphysical or religious question. It leaves the materialist free to deny any transcendental being. For the origin of space-time, he may keep the same attitude of mind he has been able to adopt for events occurring in nonsingular places in space-time. For the believer, it removes any attempt at familiarity with God, such as Laplace's "chiquenaude" or Jeans' finger. It is consonant with the words of Isaiah when he spoke of the "Hidden God," hidden even in the beginning of creation. (Lemaître 1958)

Here, Lemaître referred to Laplace and Jeans, who described God's interference in the natural course of events, as a "chiquenaude initiale" and as a "finger of God agitating the ether," respectively. Even from the theological point of view, God could be reduced to the role of a scientific hypothesis.

However, Lemaître warned that "It does not mean that cosmology has no meaning for philosophy." Philosophy and theology, when kept in isolation from scientific thought, either change into outdated self-closed systems or become dangerous ideologies.[2]

Notes

[1] Archives Lemaître, Institut d'Astronomie et de Géophysique, Université Catholique de Louvain.

[2] For more details, see Godart 1984 and Godart and Heller 1979a, 1979b.

References

De Sitter, Willem (1917). "On the Relativity of Inertia." *Koninklijke Akademie van Wetenschappen te Amsterdam. Section of Sciences. Proceedings* 19: 1217–1225.

Eddington, Arthur S. (1931). "The End of the World: From the Standpoint of Mathematical Physics." *Nature* 127: 447–453.

Einstein, Albert (1917). "Kosmologische Betrachtungen zur allgemeinen Relativitätstheorie." *Königlich Preussische Akademie der Wissenschaften* (Berlin). *Sitzungsberichte*: 142–152.

Friedmann, Aleksander A. (1922). "Über die Krümmung des Raumes." *Zeitschrift fur Physik* 10: 377–386.

Godart, Odon (1984). "Monseigneur Lemaître, sa vie, son oeuvre." *Revue des Questions Scientifiques* 155: 155–182.

Godart, Odon, and Heller M. (1979a). "Einstein–Lemaître: Rencontre d'idées." *Revue des Questions Scientifiques* 150: 23–43.

—— (1979b). "Les relations entre la science et la foi chez G. Lemaître." *Pontificia Academia Scientiarum. Commentationes* 3: 1–12.

Hawking, Stephen W. (1966). "The Occurrence of Singularities in Cosmology." *Royal Society of London. Proceedings A* 294: 511–521.

Heller, M. (1979). "Question to an Infallible Oracle." In *Physics of the Expanding Universe. Proceedings of the Cracow School in Cosmology, held at Jodlowy Dwór, Poland, 1978.* Lecture Notes in Physics, 109. Marek Demiański, ed. Berlin and New York: Springer-Verlag, 1979, pp. 199–210.

Lemaître, Georges (1924). "The Motion of a Rigid Solid According to the Relativity Principle." *Philosophical Magazine.* 48: 164–176

—— (1925). "Note on De Sitter's Universe." *Publication du Laboratoire d'Astronomie et de Géodésie de l'Université de Louvain* 2: 37–41. *Journal of Mathematics and Physics* 4: 188–192.

—— (1926). "La théorie de la relativité et l'expérience." *Revue des Questions Scientifiques* 9: 364–374.

—— (1927a). "A propos d'une note du P. Schaffers sur la relativité." *Société Scientifique de Bruxelles. Annales B* 47: 15–21.

—— (1927b). "Le mouvement varié d'un solide d'après la théorie de la relativité." *Société Scientifique de Bruxelles. Annales A* 47: 103–109.

——— (1927c). "Un univers homogène de masse constante et de rayon croissant, rendant compte de la vitesse radiale des nébuleuses extragalactiques." *Société Scientifique de Bruxelles. Annales A* 47: 49–59.

——— (1931a). "The Beginning of the World from the Point of View of Quantum Theory." *Nature* 127: 706.

——— (1931b). "Contribution to a British Association Discussion on the Evolution of the Universe." *Nature* 128: 714–706.

——— (1931c) "A Homogeneous Universe of Constant Mass and Increasing Radius Accounting for the Radial Velocity of Extra-Galactic Nebulae." *Royal Astronomical Society. Monthly Notices* 91: 483–490. [Slightly amended English translation of Lemaître 1927c].

——— (1932). "L'univers en expansion." *Publication du Laboratoire d'Astronomie et de Géodésie de l'Université de Louvain* 9: 171–205; also in *Société Scientifique de Bruxelles. Annales A* 53 (1933): 51–85.

——— (1934). "Evolution of the Expanding Universe." *National Academy of Science USA. Proceedings* 20: 12–17.

——— (1958). "The Primaeval Atom Hypothesis and the Problem of the Clusters of Galaxies." In *La Structure et l'Évolution de l'Univers. Report of the Onzième Conseil de Physique Solvay.* Bruxelles: Stoops, pp. 1–25.

Schrödinger and Cosmology

Helmuth Urbantke

In the field of relativity and cosmology, Erwin Schrödinger did not achieve the kind of breakthrough that would compare with his development of nonrelativistic wave mechanics, the physics of atoms and molecules. However, if one considers larger parts of his work, one recognizes connections that, while nonexistent in nature as we know it today, were unconsciously operative in his mind, thus contributing to work in apparently unrelated areas. This aspect makes an analysis of his ideas in cosmology quite rewarding.

In this chapter, I rely heavily on material and information given to me by the following persons: J.L. Synge (Dublin), Dr. R. Mansouri (Teheran), R. Sexl, W. Thirring, Dr. A. Dick, and Drs. G. and W. Kerber (Vienna). Finally, a paper by Alexander Rüger (Rüger 1988) that evaluates Schrödinger's correspondence as far as our subject is concerned must be cited. I refer to his work for a far more complete list of references.

Schrödinger entered the field of cosmology in 1918 with a paper (Schrödinger 1918) in which he only reinterpreted the solution of the relativistic gravitational field equations that Einstein had given in "Kosmologische Betrachtungen zur allgemeinen Relativitätstheorie." In retrospect, Schrödinger's paper is not an important one, but it gives some insight into his world of ideas. It is necessary to be aware of the status of empirical cosmology during that time in order to appreciate the impression Einstein's theory made on Schrödinger. Confirmed knowledge did not reach beyond our own galaxy; the interpretation of the nebulae as distant galaxies was no more than a conjecture made by a few people. There was, then, no indication whatsoever that the average matter distribution in the universe could be homogeneous. Then, Einstein found new ways of thinking about the universe as a result of contemplating a few principles: the principles of relativity, covariance, and Mach. Used to working with detailed empirical material, Schrödinger must have been very impressed. At the same time, he acquainted himself with the content of the Upanishads, and the idea of unity in the world never left him. Now, in his first cosmological paper, Schrödinger did nothing but add the so-called cosmological term to the dust matter term and interpret the sum as a new matter term, corresponding to matter with stresses of the order density × speed of light. This he did in favor of an interpretation that to him seemed more Machian in the sense of "relativity of mass," thus coming closer to "general relativity" in its truest sense. It is remarkable that in the margins of a reprint of his paper we see that he inserted numbers into his equations (which had meanwhile become available) to estimate these stresses!

A further decisive impression upon Schrödinger came from H. Weyl's attempt at geometrizing electrodynamics. The first version of Weyl's gauge theory failed on empirical grounds because it implied, in addition to the "first clock effect, " the "twin paradox" of relativity, a second clock effect on the clock rate under transport around a closed orbit, which is empirically absent in atoms. Weyl tried to explain this by a remark that is unimportant except for the phrase "microphysics from a detour through cosmology," which became firmly rooted in Schrödinger's mind. Now, after a period of thinking about how the principle of geodesic motion was realized for charged particles in Weyl's theory (remember that at this time no neutral massive elementary particles were known!), Schrödinger almost saved the theory by a detour into microphysics. He noted (Schrödinger 1922) that the logarithm of the clock rate correction factor, multiplied by a suitable constant, became integer valued on Bohr's quantum orbits of the hydrogen atom; thus, including a factor $\sqrt{-1}$ and exponentiating, unity would be obtained. It did not come to him, however, at the time that this inclusion of $\sqrt{-1}$ could be interpreted as changing a clock rate correction factor into a phase correction of a complex wave amplitude. Thus, for the development of wave mechanics, he had to learn about the Bose–Einstein gas theory and de Broglie waves, and it is only in the last "Mitteilung" that he introduced the time-dependent equation that requires considering essentially complex wave functions (Schrödinger 1926b).

The correct wave-mechanical interpretation of the phase correction factor was given by others (London, Fock, Weyl, etc.). After this, the idea of links between gravitation, electromagnetism, world geometry, cosmology, and quantum theory became stronger and stronger in Schrödinger's mind. In the early 1930s, he formally discussed this new geometrization of electromagnetism in his paper on Dirac's electron in the gravitational field (Schrödinger 1932). But, in retrospect, what he considered to be an achievement turned out to block a view into the future: for reasons of apparent mathematical simplicity, he suppressed, as far as possible, the group theoretical aspect of the problem, not knowing that the proper tool toward simplification was the apparatus of differential forms. Thus Schrödinger did not grasp the idea of paralleling the symmetry group that arose later in the context of nuclear forces with the symmetry groups arising in gravitation and electromagnetism, although the former was being studied at the Dublin Institute, where he was striving for the unification of all three types of forces on a very different 4-dimensional geometrical basis, which finally was not successful. In his wave-mechanical papers, he stressed that for him, understanding was in terms of "Denkformen," (Schrödinger 1926a, pp. 508–509) which had to be rooted in space and time; therefore, the hybrid type object of a space-time spinor was closer to him than isotopic spin, and in his paper, he discussed tying the spinors even closer to space-time.

In the sequel, we see Schrödinger playing with the Dirac equation, written in various cosmological spaces, and having exchanges with the Princeton school (Veblen, von Neumann, Taub) on this subject. But, only after 1936, when he started studying the work of Eddington did he seem to grasp the physical significance of this game.

In a talk at the Galvani Bicentenary celebration, held in 1937 in Bologna, Schrödinger presented Eddington's ideas (Schrödinger 1938a). These were to explain microphenomena on the basis of cosmology and visa versa, via eigenvalue and self-consistency considerations in a closed universe. In particular, the fundamental dimensionless numbers in physics, such as mass ratios of elementary particles, the dimensionless number associated with the elementary charge, and the number of particles in the universe, were to be explained by pure reasoning! (It should be mentioned that leading theorists were able to imagine a removal of various divergence difficulties plaguing quantum electrodynamics only as concurrent with an explanation of the value of the fine structure constant!) Soon after, he reviewed Eddington's book on these ideas quite enthusiastically for *Nature* (Schrödinger 1937). The basic idea is that of a completely degenerate relativistic Fermi gas "enclosed" in a closed universe: the geometry determines the energy levels, and the gas determines the geometry such that everything becomes self-consistent. It is interesting to note that Schrödinger imagined a further unification by thinking of the matter waves constituting the Fermi gas as finally reducible to gravitational waves. (This is found only in his notebooks as "cum grano salis": see also a letter to Einstein, quoted in Rüger 1988, p. 394. The universal connection between spin and statistics does not seem to have been accepted at this time!)

His enthusiasm for Eddington's grandiose conception was not hampered at this time by his knowledge about the many difficulties in the details. But, he had, even before the Bologna conference, a long correspondence with Eddington, mostly concerning the famous relation $mc^2 = (hc/R)N^{1/2}$ between the proton mass, the radius of the universe, and the number of particles in the universe. Schrödinger pointed out that from correctly handling Fermi statistics one arrives at the empirically wrong version $mc^2 = (hc/R)N^{1/3}$, which finally led him to discard Eddington's theory. However, it took him three years to give up his attempts to justify $N^{1/2}$ (Schrödinger 1940). Eddington could not be convinced that here was, on his part, a fundamental neglect or misunderstanding of degenerate relativistic Fermi gases. (S. Chandrasekhar had been victim of this to some extent.) Schrödinger presented the $N^{1/2}$ version in Bologna in front of Pauli, Bohr, Heisenberg, and others without further comment, thus intentionally maneuvering himself into a difficult position that added to the position of outsider, which he already had by his nonacceptance of the Copenhagen interpretation of quantum mechanics. One can appreciate from this just how much unity in physics meant to Schrödinger.

During 1937–1940, he did some detailed calculational work (Schrödinger 1938b) to "penetrate the rich, but difficult to grasp, treasure of ideas of A. S. Eddington." First, he calculated the proper modes of a scalar and a Dirac field in the static Einstein universe, as a "not indispensable, but still desirable preparation," his "modest . . . diligence" being rewarded by insights into geometry (Clifford cylinders) and wave propagation on a 3-sphere. The same was done by a student of his (Müller 1939) for a static De Sitter space. Curiously, the question of a possible discrete mass spectrum was not even raised here, whereas it was in independent work by Podolsky and Branson (Podolsky and Branson 1940) in the Unites States. Technically, their investigation is owed to A.H. Taub and, conceptually, to Eddington, whose invention of anti-De Sitter space with periodic time but noncompact space led them to a discrete, albeit numerically unacceptable, mass spectrum.

Actually, Eddington's program was not really promoted by this work, nor was it by subsequent investigations (Schrödinger 1939) into an expanding universe. But, here he made a surprising discovery: he found that the expansion might create particle–antiparticle pairs, the effect being possibly significant during critical phases of accelerated expansion.

He called these "alarming phenomena," but it is odd what he did about them!

In the paper, he convinced himself that during unaccelerated expansion nothing dramatic happens. Then he put this discovery aside, together with the whole of Eddington's theory! He turned to nonlinear electrodynamics and unified theories, never taking it up again. He did not even mention it in his booklet *Expanding Universe* (Schrödinger 1956), although there he developed all the necessary mathematics needed to discuss it. When his colleagues at the Dublin Institute took up studying the new steady-state model of cosmology in which particle creation was postulated phenomenologically and which is expanding exponentially, he did not react, and nobody seemed to know about his work!

Let me, to conclude this chapter, devote a few words to the later development of this discovery, although it proceeded unconnected to Schrödinger, therefore being retarded by more than 20 years. Schrödinger's precedence, together with exact references, seems to have been pointed out first by L. Halpern only in the early 1970s (Halpern 1972)! First, one has to observe that all subsequent work was done in the language of second quantization (field quantization). However, the decisive calculation is always one à la Schrödinger, unless one attacks the problem by a perturbative method, starting from flat space. Considering the latter method, the effect seemed obvious to many people, and the important thing was to study the effect in relevant situations. As regards the nonperturbative approach, there were two main problems to be solved: Schrödinger was aware of only one of them, namely, the ambiguity involved in the definition of positive frequencies, i.e., of particles versus antiparticles. The other one, the problem of how to regularize the expectation of the energy-momentum tensor, was not mentioned. It had to be solved in order to proceed with a back reaction calculation in the semi-classical approximation and was first recognized by de Witt (de Witt 1953).

To my knowledge, W. Thirring was the first to relate Schrödinger's effect to the steady-state theory, and he acquainted me in 1962 to this question. [Although Thirring had worked at the Dublin Institute in 1950 for half a year under the guidance of Schrödinger (on field equations of higher order), Schrödinger had never mentioned his work on the particle production to him. In 1953, Thirring suggested the idea of particle production by the deep well representing the effective potential of Schwarzschild's metric to Einstein, who rejected it, however. Of course, the global properties of the Schwarzschild horizon were unknown to both at that time.] In 1966, M. Fierz closed a letter in reply to my letter of results: "Has not Schrödinger worked on this problem?", giving me no hint as to where to find his work. In 1965, due to the discovery of cosmic background radiation, big bang cosmologies became favored, and the late R. Sexl related their singular initial state to particle creation, envisaging a self-consistent explosion. "The question whether the particles creating the gravitational field or the field creating the particles came first is meaningless because of the time scale involved," he wrote (Sexl and Urbantke 1969). During the same time, S. Coleman introduced his student L. Parker to the problem of cosmic particle creation (Parker 1968).

Around 1970, Ya. B. Zeldovich et al. in the Soviet Union and C.W. Misner et al. in the United States tried to use the effect to explain the isotropization of an initially anisotropic homogeneous universe. But, it was only in 1974 that the effect moved toward the center of interest when S.W. Hawking showed in a famous paper (Hawking 1974) that black holes should radiate thermally. This radiation is not completely uncontroversial, but it is a missing link for a formal classical thermodynamics of black holes, according to which these objects should have some nonzero entropy at temperature zero. When Hawking in 1975 made it plausible that these quantum properties of black holes could also be derived directly from quantum considerations of the gravitational field itself, the search for a correct quantum theory of gravity became more intense. We still have not had success, but it is suspected that success can only come from including all of the presently known elementary particle theories into the process: unification is a major goal in these days.

References

de Witt, Bryce S. (1953). "Pair Production by a Curved Metric." *Physical Review* 90: 357.

Halpern, Leopold (1972). "Virtual Particles as Sources of the Gravitational Field." *General Relativity and Gravitation* 3: 405–407.

Hawking, Stephen W. (1974). "Black Hole Explosions." *Nature* 248: 30–31.

Müller, Paul O. (1939). "Über die Eigenschwingungen der zylindrischen und sphärischen Welt." *Physikalische Zeitschrift* 40: 366–384.

Parker, Leonard (1968). "Particle Creation in Expanding Universes." *Physical Review Letters* 21: 562–564.

Podolsky, Boris, and Branson, Hermann (1940). "On the Quantization of Mass." *Physical Review* 57: 494–500.

Rüger, Alexander (1988). "Atomism from Cosmology: Erwin Schrödinger's Work on Wave Mechanics and Space Time Structure." *Studies in History and Philosophy of Science* 18: 377–401.

Schrödinger, Erwin (1918). "Über eine Lösungssystem der allgemein kovarianten Gravitationsgleichungen." *Physikalische Zeitschrift* 19: 20–22.

——— (1922). "Über eine bemerkenswerte Eigenschaft der Quantenbahnen eines einzelnen Elektrons." *Zeitschrift für Physik* 12: 13–23.

——— (1926a). "Quantisierung als Eigenwertproblem. Zweite Mitteilung." *Annalen der Physik* 79: 489–527.

——— (1926b). "Quantisierung als Eigenwertproblem. Vierte Mitteilung." *Annalen der Physik* 81: 109–139.

——— (1932). "Dirasches Elektron in Schwerefeld, 1." *Königlich Preussiche Akademie der Wissenschaften* (Berlin). *Sitzungsberichte*: 105–128.

——— (1937). "World Structure." *Nature* 140: 742–744.

——— (1938a). "Sur la théorie du monde d'Eddington." *Nuovo Cimento* 15: 246–254.

——— (1938b). "Eigenschwingungen des Sphärischen Raumes." *Pontificia Academia Scientiarum. Commentationes* 2: 321–364.

——— (1939). "The Proper Vibrations of the Expanding Universe." *Physica* 6: 899–912.

——— (1940). "The General Theory of Relativity and Wave Mechanics." *Wis- en Natuurkundig Tijdschrift* 10: 2–9.

——— (1956). *Expanding Universes.* London and New York: Cambridge University Press.

Sexl, Roman U., and Urbantke, Helmuth K. (1969). "Production of Particles by Gravitational Fields." *Physical Review* 179: 1247–1250.

French Summaries

Julian B. Barbour

Einstein et le principe de Mach

Dans une perspective historique, on pose la question du caractère "machien" de la relativité générale. On soutient qu'Einstein n'était pas clairement conscient des idées de Mach pouvant et devant être réalisées; mais que, au niveau plus profond de la structure spécifique de la théorie de la relativité générale, il réussit vraiment à créer un cadre qui soit aussi "machien" que possible.

Gunnar Berg

Sur l'origine du concept d'espace d'Einstein

Afin de montrer l'influence de la théorie de la relativité générale sur le développement subséquent de la géométrie différentielle, on décrit l'émergence du concept d'espace d'Einstein.

Peter G. Bergmann

La quantification du champ gravitationnel, 1930–1988

Il s'agit d'une perspective concernant la quantification du champ gravitationnel depuis les travaux de Léon Rosenfeld en 1930 aux travaux les plus récents de l'école de Syracuse.

Thibault Damour et Gerhard Schäfer

Levi-Civita et le problème du mouvement en relativité générale

Il s'agit d'une analyse technico-historique des contributions de Levi-Civita au problème du mouvement des deux corps en relativité générale au premier niveau d'approximation post-newtonienne; on s'intéresse en particulier aux questions de l'avance du périastre et de l'accélération du centre de masse, travaux qui font montrer de contradictions internes remarquables.

Odon Godart

Les contributions de Lemaître à la relativité générale(1922–1934)

Il s'agit d'un exposé des travaux de Georges Lemaître concernant la cosmologie relativiste pendant la période 1922–1933. L'auteur est un ancien élève de Lemaître.

Hubert F. M. Goenner

La réception de la théorie de la relativité en Allemagne à travers les ouvrages publiés entre 1908 et 1945

Dans cette contribution, on trouve une discussion des statistiques concernant les ouvrages sur la relativité publiés en Allemagne entre 1908 et 1945. Le maxima de l'activité de publication en 1921 est influencé à la fois par la fin de la première guerre mondiale et par l'annonce des mesures de déflexion de la lumière en Novembre 1919. Une importante bibliographie des livres sur la relativité publiés en Allemagne à cette époque est incluse.

Joshua N. Goldberg

Le soutien de l'US Airforce à la relativité générale: 1956–1972

De 1956 à 1972, un groupe de recherches de la Wright-Patterson Air Force Base dans l'Ohio a été engagé dans l'étude de la relativité générale. Ce groupe avait aussi la responsabilité d'une aide à la recherche sur la relativité générale. Cet article vise à une brève histoire de ce groupe, de ses propres recherches et des projets aidés.

Gennady E. Gorelik

Les premiers pas de la gravité quantique et les valeurs de Planck

On fait ici l'histoire des "valeurs de Planck," combinaison des trois constantes naturelles *c*, *G* et *h*. Aujourd'hui, les "valeurs de Planck" sont liées aux limites quantiques à l'applicabilité de la relativité générale, limites qui furent découvertes dès 1935 par le théoricien soviétique Matvey P. Bronsteyn.

Don Howard

Einstein et "l'univocité": un thème négligé dans les fondements philosophiques de la relativité générale

On explore ici le rôle de la pensée d'Einstein pour ce qui concerne le principe "d'univocité," un thème fondamental bien que très négligé dans la philosophie de la science allemande du début du siècle. L'argument "du trou" représente l'illustration la plus frappante de l'importance du principe d'univocité dans les travaux d'Einstein.

József Illy

La correspondance entre Albert Einstein et Gustav Mie, 1917–1918

La correspondance entre Albert Einstein et Gustav Mie consiste en cinq lettres d'Einstein et quatre lettres de Mie écrites entre 1917 et 1918. Elle concerne avant tout la question de l'existence d'un système de référence "préférentiel."

Michel Janssen

La tentative de Lorentz pour donner à la relativité générale une formulation intrinsèque

Pendant l'année 1916, H. A. Lorentz soumit une série de quatre articles sur la théorie de la relativité générale à l'Académie des Sciences d'Amsterdam. Il s'agit de la première tentative de parvenir à une formulation géométrique indépendante des coordonnées de la formulation de la théorie de la relativité générale.

S. Kichenassamy

L'équation de Dirac en espace courbe

Deux théories peuvent actuellement prétendre à la description de l'électron de Dirac dans un champ de gravitation: l'une dans le cadre de la relativité générale et l'autre dans celui de l'espace-temps d'Einstein–Cartan avec torsion. Le développement historique de ces théories fut confronté à des problèmes techniques dont la résolution doit beaucoup à Hermann Weyl.

Ludwik Kostro
Esquisse de l'histoire de la conception einsteinienne de l'éther relativiste

Entre 1916 et 1918, Einstein publia plusieurs articles dans lesquels les modèles d'espace-temps de la théorie de la relativité sont interprétés comme modèles d'un éther relativiste. Sur la base des articles publiés ou non et de lettres d'Einstein on présente ici l'histoire de l'éther relativiste d'Einstein.

A. J. Kox
La relativité générale aux Pays-Bas, 1915–1920

Cet article présente un large panorama de la réception et du développement de la relativité générale aux Pays-Bas ainsi que la description des travaux des relativistes et des anti-relativistes néerlandais.

André Lichnerowicz
Mathématique et relativité générale: un témoignage

L'auteur analyse la situation de la recherche mathématique en relativité générale dans les années 1937–1955. Il replace ses principaux travaux en relativité et les premières analyses concernant les propagateurs d'un champ physique sur un espace-temps courbe dans le contexte du moment.

David Lopes Gagean et Manuel da Costa Leite
La relativité générale et le Portugal: quelques remarques à propos des études sur la réception

Cet article vise à contribuer à une étude historique comparative de la relativité et à la compréhension de son processus de diffusion; certains aspects de l'histoire de la relativité générale au Portugal après les années cinquante sont discutés.

André Mercier

La relativité générale au tournant du renouveau

Ce texte est à la fois une description des origines spirituelles des réunions internationales sur la relativité générale (GRG) qui ont eu lieu dés 1955 et un témoignage de l'auteur sur ces mêmes origines.

Arthur I. Miller

L'article d'Albert Einstein dans le Jahrbuch *de 1907: le premier pas de la relativité générale restreinte à la relativité générale*

L'article d'Einstein de 1907, publié dans le *Jahrbuch* est analysé dans son contexte historique et scientifique. Le coup de maître d'Einstein fut de réaliser là l'équivalence entre les phénomènes de gravitation et d'accélération. On propose ici une analyse de cette question en recherchant les liens avec les concepts pertinents quant à l'invention de la théorie de la relativité de 1905.

John Norton

Le contenu physique de la covariance générale

Einstein a toujours insisté sur le caractère fondamental de la covariance générale comme principe physique donnant une expression mathématique à un principe de relativité complètement généralisé. Le point de vue d'Einstein quant au contenu physique de la covariance générale est analysé ainsi que les objections de Kretschmann.

José M. Sánchez-Ron

La réception de la relativité générale parmi les physiciens et les mathématiciens anglais (1915–1930)

Cet article concerne la réception et le développement de la relativité générale en Grande-Bretagne durant la période 1918–1930. Le rôle et la contribution des principaux physiciens, mathématiciens et philosophes anglais est exposé et analysé.

Bronisław Średniawa

Les travaux de Myron Mathisson et de Jan Weyssenhoff sur le problème du mouvement en relativité générale

Les contributions de Myron Mathisson, relativiste polonais de l'entre-deux-guerres sont exposées et analysées. Elles concernent essentiellement la question du mouvement en relativité générale.

John Stachel

Le problème de Cauchy en relativité générale—les premières années

Après avoir rappelé les travaux de Cauchy sur les systèmes d'équations différentielles partielles, on analyse en détail les travaux de David Hilbert qui le premier aborda le problème de Cauchy pour les équations d'Einstein de la relativité générale. Les contributions de Georges Darmois, de Kornel Lanczos et de Karl Stellmacher sont ensuite analysées.

Helmuth Urbantke

Schrödinger et la cosmologie

On expose l'intérêt permanent de Schrödinger quant à la quête pour l'unité de la Nature qui à partir de la relativité générale et à travers la théorie quantique le conduisit à des considérations quantiques concernant la théorie relativiste.

Christiane Vilain

Les coordonnées sphériques en relativité générale de 1915 à 1960: une interprétation physique

On discute l'intérêt physique de certains systèmes de coordonnées employés avant 1960 pour l'étude de l'espace vide à symétrie sphérique. On analyse en particulier l'argumentation d'Eddington, de Robertson et de Synge.

List of Participants

Julian B. Barbour
Gunnar Berg
Margot Bergmann
Peter G. Bergmann
Michel Biezunski
Manuel da Costa Leite
François de Gandt
Nathalie Deruelle
Charles Duval
John Earman
Jean Eisenstaedt
Odon Godart
Hubert F. M. Goenner
Joshua N. Goldberg
Peter Havas
Don Howard
József Illy
Michel Janssen
S. Kichenassamy
Ludwik Kostro
A. J. Kox
André Lichnerowicz
André Mercier
Arthur I. Miller
John North
John Norton
Achille Papapetrou
M. Perrin
Jürgen Renn
Jim Ritter
Jean-François Rober
Solange Rutz da Fonseca
José M. Sánchez-Ron
Skuli Sigurdsson
Bronisław Średniawa
John Stachel
G. Tuynman
Helmuth Urbantke
Christiane Vilain

Contributors

Julian B. Barbour, College Farm, South Newington, Banbury, Oxon OX15 J4G, Great Britain

Gunnar Berg, Department of Mathematics, Uppsala University, Thunbergsvagen 3, S-75238 Uppsala, Sweden

Peter G. Bergmann, Department of Physics, New York University, 4 Washington Place, New York, NY 10003, USA

Manuel da Costa Leite, Universidad Nova de Lisboa, Faculdad de Ciències e Tecnologia, 2825 Monte da Caparica, Portugal

Thibault Damour, Institut des Hautes Études Scientifiques, 91149 Bures-sur-Yvette, France and Départment d'Astrophysique Relativiste et de Cosmologie, CNRS, Observatoire de Paris, 92195 Meudon Cédex, France

Odon Godart, Institut d'Astrophysique et de Géophysique Georges Lemaître, Université Catholique de Louvain, Chemin du Cyclotron 2, B-1348 Louvain-la-Neuve, Belgium

Hubert F. M. Goenner, Institut für Theoretische Physik, Universität Göttingen, Bunsenstrasse 9, 3400 Göttingen, Germany

Joshua N. Goldberg, Department of Physics, Syracuse University, 201 Physics Building, Syracuse, NY 13244, USA

Gennady E. Gorelik, Institute of History of Science, Academy of Sciences USSR, Staropansky per. 1/5, Moscow 103012, USSR

Don Howard, Department of Philosophy, University of Kentucky, Lexington, KY 40506, USA

József Illy, Institute of Isotopes, The Hungarian Academy of Sciences, H-1525 Budapest, Hungary

Michel Janssen, Department of History and Philosophy of Science, 1017 Cathedral of Learning, University of Pittsburgh, Pittsburgh, PA 15260, USA

S. Kichenassamy, Laboratoire de Physique Théorique, Institut Henri Poincaré, 11, Rue Pierre et Marie Curie, 75231 Paris Cédex 05, France

Ludwik Kostro, Institute of Experimental Physics, University of Gdansk, Ul. Wita Stowosza 57, 80-952 Gdansk, Poland

A. J. Kox, Instituut voor Theoretische Fysica, Universiteit van Amsterdam, Valckenierstraat 65, 1018 XE Amsterdam, The Netherlands and The Collected Papers of Albert Einstein, Boston University, 745 Commonwealth Avenue, Boston, MA 02215, USA

André Lichnerowicz, Collège de France, 6, Place Marcelin Berthelot, 75231 Paris Cédex 05, France

David Lopes Gagean, Gabinete de Filosofia do Conhecimento, Rua Virginia Vitorino n.9 5 Dto, 1600 Lisbon, Portugal

André Mercier, Institut für Theoretische Physik, Sidlerstrasse 5, 3012 Bern, Switzerland

Arthur I. Miller, Department of History and Philosophy of Science, University College London, Gower Street, London WC1E 6BT, Great Britain

John Norton, Department of History and Philosophy of Science, 1017 Cathedral of Learning, University of Pittsburgh, Pittsburgh, PA 15260, USA

José M. Sánchez-Ron, Departamento de Física Teórica C-XI, Universidad Autónoma de Madrid, Canto Blanco, 28049 Madrid, Spain

Gerhard Schäfer, Max Planck Institut für Physik und Astrophysik, Institut für Astrophysik, Karl Schwarzschild Strasse 1, 8046 Garching bei München, Germany

Bronisław Średniawa, Institute of Physics, Jagellonian University, Reymonta 4, 30-318 Cracow, Poland

John Stachel, Department of Physics, Boston University, 590 Commonwealth Avenue, Boston, MA 02215, USA

Helmuth Urbantke, Institut für Theoretische Physik, Universität Wien, Boltzmanngasse 5, A-1090 Wien, Austria

Christiane Vilain, Département d'Astrophysique Relativiste et de Cosmologie, CNRS, Observatoire de Paris, 92195 Meudon Cédex, France